ᘔ͢S

HANDBOOK OF PHYCOLOGICAL METHODS

ECOLOGICAL FIELD METHODS:
MACROALGAE

Also sponsored by the Phycological Society of America

Handbook of phycological methods
Culture methods and growth measurements
Edited by Janet R. Stein
(published 1973)

Handbook of phycological methods
Physiological and biochemical methods
Edited by Johan A. Hellebust and J. S. Craigie
(published 1978)

Handbook of phycological methods
Developmental and cytological methods
Edited by Elisabeth Gantt
(published 1980)

HANDBOOK OF
PHYCOLOGICAL METHODS

ECOLOGICAL FIELD METHODS:
MACROALGAE

EDITED BY

MARK M. LITTLER AND DIANE S. LITTLER
CURATOR OF BOTANY   RESEARCH ASSOCIATE

NATIONAL MUSEUM OF NATURAL HISTORY
SMITHSONIAN INSTITUTION, WASHINGTON, D.C.

SPONSORED BY THE
PHYCOLOGICAL SOCIETY OF AMERICA, INC.

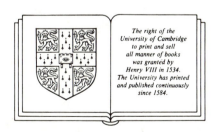

The right of the
University of Cambridge
to print and sell
all manner of books
was granted by
Henry VIII in 1534.
The University has printed
and published continuously
since 1584.

CAMBRIDGE UNIVERSITY PRESS
CAMBRIDGE
LONDON • NEW YORK • NEW ROCHELLE
MELBOURNE • SYDNEY

Published by the Press Syndicate of the University of Cambridge
The Pitt Building, Trumpington Street, Cambridge CB2 1RP
32 East 57th Street, New York, NY 10022, USA
10 Stamford Road, Oakleigh, Melbourne 3166, Australia

First published 1985

Printed in the United States of America

Library of Congress Cataloging in Publication Data
(Revised for vol. 4)
Main entry under title:
Handbook of phycological methods.
Vol. 2-4 published by Cambridge University Press,
Cambridge, New York.
Includes bibliographies and indexes.
Contents: [1] Culture methods and growth measurements,
edited by J. R. Stein – [2] Physiological and biochemical
methods, edited by J. A. Hellebust and J. S.
Craigie – [etc.] – [4] Ecological field methods,
edited by Mark M. Littler and Diane S. Littler.
1. Algology – Methodology – Collected works.   I. Stein,
Janet R.   II. Hellebust, J. A.   III. Phycological
Society of America.
QK565.2.H36   589.3'028   73-79496
ISBN 0 521 20049 0 hard covers   0 521 29747 8 paperback (v. 1)
ISBN 0 521 21855 1 (v. 2)
ISBN 0 521 22466 7 (v. 3)
ISBN 0 521 24915 5 (v. 4)

Contents

III ECOLOGICAL ENERGETICS

IV BIOLOGICAL INTERACTIONS

Contributors

Abbott, Isabella A., Department of Botany, University of Hawaii, Honolulu, Hawaii 96822 (Chapter 4)

Andrews, John H., Department of Plant Pathology, University of Wisconsin, Madison, Wisconsin 53706 (Chapter 27)

Arnold, Keith E., Department of Biological Sciences, California State Polytechnic University, Pomona, California 91768 (Chapters 17, 18)

Belbeoch, G., COB (CNEXO), B.P. 337, 29273 Brest Cédex, France (Chapter 9)

Belsher, T. Antenne COB (CNEXO), Station Biologique, 29211 Roscoff, France (Chapter 9)

Brinkhuis, Boudewijn H., Marine Sciences Research Center, State University of New York, Stony Brook, New York 11794 (Chapter 22)

Browse, John A., Plant Physiology Division, Department of Scientific and Industrial Research, Palmerston North, New Zealand (Chapter 19)

Buggeln, Richard, G., Marine Sciences Research Laboratory, Memorial University of Newfoundland, St. Johns, Newfoundland, Canada A1C 5S7 (Chapter 20)

Carefoot, Thomas H., Department of Zoology, University of British Columbia, Vancouver, British Columbia, Canada V6T 2A9 (Chapter 23)

Chapman, A. R. O., Department of Biology, Dalhousie University, Halifax, Nova Scotia, Canada B3H 4J1 (Chapter 12)

Cheney, Donald P., Department of Biology, Northeastern University, 360 Huntington Ave., Boston, Massachusetts 02115 (Chapter 5)

Dawes, Clinton J., Department of Biology, University of South Florida, Tampa, Florida 33620 (Chapter 16)

Dayton, P. K., Scripps Institution of Oceanography, University of California, San Diego, La Jolla, California 92037 (Chapter 25)

Dean, Thomas A., Kelp Ecology Project, 531 Encinitas Blvd., Encinitas, California 92024 (Chapter 10)

Denley, E. J., Scripps Institution of Oceanography, University of California, San Diego, La Jolla, California 92037 (Chapter 25)

Denny, Mark W., Hopkins Marine Station, Stanford University, Pacific Grove, California 93950 (Chapter 1)

De Wreede, Robert E., Department of Botany, University of British Columbia, Vancouver, British Columbia, Canada V6T 2B1 (Chapter 7)

Deysher, Larry E., Kelp Ecology Project, 531 Encinitas Blvd., Encinitas, California 92024 (Chapter 10)

Druehl, Louis D., Department of Biological Sciences, Simon Fraser University, Burnaby, British Columbia, Canada V5A 1S6 (Chapter 15)

Earle, Sylvia A., California Academy of Sciences, San Francisco, California 94118 (Chapter 11)

Fenical, William H., Institute of Marine Resources, Scripps Institution of Oceanography, University of California, San Diego, La Jolla, California 92093 (Chapter 6)

Foottit, Robert G., Department of Biological Sciences, Simon Fraser University, Burnaby British Columbia, Canada V5A 1S6 (Chapter 15)

Foster, Michael S., Moss Landing Marine Laboratories, P.O. Box 223, Moss Landing, California 95039 (Chapters 10, 13)

Goff, Lynda J., Center for Coastal Marine Studies, University of California, Santa Cruz, California 95064 (Chapter 27)

Harlin, Marilyn M., Department of Botany, University of Rhode Island, Kingston, Rhode Island 02881 (Chapter 24)

Kinsey, Donald W., Australian Institute of Marine Science, Townsville, Queensland 4810, Australia (Chapter 21)

Koehl, M. A. R., Department of Zoology, University of California, Berkeley, California 94720 (Chapter 14)

Littler, Diane S., Department of Botany, National Museum of Natural History, Smithsonian Institution, Washington, D.C. 20560 (Chapter 8)

Littler, Mark M., Department of Botany, National Museum of Natural History, Smithsonian Insitution, Washington, D.C. 20560 (Chapters 8, 17, 18)

Loubersac, L., COB (CNEXO) B.P. 337, 29273 Brest Cédex, France (Chapter 9)

Norris, James N., Department of Botany, National Museum of Natural History, Smithsonian Institution, Washington, D.C. 20560 (Chapter 6)

Ramus, J., Botany Department and Marine Laboratory, Duke University, Beaufort, North Carolina, 28516 (Chapter 2)

Sousa, Wayne P., Department of Zoology, University of California, Berkeley, California 94720 (Chapter 13)

Tsuda, Roy T., Division of Bioscience and Marine Studies, University of Guam, P.O. Box EK, Agana, Guam 96910 (Chapter 4)

Vadas, Robert L., Departments of Botany and Plant Pathology and Zoology, University of Maine, Orono, Maine 04469 (Chapter 26)

Wainwright, Stephen A., Department of Zoology, Duke University, Durham, North Carolina 27706 (Chapter 14)

Wheeler, Patricia A., School of Oceanography, Oregon State University, Corvallis, Oregon 97331 (Chapters 3, 24)

Editors' preface

In 1967, a special editorial committee constituted by the Phycological Society of America proposed a four-volume handbook series that would treat culture methods and growth measurements, biochemical and physiological determinations, cytological procedures, and field-oriented techniques. The first volume of the proposed series, *Culture Methods and Growth Measurements*, was edited by Janet R. Stein (University of British Columbia), the second volume, *Physiological and Biochemical Methods*, by Johan A. Hellebust (University of Toronto) and James S. Craigie (National Research Council of Canada), and the third volume, *Developmental and Cytological Methods*, by Elisabeth Gantt (Smithsonian Institution). The three texts were published by Cambridge University Press in 1973, 1978, and 1980, respectively. All three treatments have been well received by the phycological community as valuable reference sources.

This handbook on ecological methods for macroalgae, with its field-oriented perspective, is the fourth volume in the series and follows the original concept proposed by the first editorial committee of the Phycological Society. The inclusion of planktonic microalgal techniques was contemplated, but the prior publication in 1978 of an excellent and thorough manual on ecological methods for phytoplankton research (edited by A. Sournia) fulfilled this requirement. The editing of the present volume began with the receipt of the first manuscript draft in 1981 and progressed slowly but consistently. We sincerely appreciate the cooperation and valuable help of the editorial committee throughout the time-consuming process of reviewing contributions to the volume, which augmented our own efforts considerably. The committee's suggestions of topics and contributors were instrumental in broadening the overall coverage, and their critiques, along with those of additional reviewers selected from among the phycological community, resulted in numerous significant improvements.

The final product, with 27 chapters contributed by 37 scientists, was possible only because of the synergistic cooperation among many individuals. We are extremely grateful for the enthusiastic efforts of the authors and for the high quality of their chapters. We regret that, because of space and time constraints, many excellent potential contributors could not be invited to participate.

Gratitude is extended for the financial support of the Phycological Society of America. We also acknowledge the Department of Botany, Smithsonian Instituion, for considerable logistic support and Cambridge University Press for patience, cooperation, and help.

The scholarly opinions expressed in this work are those of the authors, who were asked to cite materials and equipment that they currently use. Equipment items are listed for reference only, and their inclusion should not be construed as an endorsement over other materials available at the time of writing or developed since. Lists of suppliers and their current addresses are updated annually for the United States in *Science* (American Association for the Advancement of Science, 1515 Massachusetts Ave., N.W., Washington, D.C. 20005) and for Canada in *Research and Development* (MacLean Hunter, 418 University Ave., Toronto 101, Ontario, Canada) and in *Laboratory Products News* (Southern Business Publications Ltd., 1450 Don Mills Rd., Don Mills, Ontario, Canada).

Mark M. Littler
Diane S. Littler

Department of Botany
National Museum of Natural History
Smithsonian Institution
Washington, D.C.

Editorial Committee

Keith E. Arnold, Department of Biological Sciences, California State Polytechnic University, Pomona, California, 91768

Boudewijn H. Brinkhuis, Marine Sciences Research Center, State University of New York, Stony Brook, New York, 11794

Robert E. De Wreede, Department of Botany, University of British Columbia, Vancouver, B.C., Canada V6T 2B1

Michael S. Foster, Moss Landing Marine Laboratory, P.O. Box 223, Moss Landing, California 95039

Janet R. Stein, Department of Botany, University of British Columbia, Vancouver, B.C., Canada V6T 2B1

General references

Gantt, E. (ed.). 1980. *Handbook of Phycological Methods: Developmental and Cytological Methods.* Cambridge University Press, Cambridge. 425 pp.

Hellebust, J. A., and Craigie, J. S. (eds.). 1978. *Handbook of Phycological Methods: Physiological and Biochemical Methods.* Cambridge University Press, Cambridge. 512 pp.

Sournia, A. (ed.). 1978. *Phytoplankton Manual.* UNESCO, Paris. 337 pp.

Stein, J. R. (ed.). 1973. *Handbook of Phycological Methods: Culture Methods and Growth Measurements.* Cambridge University Press, Cambridge. 448 pp.

Introduction

Marine macroalgae are a diverse group of organisms that have evolved an astounding variety of life histories, external morphologies, internal anatomical features, biochemical constituents, and metabolic activities. Although most macroalgae are restricted to a relatively small portion of the world's oceans, their concentrated biomass, high primary productivity, and role in coastal detrital and herbivore food webs make them important contributors to continental borderlands, deep sea benthic communities, and planktonic ecosystems. Yet despite their ecological significance and diversity, macroalgae have been largely overlooked as experimental organisms for the examination of selective forces that may not be operable or obvious in terrestrial and planktonic habitats or that may not have been considered in the predominantly animal-oriented studies of benthic marine systems. This treatment is especially timely because of recent technological advances and an increased awareness of the potentialities and amenability of macroalgal systems to innovative ecological experimentation.

Because of space constraints, the individual chapters are not intended to be comprehensive. References to more technical and specialized methods have been provided by all authors. The diverse audience for whom this volume is intended includes novice as well as seasoned researchers. In general, each section is concerned primarily with the method itself rather than its theoretical or historical development. Modern ecological research represents a quantitative discipline designed to produce statistically sound data bases, which are beyond the capabilities of subjective visual surveys, arbitrary scales, and other anecdotal procedures. Consequently, such approaches are not included in this handbook. The procedures presented here, although state of the art, usually are quite project specific and should be utilized as a guide and modified according to the questions being asked, resources available, and systems used. All have room for further development and improvement depending on the individual need, local circumstances, and backgrounds of those using the techniques.

An attempt has been made to present the methods in a reasonably consistent fashion; however, because the subject matter is not uni-

[1]

form, natural variations in content and format are to be expected. Inconsistencies occur among authors in controversial nuances of technique (e.g., optimal temperatures for obtaining dry weights), and some redundancies have been retained to avoid cumbersome cross-referencing. In most cases, the authors have described the limitations of the various techniques and have included references and discussions of how procedures may be adapted to suit various habitats, different algal systems, or other conditions.

Creativity and originality are essential to the experimental field ecologist. Consequently, we thought that a constraining format structure requiring either a "how to" viewpoint, literature survey approach, or theoretical/philosophical perspective, for example, might unnecessarily restrict the potential quality of the authors' contributions. Therefore, to the benefit of all concerned, a spectrum of appropriate tactics was encouraged to various degrees. The content ranges from stepwise descriptions of routine and standardly applied techniques to pioneering methods that are still in a rapid state of conceptual or technological development. Most of the procedures are readily adaptable for general use by the majority of scientists, whereas others require considerable sophistication, are still prohibitively expensive in their application, or have significant future potential.

In recent years there has been a dramatic advance in benthic algal ecology owing largely to the synthesis of traditional and empirical (i.e., observational and correlative) approaches and mechanistic or causative (i.e., experimental) studies, although vital, descriptive studies dominated previous research on algal ecology. Philosophically, this is an important point, because the products of such studies are usually empirical correlations based on habitat- or organism-oriented descriptions that are too often repeated from one algal system to another. We do not hesitate to emphasize that no good substitutes exist for thorough understanding of natural history based on careful field observations. The importance of preceding and supplementing experimental work with a generous amount of descriptive information and common sense should be underscored. Furthermore, ecological research frequently requires time spans of more than one cycle of seasons so that unusual macro- and microclimatic fluctuations will not exert a disproportionate influence on the outcome or interpretations.

However, properly controlled, experimental, hypothesis-testing approaches that lead to predictive understandings of causal phenomena generally have been relatively few, although the number of such studies on biological interactions is increasing rapidly. Manipulative investigations are leading to improved theory at the physiological, populational, and community levels. Many significant advances, concerning algal growth, productivity, distribution, succession, and especially algal–algal and algal–animal interactions, have been forthcoming from experiments and controlled perturbations performed under natural field conditions.

Another reasonable approach, which recently has garnered renewed attention, involves searching for convergent evolutionary patterns within macroalgal systems by indirect means, taking advantage of natural experiments, successional events, or developmental sequences. This technique has been designated "postdictive," rather than predictive, since the focus is on attempting to decipher the events of the past leading to present results. However, this viewpoint does contain a strong element of prediction, because hypotheses are generally of the form, "If selection has acted in the following way over evolutionary time, then we would expect nature to have the following structure."

In addition, important methodological developments have resulted in the acquisition of new information, greater standardization, and improved consistency. The recognition of the importance of both physiological stress and physical disturbance has led to major advances in our ability to understand the effects of natural and anthropogenic factors on seaweed community function, stability, and diversity. General ecological theory is becoming increasingly influenced and revitalized by studies of macroalgal ecology, and a broader awareness of the amenability and advantages of seaweeds as experimental systems for the elucidation of ecological and evolutionary mechanisms promises exciting prospects for the future. The methodology included herein contains considerable potential for developing approaches that will shed new light on ecological and evolutionary processes that may be quite widespread throughout the vast oceanic realm of the biological world. Recognition of the great importance of macroalgae as ecological research tools and their roles in marine ecosystems is long overdue.

Section I

Environmental sampling and monitoring (major parameters)

1: Water motion

MARK W. DENNY

Hopkins Marine Station, Stanford University, Pacific Grove, California 93950

CONTENTS

I. Introduction

The flow of water affects macroalgae in many ways: (1) The velocity and acceleration of the fluid impose forces on plants. In areas of rapid water motion, such as wave-swept rocky shores, these forces are substantial and may even break the plant or dislodge the holdfast. (2) Metabolism requires that inorganic nutrients and CO_2 be taken up from the water surrounding the plant and that wastes be expelled into the water. If the water is not moving, the rate of diffusion sets a limit to the rate at which metabolic processes may occur. This limit is increased substantially if the water is moving relative to the plant. (3) Many macroalgae depend on water movement to transport gametes and to disperse spores and propagules. The movement of water not only affects how far and in what direction spores will disperse, but may also determine which areas are hydrodynamically suitable for settlement.

Precise measurement of water motion is a difficult process. At some time most of us have mused over a cup of coffee, gradually stirring in cream and watching the pattern of swirls and eddies that results. This simple act gives one an intuitive grasp of how difficult it is to describe with any precision where each particle of fluid is going at one time, much less to describe how the pattern changes with time. The flow pattern in any natural setting is much more complicated than the flow in a coffee cup. A brief examination of the force exerted on an object by flowing fluid will serve to introduce the flow parameters that must be measured to describe a natural flow regime accurately.

Consider a sphere, anchored in space, with water flowing past it. If the water is flowing at a constant rate, the force on the object (in this case a drag force) is accurately described by

$$\text{drag force} = \tfrac{1}{2}\rho C_d \pi R^2 U^2 \tag{1}$$

where ρ is the density of seawater (~ 1024 kg $\cdot$ m^{-3}), πR^2 is the projected area of the sphere of radius R, U is the water velocity, and C_d is the drag coefficient (Vogel 1981). The drag coefficient of a

[8]

sphere varies as a function of the radius and the water velocity. It is advantageous to incorporate radius and velocity into a dimensionless number, which can be used to scale a particular flow pattern. This number is the Reynolds number, Re,

$$Re = \rho 2RU/\mu \qquad (2)$$

where μ is the viscosity of the fluid (1.072×10^{-3} kg $\cdot$ m^{-1} $\cdot$ s^{-1} for seawater). As long as the Reynolds number is held constant, the flow pattern around a sphere is the same, and consequently the drag coefficient is the same. The calculation of velocity from force is simplified if C_d is constant, and the drag coefficient for a sphere in steady flow is reasonably constant at 0.47 for Re between 10^2 and 10^5 (Hoerner 1965).

Equation 1 is a complete description of the force on a spherical object only if the water velocity is constant. If the water movement is unsteady, that is, the fluid is accelerating or decelerating, an additional force arises. The equation describing force in unsteady flow is generally accepted to be

$$force = (\tfrac{1}{2}\rho C_d \pi R^2 U^2) + (\rho C_m \tfrac{4}{3}\pi R^3\, dU/dt) \qquad (3)$$

(Sarpkaya and Isaacson 1981). The first expression on the right-hand side of the equation is the drag force as expressed in Equation 1. The second expression represents an inertial force in that it is proportional to the water's acceleration, dU/dt. This second expression also differs from the drag force in being proportional to the object's volume ($\tfrac{4}{3}\pi R^3$) rather than its projected area. The coefficient C_m is the coefficient of inertia and is dimensionless. In unsteady flow the drag coefficient C_d and the coefficient of inertia C_m vary both with the Reynolds number and with another dimensionless number, the period parameter K, as defined by Keulegan and Carpenter (1958):

$$K = U_m T/2R \qquad (4)$$

Here, U_m is the amplitude of the velocity fluctuations, and T is the period of the fluctuations. If the amplitude or period of the fluctuations varies, a time-averaged period parameter can be calculated using average values for T and U_m.

For Re between 100 and 100,000 and for period parameters greater than 20, C_d and C_m for a sphere are nearly constant at 0.72 and 1.07, respectively (Sarpkaya 1974). The C_d is higher in this case than for steady flow due to the unsteady nature of the fluid motion. If the sphere is within one to two diameters of the water's surface, the value of C_m decreases, but this is generally not the case.

The a priori specification of what would seem to be a simple parameter – the instantaneous force on an object – requires a knowledge of the following:

1. The instantaneous fluid velocity
2. The instantaneous fluid acceleration
3. The period and amplitude of the fluctuations in water velocity (for the calculation of K)
4. The size and shape of the object (for calculating Re and K)
5. C_d as a function of Re and K
6. C_m as a function of Re and K

This is a sizable list, one that is very difficult to compile for natural objects in natural flow regimes. Many of the same parameters must be known for other aspects of water motion to be understood. For instance, the thickness of a boundary layer (a layer of fluid moving at reduced velocity near the surface of an object) can depend on velocity, acceleration, and the shape and size of the object (Schlichting 1979).

It may appear that any measurement of water motion is a hopelessly complex task. It is not, but one must be willing to accept limitations. The complexity of measurement is a result of the rigor with which the environment must be measured, and the required rigor varies with the question asked. For example, as shown later, it is much easier to measure the average water velocity than it is to measure an instantaneous value. However, what one gains in ease of measurement one loses in the amount of information acquired; for example, from a continuous record of instantaneous velocities the average velocity can be calculated, whereas given the average alone, no information can be gleaned about instantaneous values.

One is advised to accept this situation and, rather than attempt to measure flow exactly, concentrate on measuring those aspects of flow that are of importance in a given situation. Different aspects of flow are measured by different methods, and it is the objective of this chapter to describe these varied techniques. The following is not intended to be an exhaustive review of flow measurement techniques; emphasis is placed on describing those methods that are (or soon will be) appropriate for field use on wave-swept shores.

II. Measurement of cumulative water motion

At the most basic level one may simply want some means to compare the cumulative water motion among areas over some extended period of time. Methods for such measurements are appropriate only if information concerning short-term phenomena is unimportant. For example, if data regarding maximum velocity, range of velocity, or direction of motion are required then these methods are inappropriate.

A. *Plaster spheres*

Muus (1968) measured cumulative water motion by determining the rate at which plaster of Paris dissolved from standard plaster spheres. Balls are cast in spherical ice-cube molds (the head of a nail being incorporated into each sphere) and allowed to set for a standard period (~4 wk). Spheres are then placed in the field for a known period, recovered, and dried, and their weight loss measured. Each set of spheres is calibrated by being placed in a flow of known velocity for a period of time and the decrease in dry weight measured.

These spheres have been used with some success to measure average water velocity over a tidal sand flat (Muus 1968), and a variant of the technique (plaster "clod cards") has been used in various coral reef environments (Doty 1971).

These methods have several drawbacks:

1. The rate at which plaster dissolves at a given water velocity is temperature dependent (Muus 1968). Thus, separate calibrations must be made for each temperature encountered in the field. If the temperature in the field varies during the course of a test or among sites, the accuracy of the results is reduced. Temperature measurements should be made simultaneously with the flow measurements, and the weight loss suitably corrected.

2. Particulate matter and grazing animals may abrade the plaster, resulting in aberrant readings.

3. Variation in the rate of weight loss is substantial among spheres at one water velocity (Muus 1968). As a result, average water velocity can be measured with an accuracy of only approximately ±20%.

4. The "clod cards" used by Doty (1971) are not radially symmetric; flow patterns around the clod will vary with flow direction. Weight loss may consequently vary as a function of flow direction.

5. It is unlikely that the design described by Muus (1968) can withstand the wave forces present on exposed shores.

B. *Plate and friction fitting*

Harger (1970) measured cumulative wave forces using a flat plate attached to a nail by a friction fitting (Fig. 1–1). The nail is pounded into the substratum of the shore such that the plate lies parallel to the substratum. Each wave breaking on the plate imposes a force, and each force (above some threshold) slides the plate a distance down the nail. Thus, the distance moved by the plate is an indication of the cumulative number or average severity of the forces encountered. The design can be modified to work in a variety of habitats. The size of the plate and the frictional resistance of the sliding fitting are adjusted so that the distance moved during the course of a test

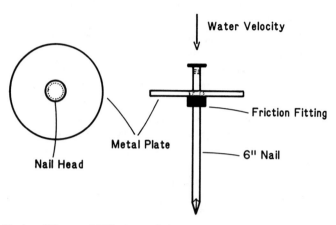

Fig. 1–1. Device of Harger (1970). A metal plate with a hole drilled through the center is held in place on a nail by a C-clip friction fitting. Water velocity directed perpendicular to the plate forces the plate down the nail. Redrawn from Harger (1970) courtesy of the *Veliger*.

is measurably large but less than the total length of the nail. The minimum force required to move the plate is determined for each device by placing weights on the plate or by pulling the plate with a spring scale. Care must be taken to ensure that each nail is free of burrs, and friction fittings must be changed frequently if reproducible results are to be obtained.

The device has been used to measure the relative exposure to wave forces of various mussel beds (Harger 1970).

Data taken with this device must be used with caution:

1. The device responds only to forces acting normal to the plane of the plate. Thus, shear forces (forces acting parallel to the plane of the substratum), even if quite large, are not recorded.

2. The response time of the device is difficult to determine. It is quite possible that the plate moves the same distance in response to a small force applied steadily for 1 to 2 s as it does in response to a very large force (such as a wave impact) applied for 0.1 s or less. As a consequence, exactly what flow parameter the device measures in a natural flow regime is uncertain. It is difficult to translate readings made with the device into a quantity of importance to algae, such as maximum or average water velocity.

III. Measurement of maximum force

In many situations it is much more useful to know the *maximum* force or velocity an organism encounters than to know the *average* force

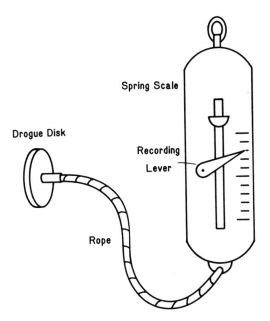

Fig. 1–2. Recording spring scale. A lever is bolted to a standard spring scale to record the maximum excursion of the spring and thereby measure the maximum force. A drogue disk is attached to the scale by a rope. The entire apparatus is attached to the substratum by a swivel fitting. Redrawn from Jones and Demetropoulos (1968) courtesy of the *Journal of Experimental Marine Biology and Ecology.*

or velocity. For example, it is the maximum water velocity, rather than the average, that breaks a plant. Two devices constructed to measure maximum water velocity are described in the following two sections.

A. *Recording spring scale with drogue disk*

Jones and Demetropoulos (1968) used a recording spring scale to measure the maximum drag force exerted on a small drogue disk held perpendicular to the flow (Fig. 1–2). These force measurements were used to calculate the maximum water velocity occurring in various habitats for comparison with the observed species distribution.

The recording spring scale is constructed in a simple manner from a standard spring scale, a lever recording the maximum excursion of the scale marker (Fig. 1–2). The scale is anchored by a swivel fitting to the substratum, and a drogue disk is attached to the scale by a rope. As water flows over the device, a drag force is placed on the drogue, and the force is recorded by the scale. Records of maximum force can be used to estimate water velocity. From Equation 1,

$$\text{velocity} = [(2 \text{ drag force})/(\rho\pi R^2 C_d)]^{1/2} \tag{5}$$

where R is the radius of the drogue disk. All variables must be expressed in compatible units (e.g., velocity in meters per second, force in newtons, and radius in meters); the drag coefficient is dimensionless. Hoerner (1965) gives values of the drag coefficient for a disk over a wide range of Reynolds numbers and shows that the drag coefficient is constant within a certain range. For example, a drag coefficient of 1.2 is appropriate for a disk 2 cm in radius (as used by Jones and Demetropoulos 1968) at velocities from 0.1 to at least $100 \text{ m} \cdot \text{s}^{-1}$. This range easily spans the velocities likely to be encountered on wave-swept shores. The size of the disk can be varied within reasonable limits to adjust the drag force so that the range of the spring scale is not exceeded.

Four problems arise when this device is used:

1. The calculation of water velocity assumes that the drogue disk is subjected to mainstream flow. However, the disk always lies in the wake created by the spring scale and, as Vogel (1981) explains, actually encounters flows considerably slower than those present in the mainstream. This problem can be circumvented to a certain degree by calibration of the device.

2. The second problem is less easily remedied. The response time of the spring scale is ~0.1 s (Denny 1982). Thus, large forces applied for a short period (<0.1 s) do not result in full deflection of the recording lever, the end result being an underestimation of the maximum water velocity. The magnitude of the underestimation increases as the duration of the force decreases, and for forces such as wave impact (which may last only a few milliseconds) this underestimation may be substantial. The response time of the apparatus is further increased by the flexibility of the device. After a change in the direction of flow, the entire device must realign itself before a new force can be measured. For a more complete discussion of the effects of response time, see Alexander (1968) or Thomson (1981).

3. Equation 5 does not take into account any component of the force due to the acceleration of the water. If acceleration forces exist, Equation 5 tends to overestimate water velocity.

4. The device does not measure the direction from which the force is applied and responds only to shear forces.

Despite these caveats, the device is useful and as a possible analog of a flexible alga, such as *Alaria* or *Lessoniopsis*, may give accurate answers regarding the magnitude of forces exerted on these genera.

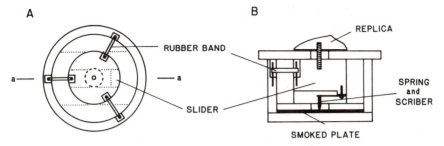

Fig. 1–3. Device for recording the direction and magnitude of the maximum shear force exerted on an object. A Teflon slider is suspended by rubber bands inside a plastic housing. A shear force on the test object (here a replica of a limpet) causes the slider to be deflected. The deflection is recorded as a scratch on a smoked-glass slide. (A) Top view with the top plate removed. (B) Side view cross section through a–a of A. Reproduced from Denny (1983) courtesy of the American Society of Limnology and Oceanography.

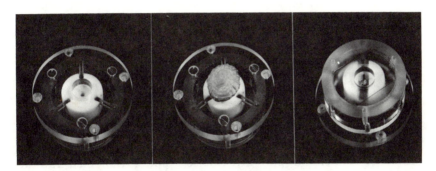

Fig. 1–4. Maximum-force recorder. Left: top view with the test object removed. Middle: top view with the test object in place. Right: bottom view showing the scriber and the groove into which the smoked-glass slide fits.

B. Direction and maximum-force recorder

Denny (1983) designed a device to record directly the direction and magnitude of the maximum force exerted on intertidal organisms (Figs. 1–3 and 1–4). A Teflon slider is sandwiched between two parallel plates in the housing of the device and is held centered by three rubber bands. The organism on which force is to be measured is attached to the slider by a bolt passing through a hole in the top plate. Any hydrodynamic shear force applied to the organism causes the slider to move in the direction of force and to a distance directly proportional to force. This deflection is recorded by a spring-tensioned scriber that extends through a hole in the bottom plate and writes on a smoked-glass microscope slide. The device is mounted

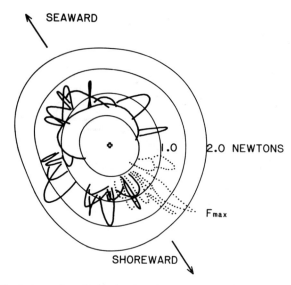

Fig. 1–5. Typical results obtained with the maximum-force recorder. The cross designates the centered, zero-force position of the scriber. The solid and dotted lines were obtained at two different sites. The calibration "bull's-eye" is superimposed on the records. The maximum force (about 2 N) was directed in a generally shoreward direction. Reproduced from Denny (1983) courtesy of the American Society of Limnology and Oceanography.

in prepared sites on the shore. A hole (approximately 6 cm in diameter and 10 cm deep) is drilled into the rock and a sleeve (a section of 5.08-cm-i.d. polyvinyl chloride pipe) is cemented into the hole (for details see Denny 1982). The device fits snugly into the sleeve, and the top plate is bolted to the sleeve. A freshly smoked microscope slide is inserted in the device, and the apparatus installed on the shore during low tide. The smoked-glass slide is recovered at a subsequent low tide, and the record examined with a measuring microscope or photographed for later examination. Typical results are shown in Fig. 1–5.

One calibrates the device by turning the housing on its side at the edge of a table and hanging a known weight from a bolt inserted into the slider. The device is then rolled along the table. The procedure is repeated for several weights, the result being a bull's-eye pattern (Fig. 1–5). This pattern is superimposed over the force record to estimate the maximum force. Construction of the device is straightforward, although the machining must be reasonably precise. A complete description is given by Denny (1983).

The device can be used to measure forces on a wide variety of organisms and in a wide variety of flow regimes. The minimum force that can be measured accurately is limited by the frictional resistance of the slider within the housing. The rubber bands must be stiff enough to overcome this resistance and accurately return the slider to its centered position when no force is applied. For the device constructed by Denny (1983) this minimum measurable force was ~0.2 N (the weight of a 20-g mass). The upper limit of force is set by the stiffness of the rubber bands available. In practice, the stiffness of the rubber bands is adjusted by trial and error so that the deflection recorded for a particular organism in a particular flow regime is large enough to be read but does not go offscale.

The object attached to the device may be either the test plant itself, suitably attached to a bolt, or a replica constructed from plastic sheeting, rubber, or epoxy. Because the force measured is in part a function of the exact shape and flexibility of the test object, care should be taken in interpreting results gathered using replica organisms.

Although this device is designed primarily to measure directly the maximum forces exerted on organisms, it may be used to estimate maximum water velocity. The calculation is identical to that described for the device of Jones and Demetropoulos (1968), but in contrast to the case of the drogue disk, the drag coefficient of most biological objects is not known. Consequently, if the device is to be used to estimate water velocity, either the drag coefficient of the test organism must be determined empirically or a suitable nonbiological object (e.g., a sphere or cylinder of known C_d) should be used. As with the device of Jones and Demetropoulos (1968), the water velocities calculated in this fashion must be considered rough estimates, because the possible effects of inertial forces have been neglected. Furthermore, Equation 5 assumes that the object is stationary; thus, it is not strictly applicable to objects that, when bending, move with the flow.

Two problems are associated with the use of this device:

1. The slider and rubber bands form a spring-and-mass system with a resonant period of oscillation. Consequently, when a force is applied, it takes a finite time for the apparatus to respond with the appropriate deflection. As explained by Denny (1983) this response time is approximately equal to the resonant period of the device and, in this case, is ~25 ms. The response time of each individual device depends on (and is directly proportional to) the ratio of the moving mass (slider plus organism) to the restoring force (a function of the stiffness of the rubber bands). There is a tradeoff between response time and sensitivity to force. A device with stiff rubber bands has a fast response but is relatively insensitive to small forces, whereas a device with very flexible rubber bands is slower to respond but more sensitive to force. The response time of Denny's device is considerably

better than that of Jones and Demetropoulos (1968) but may still be too slow to record accurately the most quickly applied forces and changes in water velocity.

2. The stiffness of rubber bands is directly proportional to the absolute temperature. Thus, the calibration of the device varies with the ambient temperature (~0.3% per 1.0°C) and calibration should be carried out at the temperature expected in the field. The stiffness of stretched rubber bands also decreases with time. The calibration of the device should be checked once per week and rubber bands changed every 2–3 wk.

IV. Continuous measurement of water velocity

For many applications, a knowledge of maximum velocity and direction is not sufficient. For example, effective settlement in one microhabitat may depend on the water velocity being within a specific range. It is of interest to know the fraction of time during which the water velocity in that habitat falls within this range. To accomplish this, the device used must be capable of continuously measuring instantaneous water velocity. Many other examples exist in which a continuous recording of water velocity is required. Five methods are commonly used for this purpose.

A. Propeller devices

A propeller placed in flowing fluid turns at an angular velocity proportional to the velocity of the water. This principle has been used in constructing a number of flow meters. The primary requirement for meters of this type is that the rate of rotation of the propeller be monitored without interfering with the motion of the propeller. The rate of rotation can be sensed by a number of methods, but most commercial devices use an optical system. Once during every revolution each propeller blade passes a light detector, preventing light from reaching the detector. This change in light intensity is sensed electronically, and a voltage signal noting the passage of a blade is transmitted via an electrical cable to the velocity meter. Thus, the number of signals per unit time is a measure of water velocity. Calibrating meters of this type involves placing them in known flows and noting the rate of rotation.

The size of propeller flow meters ranges from those with a propeller <0.2 cm in radius to those with propellers >10 cm in radius. Smaller propellers are preferable for two reasons:

1. The spatial resolution of the device is controlled by the size of the propeller, since the device essentially measures the average flow

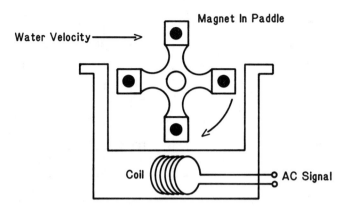

Fig. 1–6. Simple paddle-wheel flow meter. The axle of the paddle wheel lies perpendicular to the plane of the page. Water motion causes the wheel to rotate, and the magnet in each paddle induces an electric current in the coil.

through a circular area equal in diameter to the propeller. Thus, if the fine structure of flow is to be sensed, the propeller must be small. This need arises, for example, in the measurement of accelerations associated with small, turbulent eddies and the inertial forces on small objects ($\sim$1–50 cm).

2. The inertia of the propeller determines the response time. A propeller rotating at a certain angular velocity requires some time to speed up if subjected to a sudden increase in water velocity; similarly, a propeller continues its rapid rotation for a period after flow velocity decreases. The response time is minimized by minimizing the rotational inertia of the propeller. In other words, one makes the propeller as small as possible and of a material with a low density. A rough estimate of the response time of a propeller can be obtained by thrusting a nonrotating propeller into a steady flow and noting the time required to achieve a constant angular velocity.

Commercial propeller flow meters are accurate and reliable (Swoffer Marine Instruments, Inc.). However, problems are associated with their use on wave-swept shores. The tiny propellers of many meters are very delicate and are easily damaged on impact with solid objects in the moving water. There may be no simple correlation between the direction of flow relative to the axis of the propeller and the rate of rotation. If a propeller flow meter is to be used to record flow direction as well as speed, the directional response of the meter should be carefully determined.

A simple version of the commercial meters can be constructed from the paddle-wheel-type velocity meters sold for sailboats (Fig. 1–6). Each arm of the paddle wheel carries a magnet, and rotation

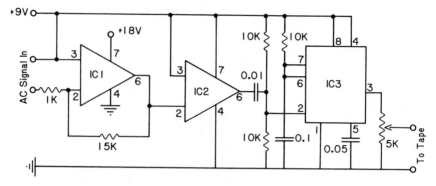

Fig. 1–7. Simple circuit for conditioning the signal from the paddle-wheel flow meter. IC1 (a microampere 741 op amp) amplifies the AC signal from the meter; IC2 (a microampere 741 op amp) acts as a comparator. When the output from IC1 goes above 9 V, IC2 quickly drops the voltage at its output (pin 6). This decrease in voltage triggers the microampere 555 timer (IC3). The amplitude of the resulting square-wave pulse is adjusted by the 5000-Ω potentiometer and is recorded on a cassette tape recorder. All resistors are 0.25 W, 10%. Capacitance values in microfarads. Current drain is ~8 mA.

of the wheel causes an alternating electric current to be induced in a wire coil mounted in the base of the housing. This current is carried by cable to the simple circuit shown in Fig. 1–7. This circuit detects the voltage signal from the passage of each paddle and converts the signal to a 1-ms square-wave pulse suitable for recording on a cassette tape recorder. Data recorded in the field are returned to the laboratory, and the tape record is placed onto a chart recorder for analysis. Taped records can also be played into the frequency-to-voltage converter described by LaBarbera and Vogel (1976), in which case a DC voltage proportional to water velocity is the output. The paddle wheel of this device is relatively massive; consequently, the response time is quite long (~1 s). As a result, the device does not accurately record rapid changes in velocity. Below a speed of 0.2 to 0.3 m · s^{-1} the rotation of the paddle wheel is erratic, but at higher water velocities the device responds accurately to flow.

B. Drag sphere devices

This method of flow detection is based on the same principle as the devices of Jones and Demetropoulos (1968) and Denny (1983); the force imposed on an object is a function of the velocity and acceleration of the water. The problems due to unmeasured accelerational effects encountered in applying this premise to the devices described earlier can be circumvented by proper design. Given the possible variations in C_d and C_m, the equation describing the total force on an object (Equation 3) can be applied most effectively if Re and K are in the range where C_d and C_m are constant. The size of the required sphere can be calculated.

The density and viscosity of seawater are fixed. From the data of M. W. Denny (unpublished), a typical set of flow parameters encountered on a wave-swept shore is estimated to be $U = 10$ m $\cdot$ s^{-1}, $U_m = 5$ m $\cdot$ s^{-1}, and $T = 0.1$ s. With respect to Re, the drag coefficient and inertia coefficient are constant if $200 < \rho 2RU/\mu < 10^5$ and 10^{-5} m $< R < 5 \times 10^{-3}$ m. With respect to the period parameter, C_d and C_m are constant if $U_m T/2R > 20$, that is, $R < 1.25 \times 10^{-2}$ m.

For a sphere of radius less than 0.5 cm and greater than 10 μm, Equation 3 can be applied without correcting for variation in C_d and C_m. In general, unless the acceleration is very large, the inertial component of the force is small compared with the drag force. For example, using a sphere 0.5 cm in radius at a velocity of 10 m $\cdot$ s^{-1} and an acceleration of 50 m $\cdot$ s^{-2}, one finds that the inertial force is only ~1% of the total force and can safely be neglected; however, much higher accelerations are possible. M. W. Denny (unpublished) has measured accelerations of greater than 400 m $\cdot$ s^{-2}. In this case, the inertial force may be 10% of the total force and should be taken into account. The inertial force is included in the calculation of velocity by numerically solving Equation 3 in the rearranged form

$$dU/dt = [F(t) - \tfrac{1}{2}\rho C_d \pi R^2 U^2]/(\rho C_m \tfrac{4}{3}\pi R^3) \qquad (6)$$

where $F(t)$ is the force at time t. The numerical solution is easily obtained by a computer using the Runge–Kutta method (e.g., see Dorn and McCracken 1972) with the boundary condition that at zero time all force is due to the acceleration reaction. The velocity and acceleration can be computed to the accuracy permitted by the assumption that the drag coefficient and inertia coefficient are constant. The use of this theory requires a continuous measurement of the force acting on the drag sphere. Two devices have been constructed for this purpose.

Denny (1982) used three separate transducers to measure the components of force exerted on replicas of limpets and barnacles (Fig. 1–8). The sensing element of each transducer is a double-cantilever beam milled from a single piece of acrylic plastic. The base of the beam is bolted firmly to the transducer housing, and the free end of the beam is connected to the drag sphere by a bolt extending through a hole in the top plate of the housing. Force on the sphere results in a bending of the beam; the amount of deflection is proportional to force. A double cantilever of this sort is sensitive to

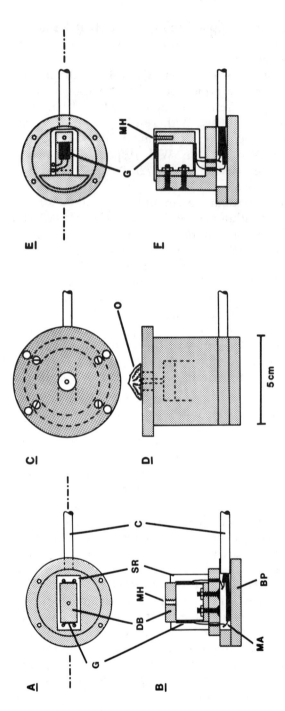

Fig. 1–8. Force transducers for a telemetry system. The sensing element is a double beam (DB) on which strain gauges (G) are mounted. A, B Top and side view, respectively, of a transducer for measuring shear forces; outer housing removed. C, D Top and side view, respectively, of the assembled transducer. E, F Top and side view, respectively, of a transducer for measuring forces directed normal to the substratum. BP, Base plate; C, cable; MA, milled area to accommodate gauge and cable connections; MH, mounting hole for test object; O, test object; SR, silicone rubber to seal and damp the beam. Reproduced from Denny (1982) courtesy of the American Society of Limnology and Oceanography.

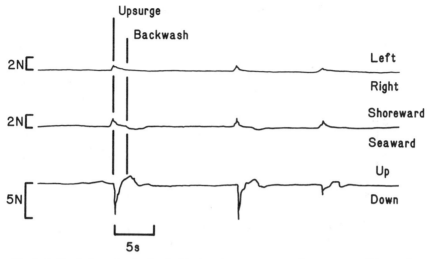

Fig. 1–9. Typical results obtained with the telemetry system: force on a small barnacle. The onset of the wave upsurge and backwash are noted. Redrawn from Denny (1982) courtesy of the American Society of Limnology and Oceanography.

force only in the direction parallel to a line joining the tips of the cantilevers. Forces in other directions cause relatively little deflection. The bending of the beam is sensed by a pair of foil strain gauges, one attached to each side of the beam. A force applied to the drag sphere thus causes the extension of one gauge and the compression of the other with a corresponding change in the electrical resistance of the gauges. The gauges are arranged as one arm of a Wheatstone bridge, and the resulting voltage signal is carried by a shielded cable.

Three transducers, each measuring force in a different direction, are implanted in the rock surface as described for the maximum-force recording device. The cables from the transducers are anchored to the rock and connect the transducers to a telemetry system placed well above the surf zone. The voltage signal from each transducer is changed to a frequency-modulated audio signal. The audio signals from the three transducers are mixed and transmitted as an FM radio signal. The radio signal is received by a standard FM tuner, and the reconstituted audio signal recorded on magnetic tape for later decoding and analysis. The range of the telemetry system used by Denny (1982) is ~1 km. Details of the construction of the transducers and telemetry system are given in Denny (1982).

A typical example of the data recorded by this system is shown in Fig. 1–9. From the three force versus time signals, the three spatial

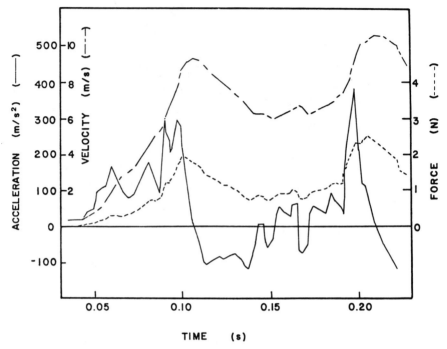

Fig. 1–10. Typical force trace analyzed according to Equation 6 to give instantaneous values of one-directional component of velocity and acceleration.

components of water velocity and acceleration can be calculated as outlined earlier. A typical result is shown in Fig. 1–10.

Two problems are encountered in the use of this device:

1. The measurements of the components of force are made at separate locations. Even though the three transducers may be only 10–15 cm apart, they do not sample the same water, and as a consequence it is inappropriate to perform a vector summation of the recorded forces to arrive at the instantaneous overall force.

2. The response time of the apparatus is a function of the ratio of moving mass (organism plus cantilever beams) to the restoring force (a function of the stiffness of the beam). Thus, as noted for all devices previously described, there is a tradeoff between sensitivity to force (the less stiff the beam, the more sensitive) and the response time (the more stiff the beam, the shorter the response time). The device of Denny (1982) is sensitive to forces greater than ~0.1 N with a response time of ~8 ms.

Donelan and Motycka (1978) used a drag sphere mounted on a triaxial force transducer to measure turbulent water velocities beneath

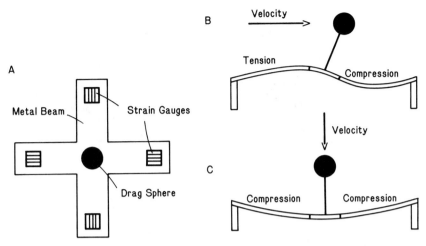

Fig. 1–11. Three-axis force transducer. (A) Top view. The drag sphere is attached to a cruciform beam. A strain gauge is glued to each arm to detect tension or compression. The strain gauges are sensitive to changes in beam dimensions only along the lines shown and are insensitive to twisting. (B) Side view. When the drag sphere is exposed to a horizontal force, the opposite arms of the beam are deformed in opposite directions; the upper surface of the left arm is in tension, that of the right arm is in compression. The other arms of the beam (those extending into and out of the plane of the page) are twisted. (C) Side view. A vertical load deforms all arms of the beam in the same direction. The electronics monitoring the strain gauges compare the deformation of all arms to yield a measure of the direction (in three dimensions) and magnitude of the force applied to the drag sphere. In the actual transducer, only the drag sphere is exposed to flow. Drawn from the description presented by Donelan and Motycka (1978).

waves. The drag sphere (0.215-cm radius) is attached by a rigid rod to a cruciform steel beam. Force imposed on the sphere in any direction results in a complex deformation of the beam (Fig. 1–11). This deformation is sensed by strain gauges fixed to the four arms of the beam. Signals from the four arms are electronically combined to yield the three components of force acting on the sphere. The device of Donelan and Motycka (1978) is connected by cable directly to the electronics responsible for deciphering and recording the signal. It would be possible, however, to operate the device with the telemetry system described by Denny (1982), and the transducer could be mounted on the shore as described by Denny (1982).

The device of Donelan and Motycka (1978) is not as easily constructed as that of Denny (1982) but has the advantage of measuring the components of force on a single drag sphere. The reported response time of the device is ~13 ms and can be adjusted by varying the stiffness of the cruciform beam. A method for

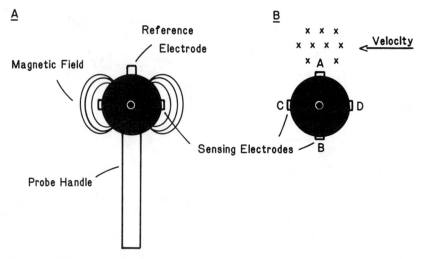

Fig. 1–12. Electromagnetic flow meter. *A* Side view of the flow probe. A time-varying magnetic field is produced by a coil in the probe. *B* The flow of water (a conductor) through the magnetic field (symbolized for the region near electrode A by the crosses) induces a voltage difference between A (or B) and the reference electrode. For velocity in the direction shown, no voltage is induced between electrodes C and D.

calculating the spatial resolution of a drag sphere probe is presented by Donelan and Motycka (1978). The spatial resolution for their sphere (0.215-cm radius) has a length scale of ~287 cm. This value is calculated treating the inertial force as "noise" in the system. If the inertial force is included in the velocity calculation (as explained in Equation 6), the spatial resolution improves by a factor of ~10.

All of the methods discussed so far rely on measuring the force imposed on some object as an indicator of water velocity. However, it is possible to sense water velocity from a number of other factors.

C. Electromagnetic flow meters

The movement of a conductor through a magnetic field creates a voltage. This principle of induction is the means by which electric generators work and can be applied to the measurement of water velocity. Seawater is a conductor. Thus, if seawater flows through a magnetic field, a voltage must result. A practical configuration for such an apparatus is shown in Fig. 1–12. By placing two pairs of voltage-sensing electrodes perpendicular to each other, one can measure simultaneously the components of velocity along two axes. The voltage produced by such an apparatus is quite small, and the electronics required to sense it are consequently sophisticated. This places the method beyond the do-it-yourself capabilities of most ecologists. However, commercially constructed devices suitable for field use are available; model 511 of Marsh-McBirney Instruments, Inc., is used here as a typical example.

The polarity of the magnetic field produced by this device is rapidly alternated (30 Hz). Consequently, the voltage signal proportional to flow is an AC signal. This signal is amplified and low-pass-filtered to provide a DC voltage varying in response to flow. The process of filtering limits the response time of the device to 50 to 100 ms. The device measures a weighted average of the flow passing through the magnetic field, and the effective magnetic field extends well beyond the sensing apparatus; for the standard spherical sensor (3.7-cm diameter) the flow measured is the average flow through a torus of ~10-cm major diameter. The sensitivity of the device is altered by nearby solid structures (especially metal structures), and reasonable care must be taken to ensure that the device is suitably isolated or that it is calibrated in place. This device measures the directional components of velocity along two axes.

The voltage signal output of the machine is suitable for recording on any standard chart recorder or can be monitored visually from dials on the control panel. The probes are rugged and have been used with great success to measure flow in breaking waves (e.g., Thornton et al. 1976).

D. Hot-wire, hot-film, and thermistor flow meters

These devices use the rate of transfer of heat from a solid object to the moving fluid as a measure of fluid velocity. The principle is quite simple: Imagine yourself holding a spoonful of soup. To cool the soup you blow a stream of cool air across it; the harder you blow, the faster the soup cools. In practice, a resistive element (a wire, a metallic film, or a thermistor bead) is heated by an electric current to a temperature several degrees above ambient temperature. As water moves past the heated element, the temperature of the element has a tendency to drop. The circuitry of the device counteracts this tendency by passing more electric current through the resistive element. As a result, the element is constrained to stay at a constant temperature and the electric current required to maintain the temperature can be used as a measure of the flow velocity.

A wide variety of hot-wire and hot-film probe configurations are available commercially. Depending on the probe structure, these can measure flow along one, two, or three axes simultaneously. The spatial resolution of these probes is quite good (for a heated wire, essentially the length of the wire, 1–5 mm), and their response time can be 1 ms or less. There are, however, several problems with the use of hot-wire or hot-film probes on wave-swept shores:

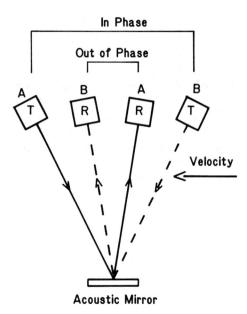

Fig. 1–13. Scheme for an acoustic current meter. Sound is produced by the two transmitters (T) and received by the two receivers (R). The sound waves are in phase as they leave the transmitters. Sound traveling from transmitter A to receiver A (via the mirror) is retarded due to the water velocity, whereas sound traveling from transmitter B to receiver B is advanced. As a consequence, the sounds received are out of phase by an amount proportional to the water velocity.

1. Hot-wire probes are exceedingly delicate. The impact of even a small solid particle will break the wire and destroy the probe. Hot-film probes are much less delicate, but at present, if more than one directional component of velocity is to be measured, there must be one probe per direction.

2. The response of these probes is not a linear function of velocity. The sensitivity to flow decreases as the velocity increases. If the output of the flow meter is to be conveniently analyzed, it must first be linearized. Electronic devices are available to carry out this linearization (e.g., the polynomial linearizers produced by TSI Inc.).

3. Both the hot-wire and hot-film probes are sensitive to bubbles in the water. If bubbles coat the probe surfaces, the heat transfer coefficient of the probe is drastically altered and the response becomes inaccurate. Commercial probes have been carefully designed to avoid such contamination problems; however, in flow regimes with substantial amounts of entrained air (e.g., breaking waves) bubbles may still cause problems.

Although the technology of hot-wire and hot-film water velocity meters is such that their construction is beyond the abilities of the typical ecologist, a simplified version of these devices has been designed by LaBarbera and Vogel (1976). A glass-encapsulated thermistor bead is used as the sensing device. The associated electronics are simply and inexpensively constructed. The probes are easy to build and have proved to be rugged enough for field use. The response time of the device is ~0.1 s, and the spatial resolution of the order of 1 mm. Details regarding the construction of the device are given by LaBarbera and Vogel (1976). A correction of the schematic circuit diagram is noted by Vogel (1981).

This thermistor flow meter has one major drawback that hot-wire and hot-film probes do not: Above ~0.5 m · s^{-1}, it is insensitive to any increase in water velocity. This device is suitable for flow measurements in environments where velocities are relatively low but is inappropriate for measurements on wave-swept shores.

Hot-wire, hot-film, and thermistor flow meters are calibrated by exposure of each probe to known flow velocities. A scheme for providing these velocities is discussed by LaBarbera and Vogel (1976).

E. Acoustic and optical methods

The methods described so far are all less than ideal in one common respect: They require that some physical structure be placed in the flow in order to sense velocity. Two methods of flow measurement that avoid this problem have been devised.

1. Acoustic current meter. The time it takes a sound wave to travel from one point to another is affected by the motion of fluid between the two points. Point 1 is at distance D from point 2. Sound in water travels with velocity c. In still water, a sound emitted at point 1 takes D/c seconds to reach point 2. If the fluid is moving from 1 to 2 with velocity U, the time it takes sound to travel from 1 to 2 is $D/(c + U)$. This decrease in time means that the sound passing through moving water is shifted in phase when it reaches point 2 relative to the sound traveling through a stationary fluid. The phase shift is directly proportional to the fluid velocity and can be electronically detected. The advantage of a device based on this principle is that it is inherently calibrated.

Commercial instruments of this type are available – for example, model DRCM-2 of Neill Brown Instrument Systems, Inc. This two-axis instrument is used here as a typical example. At present, the device is physically configured for measuring flow in open water rather than near a solid boundary (Fig. 1–13). The spatial resolution is not particularly good (the transmitter and receiver are ~10 cm apart) and the response time is 0.2 s. In principle, however, such devices could be configured for use near the substratum on rocky shores.

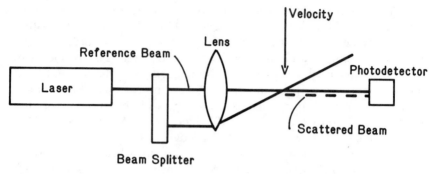

Fig. 1–14. Typical optical configuration for a laser Doppler flow meter: the single-lens heterodyne configuration. The reference beam travels straight through the apparatus to the photodetector. A second beam is focused on the test volume at an angle to the reference beam. From this second beam only that light scattered by particles in the fluid reaches the photodetector. This light is frequency-shifted relative to the reference beam by an amount proportional to the velocity of fluid along the second beam. This change in frequency is detected as a periodic change in the intensity of light reaching the photodetector.

2. Laser Doppler techniques. The ultimate technique presently available for flow measurement is an optical method in which the frequency shifting of light by waterborne particles is an indication of velocity. A typical configuration is shown in Fig. 1–14. Light emitted by a laser is split into two beams and focused on a small test volume of moving fluid. One beam (the reference beam) passes straight through the fluid and onto a photodetector. The second beam arrives at the test volume from an angle different from the reference beam. As a consequence, the only light from the second beam to reach the photodetector is that which has been scattered by particles in the fluid. If these particles have a component of velocity in the direction of the second beam, the scattered light they emit has a slightly different frequency than the light in the reference beam; the greater the particle velocity, the greater is the frequency shift. The difference in frequency between the reference and scattered beams causes the beams to interfere, and the intensity of light received by the photodetector fluctuates. This fluctuation occurs at a beat frequency equal to the difference in frequency between the reference and scattered light. By measuring the frequency of intensity fluctuation, one can directly determine the speed in the test volume. The particles required to scatter the light can be extremely small (10 μm or less) and can be assumed to move precisely with the water in which they are dispersed.

The spatial resolution of the device is determined by the size of the volume on which the beams are focused and therefore can be made microscopically small. The response time of the devices may easily be as short as 1 μs. The optical design can be arranged so that flow can be measured simultaneously along multiple axes. As with the acoustic apparatus, this device is inherently calibrated.

Although laser Doppler techniques present superb potential for flow measurement, practical problems have limited their use in field situations:

1. The optical configuration most often used (similar to that shown here) requires that the photodetector be on the opposite side of the test volume from the laser, a situation that would be inconvenient for a field device.

2. This configuration is subject to disturbance from bubbles in the water. Too many bubbles or even one big bubble filling the test volume would effectively block both beams from reaching the photodetector. This problem can be circumvented to a certain degree by operating the device in a backscatter mode in which the photodetector is on the same side of the test volume as the laser. However, the intensity of light scattered back toward the source is several orders of magnitude lower than that scattered forward. In this configuration, a more powerful laser must be used to generate a detectable signal. A device operating in the backscatter mode has been marketed in which the beam splitter and lenses are connected to the laser and photodetector by fiber optics (DISA Electronics). The flexible "probe" is prealigned and, if a suitable waterproof housing can be devised, offers promise for field use.

3. All laser Doppler devices must be very accurately aligned, and this alignment would be difficult to maintain on a rocky shore.

4. At present, laser Doppler devices are very expensive (more than $10,000).

If these practical problems can be overcome, laser Doppler devices may prove to be invaluable field instruments.

V. Acknowledgments

I thank S. Gaines, T. Daniel, W. Magruder, and S. Denny for their help and guidance in preparing the manuscript. The paddle-wheel flow meter was designed as part of a project supported by the Office of Naval Research (contract no. N00014-79-C-0611). I thank R. T. Paine and the U.S. Coast Guard for the opportunity to test this meter on Tatoosh Island, Washington.

VI. References

Alexander, R. M. 1968. *Animal Mechanics*. University of Washington Press, Seattle. 3467 pp.

Denny, M. W. 1982. Forces on intertidal organisms due to breaking ocean waves: design and application of a telemetry system. *Limnol. Oceanogr.* 27, 178–83.

Denny, M. W. 1983. A simple device for recording the maximum force exerted on intertidal organisms. *Limnol. Oceanogr.* 28, 1269–74.

Donelan, M. A., and Motycka, J. 1978. Miniature drag sphere velocity probe. *Rev. Sci. Instr.* 49, 298–304.

Dorn, W. S., and McCracken, D. D. 1972. *Numerical Methods with Fortran IV Case Studies*. Wiley, New York. 447 pp.

Doty, M. S. 1971. Measurement of water movement in reference to benthic algal growth. *Bot. Mar.* 14, 32–5.

Harger, J. R. E. 1970. The effect of wave impact on some aspects of sea mussels. *Veliger* 12, 401–14.

Hoerner, S. F. 1965. *Fluid-Dynamic Drag*. Hoerner Fluid Dynamics, Brick Town, N.J. 454 pp.

Jones, W. E., and Demetropoulos, A. 1968. Exposure to wave action: Measurements of an important ecological parameter on rocky shores of Anglesey. *J. Exp. Mar. Biol. Ecol.* 2, 46–53.

Keulegan, G. H., and Carpenter, L. H. 1958. Forces on cylinders and plates in an oscillating fluid. *J. Res. Natl. Bur. Stand.* 60, 423–40.

LaBarbera, M., and Vogel, S. 1976. An inexpensive thermistor flowmeter for aquatic biology. *Limnol. Oceanogr.* 21, 750–6.

Muus, B. J. 1968. A field method for measuring "exposure" by means of plaster balls. *Sarsia* 34, 61–8.

Sarpkaya, T. 1974. *Periodic Flow about Bluff Bodies*, part 1: *Forces on Cylinders and Spheres in a Sinusoidally Oscillating Fluid*. Technical Report NPS-595674091, Naval Postgraduate School, Monterey, Calif. 89 pp.

Sarpkaya, T., and Isaacson, M. 1981. *Mechanics of Wave Forces on Offshore Structures*. Van Nostrand Reinhold, New York. 651 pp.

Schlichting, H. 1979. *Boundary Layer Theory*, 7th ed. McGraw-Hill, New York. 830 pp.

Thomson, W. T. 1981. *Theory of Vibration with Applications*, 2nd ed. Prentice-Hall, Englewood Cliffs, N.J. 493 pp.

Thornton, E. B., Galvin, J. J., Bub, F. L. and Richardson, D. P. 1976. Kinematics of breaking waves. In *Proceedings of the 15th Coastal Engineering Conference*, pp. 461–76. American Society of Civil Engineers, Honolulu.

Vogel, S. 1981. *Life in Moving Fluids*. Willard Grant Press, Boston. 352 pp.

2: Light

J. RAMUS

Botany Department and Marine Laboratory, Duke University, Beaufort, North Carolina 28516

CONTENTS

I. Introduction

Light has several properties to which algae respond, namely magnitude (quantity), spectral distribution (color), spatial distribution, and variability in time. The appropriate measure of these properties includes one or more of three parameters: light available, light absorbed, and light utilized. Furthermore, the quantitative measurement of biologically useful light requires a thorough understanding of available instruments and the specific purpose of the data acquisition. However, light measurements need not be unnecessarily complex. The contributions of physical scientists to the biological sciences in the past several decades have greatly revolutionized and reduced the technical problems. The electronics industry responded to the demands with relatively simple, low-cost, reliable instruments designed for biological application. The appropriate choice and use of these instruments are addressed here.

A particularly important constraint in ecology is *scale*, the match in time and space between biological response and the environmental forcing function. For example, when an algal cell has the potential to divide one to several times per day, then knowledge of temporal scales of light availability becomes critical for understanding growth rates. Light incident (I_0) to the hydrosphere is quite variable in time, the greatest variability occurring within a day (Fig. 2–1) and between days (Figs. 2–1 and 2–2). This observation, among others, has caused some investigators (e.g., Harris 1980) to challenge equilibrium ($dX/dt = 0$) solutions to ecological models. Although mathematically tractable, the assumption of steady-state (equilibrium) conditions is not always appropriate. As demonstrated here, I_0, at least at the temporal scale of physiological response (~ 1 d), is seldom steady state; rather, it fluctuates. Therefore, the researcher must make measurements at the appropriate time scale.

The transmission of light through water is a function of the kinds and concentration of dissolved materials and the discreteness of suspended materials. Thus, lotic, riverine, lacustrine, estuarine, neritic, and oceanic waters are expected to have their own peculiar transmission characteristics. Anomalies occur within a water type,

[34]

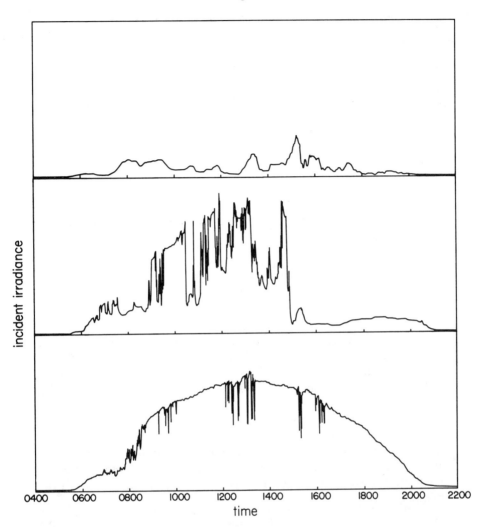

Fig. 2–1. Continuous recordings of incident irradiance (relative units) at Beaufort, North Carolina, on three successive days in August 1979 by an Eppley 8-48 pyranometer.

for example, density stratification, accumulation of suspended particles at a pycnocline, or resuspension of particles at the sediment–water interface. The challenges encountered in measuring optical properties vary as well, for example, those imposed by wind velocity, wave height, platform stability, current velocity, and depth of the water column.

Absorption and scattering are fundamental processes that determine the propagation of light in natural waters. Absorption is a

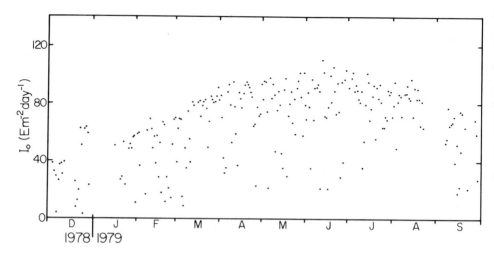

Fig. 2–2. Day-rate integrals for incident PAR quantum irradiance at Beaufort, North Carolina, from December 1978 through September 1979 using a Li-Cor 193-SB spherical quantum sensor coupled to a Li-Cor 550 printing integrator.

change of light energy to other forms, whereas scattering is a change in the direction of light without loss of energy. In a pure absorbing (homogeneous) medium, the loss of light due to absorption in a collimated (made parallel) beam of monochromatic light is given by the Beer–Lambert law,

$$I_z = I_0 e^{-az} \tag{1}$$

where I_0 is the light incident, I_z is the light at distance z, and a is the absorption coefficient. Similarly, in a pure scattering (heterogeneous) medium, the light redirected from a collimated beam of monochromatic light is given by

$$I_z = I_0 e^{-bz} \tag{2}$$

where b is the volume scattering coefficient. Attenuation is the sum of absorption and scattering; thus, $a + b = k$, where k is the beam attenuation coefficient, here in units of reciprocal meters (m^{-1}). The light lost from a collimated monochromatic beam of light in a scattering and absorbing medium is

$$I_z = I_0 e^{-(a+b)z} = I_0 e^{-kz} \tag{3}$$

This can be rewritten as

$$T_z = I_z I_0^{-1} = e^{-kz} \tag{4}$$

where T_z is the transmittance over a distance z in meters. Thus, in natural waters, sunlight is attenuated as

$$I_z = I_{-0}e^{-kz} \qquad (5)$$

where I_{-0} is the irradiance just below the surface, and the attenuation coefficient k includes the effects of both scattering and absorption over all wavelengths. The difference $I_0 - I_{-0}$ accounts for the light lost at the air–water interface by albedo, the sum of reflectance and backscatter. The value of albedo $(I_0 - I_{-0})$ is difficult to predict because it depends on the solar elevation, degree of overcast (diffusivity), particulate matter load, and sea state. It is least, however, when the sun is at its zenith and increases exponentially when the elevation of the sun declines toward the horizon. The total value of albedo may result in a 75% surface energy loss (Weinberg 1976).

Scattering is the result of three physical phenomena: diffraction, refraction, and reflection. Scattering in the water column has two entirely different components, namely, the scattering produced by pure water (described by the Rayleigh theory) and scattering produced by suspended particles (described by the Mie theory). The scattering by pure water shows relatively small variations, affected only by changes in temperature and pressure. However, the total scattering coefficient is wavelength dependent and varies approximately as λ^{-4}; that is, it increases from long to short wavelengths. Variation in particle scattering, on the other hand, is dependent primarily on particle size or, in natural waters, on particle size distribution. With increasing particle size, forward scattering is intensified relative to backscattering. Particle scattering in surface waters is produced chiefly by large particles (>2 μm), and the effect is virtually independent of wavelength (Jerlov 1976).

The scattering process leads to a change in the distribution of light, which has far-reaching consequences for algae. The angular distribution of light incident to an alga is affected by its position in the water column. For neuston (e.g., floating cyanobacteria), the principal component will be from above and largely unaffected by scattering. For benthos (e.g., understory kelp), the principal components will be downwelling and sidewelling. For plankton (e.g., diatoms), the principal components will be downwelling, sidewelling, and upwelling.

Pure water, dissolved substances, and suspended particles all absorb light, and the absorption is wavelength (λ) dependent as well. Hence, the value of k varies with λ and for the sake of the ensuing discussion is referred to as the specific attenuation coefficient k_λ. Furthermore, the value of k_λ is nonlinear with the "discreteness" of particles in suspension; that is, k_λ is the product of the size of the particle (d) and the absorption coefficient of the cell material (a_{cm}). The limiting

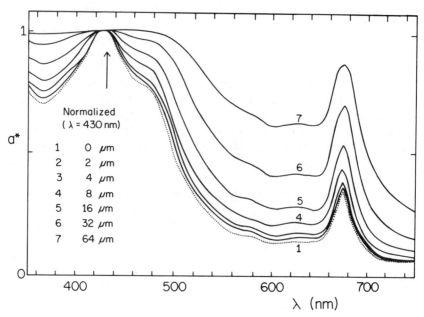

Fig. 2–3. Change in spectral absorption values with cells of variable diameter but similar composition. The spectral absorption values of this material, somewhat arbitrarily adopted, are shown as the dotted plot. All curves are normalized at 430 nm to show progressive deformation. Intracellular chlorophyll *a* concentration is assumed to be a constant 2.86×10^6 mg·m^{-3} (a^* is the specific absorption coefficient). Redrawn from Morel and Bricaud (1981).

value of this function ($d \rightarrow 0$) describes the case of a solution of like material, whereas variations in d and a_{cm} imply that k_λ is variable in magnitude and in spectral behavior. Consequently, the Beer–Lambert law, which rests on the existence of a constant k_λ, cannot generally apply when seston, especially phytoplankton populations, intervene in the canopy. In short, the value of k_λ varies with the size of the phytoplankton cells (Fig. 2–3) and their pigment content. The fact that light-harvesting capacity per unit pigment depends on cellular architecture, cell density, and pigment composition, the so-called packaging effect of Bannister (1974), has been explored in detail by Kirk (1975), Platt and Jassby (1976), Ramus (1978), and Morel and Bricaud (1981).

The wavelength dependence and nonlinearity of k change the spectral distribution of transmitted light with depth (Fig. 2–4). It might be expected that each natural water column has a characteristic set of optical properties. However, by synthesizing available data on spectral transmittance of downward irradiance at high solar latitudes, Jerlov (1951) proposed a classification of optical water types: oceanic

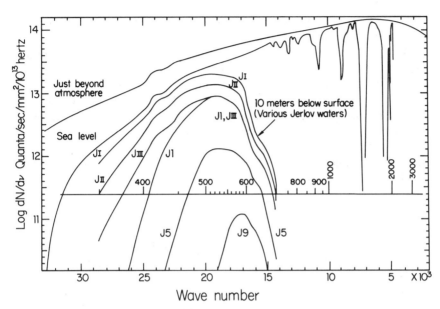

Fig. 2–4. Solar spectral irradiances outside the earth's atmosphere, at sea level, and 10 m below the surface of Jerlov-type oceanic (I, II, and III) and neritic (1, 5, and 9) waters. Irradiances (the ordinate) are scaled as log quantal flux (dN/dv), that is, as quanta per unit time per unit area per unit frequency interval. Frequency (the abscissa) is scaled as wavenumber, that is $1/\lambda_{cm}$. Dartnall (1975) prefers a frequency scale (rather than a wavelength scale) because λ_{max} values appear most clearly when graphed in this form. Redrawn from Dartnall (1975).

types I, II, and III and coastal types 1, 3, 5, 7, and 9. The extremes are represented by the clearest oceanic type I and the most turbid coastal type 9. The downward irradiance attenuation coefficient k_d increases from type I to 9; the wavelength of lowest K_d (greatest transmittance) for type I is 475 nm and for type 9 is 575 nm. Thus, at the limits of the photic zone (arbitrarily defined as 0.01 I_0) in oceanic type I water the light field appears to be blue, whereas in coastal type 9 the light field appears to be yellow. The ecological significance of spectral distribution to algae is the subject of debate (see Dring 1981; Lüning 1981; Ramus 1981, 1982).

A detailed discussion of terminology, units, and conversion factors germane for the phycologist was written by Lüning (1981) and is not repeated here save for a few conventions recommended by the SCOR Working Group 15 on photosynthetic radiant energy (Tyler 1975). The fundamental quantity to be measured is radiation, in this case light, incident or absorbed per unit area per unit time, specifically radiation flux density, or irradiance. This term is not to be confused with "intensity," which is a property of the light source. Irradiance can be expressed in units of energy, power, or quanta (photons). However, most, if not all, algal responses to light are photochemical in nature, and the primary act in photochemical reactions depends on the number of quanta absorbed. Therefore, it is most appropriate to express irradiance in terms of the number of quanta available or absorbed per unit area per unit time. Many authors now favor microeinsteins per square meter per second ($\mu E \cdot m^{-2} \cdot s^{-1}$), where an einstein is a mole quantum (Avogadro's number, 6.02×10^{23} quanta). For photosynthetic or productivity studies, the researcher should limit spectral distribution to photosynthetically active radiation (PAR), namely, 400–700 nm.

II. Radiometers

Radiometry is the measurement of the properties of radiant energy with a sensor of uniform spectral sensitivity, whereas in photometry the spectral sensitivity coincides with that of the standard (CIE, Commission Internationale de L'Eclairage) human eye. Thus, photometry is of little use to the photobiologist, and only state-of-the-art radiometry is discussed here.

A. Sensors

Two basic types of radiant-flux-receiving devices are in common use: thermoelectric and photoelectric. Encased thermoelectric detectors, known as thermopiles, pyranometers, and pyroheliometers, degrade radiant flux to heat energy on a blackened receiver. The heat energy causes the temperature of the receiver to rise as a measure of the radiant flux received. The temperature change is determined by the electric resistance of the receiver or by an electromotive force (EMF) or current produced in the thermocouple junctions of a thermopile. Thermoelectric devices are excellent for measuring radiant energy flux over the entire insolation spectrum (300–2800 nm). Presently, the Smithsonian Radiation Biology Laboratory monitors solar radiation incident to four geographical locations: Barrow, Alaska; Rockville, Maryland; Washington, D.C.; and the Panama Canal Zone. The detectors are double-domed thermopile pyroheliometers with circular black receivers produced by Eppley Laboratory. Six detectors, each equipped with a differently colored outer dome (long-wavelength cutoff filters), monitor ~100-nm bandpass regions for spectral distribution between the near-ultraviolet and the near-infrared regions. Without such an elaborate pyranometer array, the langleys per minute ($ly \cdot min^{-1}$) data are theoretically not easily converted to PAR data. An Eppley black and white pyranometer (model 8-48) was used to collect radiation simultaneously with a Li-Cor 193 SB spherical quantum sensor for 87 d in Beaufort, North Carolina. A regression analysis of the pyranometer output (langleys per day) on the quantum sensor (PAR, microeinsteins per square meter per day) gave $r^2 = 0.960$. Therefore, rough conversions from pyranometer to PAR data can indeed be made with a pyranometer calibrated to a PAR meter, at least for solar radiation day-rate integrals.

The photoelectric sensor in most general use is the photovoltaic cell, a semiconductor film that is located between two electrodes and generates an EMF upon absorption of photons. The voltage generated is sufficient to power a microammeter directly. The most popular semiconductor material forming the barrier layer in the past was selenium but is now mostly silicon. The silicon photodiode produces a higher current, has better linearity, and is less susceptible to fatigue than the selenium photodiode. A sensor that responds equally to photons (i.e., "counts" photons) should show a linear energy response with wavelength. The energy of a photon is inversely proportional to its wavelength:

$$E = h\nu = hc/\lambda \qquad (6)$$

Therefore, the ideal PAR quantum sensor should have an energy response slope between 400 and 700 nm of 1% per 7 nm normalized at 700 nm. A silicon photodiode gives this ideal quantum response when fitted with a visible bandpass interference filter (for a sharp cutoff at 700 nm) coupled with colored optical glass filters (Fig. 2–5).

B. Collectors

Irradiance is, strictly speaking, the radiant flux incident on an infinitesimal element of a surface containing the point under consideration divided by the area of that element (Morel and Smith 1982). Biologists are concerned primarily with incident, downward, and upward irradiance, all defined for horizontal surfaces. When a parallel beam of radiation of given cross-sectional area spreads over a flat surface, the area that it covers is proportional to the cosine of the angle (θ) between the beam and a plane normal to the surface. This is the Lambert cosine law, which is written

$$I_0 = I \cos \theta \qquad (7)$$

A flat plate sensor (surface area SA $= \pi r^2$) will conform to the law only if fitted with a cosine collector, that is, a translucent plate (often an opal glass diffuser) capable of transmitting radiant energy from different angles. For a detailed treatment of cosine collectors, see Smith (1969).

However, cosine-corrected sensors measure only horizontal irra-

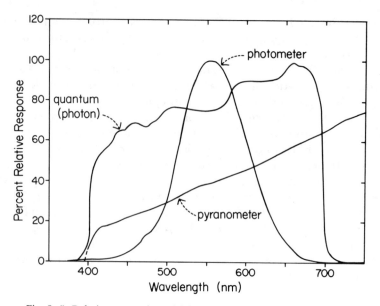

Fig. 2–5. Relative spectral sensitivity of light sensors in common use.

diance (Fig. 2–6), which may be totally inadequate insofar as most algae live in a three-dimensional, and more importantly, a diffuse-light environment. Therefore, for benthic algae, hemispherical collectors (SA = $2\pi r^2$, Fig. 2–6) are preferable, and for pelagic algae, spherical collectors (SA = $4\pi r^2$, Fig. 2–6) are preferable. These collectors (diffusers) more completely measure the irradiance in the ambient light field. It should be noted that, with a spherical collector, the integral of the radiance distributed around the sensor is measured, a quantity called scalar or spherical irradiance. If the spectral sensitivity of the sensor is PAR quantum irradiance (photosynthetic photon flux density, PPFD), then the quantity measured is called photosynthetic photon flux fluence rate (PPFFR) by crop scientists (Shibles 1976) and quantum scalar irradiance by oceanographers and limnologists. Note that there is no unique relationship between PPFD and PPFFR. An ideal PPFFR sensor placed in a uniform radiance distribution (perfectly diffuse radiation) would indicate a PPFFR that is four times higher than the PPFD measured by an ideal cosine-corrected sensor also placed in such a uniform radiance distribution, namely, $4\pi r^2/\pi r^2$. In practical situations, the ratio is somewhere between 1 and 4. To avoid confusion for making comparisons of available light, the type of collector must be specified.

In highly turbid waters, spherical sensors will indicate high PPFFR values due to the displacement of water by the sensor sphere volume.

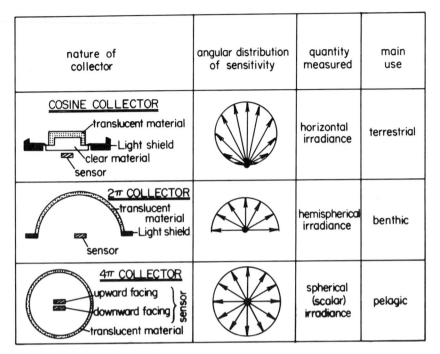

nature of collector	angular distribution of sensitivity	quantity measured	main use
COSINE COLLECTOR		horizontal irradiance	terrestrial
2π COLLECTOR		hemispherical irradiance	benthic
4π COLLECTOR		spherical (scalar) irradiance	pelagic

Fig. 2–6. Various light collectors, their angular distribution of sensitivity, quantity measured, and primary application. Redrawn from Weinberg (1976).

This (the displacement error) is because the point of measurement is taken to be at the center of the sphere, but the attenuation that would have been provided by the water within the sphere is absent. The error is typically ±6% for water with an attenuation coefficient k of 3 m^{-1}, that is, highly turbid water.

A cosine-corrected sensor will have an immersion effect when immersed in water. Radiation entering the diffuser scatters in all directions within the diffuser, with more radiation lost through the water–diffuser interface than the air–diffuser interface. Most manufacturers supply immersion correction factors for underwater sensors.

C. Measuring systems

Of course, the sensor/collector package must be interfaced with other components, the sophistication of which depends on the requirements and resources of the investigator. The least requirement is a readout instrument(s), as a meter, recorder, and interval printer. Most readout instruments have the capacity for integrated measurements, time scaling from seconds (usually the time constant of the instrument)

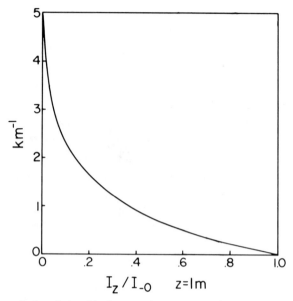

Fig. 2–7. Hyperbolic relationship between beam attenuation coefficient k and water transparency as measured by the ratio I_z/I_{-0}.

to solar days. Sensitivity range is also of paramount importance, especially in making time-scaled or underwater measurements, when at least seven orders of magnitude might be required. Some systems are equipped with microcomputer control and memories. The capacity for expansion is often necessary to interface the systems with compatible cassette tape recorders, tape readers, terminals, and computers. For field applications, portability is of prime concern. The system should have low power requirements, a self-contained power supply, weatherproofing, humidity control, durability, and modular electronics for easy repair.

If light versus depth profiles or attenuation coefficients k are required, there are two exacting requirements: a deck cell and a lowering frame. Insofar as the incident irradiance fluctuates over short time intervals (Fig. 2–1), I_{-0} and I_z must be measured simultaneously, unless the investigator opts for flawless insolation conditions. The lowering frame must be constructed so it produces no anomalies as it rotates about its vertical axis. It must be sufficiently weighted to keep the vertical axis normal to the horizon in currents. Ideally, the angular response of spherical (4π) sensors is uniform in all directions (Fig. 2–6). However, the geometric constraints of placing a single sensor at the base of an integrating sphere, the usual construction, produces an angular response that is not uniform.

Specifically, these sensors collect less upward than downward irradiance.

Culture studies require the measure of irradiance in confined spaces, such as flasks, chemostats or turbidostats, and outdoor tanks. Size-scaled sensor probes are available for this purpose (e.g., the Biospherical Instruments QSL 100 laboratory quantum scalar irradiance meter, which features a 2-cm-diameter, Teflon sphere mounted on a probe to collect light).

It is necessary to calibrate measuring systems. This is usually done by the manufacturer, but fatigue, spatial, or relative spectral response errors eventually affect the calibration. Several alternatives exist for recalibration. Most manufacturers offer a recalibration service for their own instruments, but this usually requires time and the hazards of shipping. Some measuring systems are equipped with an internal calibration source (lamp) traceable to the national Bureau of Standards (NBS). Finally, one can have on hand a calibration source traceable to the NBS, which, although preferable, is the most expensive alternative. Required are a high-quality lamp, a stabilized power source, and an optical bench.

III. Transparency

Dissolved and suspended materials determine the rate of light attenuation (k in Equation 5) in the water column, and the determination of k requires accurate and precise measurements. The attenuation coefficient varies from 0 to ∞ as a hyperbolic function of transparency (Fig. 2–7). In a given water column, transparency will vary in time, space, and wavelength bandwidth; hence, the notion of scaling light measurements to biological rate processes is again presented. For example, temporal variability might be expected in tidal estuaries, shallow systems acutely influenced by terrigenous, oceanic, and atmospheric factors. In fact, for the Newport River estuary, North Carolina, turbidity measured on an hourly basis over a 2-wk period was highly variable, the time constant for change was small, and the variability was patterned (Fig. 2–8).

A. Secchi disk

A time-honored device used to determine k is a standard 30-cm disk with white or alternating black/white or red/white quadrants. The disk is lowered into the water by means of a line until it disappears from the observer, which is the Secchi depth (D_{sd}) in meters. The attenuation coefficient can be calculated from the relation (Poole and Atkins 1929)

$$k = q(D_{sd}^{-1}) \tag{8}$$

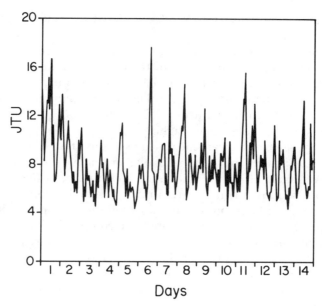

Fig. 2–8. Turbidity (in Jackson turbidity units, JTU) measured every hour for 14 d in the Newport River estuary, North Carolina, using a Turner Designs model 40 nephelometer.

However, reported values for the constant q vary, for example, 1.24 off Corsica (Weinberg 1976), 1.39 in the Mediterranean (Weinberg 1975), 1.41 in the North Sea (Gall 1949), 1.44 off California (Holmes 1970), and 1.70 in the English Channel (Poole and Atkins 1929). The value of D_{sd} is subject to considerable error, depending on sea surface state, position of the sun, meteorological conditions, distance of the observer from the sea surface, and visual acuity of the observer. Notwithstanding these sources of error, the Secchi disk was shown by Holmes (1970) and Weinberg (1976) to be a reasonably accurate device for determining the value of k, although the latter author proposed a new empirical hyperbolic relationship between k and D_{sd}:

$$k = 2.6/(D_{sd} + 2.5) - 0.048 \qquad (9)$$

B. Submersible radiometric sensors

The use of this class of instrument eliminates much of the error to which Secchi disk measurements are subject and in addition makes possible vertical profiles of light distribution. Water columns are seldom so uniformly mixed that they are homogeneous in the vertical axis (z); turbidity often is greatest at pycnoclines and sediment–water interfaces. Major sources of error, however, are fluctuations in I_0

during the course of I_z measurements. The solution is a "deck" sensor that measures I_0 or, better yet, one that is positioned at the air–water interface to measure I_{-0}. The circuitry of the meter should be designed to allow rapid comparisons of I_{-0} and I_z Another problem is encountered when one is taking measurements from a pitching vessel in turbid waters, namely, fluctuating I_z values. Thus, an integrating circuit in the metering system is useful, with time constants approximating the pitch frequencies of the vessel. Most commercial instruments are woefully inadequate in these two respects, requiring the user to make a large number of measurements and integrate the variability manually.

C. Transmissometers

In place of submersible radiometers, in situ transmittance/attenuance meters may be used to measure beam transmission and the concentration of suspended matter. Called transmissometers, these instruments allow rapid and accurate profiling in space and time. In these meters, a collimated beam of light emitted by a standard source passes through the column of water to be characterized and is measured by a detector. For example, the Sea Tech 25-cm transmissometer uses a light-emitting diode (λ_{max} = 660 nm, spectral line half-width = 20 nm) and a silicon photovoltaic detector. Collimation is produced by achromatic lenses and aperatures mounted next to the light source and receiver. Baffles are mounted to eliminate sunlight from the optical path. The optical path length may vary from 0.25 to 10 m, the longer path lengths created with reflecting mirrors without seriously affecting the size of the instrument package, which is generally mounted in a pressure housing. For horizontal profiling, the transmissometer is towed by a moving vessel and transmittance correlated with position. For vertical profiling, the instrument is raised and lowered from a stationary vessel, and for temporal profiling the instrument is moored in place. Calibration of the instrument for particle volume or weight must be done experimentally. The attenuation properties of a collection of particles depend on size, shape, and index of refraction; thus, the relationship between attenuance and particle volume or weight varies. This is especially true of phytoplankton blooms. Nevertheless, correlation coefficients for the two parameters of 0.80 to 0.98 are routinely obtained.

D. Nephelometers

Transmittance/attenuance and particle volume/weight can also be measured from water samples with a nephelometer. In this instrument, an intense beam of light is passed through the sample; light scattered to right angles of the beam is measured by a photocell, and the resulting electrical signal is amplified. The instrument is calibrated in nephelometric turbidity units (NTU), which are equivalent to Jackson turbidity units (JTU), used frequently by limnologists and oceanographers. Experimental calibration is required for conversion of turbidity units to attenuance/transmittance or particle volume/weight. Correlation coefficients r between JTU and D_{sd} for the Newport River estuary, North Carolina, for February, May, and August 1982 (Fig. 2–8) were $-.68$, $-.83$, and $-.82$, respectively, for ~200 degrees of freedom ($p < .001$). For the determination of attenuance/transmittance and particle volume/weight, nephelometry is probably the most accurate method at the least expense.

IV. Spectral distribution

The beam attenuation coefficient k includes absorption and scattering by pure water (k_w), by dissolved materials (k_s), and by particulate suspended matter (k_p); hence, $k = k_w + k_s + k_p$. Dissolved substances are primarily terrigenous humic materials that are yellow in color (*Gelbstoff*) insofar as they absorb in the blue region. As pointed out previously, k is wavelength dependent, and $k\lambda$ varies in space and time. If the distribution of photons in the environment becomes a matter of biological concern, then spectral irradiance (Fig. 2–9), as contrasted with total irradiance, must be measured. The instrument for this purpose is the spectroradiometer. Although technically available for some time, most spectroradiometers were custom fabrications, one-of-a-kind instruments. Examples are the venerable Scripps instrument (Tyler and Smith 1970) and the Gamma Scientific 3000R (Munz and McFarland 1973; McFarland and Munz 1975), which were used for the spectral characterization of underwater light fields. Mass-produced (here a relative term) instruments have only recently become available and are still in the development stage. The design features of spectroradiometers are similar and include (in line) a collector, monochromator, detector, and amplifier. This basic configuration is accomplished in at least three ways. The first design utilizes a diffraction grating or prism that separates the spectrum into component colors. The dispersive element is moved mechanically, bringing the spectral components sequentially to a photomultiplier or a silicon photodiode. The second type uses a fixed element to disperse the spectrum over the detector, such as a linear photodiode array or a video camera tube. The spectrum can then be electronically scanned at greater speeds than mechanical scans. A third type uses a series of silicon photodiodes, each with its own interference filter and electrometer amplifier to form an array that is electronically

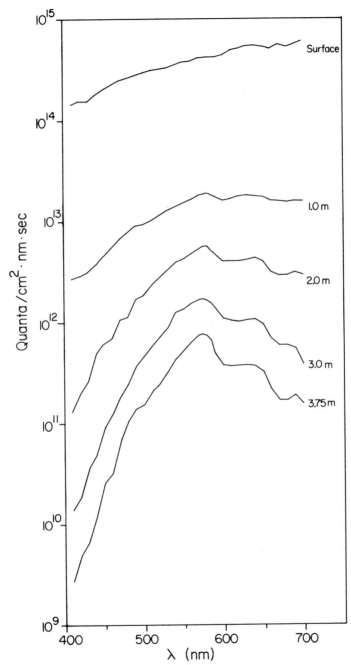

Fig. 2–9. Solar spectral irradiance taken at the surface, 1.0, 2.0, 3.0, and 3.75 m (bottom), Gallants Channel, Beaufort, North Carolina, with a Gamma Scientific 3010 underwater scanning spectroradiometer in August 1982.

scanned by an autoranging amplifier. For underwater use, packaging requires a pressure housing and provisions for remote activation, data acquisition, data storage, and data retrieval. Microelectronics allow miniaturization of most of these functions.

New products are available for above- and below-surface scans. The Biospherical Instruments MER 1000 is based on a prototype described by Booth and Dustan (1979) and used on Caribbean reefs (Dustan 1982). It is fitted with a silicon photodiode-interference filter array giving 12 optical channels from 400 to 700 nm. The manufacturer claims 10-ms scans, which is of considerable advantage. The Li-Cor LI-1800 UW uses a holographic monochromator and silicon photodiode. Scan time for 400 to 700 nm is ~10 s, but the trace will be continuous.

V. Acknowledgments

I wish to acknowledge the unpublished data contributed to this text by colleagues at the Duke University Marine Laboratory: R. T. Barber and J. Kogelschatz for the pyroheliometer recordings in Fig. 2–1; Gene Rosenberg for the quantum irradiance data in Fig. 2–2; C. S. Duke and R. W. Litaker for the turbidity readings in Fig. 2–8; and R. B. Forward for spectral irradiance on depth in Fig. 2–9. This chapter was written during the tenure of NSF Research Grant OCE81-13328.

VI. References

Bannister, T. T. 1974. A general theory of steady state phytoplankton growth in a nutrient saturated mixed layer. *Limnol. Oceanogr.* 19, 13–30.

Booth, C. R., and Dustan, P. 1979. Diver-operable multiwavelength radiometer. *Proc. Soc. Photo-opt. Instr. Eng.* 196, 33–9.

Dartnall, K. J. A. 1975. Assessing the fitness of visual pigments for their photic environments. In Ali, M. A. (ed), *Vision in Fishes*, pp. 543–63. Plenum Press, New York.

Dring, M. J. 1981. Chromatic adaptation of photosynthesis in benthic marine algae: an examination of its ecological significance using a theoretical model. *Limnol. Oceanogr.* 26, 271–84.

Dustan, P. 1982. Depth-dependent photoadaption by zooxanthellae of the reef coral *Montastrea annularis*. *Mar. Biol.* 68, 253–64.

Gall, M. H. W. 1949. Measurements to determine extinction coefficients and temperature gradients in the North Sea and English Channel. *J. Mar. Biol. Assoc. U.K.* 28, 757–80.

Harris, G. P. 1980. Temporal and spatial scales in phytoplankton ecology: mechanisms, methods, models and management. *Can. J. Fish. Aquat. Sci.* 37, 877–900.

Holmes, R. W. 1970. The Secchi disc in turbid coastal waters. *Limnol. Oceanogr.* 15, 688–94.

Jerlov, N. G. 1951. Optical studies of ocean water. *Rep. Swed. Deep Sea Exped.* 3, 1–59.

Jerlov, N. G. 1976. *Marine Optics*. Elsevier, Oxford. 231 pp.

Kirk, J. T. O. 1975. A theoretical analysis of the contribution of algal cells to the attenuation of light within natural waters. 1. General treatment of suspensions of pigmented cells. *New Phytol.* 75, 11–20.

Lüning, K. 1981. Light. In Lobban, C. S., and Wynne, M. J. (eds.), *The Biology of Seaweeds*, pp. 326–55. Blackwell, Oxford.

McFarland, W. N., and Munz, F. W. 1975. The photic environment of clear tropical seas during the day. *Vision Res.* 15, 1063–70.

Morel, A., and Bricaud, A. 1981. Theoretical results concerning light absorption in a discrete medium, and application to specific absorption of phytoplankton. *Deep-Sea Res.* 28A, 1375–93.

Morel, A., and Smith, R. C. 1982. Terminology and units in optical oceanography. *Mar. Geodesy* 5, 335–49.

Munz, F. W., and McFarland, W. N. 1973. The significance of spectral position in the rhodopsins of tropical marine fishes. *Vision Res.* 13, 1829–74.

Platt, T., and Jassby, A. D. 1976. The relationship between photosynthesis and light for natural assemblages of coastal marine phytoplankton. *J. Phycol.* 14, 352–62.

Poole, H. H., and Atkins, W. R. G. 1929. Photo-electric measurements of submarine illumination throughout the year. *J. Mar. Biol. Assoc. U.K.* 16, 297–34.

Ramus, J. 1978. Seaweed anatomy and photosynthetic performance: the ecological significance of light guides, heterogenous absorption and multiple scatter. *J. Phycol.* 14, 352–62.

Ramus, J. 1981. The capture and transduction of light energy. In Lobban, C. S., and Wynne, M. J. (eds.), *The Biology of Seaweeds*, pp. 458–92. Blackwell, Oxford.

Ramus, J. 1982. Engelmann's theory: the compelling logic. In Srivastava, L. (ed.), *Synthetic and Degradative Processes in Marine Macrophytes*, pp. 29–46. De Gruyter, Berlin.

Shibles, R. 1976. Committee report: terminology pertaining to photosynthesis. *Crop Sci.* 16, 437–9.

Smith, R. C. 1969. An underwater spectral irradiance collector. *J. Mar. Res.* 27, 341–51.

Tyler, J. E. 1975. SCOR Working Group 15 (with UNESCO and IAPSO). Photosynthetic radiant energy: recommendation. *Proc. Sci. Commun. Occ. Res.* 10, 37–42.

Tyler, J. E., and Smith, R. C. 1970. *Measurements of Spectral Irradiance Underwater*. Gordon and Breach, New York. 103 pp.

Weinberg, S. 1975. Ecologie des octocoralliennes communs du substrat dans la region de Banyuls-sur-Mer: essau d'une méthode. *Bydr. Diek.* 45, 50–70.

Weinberg, S. 1976. Submarine daylight and ecology. *Mar. Biol.* 37, 291–304.

3: Nutrients

PATRICIA A. WHEELER

School of Oceanography, Oregon State University, Corvallis, Oregon 97331

CONTENTS

I. Objectives of measurements

A. Major nutrients

There are two primary objectives of environmental sampling and monitoring of nutrients: (1) determination of the extent to which nutrient availability affects macroalgal productivity and (2) determination of macroalgal nutrient uptake at environmental levels. The most extensive work to date has been concerned primarily with macronutrients that supply the nitrogen and phosphorus necessary for plant growth. The major nitrogenous nutrients usually considered are NO_3^- and NH_4^+; NO_2^- is also utilized (see DeBoer 1981) but is generally present in low quantities and not analyzed separately from NO_3^-. Inorganic phosphorus, PO_4^{3-}, is the only form of phosphorus that is routinely monitored.

B. Spatial and temporal variations

In most cases, the monitoring of environmental nutrient levels must be conducted by periodic discrete sampling schemes designed to give a representative coverage of expected spatial variations in the specific environment of interest. Obvious spatial dimensions for aquatic environments are depth in the water column and distance from major current patterns.

Temporal variations of major interest range from seasonal to tidal, although both larger and smaller temporal variations may be considered. In any given case, the investigator must make some judgment concerning the most relevant spatial and temporal scales.

C. Flux determinations

Although monitoring of nutrients provides some information on nutrient availability, it gives no indication of the actual rate of nutrient utilization or supply (i.e., flux). In order to assess the flux of nutrients under any given set of environmental conditions, it is necessary to know both the concentration of nutrients and the rates at which they are utilized. The most common approach to obtaining the latter estimate is kinetic analysis of short-term uptake rates (see

[54]

Harlin and Wheeler, Chap. 24). The importance of flux determination is illustrated by Topinka's (1978) study of nitrogen utilization by *Fucus*. He found that, although NH_4^+ and NO_3^- concentrations ranged from 1.2 to 2.1 and 1.6 to 12.6 μg-atom nitrogen per liter, respectively, NH_4^+ accounted for ~40% of the environmental nitrogen utilized by *Fucus*. Thus, it must be stressed that monitoring nutrient levels provides very limited information on the relationship between environmental supply rates and macrophyte productivity. Whenever possible, nutrient levels should be monitored in conjunction with some assessment of rates of utilization.

II. Equipment and supplies

A. *Equipment*

The suite of nutrient analyses for the major macronutrients, $NO_3^- + NO_2^-$, NH_4^+, and PO_4^{3-}, can be most conveniently performed by standard spectrophotometric techniques. A prism or grating spectrophotometer, operable from 543 to 885 nm and equipped to accommodate absorption cells, of both 1- and 10-cm path length, is required. If NO_3^- levels are high, 1-cm cells can be used. For low NO_3^- levels and for NH_4^+ and PO_4^{3-} determinations, 10-cm cells are necessary.

For nitrate analysis, cadmium–copper reduction columns are required. The columns can be easily fabricated, and the size and design are somewhat flexible (for an example see Fig. 3–1). To prepare the reduction columns, the cadmium granules should be freed of oxides by washing with 2 N HCl and then stirred for several minutes with 2% (w/v) of copper sulfate ($CuSO_4 \cdot 5H_2O$). The copper sulfate should be poured off and the granules rinsed well with deionized or glass-distilled water. A small plug of copper turnings should be inserted into the column, which can then be filled with dilute ammonium chloride (Sec. II.B) and the cadmium–copper mixture. Packing of the column is facilitated by gentle tapping during additions. When nearly full, another plug of copper turnings should be inserted. Between uses, the column should be kept filled with dilute ammonium chloride. Grasshoff (1976) recommends activating the reducing columns whenever they have been out of use for several days by pouring through dilute ammonium chloride containing 100 μg-atoms NO_3^-–N per liter followed by a thorough rinse with dilute ammonium chloride. The columns should be regenerated after about 100 determinations. This is done by emptying the cadmium–copper mixture into a beaker and repeating the HCl rinsing and copper plating as described earlier. Strickland and Parsons (1972), Eppley (1978), and Grasshoff (1976) can be consulted for additional detail.

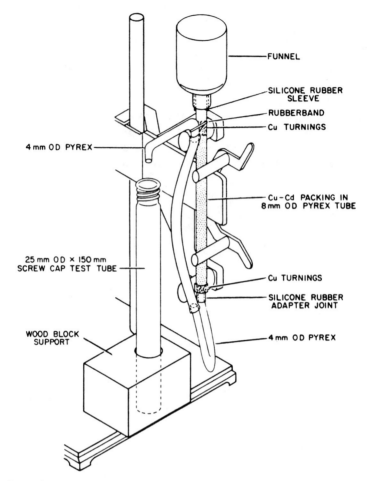

FUNNEL

SILICONE RUBBER
SLEEVE

RUBBERBAND

Cu TURNINGS

4 mm OD PYREX

Cu–Cd PACKING IN
8 mm OD PYREX TUBE

25 mm OD × 150 mm
SCREW CAP TEST TUBE

Cu TURNINGS

SILICONE RUBBER
ADAPTER JOINT

WOOD BLOCK
SUPPORT

4 mm OD PYREX

Fig. 3–1. Nitrate reduction column. The length of the column is 15 cm overall, with an 11-cm length of Cu–Cd packing. The funnel is a 50-ml plastic bottle cut off near the base and inverted. A short length of silicone rubber tubing serves as a sleeve to effect a good seal between the neck of the funnel and the column. The liquid level in the column (it must not be allowed to go dry) is maintained by the position of the 4-mm Pyrex outlet tube. A length of flexible tubing such as Tygon, attaching the outlet tube to the column, allows easy positioning. From Eppley (1978).

Other materials necessary for routine nutrient sampling are polyethylene collection bottles and glass reaction vessels. Erlenmeyer flasks (125 ml) are suitable for all analyses, but in most cases (see Strickland and Parsons 1972), screw cap test tubes are more convenient because they use less space.

B. Reagents and supplies

1. NO₃⁻ + NO₂⁻ analysis

Reduction columns:

1. Granulated cadmium (99.9% purity) between 0.5 and 2.0 mm in longest dimension or between 40 and 60 mesh.
2. Copper turnings (very fine copper "wool").
3. 2% (v/v) copper sulfate ($CuSO_4 \cdot 5H_2O$).
4. Concentrated ammonium chloride.
5. Dilute ammonium chloride. Dissolve 10 g ammonium chloride in ~950 ml deionized water, adjusting the pH to 8.5 with ~1.5 ml concentrated ammonium and bringing the final solution to 1-liter volume.

Analytical reagents:

1. Sulfanilamide solution. Dissolve 5 g sulfanilamide in 50 ml concentrated HCl and 300 ml deionized water; dilute to final volume of 500 ml. Solution is stable for months.
2. *N*-(1-Naphthyl)ethylenediamine dihydrochloride (NEDA) solution. Dissolve 0.5 g in 500 ml of distilled water and store in brown bottle in the refrigerator. Replace stock solution monthly or whenever a brown color develops.

2. NH₄⁺ analysis

1. High-quality deionized water. It is difficult to obtain water completely free of ammonia. We routinely use distilled water that has been passed through a Milli-Q ion-exchange system and obtain acceptable blank values. Alternatively, glass-distilled water can be run through Metex (IWT, ion-exchange columns) in series. If neither of these procedures is feasible, Grasshoff (1976) should be consulted for a distillation-purification technique.
2. Phenol solution. Dissolve 20 g of crystalline analytical-reagent-grade phenol in 200 ml of 95% (v/v) ethyl alcohol.
3. Sodium nitroprusside solution. The disodium nitroprusside dihydrate $Na_2Fe(CN_5)NO \cdot 2H_2O$, also referred to as sodium ferricyanide, should be recrystallized to remove impurities and to produce finer crystals that facilitate the weighing of small quantities. This can be achieved by dissolving ~5 g in 20 ml deionized water, adding 400 ml 95% ethanol, storing at −5 to 0°C overnight, collecting crystals on filter paper, followed by complete drying. The reagent solution can then be made by dissolving 1.0 g of recrystallized sodium nitroprusside in 200 ml of deionized water. This solution should be

stored in a refrigerator in a brown glass bottle and is stable for at least 1 mo.

4. Alkaline reagent. Dissolve 100 g of sodium citrate and 5 g of sodium hydroxide (analytical reagent) in 500 ml of deionized water. This solution is stable indefinitely.

5. Sodium hypochlorite solution. Use a solution of commercial hypochlorite (e.g., Chlorox), which should be about 1.5 N. Strickland and Parsons (1972) describe procedures for checking solution strength.

6. Oxidizing solution. Mix 100 ml of alkaline reagent and 25 ml of hypochlorite solution. Keep stoppered when not in use and prepare fresh daily.

3. PO_4^{3-} analysis

1. Ammonium molybdate solution. Dissolve 15 g ammonium paramolybdate $(NH_4)MoO_{24} \cdot 4H_2O$ (analytical reagent) in 500 ml deionized water. Store in a plastic bottle out of direct sunlight. Solution is stable indefinitely.

2. Sulfuric acid solution. Add 140 ml concentrated sulfuric acid (analytical reagent, sp. gr. 1.82) to 900 ml dionized water. Store in a glass bottle.

3. Ascorbic acid solution. Dissolve 27 g ascorbic acid in 500 ml deionized water. Store in a plastic bottle and keep frozen between uses.

4. Potassium antimonyl tartrate (tartar emetic). Dissolve in 250 ml deionized water, warming if necessary. Solution is stable for many months.

5. Mixed reagent. Mix together 100 ml ammonium molybdate, 250 ml sulfuric acid, and 50 ml potassium antimonyl tartrate solutions. Prepare for immediate use only, and discard excess. Do not store for more than 6 h.

III. Methods

A. Sample collection, filtration, and storage

Samples should be collected in polyethylene bottles that have been rinsed twice with sample water before filling. As a general procedure it is recommended that samples be filtered as soon as possible after collection. Glass fiber filters (e.g., Whatman GF/C or GF/F) will serve to remove particulate material greater than 0.6–0.8 μm and have reasonably high flow rates. Filtration removes larger particles that might produce turbidity and interfere with nutrient analyses as well as larger microorganisms, which could remove significant quantities of nutrients if the samples are not analyzed immediately. A mild vacuum (≤ 150 mm Hg) should be used to minimize disruption of fragile cells and subsequent release of intracellular nutrient pools.

It is best to analyze samples as soon as possible after collection. If any length of time passes before samples can be analyzed, bottles should be kept cold and in the dark. Samples that cannot be analyzed within 2 h should be frozen. However, frozen storage is unacceptable for precise NH_4^+ determinations (Carpenter and McCarthy 1975).

B. Analyses

1. NO_3^- + NO_2^-. Add 0.5 ml concentrated NH_4Cl solution to each 25-ml sample. Let 10 ml of this sample pass through the column as a rinse and discard. Pass a second 10-ml aliquot through the column and collect in a clean test tube; save for analysis. Pass a 10-ml volume of dilute NH_4Cl through a column before introducing another sample.

To develop color of nitrate formed, add 0.2 ml of sulfanilamide solution and 0.2 ml of NEDA solution, mixing well after each addition. Color development is complete within 10 min. Measure absorbance at 543 nm, and calculate nitrate concentrations by comparison with a standard curve (using reagent-grade KNO_3) that covers the range of nitrate values analyzed and includes a reagent blank.

2. NH_4^+. All glassware should be washed with dilute HCl and rinsed well with deionized water. Since clean glassware is essential, it is useful to burn out test tubes immediately before use by adding 10 ml deionized water and all reagents for NH_4^+ analysis. These wash samples can be discarded immediately, and tubes are ready for use after another rinse with deionized water.

Add 10 ml of sample, blank, or standard to each reaction tube. Sequentially add 0.4 ml phenol solution, 0.4 ml nitroprusside solution, and 1.0 ml oxidizing solution, mixing well after each addition.

Store samples in the dark 1–2 h for color development. Measure absorbance at 640 nm, and calculate ammonium concentrations by comparison with a standard curve (using reagent-grade NH_4Cl) that covers the range of values analyzed and includes a reagent blank.

3. PO_4^{3-}. To 10 ml of sample, add 1 ml of mixed reagent and mix immediately. After 5 min (within 30 min), measure absorbance at 885 nm, and calculate phosphate concentrations by comparison with a standard curve (using reagent-grade KH_2PO_4). A turbidity blank (absorbance of sample without reagents) should be determined and subtracted if unfiltered water is assayed.

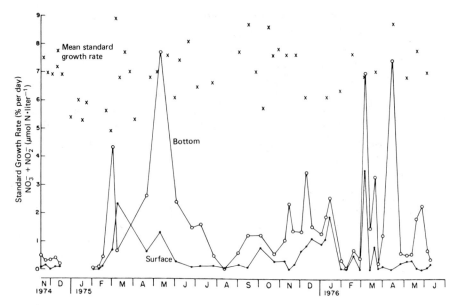

Fig. 3–2. Seasonal changes in NO_3^- + NO_2^- in a southern California kelp bed. Growth rates were determined by bimonthly measurements of frond length. Adapted from Wheeler and North (1981: Fig. 2.).

IV. Sample data

Seasonal changes in NO_3^- + NO_2^- in a southern California kelp bed are illustrated in Fig. 3–2. These data illustrate the distinct spring upwelling resulting in elevated NO_3^- concentrations, particularly in the lower portion of the water column. However, note the poor correlation between growth rate and concentration.

Short-term variations also occur, the most notable of which is the effect of tidal cycles. Hourly monitoring of the thermocline shows distinct tidal cycles (Fig. 3–3). In the absence of extensive mixing activity, temperature–nutrient relationships typical of stratified water (Fig. 3–4) can be established. As a result, by knowing the temperature history of a particular fixed location, one can calculate the amount of nutrient (nitrate) available to attached algae.

V. Critical evaluation

The procedures referred to in this chapter are the most sensitive and reliable spectrographic methods for major nutrient analysis currently in use. If reasonable care is taken in sample collection and sample storage is kept to a minimum, a precise monitoring of

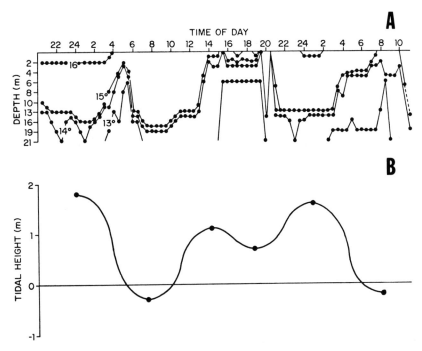

Fig. 3–3. (A) Surfaceward excursions of cold (nutrient-rich) water on a typical day off the coast of Santa Catalina Island, California. (B) Tidal height for same period. Unpublished data courtesy of R. Zimmerman.

nutrients levels should be achieved for the temporal and spatial scales chosen for analysis.

A potential problem with the use of a discrete sampling strategy is that important short-term and small-scale spatial variations may be overlooked. The degree to which this may be a general problem is difficult to evaluate; nonetheless, it should not be ignored completely.

A more serious problem is the common assumption that the nutrient in highest concentration is the predominant form that supports macrophytic growth. A case in point is the relative utilization of NH_4^+ and NO_3^-, which is not in direct proportion to external concentration. This illustrates the need to estimate the dynamics of nutrient use rather than simple correlations between ambient levels and some measure of algal production (see Fig. 3–1). A combination of techniques (e.g., nutrient monitoring and short-term uptake experiments) is essential to determine the dynamic relation between nutrient supply and algal production.

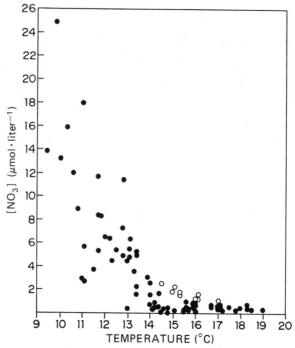

Fig. 3–4. Concentration of NO_3^- vs. temperature for waters off the coast of Santa Catalina Island, illustrating the relationship obtained for typical stratified water. Unpublished data courtesy of R. Zimmerman.

VI. Alternative techniques

A. Other nutrients

Only the major inorganic nutrients required by macrophytes have been considered here. Some marine macrophytes are able to utilize both inorganic and organic forms of nitrogen. In particular, those algae that grow well in eutrophic areas, such as *Ulva, Enteromorpha, Lithothrix,* and *Corallina,* can also accumulate amino acids (North et al. 1972). *Macrocystis,* however, cannot utilize either amino acids or urea (Wheeler 1979). Some organic forms of phosphorus can be utilized by macrophytes (for references, see DeBoer 1981).

Finally, in some circumstances it may be desirable to monitor trace elements. The number of essential nutrients that may play a significant environmental role in macrophyte production precludes a broad routine monitoring. However, in some cases, a specific trace-metal limitation can be identified (North 1980). In such cases environmental monitoring of micronutrient levels is warranted.

B. *Other analytical techniques*

The spectrophotometric procedures recommended here are widely used and can be readily conducted in any laboratory with a minimum amount of special equipment. Two alternative methods are possible when the necessary equipment is available. The first of these is the use of ion-specific electrodes. Ion electrodes are currently available for NO_3^- and NH_4^+ analyses and have been utilized in studies of NO_3^- uptake by freshwater phytoplankton (Tischner and Lorenzen 1979) and higher plants. However, the NH_4^+ electrodes lack the sensitivity necessary to monitor low ambient NH_4^+ concentrations, and NO_3^- electrodes do not work in seawater.

The second procedure for monitoring nutrients is the use of an autoanalyzer system (Haines and Wheeler 1978). Autoanalyzers are particularly useful if several nutrients are being assayed for each water sample. The main disadvantages are equipment costs and the requirement for skilled technical assistance.

VII. Acknowledgments

The procedures recommended in this chapter have become standardized through the efforts of J. D. H. Strickland, T. R. Parsons, and K. Grasshoff. We are all indebted to them for this great service. Significant improvements have been suggested by Eppley (1978). My own experience with optimal use of these methods results from time spent in the laboratories of W. J. North and J. J. McCarthy, who are both gratefully acknowledged. Support for the preparation of this chapter was provided by NSF Grants OCE-7826011 (JJM), OCE-8022990 (JJM), and OCE-8208873 (PAW) and funds from the School of Oceanography at Oregon State University.

VIII. References

Carpenter, E. J., and McCarthy, J. J. 1975. Nitrogen fixation and uptake of combined nitrogen nutrients by *Oscillatoria* (Trichodesmium) in the western Sargasso Sea. *Limnol. Oceanogr.* 20, 389–401.

DeBoer, J. A. 1981. Nutrients. In Lobban, C. S., and M. J. Wynne (eds.), *The Biology of Seaweeds*, pp. 386–92. University of California Press, Berkeley.

Eppley, R. W. 1978. Nitrate uptake. In Hellebust, J. A., and Craigie, J. S. (eds.), *Handbook of Phycological Methods: Physiological and Biochemical Methods*, pp. 401–9. Cambridge University Press, Cambridge.

Grasshoff, K. 1976. *Methods of Seawater Analysis*. Verlag Chemie, Weinheim, West Germany. 317 pp.

Haines, K. C., and Wheeler, P. A. 1978. Ammonium and nitrate uptake by the marine macrophytes *Hypnea musciformis* (Rhodophyta) and *Macrocystis pyrifera* (Phaeophyta). *J. Phycol.* 14, 319–24.

North, W. J. 1980. Trace metals in giant kelp, *Macrocystis. Amer. J. Bot.* 17, 1097–1101.

North, W. J., Stephens, G. C., and North, B. B. 1972. Marine algae and their relations to pollution problems. In Ruivo, M. (ed.), *Marine Pollution and Sea Life*, pp. 330–40. F. A. O. Fishing News (Books), London.

Strickland, J. D. H., and Parsons, T. R. 1972. *A Practical Handbook of Seawater Analysis*. Bulletin 167, Fisheries Research Board of Canada, Ottawa. 310 pp.

Tischner, R., and Lorenzen, H. 1979. Nitrate uptake and nitrate reduction in synchronous *Chlorella. Planta* 146, 287–92.

Topinka, J. A. 1978. Nitrogen uptake by *Fucus spiralis* (Phaeophyceae). *J. Phycol.* 14, 241–7.

Wheeler, P. A. 1979. Uptake of methylamine (an ammonium analogue) by *Macrocystis pyrifera* (Phaeophyta). *J. Phycol.* 15, 12–17.

Wheeler, P. A., and North, W. J. 1981. Nitrogen supply, tissue composition and frond growth rates for *Macrocystis pyrifera* off the coast of southern California. *Mar. Biol.* 64, 59–69.

Section II

Assessments of populations and communities

4: Collection, handling, preservation, and logistics

ROY T. TSUDA

Graduate School & Research, University of Guam, UOG Station, Mangilao, Guam 96913

ISABELLA A. ABBOTT

Department of Botany, University of Hawaii, Honolulu, Hawaii 96822

CONTENTS

[67]

I. Introduction

Good ecological field work requires good taxonomy, which is dependent on the quality of voucher specimens preserved as documentation. This chapter describes methods for qualitatively assessing the macroalgal flora of a study site, whether the site is a single bay, a coastline, or the reefs around an island. The various methods of collecting, handling, and preserving specimens in the field and laboratory for later taxonomic determination, as well as logistic considerations, are described and evaluated. Publications that provide excellent descriptions of field and laboratory methods of collecting, processing, and identification are as follows: Taylor (1950, 1960, 1962), Dawson (1966), Abbott and Hollenberg (1976), and Abbott and Dawson (1978). Key points of these treatments are highlighted and summarized in this chapter.

II. Field collecting

A. Equipment and supplies

Because the primary objective in a qualitative assessment is to collect as many species as possible, that is, to obtain a representative flora of the area, very little in the way of equipment and supplies is needed. If one plans to wade in the shallow waters near shore, foot protection is necessary. High-topped, athletic rubber-soled shoes (e.g., tennis or jogger shoes) or Japanese *tabi* are comfortable and will provide protection from jagged rocks, corals, and toxic marine animals (e.g., cone shells or the toxic-spined fishes that inhabit coral reefs). A glass-bottom "look box" or diving mask is useful. To assess the macroalgae of deeper waters, skin-diving gear, for example, face mask, snorkel, and swim fins, or even SCUBA equipment may be used (see Foster et al., Chap. 10); in this case, a small boat is very convenient as a diving platform and for depositing full bags of material.

One should use judgment about protective clothing. Some individuals who work in tropical waters prefer a wet suit with weight belt;

others wear coarse jeans or knee pads to protect against sharp organisms. Hats, shirts, or a protective lotion are mandatory to prevent sunburning. Obviously, a wet suit or dry suit is essential for diving in colder waters. Hip-height waders or wet-suit boots and trousers are necessary in temperate waters for inshore collecting. A first-aid kit is a necessary item to include on collecting trips.

In most cases, the algae can be collected by hand; however, a hammer, chisel, knife or spoon, geologist's pick, or mountain climber's ice ax are helpful for collecting smaller algae attached to hard substrata, collecting parts of larger algae (e.g., kelp), and removing holdfasts. In shallow waters, Whirl-pak bags that can be drained, sealed, and stored in larger bags or buckets are convenient. In deeper waters, a fine-mesh bag to hold the algae is best, although some collectors prefer perforated or unperforated plastic bags. Plastic vials with small holes are ideal containers for small, filamentous algae. It is very difficult to place a small specimen in an unperforated bag or vial underwater without loss of some portions of the specimen. In addition, an underwater writing slate or commercially available waterproof paper (Nalge Company) attached to a clipboard is essential for taking notes on natural history during collection. Some workers (e.g., Neushul 1965) have employed tape recorders in plastic bags or waterproof housings to take rapid floristic and ecological field notes. Tape-recorded notes are then transcribed in the laboratory.

Other supplies needed after collection are commercial Formalin or 95% ethyl alcohol (as preservatives), buckets and trays (for sorting), tweezers, plant press (for drying), bottles and vials, herbarium paper, labels (100% rag), and a permanent black ink pen.

B. Methods

In regions where the tidal amplitude is great, it is best to collect during the ebbing tide, because the water is usually clearer, especially at the period nearest mean lower low water (MLLW). Beach-drift specimens are best collected last, because fresher specimens can usually be found growing nearby and it is vastly preferable to determine the precise habitat.

After selection of the study site (collection locale), it does not matter where collecting begins as long as the algae taken from the different habitats are kept in separate containers. Each container label must accurately describe the habitat of the collection site. Experience has shown that one can collect a representative flora in a short period by simply wading or snorkeling perpendicular to the shore, whereas one can survey patchiness by swimming or wading parallel to shore. During SCUBA diving in subtidal waters, it is best to begin collecting at the deepest depth and work toward the shallower waters. Data on tidal height and zonation are very important, especially on reefs (i.e., inner reef flat, outer reef flat, reef margin, or reef front) and the deeper reef slope and terraces. The vertical ranges of species subsequently can be determined from accurate records of collection depths made on a tape recorder or with underwater writing materials, as well as photographs (see Littler and Littler, Chap. 8). The depth readings can be obtained from an accurately calibrated depth gauge during skin or SCUBA diving; time of collection must also be noted for tidal height corrections.

Ecologists frequently conduct population studies in which certain measured areas are scraped clean, and the material is sorted, weighed, acidified, dried, and reweighed (see De Wreede, Chap. 7). In sorting of the material, identification to genus is usually accomplished initially; samples should be kept for later, more specific identification. Species diversity studies also require that taxonomic vouchers of all algae be taken from measured areas. Here, it is important to have different-size classes of the same taxon, from adjacent areas if necessary, in order that a good representation of the morphological and ecological variation be understood by the person identifying the species. The small scraps of thalli frequently collected by marine ecologists in such studies are difficult to utilize even for experienced algal taxonomists.

In herbivory studies, collections within the area of the predators should be made (including crustose taxa). To reduce bulk, the gut contents of the animals may be preserved separately (using ~10% Formalin). The animals themselves are preserved differently from plants; most invertebrates require anesthetization before preservation (few zoologists will accept poorly preserved specimens for identification, and phycologists are beginning to follow suit). A preservative suitable for the whole animal is not necessarily adequate to preserve gut contents, and injection is usually required (see Vadas, Chap. 26). Time of feeding is important for collecting gut contents that can be identified. Preservation of animals varies by animal group, and *Light's Manual* (Smith and Carlson, 1975) should be consulted.

Some recommendations for collecting macroalgae are as follows. (1) When collecting epiphytes or smaller algae on solid substrata, break off the portion of the substratum on which the algae are growing; this will later provide substratum information, and there will be less chance of losing the specimens during sorting. (2) Collect the entire basal holdfast or rhizoidal portion with the specimen; these structures may greatly aid in later identification. (3) Make an effort to collect reproductive specimens, since these may also be critical for precise identification. Because the objective of the assessment and collection is to provide a qualitative checklist based on the taxonomic vouchers taken, collection of entire specimens of larger forms (e.g., kelps) is often unnecessary unless one suspects that they may represent variants or new species.

If time is limited, a considerable area can be surveyed by slowly (less than 3 knots) towing an observer/collector with a rope well behind a small boat (e.g., inflatable boat with a 15- to 25-horsepower outboard engine). Safety precautions comparable to those recommended for towing water skiers (e.g., established hand signals) should be followed. This includes an observer in the boat during towing. Wet suits or clothing worn during towing should be designed to avoid scooping water down the neck and sleeves. Fins should always be worn for maneuverability and to aid in quick exit from the water when the curious shark or barracuda inevitably appears. This method is especially effective for locating a broad range of habitats if applied safely in clear waters, especially when two divers are towed simultaneously.

In deeper, turbid, or otherwise unsafe waters, dredging may be the most practical and safest means of obtaining an algal collection; however, a ship with the capacity to dredge is seldom available for routine algal collecting. Bottom trawls that cover a considerable amount of area are much preferable to the grab sampler, which covers a very small area. As pointed out by Taylor (1960), bottom trawls are very difficult to operate in rough seas. In addition, the bottom topography must be considered, because there is danger of losing the trawl in rocky terrain. See Holme and McIntyre (1971) for detailed methods of collecting marine benthos from surface vessels.

Since some areas of a country or even the country as a whole may have special laws or permit requirements for collecting, individuals should determine these regulations and comply before entering the water. Even in areas where there may be no official laws against collecting, local customs may dictate acquiring permission from the resident population or authorities before field activities are begun.

III. Processing

A. Field processing

On a regional scale (Tsuda and Stojkovich 1980), it may be desirable to collect a set of specimens over a 1-yr period at monthly, bimonthly, or quarterly intervals to locate seasonal or ephemeral species and to document variability adequately (phenotypic plasticity). At a minimum, collections should be undertaken during each extreme season in a highly seasonal environment. Such information as the maturity and fertility of the thallus, habitat conditions, and environmental considerations (e.g., water temperature, salinity, light, and wave action) may prove to be important at a later time, to the chemotax-onomist (see Norris and Fenical, Chap. 6) as well as the ecologist and biogeographer (see Druehl and Foottit, Chap. 15).

When collecting is done over several days or weeks and laboratory facilities are unavailable for immediate processing, it is necessary to preserve the specimens after each day of collecting. One method is to preserve the specimens from each collection in plastic bags (e.g., Whirl-paks) or glass containers with either a 3–5% solution of commercial Formalin (=37% formaldehyde) or ethyl alcohol diluted to 65 to 70%. Seawater must be used to make these solutions, and a buffer, such as borax, should be added to the Formalin–seawater solution until it gives a pink reaction when tested with phenolphthal-ein.

As described by Taylor (1960), the Formalin solution is initially best for coarse species, and the specimens should be kept out of the light. Alcohol amounting to 30% added to the Formalin solution is recommended for more delicate specimens. In general, the ethyl alcohol solution is sufficient for all species, but there is a tendency for the pigments, especially chlorophyll, to extract and bleach. Some cell shrinkage occurs when any of these preservatives are used.

If only a limited amount of preservative is available, the preservative solution can be reused. The algae should be totally immersed for two or more days; then the excess preservative can be drained off, the algal specimens being left "damp-dry," and the preservative reused in new collections. The specimens can also be dried between newspaper (for coarse algae) or wax paper (for delicate algae) in a plant press or simply air-dried for a short period in the sun on a black plastic sheet and later placed in a shady area to complete the drying process. Also, hanging the algae in a mesh bag in a windy area will produce rapid drying; however, one must be careful to avoid clumping of the algae. Air-dried specimens, although brittle, can be resoaked and mounted on herbarium paper later.

Another method that can be used if no preservatives are available is the use of common rock salt. The process is quite similar to that used by fishermen in preserving fish. A large quantity of rock salt is added to seawater to form a strong brine solution. Taylor (1960:38) suggests that if "the brine first formed is drained and the algae relayered with fresh salt, many species, and even many rather delicate ones, may be preserved in this way for several years." One must quickly and thoroughly mix the salt with the specimens.

If a freezer is handy, one can freeze the algae-filled plastic bags; however, considerable cell damage may occur, especially in small, delicate species.

Each plastic bag should be numbered consecutively and keyed to its respective number in a field notebook. Short strips of herbarium paper (100% rag) make ideal labels if numbered with permanent black ink (e.g., India ink); these numbered labels are placed directly into the plastic bags. Information on the collecting site, date, habitat description, and anything else that is noteworthy should be recorded in the field notebook. It is also useful to record a general list of the visually dominant algae found in each habitat.

The plastic bags with preserved specimens and labels are then placed in a suitable shipping container (e.g., Liqui-Pak). "Ship-biscuit" aluminum 5-gal containers are also useful and can be obtained in many areas of the world. Any nonleaking container is adequate. When preparing preserved specimens for shipment (i.e., by vehicle, ship, or airplane), drain the preservative from each of the plastic bags into a suitable deep hole and bury it. Consolidate the bags into as few containers as possible. The dried specimens can be placed in plastic bags according to collection sites and shipped in cardboard boxes.

B. Laboratory processing

If the collections have been well labeled and roughly sorted as they lie in their plastic bags, they can be placed – bag and all – in large containers that have 3–5% Formalin poured over them. They should not be left in Whirl-paks that will rust over time or in containers from which the preservative evaporates. Except for certain brown algae, most marine algae will fade on standing in preservative; consequently, they should be sorted and voucher specimens prepared as soon as possible, generally within 1 to 2 mo. If previous arrangements have not been made for the care of excess materials or for their identification by colleagues and experts, it is reasonable to discard them at this time. Unless someone or some institution has announced an interest in one's algal collections, one should collect only as much as one can properly process oneself. It is a chore for someone else, and an added expense, to process collections conscientiously in order to save them rather than to study them.

1. Sorting. If the contents of plastic bags were previously rough-sorted when packed, they can be divided, for example, into groups of large and small specimens. It is usually preferable to work on the large specimens first – in order to be rid of the bulk more quickly. Place them in a bucket of seawater or fresh water, if they have been preserved in Formalin, rinsing off preservative, sand, and small animals and removing epiphytes if any are present. Many individuals are sensitive to the fumes of Formalin (it appears that such sensitivity increases with a person's age), and it may be necessary to work under

Table 4–1. *Common stains, fixatives, and preservatives used for benthic marine algae, including Corallinaceae*

	Reference
Preservatives	
3–5% formalin in seawater (treating commercial formaldehyde as 100%); ~0.06 liter borax is added to 4 liters liquid	—
35, 50, 70% ethyl alcohol, stored 70% in ethyl alcohol	—
Fixatives (for future cytological studies)	
FAA (Formalin–acetic alcohol)	Berlyn and Miksche (1976)
Karpetchenko (Papenfuss modification)	Papenfuss (1946)
Fixatives for coralline algae	
Susa solution	Suneson (1937)
Perenyi's solution	Mason (1953)
EDTA	Giraud and Cabioch (1976)
Trichloroacetic acid	Johansen (1969)
Stains (aqueous, suitable for corn syrup technique)	
Aniline blue[a]	Papenfuss (1937)
Delafield's hematoxylin	Sass (1958)
1% Fast green	Sass (1958)
1% Congo red	Sass (1958)
Crystal violet (glycerine method)[b]	Chemin (1929)

[a] Aniline blue is not a critical stain but a good general stain. It is used mainly by students of G. F. Papenfuss.
[b] Crystal violet is used by P. S. Dixon and his students. Similar results are obtained using light green or fast green in glycerine (Papenfuss 1937).

a fume hood or with specimens in running water. Use plastic or rubber gloves. Press the specimen (see Sec. III.B.2.b) or, in the case of a large, coarse brown alga such as *Sargassum* or a thick one such as *Halimeda*, place it in newspaper to dry (see Sec. III.B.2.a). Retain epiphytes and small specimens less than 4 cm tall in individual vials with appropriate labels including collection numbers and data transferred from field books. Letters are frequently added to the numbering system for individual small algae when a number of them occur epiphytically on the same host. The hosts for epiphytes should be recorded. The vials should be filled with 3 to 5% buffered Formalin (Table 4–1) or with 65 to 70% ethyl alcohol, whichever was used in the field collection. If specimens were not preserved in the field, a choice of substances is available for preservation (Table 4–

1). If specimens were previously preserved in alcohol, they are placed in a 1:1 alcohol/water solution for 15 to 20 min before pressing. Alcohol-preserved marine algae are usually less pliable and dry to a more brittle state than Formalin-preserved specimens. When a large number (50–100) of these small specimens have accumulated, put aside a few hours or a day to process them further by (1) examining under a microscope to identify at least to genus and to determine reproductive state, etc.; (2) pressing as herbarium specimens those large enough to be seen; and (3) making microscope slides of very small specimens or parts of larger specimens.

A further sorting step that will greatly facilitate retrieval can be taken at this time. In the preliminary sorting of large versus small specimens, categories of algae can be kept together by major phyla (colors of pigments) and thus taken up for identification in similar taxonomic groups. Blue-green algae (Cyanophyta), which usually constitute a small number, should be identified as soon as possible, because their gelatinous structures frequently disappear with further laboratory treatment, making it very difficult to identify them to species. If arrangements have been made for identification by an expert, his or her preferred method of preservation should be known before collections are made. Green algae (Chlorophyta) and brown algae (Phaeophyta) should be processed as herbarium specimens, except for very small taxa of browns, which should be preserved in Formalin or processed on microscope slides (see Sec. III.B.2.e). Red algae (Rhodophyta), which will be the most numerous, should be pressed immediately if the thalli are gelatinous or mucosoid; others will soon lose their color if left in preservative. A choice must be made, if time is short, between attractive specimens or material in a state that can be adequately studied. It is a generalization that, of all algal groups, red algae are the most difficult to identify from dried specimens. It is also true, however, that the ones bleached in preservative are not often in a condition to be shared and shown to others. The ideal method would be to remove small portions of red algae to vials with preservative while drying the remainder of the thalli as herbarium specimens.

2. Preservation as voucher specimens. Voucher specimens are those that serve as evidence or documentation that a given species *is* that species or was collected from a certain place. The literature is full of undocumented lists of algae that were supposedly collected at certain places. Workers that follow can never check earlier identifications, update the nomenclature, or evaluate old lists without voucher specimens. It is especially important for nontaxonomists to appreciate the value of taking voucher specimens – taxonomists keep specimens routinely. The easiest and best way to maintain voucher specimens is to press them on herbarium sheets, but small specimens may have to be kept on pieces of mica (the best, but increasingly difficult to purchase); pieces of clear, firm plastic sheeting; firm, good-quality cards; or microscope slides. Methods for these preparations are outlined here.

a. Placing coarse specimens in newspaper to dry. When nothing else is available, dry newspaper will serve adequately to contain specimens until other facilities are available. The newspaper must be changed daily until the specimen becomes dry. If the specimen adheres to the newspaper, it is usually because the paper was not changed frequently [in the field, one of us (IAA) changes newspaper twice each day]. If necessary, the specimen can be soaked off at a later date, the newsprint and paper fibers will brush off, and the specimen can be arranged (or rearranged) on a clean herbarium sheet. Appropriate collection numbers must be transferred each time.

b. Preparing a herbarium sheet. Herbarium paper, fitted blotters, and corrugated cardboard ventilators can be purchased (Carpenter/ Offutt Papers, Inc., or Turtox Biological Supplies) in standard herbarium shapes and sizes. In addition, pieces of cloth, cut or torn approximately 10×16 in., newspapers, wax paper, or paper towels should be available. Small paint brushes (2–5 cm wide) or an ear syringe (a rubber bulb with a pointed tip) are used to encourage branches to lie in one plane or to spread out to show their pattern. The cloth used can consist of pieces of unbleached muslin, which usually contain no sizing, are readily available, and are long lasting. Old bed sheets also make excellent herbarium cloths. If these sources are lacking, rags can sometimes be purchased in laundries or paint shops. Cloths are not mandatory; however, drying is hastened with their use.

Using a shallow pan or tray (cafeteria trays or plastic trays used by photographers) that is larger than the standard herbarium sheet (approximately 25×45 cm) and is 75% filled with water, insert a piece of plastic, galvanized iron, or hardware cloth (cut to fit) with a piece of herbarium paper above this firm base. Then float the specimen to be pressed out over the wet sheet of paper, slowly lift the base at an angle, straighten the specimen with branches laid flat, and finally lift the three pieces (base plus paper plus specimen) out of the pan and allow the water to drain off. Place the paper with specimen on a dry blotter that rests on a cardboard ventilator (or the side of a cardboard carton cut to fit). The advantage of a ventilator is that the corrugates will be running across the width, allowing circulating air or heat through the most openings over the shortest distance. The holes of a cardboard carton will run through the length of the board. Cover the specimen with a cloth or, if the specimen is slippery or mucosoid, with wax paper. Place a layer of newspaper over the wax paper or cloth if the specimen is bulky; otherwise, place a dry blotter over the wax paper, followed by a dry ventilator, then a dry blotter, which will ready the press for the next sandwich of specimens. When a stack about 30–60 cm high has been accumulated, place a plant-press frame or piece of plywood cut to fit on the top and bottom of the stack. Use a woven strap around this stack, pulled as tightly as possible. Twelve to 18 h later, change all blotters, ventilators, and cloths in the stack. If waxed paper was used, insert a new piece. The specimens that have left a large amount of water on the adjacent blotters should perhaps have two layers of blotters on each side of them when the new stack is made. A weight (lead brick, large rock, many books, or sandbags) can be placed on the stack. The stack can now be left for 24 h and then reexamined. It may be possible to eliminate cloths and blotters at this change. Dry the damp blotters, ventilators, and cloths by placing them in a standard, heated plant dryer or by hanging them on a line. In places that have room heat, they may be overlapped and stood on end. Pressing cloths should be washed every few months under normal use to remove accumulated salts.

In the case of a specimen with low water content, drying should be hastened by more frequent changes or by eliminating blotters. Very slow drying encourages growth of fungi or formation of salt crystals. On the other hand, the use of plant dryers that flowering plant taxonomists routinely employ will cause the algae to dry too quickly, resulting in shrinkage of thalli or leaving them too brittle for future handling. Thin, flat blades such as *Ulva* or delicate red algae such as *Polysiphonia* will probably be dry enough to remove after the second change. Other, more bulky algae such as species of *Chondrus, Gracilaria,* and *Colpomenia* may take 3–5 d (with changes each day). For very thick algae such as kelps, mold may develop before the specimens are completely dry. A dilute solution of Formalin (about 10%) should be dabbed on the moldy areas, and the specimens can be finished off by brief placement in a plant dryer or in an oven at low temperatures (~100°C) for about 20 min.

c. Calcified, lumpy, or odd-shaped specimens. Nonarticulated coralline algae, many of which have been removed from their coral (animal) or rock substrata, are too bulky to process as outlined in the preceding subsections. They are usually placed in 5 to 10% Formalin for 2 to 3 d and then air-dried. Erect, articulated corallines also are frequently placed in 5 to 10% Formalin overnight, then placed between folds of newspaper and dried between blotters and ventilators, as in the case of other specimens. When dried, they should be thoroughly glued to herbarium sheets. Coralline algae are frequently stored in shallow, flat boxes in order to keep the brittle fronds free of physical damage and to allow for easy removal for closer examination.

Crustose, noncalcified algae (e.g., *Ralfsia* among the brown algae and *Peyssonnelia* among the red algae) should be chipped as closely as possible from their substratum with a chisel or, better yet, completely detached. They are very difficult to study on rocky substrata since they dry irregularly while attached or pull away differentially from the substrata, and resulting dried specimens only approximate the natural cell arrangements. Generally, fleshy crusts are studied while fresh and microscope slides are made at that time; the remaining material becomes the voucher specimen after air-drying.

d. Small specimens (less than 2 cm long). Mica is purchased in small, multilayered sheets that are usually peeled off to one or at most two thicknesses of a size just larger than the specimen to be preserved. Mica is used like herbarium paper, except that it is not immersed in water. The specimen is positioned on the piece of mica and, by means of forceps and a wet paint brush, is laid flat. Waxed paper or cloth is placed over the specimen, and the usual drying procedure is followed. The field data or specimen number can be written directly on the piece of mica. Sheets of firm, clear plastic (of windowpane thickness) can be used for delicate and small red algae with success, but such plastic does not hold small blue-green, green, or brown algae satisfactorily. When dry, both mica-mounted and plastic-mounted specimens should be placed in an envelope or packet and glued to a herbarium sheet.

Good-quality cards of various sizes also can be used for mounting specimens that are too small for a large herbarium sheet. The usual method for a large sheet is employed. When dry, the card should be glued to a standard herbarium sheet. The practice of some herbaria is to glue several small cards containing the same species from the same or different collections to a single herbarium sheet, each specimen receiving a different accession number. In other herbaria, each card is placed on a different herbarium sheet. The saving of paper and space in a herbarium cabinet is obvious in the first practice, but the second is justifiable because of the possibility of identification errors when multiple specimens are on the same sheet. Unauthorized removal of specimens from herbarium sheets is illegal in most herbaria.

Finally, small specimens and sections for microscopic examination can be kept on glass slides. The prepared slides can in turn be stored in a slide collection and cross-indexed with the herbarium sheet (using the same field or collecting number in every case). Also, slides can be placed in cardboard containers with depressions that fit glass slides (Clay-Adams Co.), and these can be glued to the herbarium sheet or placed in a manila envelope, which in turn is glued to the sheet. Many endophytic and small epiphytic algae are kept on slides, since they would be lost on a large sheet or difficult to locate if retained on the host specimen. Although breakage of a glass slide is always a possibility (thus making mica or plastic more favorable), it is important to preserve small algae in a manner that facilitates study at a later time.

e. Preparation of permanent microscope slides as voucher specimens. The objectives are (1) to prepare a slide that can be made quickly and (2) to be able to use the slide as a permanent record. Preparation of slides for future cytological examination, a very different objective, is discussed separately.

The corn syrup method, introduced about 30 years ago at the University of California, Berkeley, has been tested over time and is now the favored technique for most marine phycologists. Clear corn syrup (Karo brand, available in most grocery stores), with phenol preservative added, is diluted with fresh water to form a series ranging from 35, 50, 70, to 80% in which most algae will not plasmolyze severely. The specimen to be used for this method must have been preserved earlier in Formalin or some other preservative in the field or laboratory. The alga is placed on a microscope slide (if small), the branches arranged so that they are in one plane; a drop of 0.5 to 1.0% aqueous stain (see Table 4–1) is added and fixed (if necessary) with a drop of 1% hydrochloric acid. This is rinsed off by flooding with distilled water and holding the slide at an angle over a paper towel to drain. The excess stain can also be carefully removed by blotting along the edges of the thallus with a piece of paper towel or bibulous paper (American Hospital Supply Corp.). A drop of 35% corn syrup is added, allowed to stand for a few minutes (up to 30 min), and then a drop of 50% corn syrup is added. With experience, it is possible to use 50% corn syrup directly; however, until the delicate thalli are known not to plasmolyze readily, it is safer to start at the lower concentrations of sugar. A cover slip is added carefully, one edge resting on a probe or forceps, and it is gently lowered over the syrup surface. In 8 to 12 h the slides should be inspected for plasmolysis or for air bubbles due to evaporation, and more syrup of the next higher concentration should be added to the margins of the cover slip. The slides must be kept on a flat surface. When bubbles no longer appear, the cover slips can be sealed with ordinary clear nail polish. This step, however, is not necessary, for if properly done, the preparation is self-sealing. The

GILBERT MORGAN SMITH HERBARIUM
Hopkins Marine Station of Stanford University

MARINE ALGAE OF THE MONTEREY PENINSULA CALIFORNIA

Ceramium eatonianum (Farlow) DeToni
  (tetrasporangial)
Common in mixed turf dominated by Corallina officinalis var. chilensis at +1.0 ft tide level, middle flats of Stillwater Cove, Pebble Beach.

Coll. I. A. Abbott 16153   Aug. 8, 1982
Det. ISABELLA A. ABBOTT

----Herbarium where specimen is filed
----Geographic range of species covered

----Binomial, including author(s)
----Specific ecological niche

----Collector, collector's number; date of collection
----Person who identified the specimen

---

from Hopkins Marine Station, Pacific Grove

MARINE ALGAE OF CALIFORNIA

Porphyra perforata J. Agardh

  Voucher for photosynthesis studies

+3.5 ft tide level on granite rock, beach southeast of Mussel Point, Pacific Grove
leg. Celia M. Smith   Aug. 8, 1982

Det. ISABELLA ABBOTT

---from a herbarium (=gift)
--- Geographic range of species

--- Binomial

--- Evidence or record

---Where collected

--- Collector; date of collection
--- Person who identified the specimen

ALGAE OF THE HAWAIIAN ISLANDS
Oahu Island

Lophocladia trichoclados (C. Ag.) Schmitz

New record for Hawaiian Islands

  8 km offshore in 70 m depth, using "Deepstar" on eroded and living coral shelf.

  leg. K. Agegian   Aug. 8, 1982

Isabella Abbott
Det.

--- Broad geographic range
--- Narrow geographic range

--- Binomial
--- Voucher for new record

--- Where collected

--- Collector; date of collection

--- Person who identified the specimen

Fig. 4–1. Format for several types of labels commonly used.

slides should remain flat for at least 1 wk, after which they can be packed in standard slide boxes and stored so that the slides are still parallel to the shelf. After 1 mo, they can be stored on edge. Slides made in this way – for example, some of those of E. Y. Dawson in the U.S. National Herbarium, Smithsonian Institution, that are more than 20 years old – are as usable as if made today. A few drops of phenol or thymol should be added to dropping bottles containing the corn syrup in order to discourage the growth of fungi and these bottles should be labeled POISON.

Standard microscope slides 25 × 75 mm are available with one end ground (on which one can write) or as clear glass. Cover slips come in several thicknesses, no. 1 being the thinnest. For microscopy no other thickness should be used. Cover slips may be round, square (22 × 22 mm), or rectangular. For the majority of algae, either the round or square cover slips are used. The disadvantage of the longer ones is that very little space is left for a label. Slide labels come in perforated sheets, which are convenient to use when one is traveling, or in rolls in which the labels (22 mm²) are self-sticking on one side. The collection number of the specimen, at least the genus name, and the place of collection are the acceptable minimum for the slide label.

The voucher specimen is fully prepared when

1. it is a dried, pressed, herbarium specimen that is adequately labeled, or

2. it is a dried or wet-preserved specimen that is labeled and contained, or

3. it is a small specimen that is on a labeled and prepared microscope slide.

To complete the documentation, an adequate label must be prepared to accompany the field or collecting number that was first assigned to the specimen, and necessary cross-indexing must be added to the herbarium materials. Any preserved material or slides containing a subset of the material, for example, should be noted on the herbarium sheet. Formats for labels and kinds of information on them are given Fig. 4–1. Labels should be on 100% rag paper so as to prevent insect damage and so that they can be used in liquid as well.

IV. Identification

The assignment of species names must be based on adequate voucher specimens and not on undocumented "reports" in the literature (Tsuda and Stojkovich 1980). Although the name for a species must be tied to one herbarium sheet (i.e., the type specimen), the interpretation of what the species *is* depends on a number of specimens showing a variety of morphology and, if possible, ecological differences (Abbott 1972). The latter are expressed in uniform ways in some species and in slightly to extremely variable degrees in others.

When practicable, it is advisable to have two or more complete sets of specimens, and when data on them are published, the places that they are stored should be noted, for example, the U.S. National Herbarium, Smithsonian Institution, which is abbreviated (Holmgren and Keuken 1974) US. Private individual herbarium collections should be discouraged since they are not readily available to workers from all parts of the world. When one is collecting abroad, it is not only courteous, but probably mandatory as well to leave one set of specimens for the local area. This not only fosters interest and pride, but encourages further activity and sometimes additional valuable collections.

Rough identification will facilitate the processing of the voucher specimens that have been prepared. Once they are identified to genus, the search for a species name will become easier. There are several good regional algal floras, but their effective use requires a knowledge of marine algae. However, Abbott and Dawson (1978) provide a pictured key to the most common marine algae of all coasts of continental United States, and many species that occur in Alaska and Hawaii. With a common alga in hand, making choices in the key will enable even those whose English is limited to find the correct genus. A brief assessment of useful marine floras is given in Table 4–2.

1. Procedures for the identification of individual species

1. Note the major group of algae to which the specimen might belong by judging its color: some clear shade of grass green (Chlorophyta); olive green to golden (Phaeophyta); pink, red, purple, even greenish (Rhodophyta); dark blue-green, brownish, or black (Cyanophyta).

2. Select an appropriate flora by geographical region (Table 4–2) and look up the major group (phylum or division) that you have chosen.

3. Use keys that are available in the flora. If they contain a vocabulary that is too technical, try the keys in Abbott and Dawson (1978) first; then refer again to the flora you have chosen.

4. Be sure to examine the illustrations that are referred to in the descriptions.

5. Read the description of the species that you think you have slowly and critically. Use the glossary if one is included in the book (glossaries are usually worked over extensively before they are

Table 4–2. *Marine floras helpful for species identification of common macrophytes*

Pacific West Coast of North America
Abbott, I. A., and Hollenberg, G. J. 1976 *Marine Algae of California*. Stanford University Press, Stanford, Calif. 827 pp. Keys to genera of green, brown, and red algae; keys to species; each species in the book illustrated.
Scagel, R. F. 1967. *Guide to Common Seaweeds of British Columbia*. British Columbia Provincial Museum, Victoria. 330 pp. Includes many more northerly species than those in California.
Waaland, J. R. 1977. *Common Seaweeds of the Pacific Coast*. Pacific Search Press, Seattle. Some ecological notes, 16 excellent colored photographs as well as black and white illustrations.

Mexico and Central America
Dawson, E. Y. 1941–1965. It would be impossible to do any critical taxonomic studies in this geographical area (and to the immediate north and south) without referring to the numerous publications of this important contributor to the understanding of marine algae. Do not neglect these studies because they are not in book form! For reference to Dawson's papers, see Abbott and Hollenberg (1976).

Caribbean and Florida
Taylor, W. R. 1960. *Marine Algae of the Eastern Tropical and Subtropical Coasts of the Americas*. University of Michigan Press, Ann Arbor. 870 pp. Well-illustrated, comprehensive coverage of the warm Atlantic and Caribbean. It can be added to but never replaced as a basic flora.

Northeastern United States and Canada
Taylor, W. R. 1957. *Marine Algae of the Northeastern Coast of North America*, rev. ed. University of Michigan Press, Ann Arbor. 509 pp. Timeless in its judgment and value.

All geographical areas
Humm, H. J., and Wicks, S. R. 1980. *Introduction and Guide to the Marine Blue-green Algae*. Wiley, New York. 194 pp. Very useful keys to genera and to species. Species descriptions contain the most common synonyms under which they may be found in other references.
Umezaki, I. 1961. The marine blue-green algae of Japan. *Mem. Coll. Agr. Kyoto Univ.* no. 83, 1–149. Keys to orders, genera, and species. More technical than Humm and Wicks (1980). Excellent illustrations.

published). If the generic description fits the specimen but not the species, look in other floras suggested in the book's references.

6. At this stage, it may be obvious that you need to know more about the specimen than you can observe without a microscope. Steps that can be taken are given in the next subsection.

7. The majority of ecologists work with common species. It is likely that the steps listed here have led to a name for the specimen, tentative as the identification may be. However, the specimen can

now be retrievably filed rather than left in a stack of unidentified specimens.

8. If uncertain, pencil the species name with its author(s) on the herbarium sheet. Initial the identification. When verified, the name can be added to the label.

9. Look over the other voucher specimens to see whether there are other specimens of the same species and put them together. You may now be more certain of whether your identification was correct.

2. Examination for details of structure in order to match descriptions. Thin sections of vegetative or reproductive structures must be made in order to place most algae within families, genera, or species. If preserved material has been saved, the job is relatively easy but, if not, the dried specimen can be sectioned.

1. Whether the specimen is liquid-preserved or dried, place it, or a small piece of it, under a dissecting microscope. Examine for reproduction if a brown or red alga.

2. Using a new single-edged razor blade and observing through the microscope, carefully cut a small piece (about 2 mm) of the specimen off. Place the tiny piece on a clean microscope slide and remove the remainder of the specimen from the microscope.

3. Place the slide with its specimen under the microscope again, hold the specimen down with a glass slide held at an angle close to the edge of the specimen, and cut across the specimen until the angled slide is reached. Lift it, move it back, and continue to cut sections, cut as many as possible.

4. Most anatomical descriptions are based on cross sections. The sections that have been cut must be turned on their broadest faces (i.e., so that they will present a cross section). Fine needles will help to do this.

5. If sections of dried brown or green algae have been cut, they are probably very hard and the cells lie very closely together. The sections can be expanded by the addition of a drop of water or of several drops of dilute detergent (~5 ml in 250 ml of water) or by gentle heating of the wet specimen on the slide over an alcohol lamp.

6. When the material is expanded, use the previous staining and mounting procedure employing corn syrup. If the material came from liquid preservative, add stain directly to the cut sections and follow the steps for making a permanent microscope slide.

7. In order to identify coralline algae, particularly nonarticulated species, special procedures are followed. Having been placed in fixatives designed for their preservation and decalcification (Table 4–1), they are usually sectioned with a freezing microtome or by the paraffin method of slide preparation. A botanical microtechnique

reference (see Table 4–1) should be consulted for these methods. A freezing microtome is used for many algae other than corallines, but for identification purposes, a sharp razor blade, a steady hand, and patience will do a more than adequate job.

3. Preparation of material for cytological study. Each phycologist has special techniques for treating various algae. Only a brief explanation is needed here to indicate that, although the methods described earlier are standard and routinely accepted by most workers in the field, specialized disciplines require more attention to the microtechnical aspects demanded by individual algal species. For example, details of nuclear structure, location of pyrenoids, and other organelles will not be sufficiently preserved or retained in the Formalin or alcohol preservatives. Furthermore Karpetchenko fixative or FAA (Table 4–1) is generally better than either Formalin or alcohol but inadequate for chromosome studies. Finally, none of these can be used in the preparation of samples for electron microscopy. It will be sufficient to state here that many more specialized microtechniques exist and are required for detailed studies of marine algae, but they are beyond the scope of the present treatment. The interested reader is referred to Gantt (1980) as a source of many appropriate cytological methods.

V. References

Abbott, I. A. 1972. On the species of *Iridaea* (Rhodophyta) from the Pacific coast of North America. *Syesis* 4, 51–72.

Abbott, I. A., and Dawson, E. Y. 1978. *How to Know the Seaweeds*, 2nd ed. Brown, Dubuque, Iowa. 141 pp.

Abbott, I. A., and Hollenberg, G. J. 1976. *Marine Algae of California*. Stanford University Press, Stanford, Calif. 827 pp.

Berlyn, G. P., and J. P. Miksche. 1976. *Botanical Microtechnique and Cytochemistry*. Iowa State University Press, Ames. 226 pp.

Chemin, E. 1929. Le bleu de cresyl comme réactif des iodures. *Bull. Soc. Bot. Fr.* 76, 1009–26.

Dawson, E. Y. 1966. *Marine Botany: An Introduction*. Holt, Rinehart, and Winston, New York. 371 pp.

Gantt, E. (ed.). 1980. *Handbook of Phycological Methods: Developmental and Cytological Methods*. Cambridge University Press, Cambridge. 425 pp.

Giraud, G., and Cabioch, J. 1976. Étude ultrastructurale de l'activité des cellules superficielles du thalle des Corallinacées (Rhodophycées). *Phycologia* 15, 405–14.

Holme, N. A., and McIntyre, A. D. 1971. *Methods for the Study of Marine Benthos*. International Biological Programme (IBP) Handbook no. 16, Blackwell, Oxford. 334 pp.

Holmgren, P. K., and Keuken, W. 1974. *Index Herbariorum*, part 1: *The*

Herbaria of the World. Oosthoek, Scheltems & Holkema, Utrecht, Netherlands. pp. 397.

Johansen, H. W. 1969. Morphology and systematics of coralline algae with
special reference to *Calliarthron. Univ. Calif. Publ. Bot.* 49, 1–98.

Mason, L. R. 1953. The crustose coralline algae of the Pacific coast of the
United States, Canada and Alaska. *Univ. Calif. Publ. Bot.* 26, 313–89.

Neushul, M. 1965. SCUBA diving studies of the vertical distribution of
benthic marine plants. *Bot. Gothoburg.* 3, 161–76.

Papenfuss, G. F. 1937. The structure and reproduction of *Claudea multifida,
Vanvoorstia spectabilis, and Vanvoorstia coccina. Symbol. Bot. Upsal.* 2, 1–66.

Papenfuss, G. F. 1946. Structure and reproduction of *Trichogloea requienii,*
with a comparison of the genera of Helminthocladiaceae. *Bull. Torrey Bot.
Club* 73, 419–37.

Sass, J. E. 1958. *Botanical Microtechnique.* 3rd ed. Iowa State University Press,
Ames. 228 pp.

Smith, R. I., and Carlson, J. T. (eds.) 1975. *Light's Manual: Intertidal Invertebrates of the Central California Coast.* University of California Press, Berkeley. 716 pp.

Suneson, S. 1937. Studien über die Entwicklungsgeschichte der Corallina-
ceen. *Lunds Univ. Arsskr.* New Series, Sec. 33, 1–101.

Taylor, W. R. 1950. Field preservation and shipping of biological specimens.
Turtox News 38, 42–3.

Taylor, W. R. 1960, *Marine Algae of the Eastern Tropical and Subtropical Coasts
of the Americas.* University of Michigan Press, Ann Arbor. 870 pp.

Taylor, W. R. 1962. *Marine Algae of the Northeastern Coast of North America,*
2nd ed. University of Michigan Press, Ann Arbor. 509 pp.

Tsuda, R. T., and Stojkovich, J. O. 1980. Modern approaches, status, and
future directions of algal taxonomy in Micronesia. In Abbott, I. A., Foster,
M. S., and Eklund, L. F. (eds.), *Pacific Seaweed Aquaculture,* pp. 157–63.
California Sea Grant College Program, La Jolla.

5: Electrophoresis

DONALD P. CHENEY

Department of Biology, Northeastern University, Boston, Massachusetts 02115

CONTENTS

I. Introduction

In the past 15 years, electrophoresis has become probably the most widely used technique for the study of genetic variation in natural populations. Electrophoresis is a reliable and practical tool for surveying the level of genetic variation within a population and/or the degree of genetic differentiation between populations of the same or closely related species. One of the principal advantages of the technique for the systematist and ecologist is that it enables one to make comparisons between individuals at the level of the gene products themselves (in the form of protein banding patterns) rather than secondary morphological characters, thereby providing a measure of genetic differentiation that is independent of morphological differentiation.

Applications and advantages of the technique, as well as limitations and biases, have been described at length elsewhere (e.g., Lewontin 1974; Ayala 1976; Gottlieb 1981). Among its numerous applications, electrophoresis is most commonly used in systematic and ecological studies for the following purposes (modified from Powell 1975a): (1) to estimate quantitatively the genetic variability of a population (or species) and assess its genetic structure, (2) to study geographical and temporal patterns of genetic variation within a species, as well as recognize individual variants within populations, and (3) to determine levels of genetic divergence among populations of different taxonomic levels.

The early literature on algal electrophoretic studies was reviewed by Holton (1973) and updated by Cheney and Babbel (1978) and Mathieson et al. (1981). In comparison with the great number of electrophoretic studies conducted on aquatic animals and land plants (for examples see Powell 1975a; Ayala 1976; Gottlieb 1977; Battaglia and Beardmere 1978; Nevo 1978; Hamrick et al. 1979; Burton 1983), the number of algal taxa studied to date is very small. In spite of a marked increase in electrophoretic studies dealing with populations in fairly recent years (e.g., Murphy and Guillard 1976; Cheney and

[88]

Babbel 1978; Grant and Proctor 1980; Blair et al. 1982; Gallagher 1980; Soudek and Robinson 1983; Hayhome and Pfiester 1983; Innes and Yarish 1984), there is still a total of less than 10 algal genera for which information about genetic structure and/or variability has been reported (Table 5–1). The reasons for this relatively small number of population electrophoretic studies with algae are no doubt numerous but must partly reflect inherent technical difficulties (e.g., Marsden et al. 1981), as well as the tendency of phycologists to utilize more traditional approaches to systematic and ecological research.

In spite of the relatively small number of algal species studied to date, electrophoresis has made some valuable contributions to our understanding of the genetic structure of algal populations and species, as well as to algal systematics. For example, electrophoretic techniques have demonstrated genetic differentiation among morphologically identical entities at or below the species level for a variety of algae, including isolates of the endosymbiotic dinoflagellate *Gymnodinium* (= *Symbiodinium*) *microadriaticum* from different coelenterate hosts (Schoenberg and Trench 1980), populations of the red alga *Eucheuma nudum* (= *isiforme*) from different depths (Cheney and Babbel 1978), summer and winter blooms of the diatom *Skeletonema costatum* (Gallagher 1980), coastal and oceanic isolates of the diatom *Thalassiosira pseudonana* (Murphy and Guillard 1976), and different lake isolates of the diatom *Asterionella formosa* (Soudek and Robinson 1983). In addition, electrophoretic analysis has provided some new insights into the genetic structure of natural and cultured populations of algae. For example, it has provided evidence for gene duplication via polyploidy in *Chara* (Grant and Proctor 1980), a loss of heterozygosity in clones of two diatom species after 6 mo in culture (Murphy 1978), asexual reproduction in *Codium fragile* (Malinowski 1974) and *Enteromorpha linza* (Innes and Yarish 1984), and a deficiency of heterozygotes in populations of *Eucheuma* with a high degree of vegetative reproduction (Cheney and Babbel 1978).

The purpose of this chapter is to review electrophoretic methods applicable to population studies of algae, with an emphasis on seaweeds. Several excellent sources provide general details on starch and polyacrylamide gel electrophoresis methods and applications (e.g., Brewer 1970; Oelshlegel and Stahmann 1973; Harris and Hopkinson 1976; Andrews 1981; Hames and Rickwood 1982). This chapter attempts to present a practical guide aimed primarily at the beginner, with the hope that such an approach will encourage increased application of electrophoresis to algal systematics and ecology.

Table 5–1. Survey of algal genera and seagrasses studied by electrophoresis, with an emphasis on population studies

Genus[a]	Purpose of study[b]	Isozymes[c]	References
Chrysophyta			
Bacillariophyceae			
Thalassiosira spp.	Taxon/ecol	MDH, IDH, PGI, G-6-PDH, GDH, SOD, GP	Murphy and Guillard (1976)
Skeletonema costatum, Thalassiosira pseudonana	Taxon/ecol	MDH, PHI	Murphy (1978)
Skeletonema costatum	Taxon/ecol	MDH, PGI, SOD, GDH	Gallagher (1980)
Asterionella formosa	Taxon/ecol	MDH, GDH, LAP, EST, ACPH, AIPH, GP	Soudek and Robinson (1983)
Dinophyceae			
Peridinium spp.	Taxon/ecol	MDH, LAP, GDH, EST, ACPH, AIPH	Hayhome and Pfeister (1983)
Peridinium cinctum	Ecol/biochem	ACPH, ALPH	Wynne (1977)
Symbiodinium (= Gymnodinium) microadriaticum, Heterocapsa, Crypthecodinium	Taxon/ecol	MDH, PGI, SOD, EST, GP	Schoenberg and Trench (1980)
Chlorophyta			
Chlamydomonas spp.	Taxon/ecol	MDH, GDH, LAP, EST, ADH, LDH	Thomas and Delcarpio (1971)
Chlorococcum spp., *Tetracystis* spp.	Taxon/ecol	MDH, LAP, EST	Thomas and Brown (1970a)
Protosiphon spp.	Taxon/ecol	MDH, LAP, EST	Thomas and Brown (1970b)
Chlorella vulgaris, Chlamydomonas reinhardii	Taxon/biochem	PGI, PGM, G-6-PDH	Herbert et al. (1979)
Chlorosarcinopsis spp., *Friedmannia israeliensis, Fasciculochloris boldii*	Taxon/ecol	MDH, LAP, EST, PER, GDH, GP	Thomas and Groover (1973)

Taxon	Type	Enzymes	Reference
Pandorina morum	Taxon/ecol	MDH, LAP, SOD, PGM, G-6-PDH	Fulton (1977)
Acetabularia spp.	Biochem/devel	ACPH	Keck (1961)
Acetabularia spp.	Biochem/devel	ACPH	Keck and Choules (1963)
Acetabularia spp., *Cymopolia* spp.	Taxon/ecol	MDH	Berger et al. (1974)
Derbesia marina–Halicystis ovalis	Biochem/devel	MDH	Wutz and Zetsche (1976)
Ulva mutabilis	Biochem/devel	GP	Hoxmark (1976)
Ulva mutabilis	Biochem/devel	GP	Hushovd et al. (1982)
Enteromorpha linza	Ecol	GOT, PGM, SOD, AMY	Innes and Yarish (1984)
Ulva lactuca, Enteromorpha sp., *Spongomorpha arcta*	Taxon/biochem	GP	Young (1970)
Chaetomorpha spp.	Taxon/ecol	MDH, PGI, PGM, ACPH, G-6-PDH	Blair et al. (1982)
Codium fragile	Taxon/ecol	MDH, PGI, SOD, CAT, G-6-PDH, ACPH, ALPH, GOT, XD, PEP	Malinowski (1974)
Euglenophyta *Euglena gracilis*	Biochem/devel	MDH	Peak et al. (1972)
Charophyta *Chara* spp.	Taxon/ecol	PGI, GOT, HEX, GDH, PGM	Grant and Proctor (1980)
Phaeophyta *Fucus serratus*	Biochem	MDH, PER, EST, ACPH, ALPH	Marsden et al. (1981)
Chorda filum, Chordaria flagelliformis, Fucus filiformis	Taxon/biochem	GP	Young (1970)
Rhodophyta *Acrochaetium* spp., *Rhodochorton purpureum*	Taxon/ecol	EST, GP	Richardson and Mallery (1973)
Acrochaetium sagraenum, Goniotrichum alsidii, Erythrocladia spp., *Bangia*	Taxon/biochem	EST, ACPH, AMP, GP, phycobilins	Mallery and Richardson (1971, 1972)

[91]

Table 5–1 (*cont.*)

Genus[a]	Purpose of study[b]	Isozymes[c]	References
fuscupurpurea, Porphyra leucostricta, Nemalion multifidum, Trailliella intricata, Ceramium fastigiatum, Seirospora griffithsiana, Spermothamnion turnerii			
Porphyra yezoensis	Taxon/ecol	CAT	Miura (1978)
Porphyra spp.	Taxon/ecol	CAT, SOD, MDH, GDH, PGI, PER, LAP, EST, & others	Miura et al. (1978)
Eucheuma spp.	Taxon/ecol	MDH, ACPH, SOD, PGI, PGM	Cheney and Babbel (1978)
Chondrus crispus	Ecol	MDH, ACPH, PGI, PGM	Cheney and Mathieson (1979)
Callithamnion spp.	Taxon	Phycobilins	Spencer et al. (1981)
Chondrus crispus, Polysiphonia spp., *Porphyra umbilicalis, Corallina officinalis, Rhodymenia palmata*	Taxon/biochem	GP	Young (1970)
Cyanophyta			
Schizothrix calcicola, Porphyrosiphon notarissi, Oscillatoria lutea, Microcoleus vaginatus	Taxon	MDH, LDH, GDH, GP	Baker and Holton (1973)
Phormidium spp., *Oscillatoria* spp.	Taxon	EST, LAP, GP, PHOS	Klein et al. (1973)
Anacystis nidulans, Synechococcus elongatus, Anabaena spp., *Aphanizomenon* spp., *Nostoc* spp.	Taxon	MDH	Schenk et al. (1973)
Nostoc muscorum	Biochem	MDH	Kovatcheva and Bergman (1979)

Taxa	Biochem/devel	Isozymes[c]	Reference
Anacystis nidulans, *Oscillatoria princeps*, *Nostoc muscorum*, *Gloeocapsa dimidiata*, *Cyanidium caldarium*	Biochem/devel	BRAN, POLY-SYN	Frederick (1977)
	Taxon/biochem	BRAN, PHOS	Frederick (1971)
Spermatophyta (Seagrasses)			
Zostera marina	Taxon/ecol	MDH, PGI, PGM, CAT, AAT, GDH, G-6-DPH, GP	Gagnon et al. (1980)
Halophila spp.	Taxon/ecol	MDH, PGI, PGM, GOT, PER, ACPH, GDH	McMillan and Williams (1980)
Thalassia spp., *Halodule* spp., *Syringodium* spp., *Cymodocea* spp., *Thalassodendron ciliatum*, *Halophila ovalis*	Taxon/ecol	MDH, PGI, PER, GOT, ACPH, PGM, LAP, G-6-PDH, EST	McMillan (1980)
Zostera spp., *Phyllospadix* spp., *Heterozostera* spp., *Thalassodendron* spp., *Amphibolis* spp., *Halodule* spp., *Cymodocea* spp., *Syringodium* spp., *Posidonia* spp., *Enhalus* spp., *Thalassia* spp., *Halophila* spp.	Taxon	MDH, PGI, PGM, PER, GOT, ACPH, G-6-PDH, EST, GDH, ADH	McMillan (1982)

Note: This survey was completed in November 1982.
See Holton (1973) for additional earlier references.

[a] Taxonomy is the same as that used in references; "spp." indicates that more than one species was examined.

[b] The principal goals of the study were judged to be taxonomic, ecological, biochemical, developmental, or a combination thereof.

[c] Isozyme abbreviations: MDH, malate dehydrogenase; PGI, phosphoglucose isomerase; PGM, phosphoglucose mutase; PER, peroxidase, ACPH, acid phosphatase; ALPH, alkaline phosphatase; LAP, leucine aminopeptidase; GDH, glutamate dehydrogenase; CAT, catalase; SOD, superoxide disoimutase (tetrazolium oxidase, indophenol oxidase); AAT, aspartate aminotransferase; G-6-PDH, glucose-6-phosphate dehydrogenase; GP, general protein; EST, esterase, HEX, hexokinase, PEP, peptidase; BRAN, branching enzyme; ADH, alcohol dehydrogenase; GOT, glutamate oxaloacetate transaminase; AMY, amylase; POLY-SYN, polyglucoside-synthesizing enzymes; PHOS, phosphorylase.

[93]

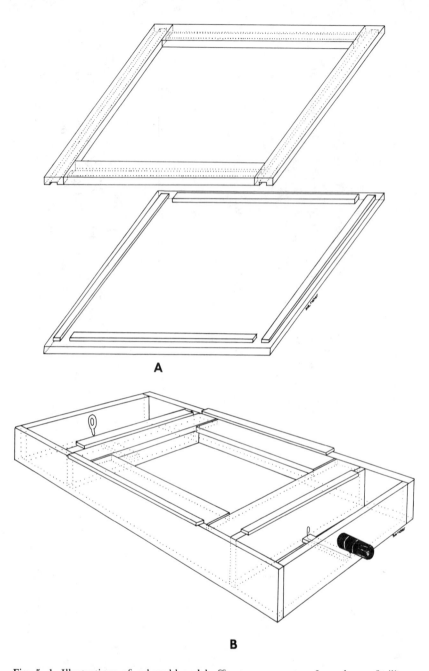

A

B

Fig. 5–1. Illustrations of gel mold and buffer tray apparatus. In order to facilitate manufacture, dimensions are provided in inches as would be required by a plastic shop. (A) The gel mold (bottom) consists of a plexiglass base ($\frac{1}{4}$ in. thick × $7\frac{3}{4}$ in. wide × $8\frac{13}{16}$ in. long) and four attached edge guides (each $\frac{1}{8}$ in. thick × $\frac{7}{32}$ in. wide × $7\frac{1}{4}$ in. long). Four grooved gel guides (top) fit on top of the edge guides; two are $\frac{9}{32}$ in. thick × $\frac{3}{4}$ in. wide × $7\frac{1}{4}$ in. long with a groove $\frac{8}{38}$ in. wide and two are similar in thickness, width, and groove width but $7\frac{7}{16}$ in. long. The slicing guides (not shown) are also

II. Equipment

Electrophoresis can be conducted using a variety of techniques and supporting media. The most common systems for population studies are horizontal starch gel electrophoresis and either horizontal or vertical polyacrylamide gel electrophoresis. Gel or slab electrophoresis is preferable to the so-called disk or tube system of electrophoresis, because it allows for the comparison of a large number of individuals simultaneously on the same gel. Although the "disk" or tube system may provide better resolution, the need for separate tubes for individual samples makes it impractical in most population studies.

The basic equipment necessary for either the horizontal or vertical gel system (using either starch or polyacrylamide) generally consists of a gel mold and tray apparatus and a high-voltage power supply. A wide variety of vertical and horizontal gel electrophoresis assemblies are commercially available (Bio-Rad, Hoefer Scientific Instruments); however, most laboratories prefer a homemade horizontal gel electrophoresis apparatus such as that illustrated in Fig. 5–1. This system consists of a plexiglass gel mold into which the molten starch or polyacrylamide is poured, and a plexiglass tray assembly, which supports the gel during electrophoresis. One advantage of such a gel mold with replaceable guides is that the gel can be sliced twice, thereby providing three slices from one gel that can be stained for different enzymes.

In addition to the electrophoresis apparatus, it is necessary to have a power supply capable of supplying between 0 and 400 V and up to 100 mA. For most studies, it is not necessary for the power supply to provide either constant voltage or constant current. However, it is desirable (or necessary, depending on the type of power supply) to use a separate power supply for each electrophoresis apparatus; therefore, the purchase of power supplies generally represents the largest initial cost. I would recommend the purchase of at least two power supplies and gel apparatuses. A number of reliable power supplies are commercially available. We have found the Heathkit 0-400 VDC regulated HV supply (either in factory-assembled or kit form) to be both very reliable and economical.

In addition to standard laboratory equipment, additional items

Caption to Fig. 5–1. (*cont.*)
grooved to fit on top of the edge guides and are $\frac{3}{16}$ in. thick $\times$ $\frac{3}{4}$ in. wide $\times$ $7\frac{7}{8}$ in. long. (B) The buffer tray apparatus is $8\frac{5}{8}$ in. wide $\times$ $13\frac{5}{8}$ in. long. It consists of a buffer chamber at each end that is $1\frac{3}{4}$ in. deep $\times$ $7\frac{13}{16}$ in. wide $\times$ 2 in. long. Platinum wire (26 gauge; e.g., A. H. Thomas no. 9924-F64) is placed along the bottom of each chamber and connected to a banana plug at each end.

that are needed include (1) magnetic stirrer–hot plate for heating starch solutions, (2) pH meter, (3) balance with accuracy to 1.0 mg, (4) vacuum and volumetric flasks, (5) refrigerator or cold room to house the gel apparatuses during electrophoresis, (6) incubator to incubate the gels during staining, (7) homemade or commercial gel slicer, and (8) camera and light box for photographing gels.

III. Methods

A. Background and general comments

Electrophoresis, as used in population studies, refers to the separation of proteins by an electric field in a medium generally of starch or polyacrylamide, coupled with histochemical staining. By means of histochemical staining procedures, specific enzymes or proteins in general can be visualized as bands on the gel. The direction and rate of migration of a protein in a gel are a function of its charge and conformation. Amino acid changes in a protein (e.g., due to point mutations) can change the charge of the protein, its conformation, or both, resulting in a protein with a different rate of migration but the same enzymatic function. Enzymes with the same biochemical function but different rates of migration are called isozymes (or isoenzymes) when coded by more than one gene locus, and allozymes when coded by different alleles of the same locus.

In the simplest case, each isozyme band represents the phenotypic expression of a single gene. However, a band can also represent an enzyme that is composed of more than one polypeptide and is coded by two or more gene loci.

If the protein products of two alleles of the same gene locus have different charges or conformation, they generally will appear as separate bands. In most cases, such bands show codominance and segregate following Mendelian expectations, making it possible to distinguish homozygous from heterozygous individuals for given loci. Generally, progeny tests are used to distinguish between those enzyme bands that represent allozymes and those that represent isozymes. This distinction is important, because it permits a quantification of the number of gene loci included in the study as well as of allele frequencies, thereby allowing an estimation of genetic variation within and divergence between populations to be made.

When undertaking an electrophoretic study, one should keep in mind that its primary strength as a systematic tool is its capacity to relate differences in electrophoretic mobility directly to genetic differences. This requires careful technique (to eliminate nongenetic causes of mobility differences) and the examination of well-defined, substrate-specific enzymes. Thus, although many of the earlier electrophoretic studies with algae utilized a general protein stain to compare species, this approach should not be used in population or systematic studies. The use of nonspecific protein stains (such as Coomassie blue) with crude tissue extracts will generally result in an uninterpretable, complex array of bands in which it is difficult at best to distinguish nonhomologous from homologous proteins. Thus, the use of only enzyme-specific stains is recommended and is described herein.

B. Starch versus polyacrylamide gel electrophoresis

Starch gel electrophoresis is generally preferred over polyacrylamide gel electrophoresis (PAGE) for plant population studies and has been the principal method used in our laboratory. Both methods have advantages and disadvantages; whichever is selected by a laboratory usually depends on past experience and specific objectives. One of the primary advantages of starch gel electrophoresis over PAGE is that the low cost and physical properties of starch make it possible to pour thick, malleable gels that can be easily sliced several times and thus stained simultaneously for separate enzyme activities. In contrast, PAGE is more expensive and toxic but provides a gel that is very transparent and therefore better adapted to quantitative studies using differential densitometry. In addition, since the polyacrylamide gel is generally cast in a sealed glass mold and run in a vertical position with samples added directly to slots at the top of the gel, smaller samples are easier to run than with starch; however, only separation of substances migrating anodally can be visualized unless the polarity is reversed. A variety of discontinuous buffer systems can be employed with PAGE (e.g., see Hames and Rickwood 1982) that involve sample loading onto a large-pore stacking gel polymerized on top of a smaller-pore resolving gel; this often results in better resolution of bands than with starch electrophoresis. If band resolution is critical, PAGE is preferable. In general, one should select the simplest system compatible with the resolution desired.

Only starch gel electrophoresis techniques are detailed in this chapter. Information on general methods of PAGE is available elsewhere (e.g., Hames and Rickwood 1982; Shields et al. 1983), as are details of its application to microscopic algae (e.g., Murphy and Guillard 1976; Gallagher 1980; Hayhome and Pfiester 1983; Soudek and Robinson 1983).

C. Gel preparation

Both Poulik and lithium hydroxide discontinuous gel systems have been successfully used with a variety of macrophytes. The composition of these and other gel systems (i.e., gel and electrode buffers) is described in Table 5–2. The basic methods of gel preparation are similar for all systems and include the following procedures in our laboratory:

1. Mix well all components of the gel electrode buffer and bring to proper pH. For the gel mold illustrated in Fig. 5–1, use 250 ml of gel buffer solution per gel. We usually make up 500 ml at a time and pour two gels simultaneously.

2. Decant ~25% of the gel electrode buffer into a graduated cylinder and keep at room temperature until mixed with the starch. Heat the remainder of the buffer in a long-necked volumetric flask to boiling.

3. Weigh out the appropriate amount of starch (Connought Starch, Fisher Scientific) and carefully pour into a large side-arm vacuum flask. Starch concentrations of ~12% but varying from 11.6 to 12.4% (w/v) have worked best for us. The optimal concentration can vary from one source of starch to another and even from one lot to another. In order to maintain a constant starch concentration (which affects protein migration) throughout a long study, sufficient starch of a single lot number should be ordered.

4. Add decanted gel electrode buffer solution to side-arm flask and mix thoroughly with the starch using a swirling motion, making certain no lumps remain on the bottom. While continuing to swirl the flask, rapidly add the remainder of the gel electrode buffer once it has just begun to boil. **Caution:** do this carefully and with gloves to avoid being burned; use of a long-necked volumetric flask that fits inside the vacuum flask is recommended.

5. Immediately apply a vacuum and continue until the starch solution appears nearly transparent and contains only small air bubbles.

6. Carefully pour the hot, evacuated starch solution into gel molds, making certain that the entire mold is evenly filled. As soon as the gel begins to harden, gently cover with a glass plate that compresses the gel to provide a uniform thickness and prevents air bubbles from forming.

7. Allow gels to harden overnight at room temperature. When necessary, gels may be hardened in a refrigerator for 1 to 4 h before use without harmful effects.

D. Sample preparation

Sample preparation is one of the most critical steps in the electro-phoresis procedure for any organism, but it can be particularly important for many macrophytes. Poorly prepared samples can result in no activity or in weak bands with poor resolution.

Table 5–2. *Compilation of commonly used grinding buffers, electrophoresis buffers, and stains*

Grinding or extraction buffers
  1. Tris–HCl–PVPP solution
       0.1 M Tris–HCl, pH 6.8
       0.001 M EDTA
       1–3% PVPP (insoluble; Sigma P-6755)

  2. Tris–Dowex solution
       0.02 M Tris–HCl, pH 7.0
       20% Dowex-1 (Sigma 1XB-50)

  3. Tris–HCl–mercaptoethanol solution[a]
       0.1 M Tris–HCl, pH 7.5
       0.02% mercaptoethanol
       1% PVP (soluble; Sigma PVP-40) or PVPP

  4. Phosphate–PVP solution[a,b]
       0.1 M phosphate buffer, pH 7.5 (1.36 g KH_2PO_4 in 25 mg H_2O
       0.02% mercaptoethanol
       0.029 M sodium tetraborate
       0.017 M sodium metabisulfite
       0.20 M L-ascorbic acid sodium salt
       0.016 M diethyldithiocarbamic acid sodium salt
       1–3% PVP

Gel and electrode buffer systems[c]
  1. Poulik buffer system
     Electrode buffer:
       0.304 M boric acid, 0.100 M NaOH, pH 8.6; 18.5 g boric acid, 4.0 g NaOH
     Gel buffer:
       0.015 M Tris, 0.003 M citric acid, pH 7.8.
       Bring 34 ml of citric acid stock (10.5 g · liter^{-1} citric acid) plus 40 ml of Tris stock (23 g·liter^{-1} Tris) to 500-ml final volume; the final pH can be altered by changing the amount of citric acid stock added; e.g., use 36 ml of citric acid stock to 40 ml of Tris for a pH of 7.5 to 7.6

  2. Lithium hydroxide buffer system
     Electrode buffer:
       0.030 M LiOH, 0.190 M boric acid, pH 8.3; 1.259 g LiOH·H_2O, 11.742 g boric acid
     Gel buffer:
       0.046 M Tris, 0.007 M citric acid, 0.003 M LiOH, 0.019 M boric acid, pH 8.3
       Add 100 ml of electrode buffer to 900 ml of Tris–citrate buffer (6.2 g Tris, 1.6 g citric acid, pH 8.4) to give 1:9 ratio; add 1.0 M NaOH to pH 8.3

  3. Tris–Glycine system
     Electrode buffer:
       0.304 M boric acid, 0.100 M NaOH, pH 8.6.
       Same as buffer no. 1

Table 5–2 (*cont.*)

Gel buffer:
  0.009 M Tris, 0.025 M glycine, pH 8.3; 1.090 g Tris, 1.878 g glycine, 1.0 M NaOH to pH 8.3

4. Tris–EDTA–borate system
  Electrode (cathode) buffer:
  0.180 M Tris, 0.005 M EDTA, 0.100 M boric acid
  Electrode (anode) buffer:
  0.0126 M Tris, 0.003 EDTA, 0.07 boric acid
  Gel buffer:
  0.045 M Tris, 0.001 M EDTA, 0.025 M boric acid
  Dilute stock solution (109 g Tris, 7.6 g EDTA, 30.9 g boric acid, pH 8.6) 1:5 for cathode buffer, 1:7 for anode buffer, and 1:20 for gel buffer

Stain solutions[d]

1. α-Acid phosphatase (ACPH; EC 3.1.3.2)[e]

Fast garnet GBC salt	50 mg
1.0 M sodium acetate buffer, pH 4.5	2.5 ml
0.1 M MgCl$_2$	0.5 ml
1% α-naphthyl acid phosphate, sodium salt[a]	1.0 ml
H$_2$O	40 ml

The amount of sodium acetate buffer added can be increased to 4 ml; the naphthyl acid phosphate should be added just before use, either from a stock solution or solid (100 mg/50 ml)

2. β-Acid phosphatase (β-ACPH; EC 3.1.3.2)

Fast garnet GBC salt	100 mg
0.1 M Tris—maleic, pH 5.0	50 ml
β-Naphthyl acid phosphate[a]	100 mg

The Tris–maleic buffer consists of 1.2 g Tris, 1.15 g maleic acid, and 7 ml 0.1 M NaOH to pH 5.0 per 100 ml; the naphthyl acid phosphate should be added just before use.

3. Catalase (CAT; EC 1.11.1.6)

A. 3% H$_2$O$_2$	5 ml
0.1 M phosphate buffer, pH 7.0	10 ml
0.06 M Na$_2$S$_2$O$_3$·H$_2$O	7 ml
H$_2$O	78 ml
B. 0.09 M KI	50 ml
H$_2$O	50 ml

Incubate in solution A at room temperature for 15–30 min; pour off solution, rinse several times with water, then add solution B; activity will appear as white bands on a dark blue background.

4. Esterase (EST; EC 3.1.1.-)

Fast blue RR salt	100 mg
1% α-naphthyl acetate[a]	1 ml
1% β-naphthyl acetate[a] dissolved in acetone	0.3 ml
0.05 M phosphate buffer, pH 6.0	50 ml

The α- and β-naphthyl acetates are dissolved first in acetone and then diluted with water to make a stock of 50% acetone; both naphthyl acetates can be added as a solid instead of from stocks; we get improved results by also adding 2 ml *n*-propanol

Table 5–2 (*cont.*)

5. Glucose-6-phosphate dehydrogenase (G-6-PDH; EC 1.1.1.49)	
0.5 M Tris–HCl, pH 7.0	4 ml
0.25 M glucose-6-phosphate, disodium salt	3–5 ml
NADP	5 mg
1% MTT[a]	1.5 ml
1% PMS[a]	0.5 ml
H_2O	40 ml
6. Glutamate dehydrogenase (GDH; EC 1.4.1.2)	
1.0 M Tris–HCl, pH 8.0	10 ml
1.5 M L-glutamic acid	15 ml
NAD	20 mg
1% MTT[a]	2.0 ml
1% PMS[a]	0.5 ml
H_2O	25 ml

Dissolve free acid or monosodium salt glutamic acid in water and adjust pH with NaOH to pH 8.0; add MTT and PMS just before use

7. Glutamate oxaloacetate transaminase (GOT; EC 2.6.1.1.)	
Fast blue BB salt	100 mg
α-Ketoglutaric acid	50 mg
L-Aspartic acid	100 mg
Pyridoxal 5′-phosphate	0.5 mg
0.1 M Tris–HCl, pH 8.0	50 ml

Add components to Tris–HCl just before staining

8. Leucine aminopeptidase (LAP; EC 3.4.11.-)	
A. 1% L-Leucine-β-naphthylamide dissolved in acetone	2 ml
0.1 M phosphate buffer, pH 6.0	48 ml
1% $MgCl_2$	1 ml
1% $MnCl_2$	1 ml
B. Black K salt	20 mg

Dissolve the black K salt in 2–3 ml H_2O and add to solution A just before use; the leucine naphthylamide is dissolved in an acetone/water (1:4) mixture; although commonly referred to as leucine aminopeptidase, this enzyme is more precisely referred to as aminopeptidase; Innes and Yarish (unpublished) use a different recipe for aminopeptidase (See enzyme no. 9)

9. Aminopeptidase (AP; EC 3.4 11.11)	
0.1 M Tris–HCl, pH 8.0	50 ml
L-Leucyl-L-alanine	20 mg
Bothrops snake venom	10 mg
o-Dianisidine-2 HCl	10 mg
0.25 M $MnCl_2$	0.5 ml
Peroxidase	10 mg
10. Malate dehydrogenase (MDH; EC 1.1.1.37)	
1.0 M Tris–HCl, pH 8.0	10 ml
1.5 M DL-malic acid, pH 7.0; adjust pH with NaOH	10–15 ml
NAD	20 mg
1% MTT (or nitro blue tetrazolium)[a]	2 ml
1% PMS[a]	0.5 ml
H_2O	30 ml

Table 5–2 (*cont.*)

The pH of the malic acid must be adjusted before use; the MTT and PMS should be added just before use.

11. Phosphoglucoisomerase (PGI; EC 5.3.1.9)

1.0 M Tris–HCl, pH 8.0	10 ml
0.1 M MgCl$_2$	10 ml
0.018 M fructose 6-phosphate[a]	1 ml
Glucose 6-phosphate dehydrogenase[a]	2.5 ml
NADP	5 mg
1% MTT[a]	0.5 ml
1% PMS[a]	0.25 ml
H$_2$O	25 ml

The glucose 6-phosphate dehydrogenase is made up as a stock, 10 units·ml^{-1}; MTT and PMS should be added just before use

12. Phosphoglucomutase (PGM; EC 2.7.5.1)

1.0 M Tris–HCl, pH 8.0	10 ml
0.1 M MgCl$_2$	10 ml
Glucose 1-phosphate	100 mg
Glucose 6-phosphate dehydrogenase[a]	1.5 ml
NADP	5 mg
1% MTT[a]	1 ml
1% PMS[a]	0.5 ml
H$_2$O	25 ml

The glucose 6-phosphate dehydrogenase stock is the same as for PGI; The MTT and PMS should be added just before use.

[a] Store stock solutions in dark bottle in refrigerator.
[b] Modifed after Soltis et al. (1980).
[c] Gram amounts for gel and electrode buffers are given for a 1-liter final volume of buffer except where noted.
[d] Stain solutions come from various sources cited in the text.
[e] Included are the common abbreviation and IUB enzyme nomenclature number (see text).

A variety of extraction procedures have been successfully utilized with algae; which one will be best for a particular species depends on the nature of the plant (i.e., unicellular or multicellular, tough or soft thallus), its relative protein concentration (which is generally low), and the presence of enzyme-inhibiting secondary compounds such as polyphenols. The detrimental effects of tannins and other polyphenols on electrophoretic analysis are well known in flowering plants and ferns (e.g., Kelley and Adams 1977; Soltis et al. 1980, 1983) and have been suggested as a major cause of electrophoretic difficulties in the Phaeophyceae (Marsden et al. 1981; R. L. Vadas, personal communication). If high concentrations of polyphenols are suspected, the use of special grinding buffers during sample preparation is probably necessary.

We generally use a crude (protein) extraction procedure in which a small amount (e.g., 100–250 mg) of tissue is macerated and ground in two to four drops of grinding buffer. The amount of grinding buffer should be kept to a minimum to avoid dilution of the protein sample. The plant tissue should be fresh and healthy. If necessary, samples can be stored in a low-temperature (e.g., $-20°C$ or below) freezer; however, some loss of enzyme activity may occur and should be evaluated. To avoid possible developmental differences in isozyme expression, plant tissue of similar age and origin (e.g., young vs. old plants, vegetative vs. reproductive parts) should be uniformly sampled.

Examples of grinding buffers are provided in Table 5–2. Common components in such buffers include mercaptoethanol (which should be used with great care) and polyvinylpyrrolidone (PVP) or polyvinylpolypyrrolidone (PVPP). The appropriate concentration of PVP (or PVPP) depends on the species. We typically use a grinding buffer containing 1–3% insoluble PVP (40,000 MW) plus a small amount of fresh PVPP powder added directly to the slurry during grinding.

Samples can be ground with an ordinary mortar and pestle; however, we prefer to use small plastic weigh boats (43 mm^2) and a handmade plexiglass pestle with a flat, rectangular bottom. Freezing the sample with liquid nitrogen greatly facilitates the grinding procedure, especially in species with tough or leathery thalli. Homogenization with a tissue grinder and subsequent centrifugation has also been used to prepare samples from some seaweeds and a variety of microalgae (e.g., Murphy and Guillard 1976; Gallagher 1980; Innes and Yarish, 1984). The latter method works well with microalgae but poorly with seaweeds rich in cell wall phycocolloids such as carrageenan, as discussed by Cheney and Babbel (1978). Marsden et al. (1981) describe an alternative method for the brown seaweed *Fucus serratus* that involves grinding the tissue in a complex buffer containing the detergent Tween 80, chelating agents, antioxidants, and PVPP, followed by centrifugation and ion-exchange chromatography on small DEAE-cellulose columns. Whichever method is used, the homogenization process must be thorough enough to provide a complete sample of the isozymes found in the cells of the tissue being sampled, including those of the plastids and mitochondria as well as the cytosol. The subcellular location of the most frequently assayed enzyme systems in plants is described in Gottlieb (1981).

Sample preparation should take place as rapidly as possible and the grinding buffer kept cold to avoid enzyme denaturation. Once the sample is ground and squashed, its cell sap is absorbed directly onto paper wicks (5 × 6 mm) precut from Scientific Apparatus electrophoresis paper or Whatman 3-mm filter paper. Saturated wicks are placed in a glass petri dish on ice and transferred to a refrigerator until loaded into the gel.

E. Gel loading

The following procedures are used in our laboratory in conjunction with the gel mold and tray apparatus illustrated in Fig. 5–1; they should be considered only a guide and modified according to the system used. Helpful illustrations of these steps can be found in several general background sources (e.g., Shields et al. 1983).

1. Carefully pry and remove the glass plate covering the gel with a large spatula. Run a small spatula along the inside edges of the gel mold and remove excess gel material.

2. Make a transverse cut ~4.5 cm from the cathodal end using a guide to make certain the cut is perfectly straight across.

3. Separate the cathodal section of the gel to facilitate wick loading. Insert wicks vertically, side by side, leaving a few millimeters of space between each wick to prevent overlapping of bands. Between 20 and 25 wicks can be inserted per gel. Some investigators soak the last wick in a front tracking dye such as 0.5% bromophenol blue.

4. Rejoin the cathodal and anodal sections of the gel. Place the gel mold into the tray apparatus (Fig. 5–1) and fill the cathodal and anodal end chambers with appropriate buffers and a sponge. We prefer to use a sponge used for household cleaning that is approximately 18 cm square by 0.5 cm thick and is available in grocery stores.

5. Cover the top 4 cm of the anodal section of the gel with the anodal sponge. Cover the entire gel with a piece of Saran Wrap down to 2 cm from the bottom of the cathodal section. Make certain that the lower edge of the Saran Wrap is straight across, since this will ensure an even front for the migration of the cathodal electrode buffer. Now place the cathodal sponge on top of the Saran Wrap so that it overlaps it by at least 2 cm. Cover the entire gel with another piece of Saran Wrap; this prevents dehydration of the cathodal sponge and allows placement of an ice tray on the gel surface for cooling.

6. Initiate migration of proteins from the wicks into the gel itself with a short electrical pulse of 200 V (75 mA) for 20 to 25 min. Wicks are subsequently removed with forceps; they should feel slightly moist but not wet. Vary the length and voltage of the pulse if the wicks are very wet or dry.

7. After removing all wicks, carefully join the cathodal and anodal sections of the gel, making sure that there is no trapped air or debris

along the cut or origin. Any debris should be removed with a cotton swab before the sections are rejoined. To prevent a gap from forming at the origin in the case of gel shrinkage, a spacer such as a narrow glass rod is sometimes inserted at the cathodal end of the gel.

8. Replace sponges and re-cover the gel with Saran Wrap. Continue the electrophoresis run at the desired voltage and amperage; we prefer a lower voltage of ~180 V and 35–50 mA. The optimal voltage and duration will depend on the buffer combination, starch concentration, and isozyme system being examined and should be determined empirically. When 180 V and a 12% starch concentration are used, the front will typically migrate 8 cm in approximately 3 to 3.5 h, providing sufficient protein separation.

9. In order to avoid protein denaturation, it is important to conduct all phases of the electrophoresis run (i.e., steps 6 and 8) at a low temperature. Typically, gels are kept in either a cold room at less than 10°C or a refrigerator during a run. In addition, we generally place an ice tray on the surface of the gel.

F. Gel slicing

After the front has migrated the desired distance, usually 8–9 cm, the gel is removed from the tray apparatus and sliced. First, the guides of the gel mold are removed, and the gel is notched at the top left corner of the anodal section to provide a method of orientation for the gel slices after staining. Slicing is accomplished by carefully drawing a taut wire horizontally through the entire length of the gel. The wire should be held against the two attached side guides of the gel mold at the same time that it is being drawn through the length of the gel. This should provide a bottom slice that is approximately 3–4 mm in thickness. Next, the two "slicing" side guides (not shown in Fig. 5–1) are placed on top of the attached side guides and the slicing procedure is repeated, a total of three slices of approximately equal thickness thereby being produced. Consistently producing slices of uniform thickness can be difficult without the aid of either a handmade or a commercial gel slicer. Plans for homemade gel slicers are available in several books; our design will be made available upon request.

G. Staining procedures

The basic principle of enzyme-specific gel staining is quite simple (for more details, see, e.g., Brewer 1970; Oelshlegel and Stahmann 1973; Harris and Hopkinson 1976). Each staining solution contains a substrate specific for the enzyme being assayed and a salt or dye that precipitates where the enzyme-catalyzed reaction occurs. This results in a colored band marking the position in the gel to which the enzyme migrated. Stains that produce such colored bands are referred to as positive stains. Negative stains also exist; when they are used (e.g., with tetrazolium oxidase), enzyme activity is evidenced by unstained bands appearing against a stained background.

The gel slices are carefully separated with a spatula and transferred by hand to separate staining solutions. A properly prepared gel will produce slices that can be easily lifted by the corners. As a rule, we prepare all the reagents for each staining solution during the electrophoresis run and mix them together only after the gel has been sliced to ensure their freshness. Each slice is placed in the stain solution with the cut surface up to improve band resolution. The middle and bottom slices generally show better staining activity than the top slice and so are placed in the stains considered most critical to the study. Because of the high costs of certain stains (e.g., those using NADP), it is important to use a staining tray that allows one to submerge the slice completely in the smallest volume possible (e.g., about 50 ml). Plastic sandwich containers with lids work well for this purpose; however, commercial stain boxes are also available (Flambeau Co.)

The stain recipes most commonly used by ourselves and other investigators are summarized in Table 5–2. Detailed information on these and other enzymes is provided in *Enzyme Nomenclature* (International Union of Biochemistry 1979), Gottlieb (1981), and Brewer (1970). It should be noted that the methods described in Table 5–2 may have to be modified for a particular species. One variable frequently experimented with is the pH of the stain. For example, Soltis et al. (1983) reported improved staining activity for glutamate dehydrogenase (GDH), glucose-6-phosphate dehydrogenase (G-6-PDH), malate dehydrogenase (MDH), phosphoglucose isomerase (PGI), and phosphoglucose mutase (PGM) in ferns by increasing the pH of the Tris–HCl buffer of the stain from 7.0 to 8.0–8.5. Likewise, the best gel and electrode buffer system used in conjunction with a stain may have to be determined experimentally, because some stains work best with certain buffer systems (e.g., β-acid phosphatase with LiOH; PGI, PGM, and MDH with Poulik). Additional helpful suggestions for improving staining results can be found in Oelshlegel and Stahmann (1973).

Although most staining procedures can be conducted at room temperature, we usually incubate our stains in the dark in a laboratory oven at 35°C, which maximizes enzyme activity and protects light-sensitive stains (e.g., those using tetrazolium salts such as dimethylthiazoldiphenyltetrazolium (MTT) and phenazine methosulfate (PMS) from inactivation. The length of time for development varies for each stain and the relative amount of protein in the sample.

Generally, most nontetrazolium stains can be left overnight to develop, whereas tetrazolium-containing stains such as MDH, PGI, and PGM develop in just 2 to 3 h. Slices should not be left in a stain any longer than necessary, because a loss of band resolution will result.

H. Gel fixation and preservation

Once a stain has developed fully, the staining solution is removed and the gel is carefully washed with distilled water and fixed inside the staining tray. Whereas most investigators use a fixative of 50% ethanol, we use 50% ethanol and 3.5% acetic acid. The fixative hardens and toughens the gel so that it can be handled more easily. After fixing overnight, gels are gently dried and wrapped in Saran Wrap; wrapped gels can be stored under refrigeration for up to several months. Analysis of banding patterns is greatly facilitated by the use of a light box. We routinely record our results photographically using black and white film and lighting the gels from underneath with either a light box or copy stand arrangement similar to that described by Sharkey and Bacci (1975). Photographic records allow for easier gel-to-gel comparisons throughout a study and are a great aid in interpreting complex banding patterns.

I. Sampling design

Although the optimal sampling design will depend on the specific purpose of the electrophoretic data, there are generally three principal considerations in any electrophoretic study involving population comparisons: (1) the nature and number of populations to be sampled, (2) the number of individuals sampled per population, and (3) the number of loci to be analyzed. It is generally desirable to sample 50 or more random individuals per population (Gottlieb 1977; Brown 1978); however, this may not always be practical or necessary (e.g., see Gorman and Renzi 1979). Often the investigator must choose between a large sample size from few locations or vice versa; which strategy is best depends on the purpose of the study and the degree of differentiation among the populations. However, no matter how many populations are being sampled, we sample a minimum of 20 individuals per population. A preliminary population sample of larger size (e.g., 50–75 individuals) can be very helpful in determining the minimum population sample size.

In systematic studies in particular, it is important to sample as many populations as possible and from as broad a range of the species' geographical and ecological distribution as possible. It is clear from past algal studies that a great deal of intraspecific genetic differentiation can be associated with habitat differences (e.g., Murphy and Guillard 1976; Cheney and Babbel 1978; Soudek and

Robinson 1983; Innes and Yarish, 1984). Studies dealing with land plants (e.g., Gottlieb 1977; Hamrick et al. 1979; Crawford 1983) suggest that, in general, greater interpopulation sampling is required for inbreeding species compared with outcrossing species. Although very little is known about breeding systems in algae, in at least three studies (Malinowski 1974; Cheney and Babbel 1978; Innes and Yarish, 1984) evidence has been reported for populations reproducing primarily by asexual reproduction.

The general rule about the number and kinds of enzymes employed is to sample as many enzymes as possible and to include both multiple- and single-substrate types. As described earlier, single-substrate enzymes tend to have fewer isozymes than multiple-substrate enzymes as well as to be more conserved among related taxa (e.g., Gottlieb 1981). Thus, one must avoid artificially influencing the degree of isozyme similarity in a systematic study by making certain to survey a mixture of enzyme types. We try to analyze a minimum of 15 to 20 loci (total) from at least five or six enzyme systems.

IV. Data recording and interpretation

After a gel has been properly stained, one or more colored bands will appear per individual, marking the position(s) to which different forms of the assayed enzyme (i.e., isozymes) migrated during electrophoresis. The total pattern of bands for an enzyme is called a zymogram.

The position of each band on a gel should be measured to the nearest millimeter and recorded. We routinely record our results on diagrams with migration distances (in millimeters) along the vertical axis and sample numbers along the horizontal axis. The actual migration distance of a band can be converted to a relative (R_f) value by dividing a band's actual migration distance by the migration distance of the front and multiplying by 100, thereby taking into account any differences in front migration from gel to gel.

In order to facilitate gel-to-gel comparisons in a large study, it is advisable to run an internal standard or reference plant in each gel, which produces one or more prominent "reference" bands for each enzyme system. Cheney and Babbel (1978), for example, maintained clones of a single reference plant in culture throughout their study for this purpose. The migration distances of such "reference" bands serve as a useful check for consistency of technique throughout a study.

A number of nomenclature systems have been used to label bands. We prefer a system that utilizes the enzyme abbreviation for labeling loci, followed by a hyphenated numeral when more than one gene

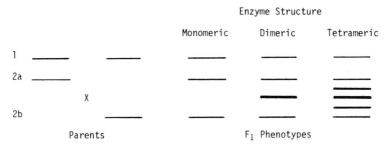

Fig. 5–2. Examples of electrophoretic patterns in parents and their offspring illustrating how the subunit structure of the enzyme determines the number of bands in the offspring. Bands are labeled as described in the text; band 1 represents a monomorphic locus, whereas bands 2a and 2b represent alleles of a polymorphic locus. Modified from Gottlieb (1977).

locus or isozyme is represented. Usually, bands are numbered in sequence, starting with the band that migrated farthest from the origin. Allelic variants are designated by a letter after the numeral. Thus, two allelic bands (allozymes) of the PGM-1 locus would be designated PGM-1a and PGM-1b. An alternative method is simply to code individual bands according to their relative migration distance (R_f value) from the origin. For example, PGM^{57} would represent a band that had a relative migration distance of 57 mm.

Electrophoretic patterns vary in complexity from one enzyme to another. Multiple-substrate enzymes such as phosphatases and esterases usually exhibit the most complex banding patterns, whereas substrate-specific enzymes such as PGI and PGM generally exhibit relatively few bands per individual.

Other factors determining the complexity of banding patterns include the quaternary structure of the enzyme assayed as well as the genotypic (homozygous or heterozygous) nature and ploidy level of the individual examined. The number of polypeptide subunits of each enzyme will determine the number of bands displayed by a heterozygous individual. For example, an individual heterozygous for a particular gene locus will exhibit two allozyme bands if the enzyme assayed is monomeric (composed of a single polypeptide), three bands if the enzyme is dimeric (composed of two polypeptides), and five bands if it is tetrameric (Fig. 5–2). In general, a heterozygous individual will exhibit one more band than the number of subunits making up the enzyme. Gottlieb (1981) provides information on the quaternary structure of the most commonly assayed enzymes in plants.

Band number is also influenced by the ploidy level of the individual. Tetraploids, for example, have been shown to display more isozymes per system than their diploid relatives and to have isozyme patterns that resemble those of heterozygous diploid individuals (Gottlieb 1981). Since diploids may exhibit more bands than haploids (when heterozygous), data for diploid and haploid individuals should not be pooled for population comparisons in order to make certain that any population differences in band frequencies are not merely due to differences in the proportion of haploid and diploid individuals sampled.

Artifacts that result from the extraction, electrophoresis, or staining procedures can also affect band number. One method of testing for staining artifacts is to delete the substrate from a stain solution or to replace the extract with water; in either case, no bands should appear. This is particularly important for dehydrogenase stains using tetrazolium salts, because artifact bands commonly called "nothing dehydrogenase" may appear.

As noted here, the causes of variation in electrophoretic banding patterns can be complex. Furthermore, to be fully useful to the systematist or ecologist, banding pattern differences among individuals must be shown to have a genetic basis and not be the result of developmental changes in isozyme expression (e.g., see Scandalios 1974), environmental effects, or experimental artifacts. Proving the genetic control of banding differences and providing a genetic interpretation of banding patterns (i.e., distinguishing between isozymes and allozymes) are generally the most difficult aspects of any electrophoretic study but necessary for taking full advantage of the electrophoretic technique (Gottlieb 1981). This is usually accomplished by formal genetic analysis through breeding studies. An alternative technique used by Weeden and Gottlieb (1979) involves comparing electrophoretic patterns of extracts from diploid somatic tissues of plants with those from haploid pollen.

For a great many (if not most) marine algae, breeding studies may be impossible or impractical to conduct due to difficulties in controlling sexual reproduction in culture. Therefore, alternative methods have been used to eliminate nongenetic sources of band variation. Gallagher (1980), for example, eliminated environmental influences by culturing her clones under several conditions and by using only those enzymes and bands with constant stability under the different conditions. Artifacts can be eliminated by running careful controls, such as described earlier. If the species being studied has morphologically distinguishable haploid and diploid generations, the approach used by Cheney and Babbel (1978) can be helpful. In the latter study, nonartifactual bands of the same enzyme assay that occurred together in haploid plants were assumed to be the product of different gene loci, whereas bands that belonged to the same zone of activity and occurred together only in diploid plants (i.e., putative heterozygotes) were judged to be alleles. D. P. Cheney and A. Mathieson (unpublished data) similarly compared the banding patterns of diploid tetrasporophytic plants with those of their haploid gametophytic offspring grown in culture to aid in interpreting banding patterns in *Chondrus*.

V. Methods of analysis

The simplest and most frequently used method of illustrating electrophoretic data in algal studies has been the use of zymograms (e.g., Murphy and Guillard 1976; Cheney and Babbel 1978; Blair et al. 1982; Soudek and Robinson 1983). Although zymograms allow for easy comparison of major differences among populations or samples, their use has limitations. Gottlieb (1977) discussed the shortcomings of simply comparing individuals or populations for the presence or absence of bands and described why differences among individuals in their number of bands are not necessarily equivalent to the number of their genetic differences. Furthermore, zymograms do not show the variation in banding patterns of individuals within populations, only a composite for the population as a whole, thereby masking potentially important information.

Whereas zymograms can demonstrate gross differences in banding patterns among samples, gene or allele frequency information is necessary for detailed comparisons of samples for genetic variability and similarity. The frequency of any gene or allele in a sample is equal to twice the number of heterozygotes with the allele (or gene) divided by twice the number of individuals in the sample. Once allele frequencies have been calculated, there are a variety of analytical measures that can be used to compare populations for their relative degree of genetic variation (e.g., see Gottlieb 1981; Hamrick et al. 1979). Two such calculations commonly used are (1) the proportion of polymorphic loci for a population (or species) as a whole and (2) the mean heterozygosity per locus. The proportion of polymorphic loci (P) is simply the proportion of all gene loci studied that are polymorphic, where a polymorphic locus is generally defined as one in which the frequency of the most common allele is no greater than 0.99. Assuming a random mating population, the expected frequency of heterozygotes (H) at a locus can be calculated directly from allelic frequencies, such that $H = 1 - \sum \underline{x_i^2}$, where x_i is the frequency of the ith allele. Mean heterozygosity ($\overline{H}$) is the average of H over all loci examined. Alternatively, one can also calculate the observed frequency of heterozygotes, simply by dividing the number of heterozygous individuals at a locus by the total number of individuals in a sample. Naturally, observed heterozygosity values may differ from expected values. Significant deviations in these two values from Hardy–Weinberg equilibrium expectations can provide valuable insight into the genetic structure of a population. See Crisp et al. (1978) for further information on testing electrophoretic results for Hardy–Weinberg equilibrium.

The most widely used method for providing a quantitative estimate of the amount of genetic differentiation or divergence among populations is the genetic identity statistic I of Nei (1972). This statistic has been used to estimate the degree of genetic differentiation at various taxonomic levels ranging from conspecific populations to distinct species in a wide variety of organisms (see reviews by Ayala 1975; Powell 1975a; Crawford 1983; Gottlieb 1977). The genetic identity between two populations at the i locus is defined as

$$I_i = \sum x_i y_i / (\sum x_i^2 \sum y_i^2)^{1/2} \tag{1}$$

where x_i and y_i represent the frequencies of the ith allele in populations x and y, respectively. For all loci in a sample, the overall genetic identity of populations x and y is defined as

$$I = I_{xy} / (I_x I_y)^{1/2} \tag{2}$$

where I_x, I_y, and I_{xy} are the arithematic means over all loci of $\sum x_i^2$, $\sum y_i^2$, and $\sum x_i y_i$, respectively. Values of I can range from 0 to 1, with $I = 1$ being the case when all allele frequencies in populations x and y are equal, and $I = 0$ when the two populations have no alleles in common. Nei's genetic identity statistic has been used in only a few algal studies (Malinowski 1974; Cheney and Babbel 1978; Gallagher 1980). Nei's (1972) genetic distance D is often calculated from Nei's I, where $D = -\ln I$, and is a measure of the genetic distance between populations x and y.

Another method of comparing populations is a modification of Hedrick's (1971) genotypic identity statistic. Hedrick's formulation has been used to compare populations with phenotype (or band) frequencies in only a few studies (e.g., Cheney and Babbel 1978; Hancock and Bringhurst 1979; Blair et al. 1982). Its one advantage is that it does not require a genetic interpretation of banding patterns (which Nei's I does) and therefore can be useful for quantifying band differences in those cases where banding patterns cannot be genetically interpreted. However, one must be careful in drawing conclusions when using such a statistic for the same reasons that one must excercise caution in using simple band counting, as described earlier.

Other methods have also been used for analyzing electrophoretic data (e.g., see Ayala 1976; Gallagher 1980; Soudek and Robinson 1983). Swofford and Selander (1981) described a helpful computer program (written in FORTRAN) for analyzing allelic variation in a variety of ways, including (1) computation of allele frequencies and genetic variability measures, (2) testing for deviation from Hardy–Weinberg equilibrium, (3) performing heterogeneity chi-square analyses, (4) calculating a variety of similarity coefficients, (5) constructing phenograms using cluster analysis, and (6) estimating phylogenies using the Wagner distance procedure.

VI. Limitations, biases, and alternative techniques

Although electrophoresis has many advantages to offer the systematist or ecologist, its use to measure genetic variation or differentiation is subject to several inherent limitations and potential sources of bias. These are discussed in detail elsewhere (e.g., Lewontin 1974; Ayala 1976; Ramshaw et al. 1979; Gottlieb 1981) and are mentioned only briefly here. One of the principal limitations of conventional methods (such as described here) is that they detect only those amino acid substitutions that result in charge differences in proteins and therefore provide an underestimate of the total allelic variation. Although migration differences among bands generally can be taken as evidence of at least one amino acid difference, band homogeneity does not necessarily indicate an identical amino acid sequence and genetic identity. Since 16 of 20 common amino acids are electrophoretically neutral in the pH range of the most commonly used buffers, substitutions involving most amino acids go undetected. In fact, the proportion of point mutations generally thought to be electrophoretically detectable is quite low and estimated to be ~0.27 (King and Wilson 1975). Thus, conventional electrophoretic techniques clearly underestimate the genetic differentiation among populations or species. One must therefore be careful in making conclusions based on a high degree of electrophoretic similarity. Furthermore, isozyme similarity (or difference) may also be caused by posttranslational modification of an enzyme by the products of other genes (e.g., Finnerty and Johnson 1979).

Another limitation of electrophoresis is that only structural genes coding for soluble proteins can be detected. Regulatory genes or those coding for nonsoluble proteins go undetected. Thus, electrophoresis does not detect a truly random sample of all loci. This is further influenced by the type of enzymes assayed.

Questions concerning the sensitivity and sample bias of electrophoresis have been discussed at length in the literature (e.g., Powell 1975b; Johnson 1977a; Ramshaw et al. 1979; Coyne 1982). Enhanced sensitivity or capacity to detect so-called hidden allelic variation has been demonstrated in several studies using more refined techniques.

These more sophisticated, high-resolution electrophoretic techniques have generally involved varying the gel pore size, pH, and temperature (e.g., Singh et al. 1976; Johnson 1977b; Coyne et al. 1979). Electrophoresis has also been combined with enzyme sensitivity to high temperature to demonstrate variation within identically migrating isozymes (e.g., Trippa et al. 1976).

Two related techniques that also exhibit increased resolution of genetic variation are two-dimensional electrophoresis (e.g., Leigh Brown and Langley 1979) and isoelectric focusing (e.g., see Hoyle 1978; Andrews 1981). Isoelectric focusing relies on fractionating molecules differing only in net charge; that is, it does not make use of "molecular sieving" as in conventional electrophoresis. By using narrow-range ampholytes in a granulated bed matrix, isoelectric focusing concentrates protein bands at their isoelectric points and can be used to concentrate quantities of an enzyme for further analysis.

VII. Concluding remarks

Despite its limitations, electrophoresis is the most practical and reliable technique available for routinely studying genetic variation and differentiation in natural populations. Taking full advantage of its applications to the study of systematic and ecological problems, however, requires careful techniques and expertise that can be gained only through experience. Before undertaking any electrophoretic study, those unfamiliar with the technique should make certain that they can make a substantial time commitment to the study and can get help from someone with electrophoretic expertise. The amount of time and effort that will be required to develop reliable assay systems for newly studied species and for genetically interpreting their banding patterns generally cannot be predicted.

VIII. Acknowledgments

The author wishes to express his gratitude to Gary Babbel for his excellent tutelage and his appreciation to D. Innes, A. Mathieson, C. Tanner, and the editorial committee for their helpful comments and review of an earlier manuscript. Special thanks go also to M. Littler and D. Littler for their patience and to Susan Reiter for her illustrations of Figs. 5–1 and 5–2.

IX. References

Andrews, A. T. 1981. *Electrophoresis: Theory, Techniques, and Biochemical and Clinical Applicatons.* Clarendon Press, Oxford. 336 pp.

Ayala, F. J. 1975. Genetic differentiation during the speciation process. *Evol. Biol.* 8, 1–78.

Ayala, F. J. 1976. *Molecular Evolution*. Sinauer Associates, Sunderland, Massachusetts. 277 pp.

Baker, A., and Holton, R. 1973. Electrophoretic analysis of proteins and malic dehydrogenase isozymes in nine oscillatorian blue-green algae. *Phycologia* 12, 83–7.

Battaglia, B., and Beardmore, J. 1978. *Marine Organisms: Genetics, Ecology and Evolution*. Plenum Press, New York. 767 pp.

Berger, L., Sandakhchiev, L., and Schweiger, H. 1974. Fine structural and biochemical markers of Dasycladaceae. *J. Microscop.* 19, 89–104.

Blair, S., Mathieson, A., and Cheney, D. 1982. Morphological and electrophoretic investigations of selected species of *Chaetomorpha* (Chlorophyta; Cladophorales). *Phycologia* 21, 164–72.

Brewer, G. 1970. *An Introduction to Isozyme Techniques*. Academic Press, New York. 186 pp.

Brown, A. 1978. Isozymes, plant population genetic structure and genetic conservation. *Theor. Appl. Genet.* 52, 145–57.

Burton, R. 1983. Protein polymorphisms and genetic differentiation of marine invertebrate populations. *Mar. Biol. Lett.* 4, 193–206.

Cheney, D., and Babbel, G. 1978. Biosystematic studies of the red algal genus *Eucheuma*. 1. Electrophoretic variation among Florida populations. *Mar. Biol.* 47, 251–64.

Cheney, D., and Mathieson, A. 1979. Population differentiation in the seaweed *Chondrus crispus*: preliminary results. *Isozyme Bull.* 12, 57.

Coyne, J. 1982. Gel electrophoresis and cryptic protein variation. In Rattazzi, M., Scandalios, J., Whitt, G., and Liss, A. (eds.), *Isozymes: Current Topics in Biological and Medical Research*, vol. 6, pp. 1–32. Liss, New York.

Coyne, J., Eanes, W., Ramshaw, J., and Koehn, R. 1979. Electrophoretic heterogeneity of α-glycerophosphate dehydrogenase among many species of *Drosophila*. *Syst. Zool.* 28, 164–75.

Crawford, D. 1983. Phylogenetic and systematic inferences from electrophoretic studies. In Tanksley, S., and Ortan, T. (eds.), *Isozymes in Plant Genetics and Breeding*, part A, pp. 257–87. Elsevier, Amsterdam.

Crisp, D., Beaumont, A., Flowerdew, M., and Vardy, A. 1978. The Hardy–Weinberg test: a correction. *Mar. Biol.* 46, 181–3.

Finnerty, V., and Johnson, G. 1979. Post-translational modification as a potential explanation of high levels of enzyme polymorphism: xanthine dehydrogenase and aldehyde oxidase in *Drosophila melanogaster*. *Genetics* 91, 695–722.

Frederick, J. 1971. Storage polyglucan-synthesizing isozyme patterns in the Cyanophyceae. *Phytochemistry* 10, 395–8.

Frederick, J. 1977. Protein and isozyme patterns of the cyanelles of *Glaucocystis nostochinearum* compared with *Anacystis nidulans*. *Phytochemistry* 16, 1571–3.

Fulton, A. 1977. Isozyme band variation within the algal species *Pandorina morum*. *Biochem. Syst. Ecol.* 5, 261–4.

Gagnon, P., Vadas, R., Burdick, D., and May, B. 1980. Genetic identity of annual and perennial forms of *Zostera marina*. *Aquat. Bot.* 8, 157–62.

Gallagher, J. 1980. Population genetics of *Skeletonema costatum* (Bacillario-phyceae) in Narragansett Bay. *J. Phycol.* 16, 464–74.

Gorman, G., and Renzi, J. 1979. Genetic distance and heterozygosity estimates in electrophoretic studies: effects of sample size. *Copeia* 1979, 242–94.

Gottlieb, L. 1977. Electrophoretic evidence and plant systematics. *Ann. Mo. Bot. Gard.* 64, 161–80.

Gottlieb, L. 1981. Electrophoretic evidence and plant populations. *Progr. Phytochem.* 7, 1–46.

Grant, M., and Proctor, V. 1980. Electrophoretic analysis of genetic variation in the Charophyta. 1. Gene duplication via polyploidy. *J. Phycol.* 16, 109–15.

Hames, B., and Rickwood, D. 1982. *Gel Electrophoresis of Proteins: A Practical Approach.* IRL Press, London. 290 pp.

Hamrick, J., Linhart, Y., and Mitton, J. 1979. Relationships between life history characteristics and electrophoretically detectable genetic variation in plants. *Annu. Rev. Ecol. Syst.* 10, 173–200.

Hancock, J., and Bringhurst, R. 1979. Ecological differentiation in perennial octoploid species of *Fragaria. Amer. J. Bot.* 66, 367–75.

Harris, H., and Hopkinson, D. 1976. *Handbook of Enzyme Electrophoresis in Human Genetics.* American Elsevier, New York. 279 pp.

Hayhome, B., and Pfiester, L. 1983. Electrophoretic analysis of soluble enzymes in five freshwater dinoflagellate species. *Amer. J. Bot.* 70, 1165–72.

Hedrick, P. 1971. A new approach to measuring genetic similarity. *Evolution* 25, 276–80.

Herbert, M., Burkhard, C., and Schnarrenberger, C. 1979. A survey for isoenzymes of glucosephosphate isomerase, phosphoglucomutase, glucose-6-phosphate dehydrogenase and 6-phosphogluconate dehydrogenase in C_3-, C_4- and crassulacean-acid-metabolism plants, and green algae. *Planta* 145, 95–104.

Holton, R. 1973. Electrophoresis and the taxonomy of algae. *Bull. Torrey Bot. Club* 100, 297–303.

Hoxmark, R. 1976. Protein composition of different stages in the life cycle of *Ulva mutabilis*, Föyn. *Planta* 130, 327–32.

Hoyle, M. 1978. *Illustrated Handbook for High Resolution of IAA Oxidase-peroxidase Isoenzymes by Isoelectric Focusing in Slabs of Polyacrylamide Gel.* Forest Service General Technical Report NE-37, Washington, D.C. 26 pp.

Hushovd, O., Gulliksen, O., and Nordby, O. 1982. Absence of major differences between soluble proteins from haploid gametophytes and diploid sporophytes in the green alga *Ulva mutabilis* Föyn. *Planta* 156, 89–91.

Innes, D., and Yarish, C. 1984. Genetic evidence for the occurrence of asexual reproduction in populations of *Entermorpha linza* (J. Agardh) (Chlorphyta) from Long Island Sound. *Phycologia* 26, 311–20.

International Union of Biochemistry. 1979. *Enzyme Nomenclature.* Recommendations (1978) of the Nomenclature Committee of the International Union of Biochemistry. Academic Press, New York. 606 pp.

Johnson, G. 1977a. Characteristics of electrophoretically cryptic variation in the alpine butterfly *Colias meadii*. *Biochem. Genet.* 15, 665–93.

Johnson, G. 1977b. Assessing electrophoretic similarity: the problem of hidden heterogeneity. *Annu. Rev. Ecol. Syst.* 8, 309–28.

Keck, K. 1961. Nuclear and cytoplasmic factors determining the species specificity of enzyme proteins in *Acetabularia*. *Ann. N.Y. Acad. Sci.* 94, 741–52.

Keck, K., and Choules, E. 1963. An analysis of cellular and subcellular systems which transform the species character of acid phosphatase in *Acetabularia. J. Cell Biol.* 18, 459–69.

Kelley, W., and Adams, R. 1977. Preparation of extracts from Juniper leaves for electrophoresis. *Phytochemistry* 16, 513–16.

King, M. C., and Wilson, A. C. 1975. Evolution at two levels: molecular similarities and biological differences between humans and chimpanzees. *Science* 188, 107–16.

Klein, S., Chapman, D., and Garber, E. 1973. Chemotaxonomy of the *Oscillatoria–Phormidium* complex. *Biochem. Syst.* 1, 173–7.

Kovatcheva, N., and Bergman, B. 1979. Some characteristics of malate dehydrogenase of the blue-green alga *Nostoc muscorum*. *Plant Sci. Lett.* 16, 189–94.

Leigh Brown, A., and Langley, C. 1979. Reevaluation of level of genic heterozygosity in natural populations of *Drosophila melanogaster* by two-dimensional electrophoresis. *Proc. Nat. Acad. Sci.* 76, 2381–4.

Lewontin, R. 1974. *The Genetic Basis of Evolutionary Change*. Columbia University Press, New York. 345 pp.

McMillan, C. 1980. Isozymes of tropical seagrasses from the Indo-Pacific and the Gulf of Mexico-Caribbean. *Aquat. Bot.* 8, 163–72.

McMillan, C. 1982. Isozymes in seagrasses. *Aquat. Bot.* 14, 231–43.

McMillan, C., and Williams, S. 1980. Systematic implications of isozymes in *Halophila* section *Halophila*. *Aquat. Bot.* 9, 21–31.

Malinowski, K. 1974. "*Codium fragile*: The Ecology and Population Biology of a Colonizing Species." Ph.D. dissertation, Yale University, New Haven, Conn. 135 pp.

Mallery, C., and Richardson, N. 1971. Disc gel electrophoresis of biliproteins. *Plant Cell Physiol.* 12, 997–1001.

Mallery, C., and Richardson, N. 1972. Disc gel electrophoresis of proteins of Rhodophyta. In Nisizawa, K. (ed.), *Proceedings of the Seventh International Seaweed Symposium*, pp. 292–300. Wiley, New York.

Marsden, W., Callow, J., and Evans, L. 1981. A novel and comprehensive approach to the extraction of enzymes from brown algae, and their separation by polyacrylamide gel electrophoresis. *Mar. Biol. Lett.* 2, 353–62.

Mathieson, A., Norton, T., and Neushul, M. 1981. The taxonomic implication, of genetic and environmentally induced variations in seaweed morphology. *Bot. Rev.* 47, 313–47.

Miura, W. 1978. Polymorphism of catalase isozyme in *Porphyra*. *Tohoku J. Agri. Res.* 29, 159–66.

Miura, W., Fujio, Y., and Suto, S. 1978. Isozymes from individual thallus of *Porphyra* species. *Jap. J. Phycol.* 26, 139–43.

Murphy, L. 1978. Biochemical taxonomy of marine phytoplankton by electrophoresis of enzymes. 2. Loss of heterozygosity in clonal cultures of the centric diatoms *Skeletonema costatum* and *Thalassoisira pseudnana*. *J. Phycol.* 12, 9–13.

Murphy, L., and Guillard, R. 1976. Biochemical taxonomy of marine phytoplankton by enzyme electrophoresis. I. The centric diatoms *Thalassiosira pseudonana* Hasle and Heimdal and *Thalassiosira fluviatilis* Hustedt. *J. Phycol.* 12, 9–13.

Nei, M. 1972. Genetic distance between populations. *Amer. Natur.* 106, 283–92.

Nevo, E. 1978. Genetic variation in natural populations: patterns and theory. *Theor. Pop. Biol.* 13, 121–77.

Oelshlegel, F., and Stahmann, M. 1973. The electrophoretic technique: a practical guide for its application. *Bull. Torrey Bot. Club* 100, 260–71.

Peak, M., Peak, J., and Ting, L. 1972. Isoenzymes of malate dehydrogenase and their regulation in *Euglena gracilis*. *Biochim. Biophys. Acta* 248, 1–15.

Powell, J. 1975a. Protein variation in natural populations of animals. *Evol. Biol.* 8, 79–119.

Powell, J. 1975b. Isozymes and non-Darwinian evolution: a re-evaluation. In Markert, C. (ed.), *Isozymes*, vol. 4: *Genetics and Evolution*, pp. 9–26. Academic Press, New York.

Ramshaw, J., Coyne, J., and Lewontin, R. 1979. The sensitivity of gel electrophoresis as a detector of genetic variation. *Genetics* 93, 1019–37.

Richardson, N., and Mallery, C. 1973. Some chemosystematic aspects of the *Audodinella* (*Acrochaetium–Rhodochorton*) complex. *Amer. J. Bot.* 60, 1051–6.

Scandalios, J. 1974. Isozymes in development and differentiation. *Annu. Rev. Plant Physiol.* 25, 225–58.

Schenk, H., Hofer, I., and Metzner, H. 1973. Molat-dehydrogenase isoenzymbaden als potentielles chemotaxomisches kriterium fur Cyanophycean species. *Biochem. Syst.* 1, 179–84.

Schoenberg, D., and Trench, R. 1980. Genetic variation in *Symbiodinium* (=*Gymnodinium*) *microadriacticum* Freudenthal, and specificity in its symbiosis with marine invertebrates. 1. Isoenzyme and soluble protein patterns of axenic cultures of *Symbiodinium microadriaticum*. *Proc. R. Soc. Lond. Ser. B* 207, 405–27.

Sharkey, J., and Bacci, C. 1975. Photography of dodecylsulfate polyacrylamide gels. *Functional Photography* November, 32–3.

Shields, C., Ortan, T., and Stuba, C. 1983. An outline of general resource needs and procedures for the electrophoretic separation of active enzymes from plant tissue. In Tanksley, S., and Ortan, T. (eds.), *Isozymes in Plant Genetics and Breeding*, part A, pp. 443–58. Elsevier, Amsterdam.

Singh, R., Lewontin, R., and Felton, A. 1976. Genetic heterogeneity within electrophoretic "alleles" of xanthine dehydrogenase in *Drosophila pseudoobscura*. *Genetics* 84, 609–29.

Soltis, D., Haufler, C., Darrow, D., and Gastony, G. 1983. Starch gel electrophoresis of ferns: a compilation of grinding buffers, gels and electrode buffers and staining schedules. *Amer. Fern J.* 73, 9–27.

Soltis, D., Haufler, C., and Gastony, G. 1980. Detecting enzyme variation in the fern genus *Bommeria*: an analysis of methodology. *Syst. Bot.* 5, 30–8.

Soudek, D., Jr., and Robinson, G. 1983. Electrophoretic analysis of the species and population structure of the diatom *Asterionella formosa*. *Can. J. Bot.* 61, 418–33.

Spencer, K., Yu, M., West, J., and Glazer, A. 1981. Phycoerythrin and interfertility patterns in *Callithamnion* (Rhodophyta) isolates. *Br. Phycol. J.* 16, 331–43.

Swofford, D., and Selander, R. 1981. Biosys-1: a Fortran program for the comprehensive analysis of electrophoretic data in population genetics and systematics. *J. Hered.* 72, 281–3.

Thomas, D. L., and Brown, R. M. 1970a. New taxonomic criteria in the classification of *Chlorococcum* species. 3. Isozyme analysis. *J. Phycol.* 6, 293–9.

Thomas, D. L., and Brown, R. M. 1970b. Isozyme analysis and morphological variation of thirty-two isolates of *Protosiphon*. *Phycologia* 9, 285–92.

Thomas, D. L., and Delcarpio, J. B. 1971. Electrophoretic analysis of enzymes from three species of *Chlamydomonas*. *Amer. J. Bot.* 58, 716–20.

Thomas, D., and Groover, R. 1973. Electrophoretic and immunological analysis of seven chlorosarcinacean algae. *J. Phycol.* 9, 289–96.

Trippa, G., Loverre, A., and Catamo, A. 1976. Thermostability studies for investigating non-electrophoretic polymorphic alleles in *Drosophila melanogaster*. *Nature* 260, 42.

Weeden, N., and Gottlieb, L. 1979. Distinguishing allozymes and isozymes for phosphoglucoisomerases by electrophoretic comparisons of pollen and somatic tissues. *Biochem. Genet.* 17, 287–96.

Wutz, M., and Zetsche, K. 1976. Zur biochemie und regulation des heteromorphen generationswechsels der grunalge *Derbesia–Halicystis*. *Planta* 129, 211–16.

Wynne, D. 1977. Alterations in activity of phosphatases during the *Peridinium* bloom in Lake Kinneret. *Physiol. Plant.* 40, 219–24.

Young, E. 1970. A comparison of the soluble proteins in various species of algae by disc electrophoresis in polyacrylamide gels. *Phytochemistry* 9, 2167–74.

6: Natural products chemistry: uses in ecology and systematics

JAMES N. NORRIS

Department of Botany, National Museum of Natural History, Smithsonian Institution, Washington, D. C. 20560

WILLIAM H. FENICAL

Institute of Marine Resources, Scripps Institution of Oceanography, University of California at San Diego, La Jolla, California 92093

CONTENTS

I. Introduction

A. Objectives

This chapter focuses on methods useful to the field biologist for identifying the presence or absence of lipid-soluble secondary metabolites in benthic marine algae. Present research indicates that ~90% of all algal secondary metabolites are lipid soluble. We also present some tests to determine whether the algal extracts or their isolated compounds are biologically active and offer a brief discussion on the use of natural products chemistry in systematics.

B. Chemistry in systematics

Chemistry played an early and important part in the systematics of algae. Harvey (1836) divided the algae into four major divisions solely on the basis of their pigment chemistry; this criterion is followed today. Harvey's use of biochemical features was the first in plant systematics (Dixon 1973).

Since then, the use of secondary metabolites in the systematics of algae has been suggested (Hegnauer 1962; Augier 1967; Dixon 1973; Lewin 1974; Boney 1978), and halogenated natural products were shown (Fenical and Norris 1975) to be useful taxonomic characters for separating some *Laurencia* species (Fig. 6–1a). Price (1974) listed phycological studies involving chromatographic methods, and Nichols et al. (1968) used pyrograms (gas–liquid chromatography) to study the systematics of several unicellular green and red algal isolates. Liaaen-Jensen (1977) developed a hypothetical evolution of the algae based on detailed structures of algal carotenoids and suggested that morphological, physiological, and biochemical criteria (both biosynthetic pathways and ultimate products) be used together in algal classification. The investigations of Glazer et al. (1982) on the use of phycoerythrins (e.g., see Gantt 1981a,b) as chemotaxonomic features supported earlier conclusions (e.g., O'hEocha 1962; van der Velde 1973) that they were probably not very useful at the ordinal or familial levels but may be beneficial at the specific and perhaps the generic levels. More recently, McCandless et al. (1982, 1983) con-

[122]

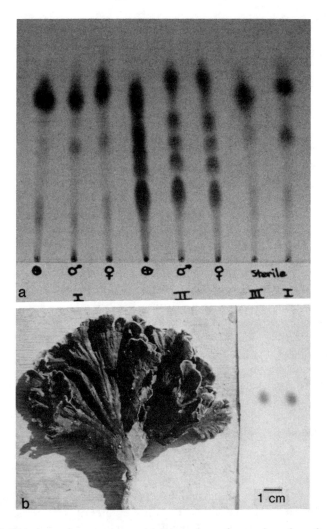

Fig. 6–1. Thin-layer chromatograms. (a) Comparison of natural products from three species of *Laurencia* in different reproductive conditions. (b) *Udotea flabellum* and a thin-layer chromatogram of its major metabolite, udoteal.

cluded that carrageenan chemistry in conjunction with taxonomic and life history features may be useful for defining the red algal families Phyllophoraceae and Gigartinaceae and certain of their genera.

Seaman and Funk (1983), using natural products chemistry (Harborne 1973; Robinson 1983) and cladistics (see Eldredge and Cracraft 1980; Wiley 1981; Funk and Brooks 1982) to study the Asteraceae, gave new insights into their classification. In marine sciences the technique of using numerical chemotaxonomy and cladistics has been successfully applied to the study of systematics and evolution of the gorgonians by Gerhart (1983). These approaches have the potential of contributing to the study of the systematics and phylogenetic relationships of marine algae as well.

C. Ecological biochemistry

In recent years, scientists have discovered numerous "secondary metabolites," many unique or novel, in various benthic marine algae in addition to the "primary metabolites" (i.e., those involved in the metabolic processes of photosynthesis, respiration, growth, and reproduction). These secondary metabolites have been identified, described, and structurally elucidated from many green, brown, red, and blue-green algae (e.g., see Fenical 1975, 1978, 1979, 1980, 1982, 1983; Baker and Murphy 1976, 1981; Faulkner 1977, 1978; Faulkner and Fenical 1977; Glombitza 1977; McEnroe et al. 1977; Maiti and Thomson 1977; Collins 1978; Moore 1978; Shield and Rinehart 1978; McConnell and Fenical 1979; Ragan and Jensen 1979; Howard and Fenical 1981; Ragan 1981; Hoppe 1982; Paul and Fenical 1983). Many of these compounds have potential application as pharmaceuticals and in agriculture or aquaculture (e.g., Baslow 1977; Colwell 1983; Fenical 1983; Rinehart et al. 1981). However, the biosynthetic pathways, biochemical processes, and function of these secondary metabolites are still largely unknown or inadequately documented.

As marine biologists have become aware of these algal natural products and the toxicity many of them exhibit, numerous challenging questions, hypotheses, and speculations on their ecological roles have emerged (e.g., Sieburth 1968; Whittaker and Feeney 1971; Burkholder 1973; Kittredge 1976; Norris and Fenical 1982). Compounds exhibiting biodynamic properties have been isolated from temperate (e.g., Ikawa et al. 1973; Bhakuni and Silva 1974; Hornsey and Hide 1974, 1976a,b; Ehresmann et al. 1977; Falkner 1978; Shield and Rinehart 1978; Henriquez et al. 1979; Biard et al. 1980), subtropical (e.g., Starr et al. 1962; McConnell and Fenical 1977; Caccamese et al. 1980), and tropical algae (e.g., Almodovar 1964; Burkholder 1973; Baker 1976; Diaz-Piferrer 1979; Norris and Fenical 1982; Fenical 1983; Paul and Fenical 1983; Sun et al. 1983).

Several of these algal compounds, which range from unpalatable to toxic, may (1) discourage herbivory (Geiselman and McConnell 1981; Norris and Fenical 1982; McConnell et al. 1982; Paul and Fenical 1983; Sun et al. 1983; Steinberg 1984), (2) possibly play a role in algal–algal (Yentsch and Menzel 1963; Chan et al. 1980; Patterson and Harris 1983) and algal–seagrass interactions (Zapata and McMillan 1979; Harrison and Chan 1980), and/or (3) be impor-

Fig. 6–2. Herbivore–algal association. *Aplysia* eating the red alga *Laurencia* in the Galápagos Islands.

tant in algal–epiphyte (Phillips and Towers 1982), algal–epizoic (Langlois 1975; Switzer-Dunlap and Hadfield 1977), algal–sessile animal (Al-ogily and Knight-Jones 1977), algal–bacterial (Sieburth 1979; Fenical 1983; Sun et al. 1983; Paul and Fenical 1983), and algal–fungal interactions (Paul and Fenical 1983).

The intricate mechanisms and biochemical interactions between an alga and its herbivores, pathogens, and competitors are likely to be both subtle and complex (e.g., see Phillips and Towers 1982; Morse and Morse 1984a). In studying marine plant–animal (Fig. 6–2) and plant–plant interactions, it might be helpful to look at those reported in higher plant–plant and plant–animal interactions (e.g., Harborne 1978, 1982; Rosenthal and Janzen 1979). The natural release of secondary compounds by algae and their interaction with their physical and biotic environment remain virtually unexplored. Coll et al. (1982) developed an underwater sampling apparatus that can isolate allelochemicals *in situ* from soft corals. N. M. Targett (personal communication) modified this apparatus so that it is totally submersible and can draw water rather than propel it through the system. A similar apparatus could be used to sample for compounds released or sloughed off by subtidal algae as well. This type of sampling apparatus now makes it possible to gather the chemical evidence needed to document those interactions that require the release of chemicals into the environment. It should be recognized that not all plant–plant or plant–animal interactions will necessarily require release of compounds into the environment.

II. Field methods

A. Collection techniques for natural products

In order to ascertain in the field which benthic algae may contain unique or unusual secondary metabolites, the investigator must begin by collecting specimens. Seaweeds can be collected in the intertidal or subtidal, using SCUBA, according to the methods presented by Tsuda and Abbott (Chap. 4; see also Abbott and Hollenberg 1976). The collections should be carefully sorted by species and then individual plants of each species divided into two groups, one for chemical extraction and the other as vouchers for taxonomic identification. Taxonomic vouchers (Fosberg and Sachet 1965) should be carefully numbered or lettered (to cross reference with the chemically analyzed specimens) with detailed notes kept on locality, habitat, depth, floral and/or faunal associations, subtratum type, collector(s), and date. They can be pressed and dried as herbarium specimens and/or liquid-preserved in a 4 to 5% buffered Formalin/seawater solution (see also Tsuda and Abbott, Chap. 4). Voucher specimens should be deposited in a major herbarium (Holmgren et al. 1981), so that in the future, if necessary, their botanical identities can be verified.

Specimens retained for chemical analyses should be blotted dry and cleaned of epiphytes or epizoa to prevent contamination. Thin-layer chromatography (see Sec. II.B.1) is then used to examine for the presence of unusual compounds. If an alga is discovered to contain interesting compounds, a minimum of ~1 liter of fresh material will be required to complete the chemical identification and biological activity tests. The collected species may be air-dried, if the compound(s) of interest are nonvolatile. Often, volatile compounds can be detected by their immediate and distinctive odors. To expedite field-drying, place material on black plastic sheets in the sun with considerable air circulation and turn it occasionally to prevent mold or bacterial growth, which may degrade desired compounds. This technique has been successfully applied to *Laurencia* and *Dictyota* species, in particular. It should be pointed out, however, that air-drying may alter the secondary metabolite composition of some species, thereby precluding chemical comparisons. An ideal method for minimizing catabolic reactions and preserving chemistry is to freeze the algae with liquid nitrogen. Alternatively, if refrigeration or an adequate supply of dry ice is available, the chemical samples of algae can also be frozen and transported for later study. However, if the compounds are known to be volatile or unstable, the algae must go directly from the sea into the extraction solvent (e.g., alcohol) in suitable storage containers (glass or Nalgene; specify linear poly-ethylene).

B. Isolation and identification

1. Thin-layer chromatography. Thin-layer chromatography (TLC) is an inexpensive, effective, and rapid method of determining whether an alga contains unusual secondary metabolites (Stahl 1969; Harborne 1973; Touchstone and Dobbins 1978; Mikês 1979). Relatively simple, this method can be readily adapted to field operations even under primitive conditions. The method requires a small quantity of algae, ordinary glassware, several chemical solvents, and TLC plates (plastic, commercially prepared types are preferred for field use), which are readily available from most laboratory or biological supply firms (see list of suppliers in Gantt 1980) and can be cut into smaller sheets (e.g., Baker Flex silica gel, 20 × 20 cm sheets). Extracts of several different species can be run on the same TLC plate at one time. Then, if warranted by the results (see Sec. II.B.1.b) additional time may be spent in the collection of requisite quantities for further biological tests and advanced chemical analyses (e.g., structural elucidation).

The solvents listed in this chapter extract only the lipid-soluble components. The separation and identification of water-soluble compounds from marine algae are more complex and variable. However, Menzel et al. (1983) successfully identified coumarins from *Cymopolia* and *Dasycladus* (Dasycladaceae; Chlorophyta); see their article for methodology. Methods for the analysis of phenolic compounds can be found in Ragan and Craigie (1978) and Geiselman and McConnell (1981).

Among the journals that can be consulted for the latest techniques are *Phytochemistry, Journal of Chromatographic Science, Journal of the American Chemical Society, Journal of Organic Chemistry, Journal of Chemical Ecology, Journal of Natural Products, Marine Chemistry, Tetrahedron Letters*, and *Analytical Biochemistry*.

a. Equipment

1. Organic solvents (**Caution:** read potential hazards on label and observe special handling if required)

    Alcohol (EtOH)
    Chloroform (CHCl$_3$)
    Diethyl ether (Et$_2$O)

2. Glassware

    Small 5-dram vials
    Glass development chamber (or suitable subtitute such as an
      empty 3-lb coffee can)

Disposable Pasteur pipettes and bulbs
Small evaporating dishes
Capillary tubes: ~10 μl (buy or make)
Glass sprayer apparatus

3. TLC plates: silica gel on plastic or on aluminum (available in different particle sizes; the larger are faster in development time, but there is a decrease in resolution between the sample components' sharpness of separation; TLC plates can be obtained from various biological and chemical suppliers, such as VWR, American Scientific Products, and Fisher.

4. Sulfuric acid (H_2SO_4): 50% (v/v) solution in water or alcohol with an appropriate sprayer [iodine crystals (I_2) can be used with similar results]

5. Heat source: electric hot plate or a white gasoline, alcohol, kerosene, or propane portable stove, e.g., backpacking or camping type (**Caution:** be extremely careful with volatile solvents and heat source)

6. Portable ultraviolet (UV) light, e.g., Multi-band light (model MSL-48) or Collecting Light Wand (no. 2813; with battery pack no. 2863) available from Ward's and Bio Quip Products, respectively

b. Method. Specimens freshly collected for chemical study (~10 g wet weight) should be blotted dry, cleaned of epiphytes, chopped into small pieces, and packed firmly in a 5-dram vial. The vial is filled with an extraction solvent such as a 2:1 chloroform/ethanol mixture. The vial is next warmed in heated water for 30 min, and then the liquid extract is decanted into an evaporating dish, where it is allowed to evaporate for a period of at least 1 h. Warming the dish may be necessary to accelerate this process. When evaporation is complete (water may remain), several milliliters of diethyl ether are added to the dish and swirled to dissolve all pigments. The ether layer is next withdrawn from the dish by tilting the dish and carefully skimming the lighter (Et_2O) solvent from the surface with a Pasteur pipette. The ether is placed in a 5-dram vial and allowed to evaporate fully, leaving a dark green residue. A few drops of ether are then added, and the concentrated extract (~30–50% by weight) is taken up in a fine capillary tube (10–50 μl) and carefully applied to the silica TLC plate with its origin marked. The extract should be applied several times to the same small spot to ensure that sufficient extract has been applied. Dark spots should be produced ~1 cm from the bottom of the plate. If more than one extract is being chromatographed, all are placed in a straight horizontal line (a pencil line being used to mark the origin), an equal distant apart (~15 mm), and a similar volume of each is applied. This plate is then placed in a glass chamber (or suitable alternative) containing the developing solvent (usually 100% Et_2O) or solvent mixture. (*Note*: be careful *not* to immerse spots in solvent.) The chamber should have a lid to prevent evaporation of the solvent. When the solvent front migrates to within ~5 cm of the top of the plate through capillary action (~15–30 min, depending on the particle size of the silica gel), the plate is removed from the chamber, the solvent front marked immediately and air-dried. **Caution:** evaporate completely all TLC solvents from the chromatogram before further analysis.

Chromatographed substances may be colorless; thus, detection requires further preparation. Spraying with a 50% sulfuric acid solution and warming to ~80°C (see point 5, Sec. II.B.1.a; if electricity is available, a hair dryer can be used) usually give immediate color reactions as the compounds begin to decompose. Colors from bright red to pink and from blue to black are commonly observed for many secondary metabolites. Of course, other detection and visualization methods are available to suit specific circumstances (e.g., see Stahl 1969, Harborne 1973, Touchstone and Dobbins 1978). For example, a portable UV light (such as the type used by geologists in field studies of minerals and rocks) can be used to assess the presence of UV-sensitive compounds displayed on the thin-layer chromotogram. A spray bottle can be purchased (e.g., Freon-driven sprayers are available commercially), or in the field a suitable sprayer can be constructed from a glass atomizer and rubber hose connected to the first stage of a SCUBA regulator and run directly off a SCUBA tank (or similar compressed air tank) or portable air compressor. Care must be taken to perform these operations under an exhaust hood or in a very well ventilated area. Sulfuric acid vapors are extremely dangerous and can cause serious burns.

Once the plates are sprayed with a fine mist of acid, they are next warmed by even heating over a very hot surface such as a hot plate, a heated inverted frying pan, or a hair dryer. Documentation is important, and the chromatogram should be photographed, drawn or traced, and cross-referenced with the appropriate voucher specimen number before it fades. The R_f values of substances should be recorded, and a photograph or tracing retains other important information (e.g., color and intensity) for comparison and analysis.

In TLC, mixtures often separate into single components. The components move from the origin and separate on the basis of their differences in affinity for the solvent and for the adsorbent. These affinites depend on the solubility, polarity, functionality, and molecular weight of the components. The distance the compound travels in relation to the distance the developing solvent travels is used to define the R_f value. It is obtained by measuring the distance from the center of the origin spot to the center of the substance spot, and this is divided by the distance between the origin and solvent front. For example, if a substance moves only one-half the distance of the solvent, its R_f value is 0.5. The R_f value and color of each spot are specific for each component.

Pigments and typical chemicals (e.g., sterols) are displayed on the chromatogram along with any novel or unexpected compounds. If one is acquainted with the R_f values and charring colors of the typical constituents, the unusual ones are generally very apparent (Fig. 6–1). The TLC readily demonstrates which samples exhibit unusual secondary metabolites by the presence of additional spots, often of brilliant and characteristic colors. Fatty acids (triglycerides) are usually the topmost spots on the finished plate. Next is chlorophyll, remaining characteristically green even after acid charring. Beneath are the sterols, characteristically grayish-black. Carotenoids usually travel only a short distance and are yellow before acid, then char black. Secondary metabolites may occur anywhere along the path, depending on their nature and polarity.

2. Preparative vacuum elution chromatography. The TLC method discussed in the preceding section is used mainly for the analytical-scale assessment of the presence of secondary metabolites in algal extracts. In many cases, the investigator may wish to purify the secondary metabolites to determine their biological and/or chemical properties. In such cases the technique of vacuum elution chromatography (VEC) is suggested since it is a rapid field-adaptable method for the separation of milligram to gram quantities of semipurified metabolites.

a. Equipment

1. Organic solvents: mixtures of one polar organic solvent and one nonpolar solvent, such as ethyl acetate/isooctane mixtures or diethyl ether/petroleum ether mixtures
2. Glassware
   Medium-porosity sintered glass filter funnel of approximately 2 to 5 cm diameter.
   Evaporating dishes: ~10
   Vacuum flask (500 ml) with rubber adapter
3. Thin-layer-grade silica gel (~300 mesh)
4. Vacuum pump (electric or hand-operated)

b. Method. Algae should be collected and the extractions made in the same manner as described for TLC. The chromatography column is first produced by filling the sintered glass funnel with dry silica gel and tapping on a hard surface to pack the column fully. The vacuum funnel should be filled to approximately 1 or 2 cm from the top to allow for solvents to be added. For 1 g of concentrated extract, a funnel of ~2 cm in diameter × 4–5 cm in height is appropriate. Ten 50-ml solvent mixtures are then made by increasing the volume percentage of the polar solvent in the nonpolar solvent. For example, if ethyl acetate and isooctane are used, the first solvent is 100% isooctane, the second 10% ethyl acetate in isooctane, the third 20% ethyl acetate in isooctane, etc. The final solvent will always be 100% of the polar solvent. The residual extract is dissolved in a minimum quantity of the first solvent (100% nonpolar), and the mixture is added to the column placed under vacuum into the vacuum flask. **Caution:** use only thick-walled glass flasks to prevent implosion under vacuum. The entire extract is added, and the mixture is allowed to run through the silica gel until the surface at the top of the funnel is dry. The filtrate (fraction 1) is collected by dismantling the apparatus, the apparatus is reconstructed, and the second solvent mixture is added under vacuum. When the second mixture has been entirely added, it too is allowed to filter through until the tip of the gel is dry. The second fraction (2) is next withdrawn as the filtrate exactly as fraction 1 was withdrawn. In this way, a minimum of 10 semipurified fractions is obtained spanning a wide range of chromatographic polarities. This method is often so efficient that pure secondary metabolites that do not require further purification can be obtained.

Note that TLC is an important tool that must be used to implement preparative VEC. Initially, TLC results can be utilized to recognize the presence of unique metabolites on the basis of their R_f values and their unique colors produced upon acid charring. Then, on the basis of the composition of the extract, a more effective series of solvent mixtures can be conceived. If the secondary metabolites in an extract are complex and relatively nonpolar ($R_f = 0.5-1.0$), solvent mixtures should be adjusted to give maximum separation of this group (e.g., see Touchstone and Dobbins 1978). A scheme could be utilized with only 5% increments to a final solvent mixture of 50% polar solvent in nonpolar solvent.

The use of TLC to assess the final outcome of a preparative chromatography must also be emphasized. Each preparative fraction should be reduced to a gummy residue and analyzed by TLC methods. Fractions that contain only one symmetric spot can be assumed to be pure for most biological testing procedures. They should be kept under refrigeration to prevent degradation. When possible, purified compounds should be used as standards and run against the isolated compounds for verification.

Isolated compounds afford more precision in bioactivity assays and field or laboratory experiments; otherwise, it is possible that there may be synergistic or antagonistic effects of the total extract on the results. In some cases, depending on the ecological or behavioral question being asked, it may be desirable to investigate various concentrations of the total extract (for the synergistic effect), the isolated compounds, and combinations of the compounds.

To identify the isolated compounds chemically, an investigator must be familiar with advanced methods of structural elucidation and have access to sophisticated chemical instrumentation. This is most easily accomplished by interdisciplinary collaboration with a qualified research chemist who has a well-equipped laboratory and is devoted to such endeavors.

III. Biological activity tests

A. Bioassays

The use of bioassays to screen marine plants for biologically active substances is commonplace. Terrestrial organisms or human pathogens have been, and still are, the usual test organisms. Although these assays have excellent merit in the search for potential pharmaceuticals and agriculture and aquaculture chemicals, there are more appropriate bioassays for ecological studies. For example, tests using marine organisms encountered in an alga's natural habitat have the prospect of providing insight into the possible ecological role of these compounds. Marine bacteria, yeasts, fungi, potential herbivores, plant competitors (i.e., algae and seagrasses) or sessile animals, pathogenic epibionts (Andrews 1976; Sieburth 1979), and fouling organisms (epiphytes or epizoa) would all be appropriate test organisms. Since organisms respond differently to the wide spectrum of secondary compounds, it is worthwhile to test as many of the diverse groups as possible to enhance the probability of discovering sensitive species.

A catalog (ATCC no. 2, 1982) of available cultures of marine bacteria and fungi can be obtained from the American Type Culture Collection (see the appendix of this volume). Cultures of algae are available from the University of Texas Culture Collection of Algae (Starr 1978): other algal culture collections are listed in Gantt (1980: 403) and Haines et al. (1982). Worldwide sources of living cultures can be found in Martin and Skerman (1972). Discussions and methods of field collection, isolation, and culture of marine bacteria can be found in Colwell and Zambruski (1972) and the "Materials and Methods" sections in Colwell and Morita (1980), of marine fungi in Johnson and Sparrow (1961), Jones (1976), Kohlmeyer and Kohlmeyer (1979), and Newell and Fell (1982), and of algae in Allen (1968), Rosowski and Parker (1971), and Stein (1973). The following

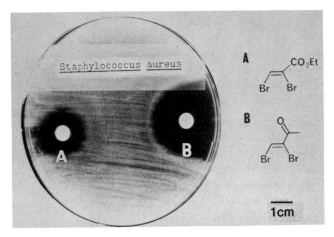

Fig. 6–3. Nutrient agar plate–disk assay. Effect of two active compounds from *Asparagopsis taxiformis* on *Staphylococcus aureus*.

are selected bioassays developed to test compounds for their biological impact on marine organisms.

1. Nutrient agar plate–disk assay. Petri plates containing suitable growth media are inoculated with the test organism (marine bacteria, yeasts, fungi, microalgae; for methods see Chan et al. 1980; also see "Culture" in index of Pelczar at el. 1977). Once the plate is grown over, a disk (such as BBL ¼ in., available from Dickerson and Co.) impregnated with the chemical compound is introduced. Adequate controls for extraction solvent(s), if present in the algal extract, are also necessary. If the algal extract is biologically active against the test organism a "zone of inhibition," where growth is suppressed or completely inhibited, will form (Fig. 6–3). Measuring the area of the inhibition surrounding the disk, which contains a measured quantity of the test compound, quantifies the strength of the activity (e.g., see Henriquez et al. 1979).

It is also possible to use the chromatogram directly to test the bioactivity of the separated chemical components. In this method, "bioautography," the chromatogram is placed directly on the culture plate of the test organism. Although less sophisticated, it will give an indication of which metabolites may be exhibiting bioactivity.

2. Comparative plate method. One problem with the preceding "disk assay" technique is that the area of inhibition is affected by the differential diffusion rates of the various compounds through the agar medium (J. J. Sims, personal communication). A method that eliminates this problem is to dissolve the algal extracts directly into the liquid agar medium, then solidify it, and compare the growth (after a specific time) on these plates with those of control plates without the algal compounds. Adequate controls for extraction chemicals, if present in the algal extract, are also needed. Growth can be quantified by a point–grid method (see Littler and Littler, Chap. 8) or determined for plants by measuring chlorophyll *a* fluorescence (Yentsch and Menzel 1963; Targett et al. 1983).

3. Mixed-culture method. Proof of allelopathy occurring in the sea may be difficult to obtain. The complexity and variability of the ecological situation make detection and collection of the algal exudate, its concentration and diffusion, measurements of microcurrents and eddies, and their interactions difficult to investigate in situ. In laboratory culture, some variables can be eliminated; it is hoped that this simplification will provide a more interpretable model of inter- or intraspecific interactions. For example, using three species in matched-pairs experiments, Russell and Fielding (1974) showed them to be competitive under a variety of environmental conditions. The macroalgae should be selected by careful field observations of algal associations or on the basis of data from quantitative community studies (see Littler and Littler, Chap. 8) via the use of computer-assisted programs, such as nearest-neighbor analysis. Experiments can be designed to test for the effects each has on the other's growth in culture, after careful experiments have initially been conducted to evaluate other factors (e.g., temperature, light, and nutrient limits; Russell and Fielding 1974). Extrapolation from the results of the mixed artificial cultures to the natural marine community must be approached with caution. If there are competitive effects, one must consider chemistry when searching for the mechanism(s) responsible. Day (1983) suggested that filamentous algae interfere with feeding of bryozoans. It would be interesting to conduct similar investigations of the role of natural products chemistry in algal–sessile animal interactions. If the release and concentration of the alga's exudate are hypothesized to be important, then their presence in culture and in the field should be documented. Procedures to detect the release of compounds from the alga into its natural habitat are being developed (for a successful method used on alcyonarians see Coll et al. 1982).

4. Sea urchin egg development. Sea urchin egg development has been extensively studied (Czihak and Peter 1975) and is frequently used as a pharmacological bioassay to screen for compounds that inhibit cell division (Cornman 1950; Hinegardner 1975; Jacobs et al. 1981). This bioassay may also have ecological significance (see Norris and Fenical 1982) if concentrations effecting response are similar to those present in nature. Eggs and sperm are collected from sea urchins after the injection of 0.1 ml of 0.5 *N* KCl into their body cavities. Sperm is then added by the drop to a solution of eggs in seawater until fertilization has occurred. The fertilized egg suspension is now added to filtered seawater, seawater containing various concentrations of the test compounds, and seawater with appropriate concentrations of dispersal compounds (if present in test compounds). Under a compound microscope, cell division is observed in the control and experimental treatments, and then compared with normal development (Czihak and Peter 1975). If the sea urchin eggs are inhibited or killed, the substances could have the potential of reducing this predator in the plant's vicinity. The alga's compounds should also be tested for effects on the settlement and development of planktonic larvae. Whether the observed effects have reality in the field may depend on the natural release of this compound at effective concentrations into the surrounding seawater in nature. If they are not released, direct contact by larvae to the alga may be important for successful development (see Morse and Morse, 1984b) or for the inhibition of further development.

B. *Behavioral bioassays*

1. Gastropod tentacle withdrawal. A behavioral bioassay developed by Targett (1979) relies on chemosensory behavior of herbivorous marine gastropods (e.g., *Nerita*) to test their reaction to the alga's chemistry (i.e., total extract or specific compounds). The mollusks are collected and maintained in aerated seawater aquaria. Individuals are placed in petri plates filled with seawater, and once they are acclimated (i.e., normal movement resumed) the algal extract is pipetted into the vicinity of one of the tentacles. If the tentacle is withdrawn, in a proper number of trials and with controls, the reaction indicates the presence of a biologically active compound. This type of bioassay could be expanded to any marine creature with an observable sensory structure reaction. The dose required to effect the mollusk's withdrawal and the amount released by the plant (undisturbed or when tasted) are important in determining if the compound affects the herbivore's behavior in the natural environment.

Targett and McConnell (1982) developed another behavioral bioassay using the periwinkle *Littorina irrorata* for the detection of secondary metabolites in macroalgae.

2. Food preferences. Chemical sensing undoubtedly has an important influence on food selection, avoidance, and preference. Determining which algae are eaten involves offering selected marine algae, either by random or haphazard placement, on suspended lines, weighted grids, or clothes pins to herbivores in the water column or on the bottom (see John and Pople 1973; Littler et al. 1983; Hay 1984). Other methods involve presenting a choice of paired algal offerings to specific herbivores, such as the captive sea urchin method of Littler et al. (1983) and the gustatory and chemoreceptive preference experiments of Vadas (1977).

Singling out the factor(s) controlling the food preference of specific fish, sea urchins, and other marine herbivores is an arduous task. Geiselman and McConnell (1981) and Bertness et al. (1983) used solidified algal homogenates to test the palatability of secondary compounds directly with the herbivorous snail *Littorina littorea*, and Valiela et al. (1979) similarly studied detritivore feeding inhibition. Algal homogenates, some including various mixtures of an alga's secondary compounds and some without them, are mixed into agar to solidify in petri plates. An equal volume of the extraction chemical(s) must be added to the control food media. These various treatments and control plates are then offered to the test organisms, in this case mollusks, which are maintained in seawater aquaria. A grid may be used to quantify the percent area of the experimental versus control medium consumed in order to detect if feeding is affected by the presence of the secondary metabolite. This type of feeding bioassay could easily be expanded to other herbivores. It is extremely important to know the biology and natural history of the organisms tested [e.g., it is necessary to establish that the herbivore is physically able to eat the test plants (Steneck and Watling 1982)].

Sun and Fenical (1979) added secondary compounds to commercial fish food that they had established would be eaten by the test fish and compared the percentage of food containing the compounds that was eaten with the percentage of food not containing the compounds that was eaten. Controls for the extraction solvents would also be necessary, if they were present in algal extracts.

Other feeding bioassays involve the direct application of secondary compounds to otherwise palatable algae (determined by preliminary preference experiments). McConnell et al. (1982) successfully applied identified pure algal compounds (nonpolar) to previously palatable algae, offered both treated and untreated specimens to sea urchins, and, after a specified time, measured the amount consumed. Upon completion of the experiment they extracted the uneaten treated ("coated") algae to document that the compounds were present throughout the experiments. Following feeding preference studies of the parrotfish *Sparisoma radians* (Lobel and Ogden 1981), N. M. Targett (personal communication) successfully applied nonpolar extracts to *Thalassia*, the preferred food of *S. radians*, to test the role of chemistry in the avoidance or preference of the fish's foraging behavior. As a modification of the agar–extract experiments of Geiselman and McConnell (1981), N. M. Targett (personal communication) used "anchored popsicles" made of agar (which the fish will eat) and conducted in situ experiments with random placement of chemically treated and untreated "popsicles." This technique allows the incorporation of water-soluble compounds in agar "popsicles" for testing as well.

3. Substrata selection by motile larvae. A planktonic stage is found in the life cycle of benthic marine invertebrates belonging to all the major phyla of the animal kingdom except the Nematoda and Gastrotricha (Crisp 1974). The planktonic larvae of many benthic invertebrates and protozoans settle and metamorphose in response to specific substances or conditions in their environment (Crisp 1974; Hadfield 1978). Some show a substratum selection for specific algae (Crisp 1974; Kriegstein et al. 1974; Langlois 1975; Switzer-Dunlap 1977; Switzer-Dunlap and Hadfield 1977; Jensen 1980). For example, planktonic abalone larvae (*Haliotis* sp.) show preferential settlement on crustose red algae (Morse, et al. 1979, 1980); direct contact of the larvae with molecules at the alga's surface induces metamorphosis (Morse and Morse, 1984a,b). Laboratory tests are performed to determine whether a secondary compound triggers specific settling and metamorphosis. Some knowledge of the selected animal's larval development (e.g., the stage and age when competent to metamorphose) is necessary. The invertebrate animal has to be cultured (Smith and Chanley 1974), and when it spawns the eggs or larvae are captured for use in experiments. A selected number of motile larvae are placed in separate dishes containing filtered seawater, some with the possible "inducer" added (at various concentrations) and others in control dishes of filtered seawater only or with similar concentrations of dispersal chemical (if used). Comparing the percentage of metamorphosis of the control groups with that of groups exposed to various concentrations of the algal extract or pure compounds (after the period of exposure) would demonstrate whether the alga's secondary metabolites were affecting larval settlement.

IV. Acknowledgments

For helpful information and critical review throughout the preparation of this chapter we are particularly grateful to Katina E. Bucher. We also thank O. J. McConnell, N. M. Targett, L. D. Coen, M. M. Littler, D. S. Littler, M. E. Hay, M. S. Foster, S. Fredericq, and V. J. Paul for their comments and review of the manuscript. This is contribution no. 142 of the Smithsonian Tropical Reef Program and the Smithsonian Western Atlantic Mangrove Program (SWAMP), supported in part by a grant from the Exxon Corporation Foundation.

V. References

Abbott, I. A., and Hollenberg, G. J. 1976. *Marine Algae of California*. Stanford University Press, Stanford, Calif. 827 pp.

Allen, M. M. 1968. Simple conditions for growth of unicellular blue-green algae on plates. *J. Phycol.* 4, 1–9.

Almodovar, L. R. 1964. Ecological aspects of some antibiotic algae in Puerto Rico. *Bot. Mar.* 6, 143–6.

Al-ogily, S. M., and Knight-Jones, E. W. 1977. Anti-fouling role of antibiotics produced by marine algae and bryozoans. *Nature* 265, 728–9.

Andrews, J. H. 1976. The pathology of marine algae. *Biol. Rev.* 51(2), 211–53.

Augier, J. 1967. Biochimie et taxonomie chez les algues. *Mem. Soc. Bot. Fr.* 1965, 8–15.

Baker, J. T. 1976. Physiologically active substances from marine organisms. *Austr. J. Pharm. Sci.* New Series, 5, 89–99.

Baker, J. T., and Murphy, V. 1976 and 1981. *Marine Products: Compounds from Marine Organisms*, vol. 1, 1976, 216 pp.; vol. 2, 1981, 240 pp. CRC Press, Boca Raton, Florida.

Baslow, M. H. 1977. *Marine Pharmacology*. rev. ed. Krieger, Huntington, N. Y. 327 pp.

Bertness, M. D., Yund, P. O., and Brown, A. F. 1983. Snail grazing and the abundance of algal crusts on a sheltered New England rocky beach. *J. Exp. Mar. Biol. Ecol.* 71, 147–64.

Bhakuni, D. S., and Silva, M. 1974. Biodynamic substances from marine flora. *Bot. Mar.* 17, 40–51.

Biard, J. F., Verbist, J. F., Le Boterff, J., Ragas, G., and Lecocq, M. M. 1980. Algues fixees de la côte Atlantique Francaise contenant des substances antibacteriennes et antifungiques. *Plant. Med,* 40(suppl.), 136–51.

Boney, A. D. 1978. Taxonomy of red and brown algae. In Irvine, D. E. G., and Price, J. H. (eds.), *Modern Approaches to the Taxonomy of Brown and Red Algae*, pp. 1–19. Academic Press, New York.

Burkholder, P. R. 1973. The ecology of marine antibiotics and coral reefs. In Jones, O. A., and Endean, R. (eds.), *Biology and Geology of Coral Reefs*, vol. 2: *Biology 1*, pp. 117–82. Academic Press, New York.

Caccamese, S., Azzolina, R., Furnari, G., Cormari, M., and Grasso, S. 1980. Antimicrobial and antiviral activities of extracts from Mediterranean algae. *Bot. Mar.* 23, 285–8.

Chan, A. T., Andersen, R. J., and Le Blanc, M. J. 1980. Algal plating as a tool for investigating allelopathy among marine microalgae. *Mar. Biol.* 59, 7–13.

Coll, J. C., Bowden, B. F., and Tapiolas, D. M. 1982. In situ isolation of allelochemicals released from soft corals (Coelenterata: Octocorallia): a totally submersible sampling apparatus. *J. Exp. Mar. Biol. Ecol.* 60, 293–9.

Collins, M. 1978. Algal toxins. *Microbiol. Rev.* 42, 725–46.

Colwell, R. R. 1983. Biotechnology in the marine sciences. *Science* 222, 19–24.

Colwell, R. R., and Morita, R. Y. (eds.). 1980. *Effect of the Ocean Environment on Microbial Activities.* University Park Press, Baltimore. 587 pp.

Colwell, R. R., and Zambruski, M. S. (eds.). 1972. *Methods in Aquatic Microbiology,* by Rodina, A. G. (revised and translated). University Park Press, Baltimore. 461 pp.

Cornman, I. 1950. Inhibition of sea-urchin egg cleavage by a series of substituted carbamates. *J. Nat. Cancer Inst.* 10, 1123–38.

Crisp, D. J. 1974. Factors influencing the settlement of marine invertebrate larvae. In Grant, P. T., and Mackie, A. M. (eds.), *Chemoreception in Marine Organisms,* pp. 177–265. Academic Press, London.

Czihak, G., and Peter, R. (eds.) 1975. *The Sea Urchin Embryo: Biochemistry and Morphogenesis.* Springer-Verlag, New York. 700 pp.

Day, R. W. 1983. Effects of benthic algae on sessile animals: observational evidence from coral reef habitats. *Bull. Mar. Sci.* 33, 597–605.

Diaz-Piferrer, M. 1979. Contributions and potentialities of Caribbean marine algae in pharmacology. In Hoppe, H. A., Levring, T., and Tanaka, Y. (eds.), *Marine Algae in Pharmaceutical Sciences,* pp. 149–64. De Gruyter, Berlin.

Dixon, P. S. 1973. *Biology of the Rhodophyta.* Hafner Press, New York. 285 pp.

Ehresmann, D. W., Deig, E. F., Hatch, M. T., DiSalvo, L. H., and Vedos, N. A. 1977. Antiviral substances from California marine algae. *J. Phycol.* 13, 37–40.

Eldredge, N., and Cracraft, J. 1980. *Phylogenetic Patterns and the Evolutionary Process.* Columbia University Press, New York. 349 pp.

Faulkner, D. J. 1977. Interesting aspects of marine natural products chemistry. *Tetrahedron* 33, 1421–43.

Faulkner, D. J. 1978. Antibiotics from marine organisms. In Sammes, P. G. (ed.), *Topics in Antibiotic Chemistry,* vol. 2, pp. 13–58. Ellis Horwood, Chichester, Great Britain.

Faulkner, D. J., and Fenical, W. H. (eds.) 1977. *Marine Natural Products Chemistry.* Plenum Press, New York. 433 pp.

Fenical, W. 1975. Halogenation in the Rhodophyta: a review. *J. Phycol.* 11, 245–59.

Fenical, W. 1978. Diterpenoids. In Scheuer, P. J. (ed.), *Marine Natural Products: Chemical and Biological Perspectives,* vol. 2, pp. 173–245. Academic Press, New York.

Fenical, W. 1979. Molecular aspects of halogen-based biosynthesis of marine natural products. In Swain, T., and Waller, G. R. (eds.), *Recent Advances in Phytochemistry,* vol. 13: *Topics in the Biochemistry of Natural Products,* pp. 219–39. Plenum Press, New York.

Fenical, W. 1980. Distributional and taxonomic features of toxin-producing marine algae. In Abbott, I. A., Foster, M. S., and Eklund, L. F. (eds.), *Pacific Seaweed Aquaculture*, pp. 144–51. California Sea Grant College Program, University of California, La Jolla.

Fenical, W. 1982. Natural product chemistry in the marine environment. *Science* 215, 923–8.

Fenical, W. 1983. Investigation of benthic marine algae as a resource for new pharmaceuticals and agriculture chemicals. In Tseng, C. K. (ed.), *Proceedings of the Joint China–U.S. Phycology Symposium*, pp. 497–521. Science Press, Beijing, People's Republic of China.

Fenical, W., and Norris, J. N. 1975. Chemotaxonomy in marine algae: chemical separation of some *Laurencia* species (Rhodophyta) from the Gulf of California. *J. Phycol.* 11, 104–8.

Fosberg, F. R., and Sachet, M.-H. 1965. Manual for tropical herbaria. *Regnum vegetabile* 39, 1–132.

Funk, V. A., and Brooks, D. R. (eds.). 1982. *Advances in Cladistics*. New York Botanical Gardens, New York. 250 pp.

Gantt, E. 1980. *Handbook of Phycological Methods: Developmental and Cytological Methods*. Cambridge University Press, Cambridge. 425 pp.

Gantt, E. 1981a. Structure and function of phycobilisomes: light harvesting pigment complexes in red and blue-green algae. *Int. Rev. Cytol.* 66, 45–79.

Gantt, E. 1981b. Phycobilisomes. *Annu. Rev. Plant. Physiol.* 32, 327–47.

Geiselman, J. A., and McConnell, O. J. 1981. Polyphenols in brown algae *Fucus vesiculosus* and *Ascophyllum nodosum*: chemical defenses against the marine herbivorous snail, *Littorina littorea*. *J. Chem. Ecol.* 7, 1115–33.

Gerhart, D. J. 1983. The chemical systematics of colonial marine animals: an estimated phylogeny of the order Gorgonacea based on terpenoid characters. *Biol. Bull.* 164, 71–81.

Glazer, A. H., West, J. A., and Chan, C. 1982. Phycoerythrins as chemotaxonomic markers in red algae: a survey. *Biochem. Syst. Ecol.* 10, 203–15.

Glombitza, K.-W. 1977. Highly hydroxylated phenols of the Phaeophyceae. In Faulkner, D. J., and Fenical, W. H. (eds.), *Marine Natural Products Chemistry*, pp. 191–204. Plenum Press, New York.

Hadfield, M. G. 1978. Chemical interactions in larval settling of a marine gastropod. In Faulkner, D. J., and Fenical, W. H. (eds.), *Marine Natural Products Chemistry*, pp. 403–13. Plenum Press, New York.

Haines, K. G., Hoagland, K. D., and Fryxell, G. A. 1982. A preliminary list of algal cultural collections of the world. In Rosowski, J. R., and Parker, B. C. (eds.), *Selected Papers in Phycology II*, pp. 820–6. Phycological Society of America (Book Division), Lawrence, Kansas.

Harborne, J. B. 1973. *Phytochemical Methods*. Chatman & Hall, London. 278 pp.

Harborne, J. B. (ed.) 1978. *Biochemical Aspects of Plant and Animal Coevolution*. Academic Press, New York. 435 pp.

Harborne, J. B. 1982. *Introduction to Ecological Biochemistry*, 2nd ed. Academic Press, New York. 278 pp.

Harrison, P. G., and Chan, A. T. 1980. Inhibition of the growth of microalgae and bacteria by extracts of eelgrass (*Zostera marina*) leaves. *Mar. Biol.* 61, 21–6.

Harvey, W. H. 1836. Part third: algae. In Mackay, J. T., *Flora Hibernica*, pp. 157–254, 277–9. Curry Jun, Dublin.

Hay, M. E. 1984. Patterns of fish and sea urchin grazing on Caribbean coral reefs: are previous results typical? *Ecology* 65(2), 446–54.

Hegnauer, R. 1962–1973. *Chemotaxonomie der Pflanzen*, 8 vols. (vol. 1: *Algae*, 1962.) Dirkhauser Verlag, Basel und Stuttgart.

Henriquez, P., Candia, A., Norambuena, R., Silva, M., and Zemelman, R. 1979. Antibiotic properties of marine algae. 2. Screening of Chilean marine algae for antimicrobial activity. *Bot. Mar.* 22, 451–3.

Hinegardner, R. 1975. Care and handling of sea urchin eggs, embryos, and adults (principally North American species). In Czihak, G., and Peter, R. (eds.), *The Sea Urchin Embryo*, pp. 10–25. Springer-Verlag, New York.

Holmgren, P. K., Keuken, W., and Schofield, E. C. 1981. *Index Herbariorum*, part 1: *The Herbaria of the World*, 7th ed. Bohn, Scheltema & Holkema, Utrecht, Netherlands. 452 pp.

Hoppe, H. A. 1982. Marine algae and their products and constituents. In Hoppe, H. A., and Lerving, T. (eds.), *Marine Algae in Pharmaceutical Science*, vol. 2, pp. 3–48. DeGruyer, Berlin.

Hornsey, I. S., and Hide, D. 1974. The production of antimicrobial compounds by British marine algae. 1. Antibiotic producing marine algae. *Br. Phycol. J.* 9, 353–61.

Hornsey, I. S., and Hide, D. 1976a. The production of antimicrobial compounds by British marine algae. 2. Seasonal variation in production of antibiotics. *Br. Phycol. J.* 11, 63–7.

Hornsey, I. S., and Hide, D. 1976b. The production of antimicrobial compounds by British marine algae. 3. distribution of antimicrobial activity within the algal thallus. *Br. Phycol. J.* 11, 175–81.

Howard, B. M., and Fenical, W. 1981. The scope and diversity of terpenoid biosynthesis by the marine alga *Laurencia*. In Reinhold, L. (ed.), *Progress in Phytochemistry*, vol. 7, pp. 263–300. Wheaton, London.

Ikawa, M. T., Thomas, V. M., Buckley, L. J., and Uebel, J. J. 1973. Sulfur and the toxicity of the red alga *Ceramium rubrum* to *Bacillus subtilis*. *J. Phycol.* 9, 302–4.

Jacobs, R. S., White, S., and Wilson, L. 1981. Selected compounds derived from marine organisms: effects on cell division in fertilized sea urchin eggs. *Fed. Proc. Amer. Soc. Exp. Biol.* 40, 26–9.

Jensen, K. R. 1980. A review of sacoglossan diets, with comparative notes on radular and buccal anatomy. *Malacolog. Rev.* 13, 55–77.

John, D. M., and Pople, W. 1973. The fish grazing of rocky shore algae in the Gulf of Guinea. *J. Exp. Mar. Biol. Ecol.* 11, 81–90.

Johnson, T. W., and Sparrow, F. K. 1961. *Fungi in Oceans and Estuaries*. Cramer, Weinheim, West Germany. 668 pp.

Jones, E. B. G. (ed.). 1976. *Recent Advances in Aquatic Mycology*. Wiley, New York. 749 pp.

Kittredge, J. S. 1976. Behavioral bioassays and biologically active compounds. In Webber, H. H. and Ruggieri, D. R. (eds.), *Proceedings of the Fourth*

Conference on Food and Drugs from the Sea, pp. 467–75. Marine Technology Society, Washington, D.C.

Kohlmeyer, J., and Kohlmeyer, E. 1979. *Marine Mycology: The Higher Fungi.* Academic Press, New York. 690 pp.

Kriegstein, A. V., Castellucci, V., and Kandel, E. R. 1974. Metamorphosis of *Aplysia californica* in laboratory culture. *Proc. Nat. Acad. Sci. U.S.* 71, 3654–58.

Langlois, G. A. 1975. Effect of algal exudates on substratum selection by motile telotrochs of the marine peritrich ciliate *Vorticella marina. J. Protozool.* 22, 115–23.

Lewin, R. A. 1974. Biochemical taxonomy. In Stewart, W. D. P. (ed.), *Algal Physiology and Biochemistry*, pp. 1–39. Blackwell, Oxford.

Liaaen-Jensen, S. 1977. Algal carotenoids and chemosystematics. In Faulkner, D. J., and Fenical, W. H. (eds.), *Marine Natural Products Chemistry*, pp. 239–59. Plenum Press, New York.

Littler, M. M., Littler, D. S., and Taylor, P. R. 1983. Evolutionary strategies in a tropical barrier reef ecosystem: functional-form groups of marine macroalgae. *J. Phycol.* 19, 229–37.

Lobel, P. S., and Ogden, J. C. 1981. Foraging by the herbivorous parrotfish *Sparisoma radians. Mar. Biol.* 64, 173–83.

McCandless, E. L., West, J. A., and Guiry, M. D. 1982. Carrageenan patterns in the Phyllophoraceae. *Biochem. Syst. Ecol.* 10, 275–84.

McCandless, E. L., West, J. A., and Guiry, M. D. 1983. Carrageenan patterns in the Gigartinaceae. *Biochem. Syst. Ecol.* 11(3), 175–82.

McConnell, O. J., and Fenical, W. 1977. Halogen chemistry of the red alga *Asparagopsis. Phytochemistry* 16, 367–74.

McConnell, O. J., and Fenical, W. 1979. Antimicrobial agents in the red algal family Bonnemaisoniaceae. In Hoppe, H. A., Levring, T., and Tanaka, Y. (eds.), *Marine Algae in Pharmaceutical Science*, pp. 403–27. De Gruyter, Berlin.

McConnell, O. J., Hughes, P. A., Targett, N. M., and Daley, J. 1982. Effects of secondary metabolites from marine algae on feeding by the sea urchin, *Lytechinus variegatus. J. Chem. Ecol.* 8, 1437–53.

McEnroe, F. J., Robertson, K. J., and Fenical, W. 1977. Diterpenoid synthesis in brown seaweeds of the family Dictyotaceae. In Faulkner, D. J., and Fenical, W. H. (eds.), *Marine Natural Products Chemistry*, pp. 179–90. Plenum Press, New York.

Maiti, B. C., and Thomson, R. H. 1977. Caulerpin. In Faulkner, D. J., and Fenical, W. H. (eds.), *Marine Natural Products Chemistry*, pp. 159–64. Plenum Press, New York.

Martin, S. M., and Skerman, V. B. D. (eds.). 1972. *World Collections of Cultures of Microorganisms.* Wiley-Interscience, New York. 560 pp.

Menzel, D., Kazlauskas, R., and Reichelt, J. 1983. Coumarins in the siphonalean green algal family Dasycladaceae Kützing (Chlorophyceae). *Bot. Mar.* 26, 23–9.

Mikês, O. (ed.). 1979. *Laboratory Handbook of Chromotographic and Allied Methods.* Halsted Press, Wiley, New York. 764 pp.

Moore, R. E. 1978. Algal nonisoprenoids. In Scheuer, P. J. (ed.), *Marine Natural Products: Chemical and Biological Perspectives*, vol. I, pp. 43–124. Academic Press, New York.

Morse, D. E., Hooker, N., Duncan, H., and Jensen, L. 1979. γ-Aminobutyric acid, a neurotransmitter, induces planktonic abalone larvae to settle and begin metamorphosis. *Science* 204, 407–10.

Morse, A. N. C., and Morse, D. E. 1984a. Recruitment metamorphosis of *Haliotis* larvae induced by molecules uniquely available at the surfaces of crustose red algae. *J. Exp. Mar. Biol. Ecol.* 75, 191–215.

Morse, A. N. C., and Morse, D. E. 1984b. GABA-Minetic molecules from *Porphyra* (Rhodophyta) induces metamorphosis of *Haliotis* (Gastropoda) larvae. *Hydrobiologia* 116/117, 155–8.

Morse, D. E., Tegner, M., Duncan, H., Hooker, N., Trevelyan, G., and Cameron, A. 1980. Indication of settling and metamorphosis of planktonic molluscan (*Haliotis*) larvae. 3: Signaling by metabolites of intact algae is dependent on contact. In Müller-Schwarze, D., and Silverstein, R. M. (eds.), *Chemical Signaling in Vertebrate and Aquatic Animals*, pp. 67–86. Plenum Press, New York.

Newell, S. Y., and Fell, J. W. 1982. Surface sterilization and the active mycoflora of leaves of a seagrass. *Bot. Mar.* 25, 339–46.

Nichols, H. W., Anderson, D. J., Shaw, J. I., and Sommerfeld, M. R. 1968. Pyrolysis–gas-liquid chromatographic analysis of chlorophycean and rhodophycean algae. *J. Phycol.* 4, 362–8.

Norris, J. N., and Fenical, W. 1982. Chemical defense in tropical marine algae. In Rützler, K., and Macintyre, I. G. (eds.), *The Atlantic Barrier Reef Ecosystem at Carrie Bow Cay, Belize 1: Structure and Communities*, pp. 417–31. Smithsonian Contributions to Marine Science, no. 12. Washington, D.C.

O'hEocha, C. 1962. Phycobilins. In Lewin, R. S. (ed.), *Physiology and Biochemistry of Algae*, pp. 421–35. Academic Press, New York.

Patterson, G. M. L., and Harris, D. O. 1983. The effect of *Pandorina morum* (Chlorophyta) toxin on the growth of selected algae, bacteria and higher plants. *Br. Phycol. J.* 18, 259–66.

Paul, V. J., and Fenical, W. 1983. Isolation of halimedatrial: chemical defense adaption in the calcareous reef-building alga *Halimeda*. *Science* 221, 747–9.

Pelczar, M. J., Jr., Reid, R. D., and Chan, E. C. S. 1977. *Microbiology*. McGraw-Hill, New York. 952 pp.

Phillips, D. W., and Towers, G. H. N. 1982. Chemical ecology of red algal bromophenols. 2. Exudation of bromophenols by *Rhodomela larix* (Turner) C. Agardh. *J. Exp. Mar. Biol. Ecol.* 58, 295–302.

Price, J. H. 1974. Advances in the study of benthic marine algae since the time of E. M. Holmes. *Bot. J. Linn. Soc.* 67, 47–102.

Ragan, M. A. 1981. Chemical constituents of seaweeds. In Lobban, C. S., and Wynne, M. J. (eds.), *The Biology of Seaweeds*, pp. 589–626. University of California Press, Berkeley.

Ragan, M. A., and Craigie, J. S. 1978. Phenolic compounds in brown and red algae. In Hellebust, J. A., and Craigie, J. S. (eds.), *Handbook of*

Phycological Methods: Physiological and Biochemical Methods, pp. 157–79. Cambridge University Press, Cambridge.

Ragan, M. A., and Jensen, A. 1979. Widespread distribution of sulfated polyphenols in brown algae. *Phytochemistry*, 18, 261–2.

Rinehart, K. L., Jr., Shaw, P. D., Shield, L. S., Gloer, J. B., Harbour, G. C., Koker, M. E. S., Samain, D., Schwartz, R. E., Tymiak, A. A., Weller, D. L., Carter, G. T., Munro, M. H. G., Hughes, R. G., Jr., Renis, H. E., Swynenberg, E. B., Stringfellow, D. A., Vavra, J. J., Coats, J. H., Zurenko, G. E., Kuentzel, S. L., Li, L. H., Bakus, G. J., Brusca, R. C., Craft, L. L., Young, D. N., and Connor, J. L. 1981. Marine natural products as sources of antiviral, antimicrobial, and antineoplastic agents. *Pure Appl. Chem.* 53, 795–817.

Robinson, T. 1983. *The Organic Constituents of Higher Plants: Their Chemistry and Interrelationships*, 5th ed. Cordus Press, North Amherst, Mass. 353 pp.

Rosenthal, G. A., and Janzen, D. H. (eds.) 1979. *Herbivores: Their Interaction with Secondary Plant Metabolites*. Academic Press, New York. 718 pp.

Rosowski, J. R., and Parker, B. C. (eds.) 1971. *Selected Papers in Phycology*. University of Nebraska, Lincoln. 876 pp.

Russell, G., and Fielding, A. H. 1974. The competitive properties of marine algae in culture. *J. Ecol.* 62, 689–98.

Seaman, F. C., and Funk, V. A. 1983. Cladistic analysis of complex natural products: developing transformation series from sesquiterpene lactone data. *Taxon* 32, 1–27.

Shield, L. S., and Rinehart, K. L., Jr. 1978. Marine-derived antibiotics. In Weinstein, M. J., and Wagman, G. H. (eds.), *Antibiotics: Isolation, Separation and Purification, Journal of Chromatography Library*, vol. 15, pp. 309–85. Elsevier, Amsterdam.

Sieburth, J. M. 1968. The influence of algal antibiosis on the ecology of marine microorganisms. In Droop, M. R., and Ferguson Wood, E. J. (eds.), *Advances in Microbiology of the Sea*, pp. 63–94. Academic Press, New York.

Sieburth, J. M. 1979. *Sea Microbes*. Oxford University Press, New York. 491 pp.

Smith, W. L., and Chanley, M. H. (eds.) 1974. *Culture of Marine Invertebrate Animals*. Plenum Press, New York. 338 pp.

Stahl, E. (ed.). 1969. *Thin-Layer Chromatography: A Laboratory Handbook*, 2nd ed. Springer-Verlag, New York. 1041 pp.

Starr, R. C. 1978. The culture collection of algae at the University of Texas at Austin. *J. Phycol.* 14(suppl.), 47–100.

Starr, T. J., Deig, E. F., Church, K. K., and Allen, M. B. 1962. Antibacterial and antiviral activities of algal extracts studied by acridine orange staining. *Texas Rep. Biol. Med.* 20, 271–8.

Stein, J. R. 1973. *Handbook of Phycological Methods: Culture Methods and Growth Measurements*. Cambridge University Press, Cambridge. 448 pp.

Steinberg, P. D. 1984. Algal chemical defense against herbivores: allocation of phenolic compounds in the kelp *Alaria marginata*. *Science* 223(4634), 405–7.

Steneck, R. S., and Watling, L. 1982. Feeding capabilities and limitation of herbivorous molluscs: a functional group approach. *Mar. Biol.* 68, 299–319.

Sun, H. H., and Fenical, W. 1979. Rhipocephalin and rhipocephanal, toxic feeding deterrents from the tropical marine alga *Rhipocephalus phoenix*. *Tetrahedron Lett.* 1979, 685–8.

Sun, H. H., Paul, V. J., and Fenical, W. 1983. Avrainvilleol, a brominated diphenylmethane derivative with feeding deterrent properties from the tropical green alga *Avrainvillea longicaulis*. *Phytochemistry*. 22, 743–5.

Switzer-Dunlap, M. 1977. Larval biology and metamorphosis of aplysiid gastropods. In Chia, F. S., and Rice, M. E. (eds.), *Settlement and Metamorphosis of Marine Invertebrate Larvae*, pp. 197–206. Elsevier/North Holland Biomedical Press, Amsterdam.

Switzer-Dunlap, M., and Hadfield, M. G. 1977. Observations on development, larval growth and metamorphosis of four species of Aplysiidae (Gastropoda: Opistobranchia) in laboratory culture. *J. Exp. Mar. Biol. Ecol.* 29, 245–61.

Targett, N. M. 1979. Gastropod tentacle withdrawals: a screening procedure for biological activity in marine macroalgae. *Bot. Mar.* 22, 543–5.

Targett, N. M., Bishop, S. S., McConnell, O. J., and Yoder, J. A. 1983. Antifouling agents against the benthic marine diatom, *Navicula salinicola*: homarine from the gorgonians *Leptogorgia virgulata* and *L. setacea* and analogs. *J. Chem. Ecol.* 9, 817–29.

Targett, N. M., and McConnell, O. J. 1982. Detection of secondary metabolites in marine macroalgae using the marsh periwinkle, *Littorina irrorata* Say, as an indicator organism. *J. Chem. Ecol.* 8, 115–24.

Touchstone, J. C., and Dobbins, M. F. 1978. *Practice of Thin Layer Chromatography*. Wiley-Interscience, New York. 383 pp.

Vadas, R. L. 1977. Preferential feeding: an optimization strategy in sea urchins. *Ecol. Monogr.* 47, 337–71.

Valiela, I., Koumjian, L., Swain, T., Teal, J. M., and Hobbie, J. E. 1979. Cinnamic acid inhibition of detritus feeding. *Nature* 280, 55–7.

van der Velde, H. H. 1973. The use of phycoerythrin absorption spectra in the classification of red algae. *Acta Bot. Neerl.* 22, 92–9.

Whittaker, R. H., and Feeney, P. P. 1971. Allelochemics: chemical interactions between species. *Science* 171, 757–70.

Wiley, E. O. 1981. *Phylogenetics: The Theory and Practice of Phylogenetic Systematics*. Wiley-Interscience, New York. 439 pp.

Yentsch, S., and Menzel, D. W. 1963. A method for the determination of phytoplankton chlorophyll and phaeophytin by fluorescence. *Deep-Sea Res.* 10, 221–31.

Zapata, O., and McMillan, C. 1979. Phenolic acids in seagrasses. *Aquat. Bot.* 7, 307–17.

7: Destructive (harvest) sampling

ROBERT E. DE WREEDE

Department of Botany, The University of British Columbia, Vancouver, British Columbia, Canada V6T 2B1

CONTENTS

I. Introduction

In this chapter destructive sampling of marine benthic macroalgae is discussed. In addition, some techniques are given for determining placement, size, and number of samples. "Standing stock" refers to the quantity of seaweed present at a particular time within a defined area. This is different from "standing crop," which refers to the repeatably harvestable (sustained yield) biomass. The great mass of seaweeds grows on rocky substratum, and thus the sampling and collection methods described here deal largely with such areas. In most cases, the same method is applicable to both intertidal and subtidal areas.

Data on standing stock can be used in many ways. To the ecologist they are one measure of the resources available to other trophic levels. To those interested in seaweeds economically, standing stock data provide information helpful for evaluating the economic viability of a harvesting operation. Also, harvested algae can provide data on resource allocation, biochemical constituents, population dynamics, and epiphytic and epifaunal associations. Any of these uses could be a primary reason for undertaking a destructive sampling program.

However, destructive sampling must be performed with caution. It is perhaps easy to rationalize the destruction and collection of seaweeds on the basis of the knowledge gained; yet before any such program is undertaken, one should carefully consider whether the information might be equally well obtained by a nondestructive sampling program (see Littler and Littler, Chap. 8). Destructive sampling of seaweeds is essential for acquiring certain kinds of information, such as biomass, and for performing some chemical content analyses. Productivity data and others are usually better obtained by methods discussed elsewhere in this manual (Littler and Arnold, Chap. 17; Arnold and Littler, Chap. 18; and Browse, Chap. 19).

The techniques described in this chapter have been obtained primarily from the literature. Some papers and books particularly

[148]

relevant to the topic of destructive sampling are the following: Gonor and Kemp (1978) on quantitative sampling procedures; sampling design by Green (1979); Bellamy et al. (1973) on estimating seaweed production on the basis of biomass data; and statistical analysis of benthic invertebrate samples by Elliott (1977). All four of these publications are well written and concise.

II. Equipment

The equipment required for destructive sampling is usually neither complex nor extremely expensive. Five categories of equipment are described here: that required to identify the site, harvest the algae, collect (contain) them, and record the appropriate data and that required in the laboratory.

A. *Site identification*

Before any sampling can begin, one must mark the appropriate sites. This is necessary in order to relocate the sites for repeated sampling or if reevaluation of the data is required. Since the physical and biological conditions in marine areas vary widely, the following materials must be chosen to fit the difficulties presented by a given locale (for additional information, see Foster et al., Chap. 10):

1. Anchor bolts: Various kinds are available from hardware supply houses. Galvanized anchor bolts will last at least a year in the intertidal zone, longer in subtidal areas. These bolts are "set" by means of a star drill and sledgehammer. A common air drill with a carbide-tipped masonry bit attached by a first stage to a SCUBA tank can serve for years if properly cleaned after use. Air drills can be obtained at most large hardware supply houses, and numerous brands are on the market. I have found this technique to be excellent for softer substrata but unsuitable for hard (granitic) bottoms, where a muffler chisel and star drill are more effective. If used under water, extra weight must be available to counteract the force needed to push on the drill. One kind of anchor bolt that has been used successfully is the Star stud bolt ($\frac{1}{4}$ S, $1\frac{3}{4}$ in., Star Expansion).

2. Pitons: These are useful when the substratum is hard and has numerous cracks and when an exact location is not necessary. Pitons will last at least 4–5 years in the intertidal and subtidal zones. A sledgehammer is required to drive them into the rock.

3. Other markers: Depending on the scale of the work, some researchers have used discarded engine blocks, barrels filled with cement, and similar heavy items to locate sites. Commercially available cement blocks (15- to 20-kg range) are suitable for calm sites but are easily moved by surge and waves.

4. Epoxy putty: Various mixtures are available, with a variety of setting times and working temperatures. A brand that I have found to work well is Sea Goin' Poxy Putty (Permalite Plastics), which has the consistency of a thick paste when mixed. Another epoxy of more liquid consistency is Cold Cure (Industrial Formulators); this can be injected into cracks and crevasses through large (50-ml) hypodermic syringes. Both will set either under water or in air. The second brand will float under water, but sand or other similar material can be added at the time of mixing to increase the density sufficiently. Various markers can be set into the epoxy while it is drying to aid in relocating the site.

5. Flagging tape: In some areas flagging tape (fluorescent yellow shows up well under water) is sufficient to mark the site but has to be replaced every 3–6 mo.

6. Lines: Nylon is the most useful material because it stretches and is long lasting. Polypropylene line can be obtained in very visible colors, but it floats, which may be a disadvantage. The line must be removed between sampling dates for, if left, it will sweep in the surge or waves and denude part of the site.

7. Spray paint: Paints can be used in the intertidal zone but should be avoided, because they are unsightly.

8. Quadrats: Quadrats made of sand-filled polyvinyl chloride (PVC) plastic or, preferably, made of metal such as brass, stainless steel, or aluminum are useful; in some cases, it is preferable that they come apart so that they can more easily be placed around the bases of the larger algae.

B. Harvesting

Harvesting requires only simple equipment, unless one heeds the advice of Gonor and Kemp (1978), who strongly advocate photographing the site before harvesting in order to have a record of the original community. The usual harvesting equipment consists of paint scrapers, diving knives, or clippers.

C. Collecting

Algae are best collected in nylon net or cloth bags. Plastic bags can serve in the intertidal zone but are awkard to use subtidally. A suction device such as a "slurp-gun" (Tanner et al. 1977) is useful for collecting smaller algae in the subtidal zone. Similarly, a PVC pipe with an airline from a SCUBA tank attached near its base can act as a suction device for gathering smaller species; a net bag at the top of the pipe collects the algae. This device is also useful in sandy or muddy substratum, because it filters the algae from the bottom material.

D. Data recording

During the harvesting process, it is necessary to record such data as site location, bag number, depth, and time; this must be done on waterproof material. Either waterproof paper (often available in dive shops) or plastic (white Lucite, Plexiglas) slates work well. Some plastics have to be sanded in order to retain pencil marks. A modified plastic slate was developed by L. Yip and T. Klinger; it consists of a PVC cylinder of sufficient diameter to fit over the hand and two back-to-back plastic report spines glued to the cylinder lengthwise. Thin flexible plastic or waterproof paper is laid along the cylinder and clipped into the report spines, and the cylinder is placed around the lower arm. This device keeps the slate out of the way when one is working. In the intertidal zone, data can also be spoken into a tape recorder. For further information on equipment, see Foster et al. (Chap. 10).

E. Laboratory equipment

Little laboratory equipment is needed to obtain standing stock data from harvested material. Depending on one's objectives, a drying oven, and perhaps muffle furnace may be needed, as will a balance, tape measure, and usually dissecting and compound microscopes.

III. Methods

The methodology of a destructive harvest will vary with the purpose of that harvest. For example, measuring the variation in standing stock over a given area at one time will require different procedures than measuring it for a particular site over many months. In addition, there are limits imposed by equipment, time, and money. Ways of optimizing information obtained for a given amount of funds are given by Poole (1974: ch. 5) and Wiegert (1962). Regardless of the purpose, certain basic decisions must be made about sampling, such as where and how frequently the sampling must be done and how many samples of what size must be taken.

A. Location of samples

All sampling programs should start with a survey of the potential sampling sites, preferably at several seasons of the year. Such a procedure will provide information on accessibility, distribution, and seasonality of the plant(s) of interest.

The next step is to locate the specific sampling sites within the general study area. Samples can be placed in either a random (not haphazard) or regular arrangement. Elliott (1977: 134) states: "If the object of the investigation is to determine the mean and variance of the population, random sampling is essential. If the object is to determine numbers in relation to position within the ecosystem, systematic sampling may be preferable." Both Lewis (1976) and Greig-Smith (1964) discuss further the advantages of regular sampling.

One advantage of random sampling is that it enables one to use statistical analyses on the data. The use of statistics in sampling is discussed by numerous authors (Green 1979: Gonor and Kemp 1978; Elliott 1977). Elliott recommends stratified random sampling (random samples placed within previously defined areas) to circumvent the problem of all random samples falling in one part of the study site. Gonor and Kemp (1978) discuss the problem of nonreplaceable sites and its effect on statistical analyses.

B. Sampling unit size and sample number

In general, it is a good idea to take as many samples as possible (Green 1979). Equal numbers of samples in either different areas or at different times are best for subsequent statistical analyses.

1. Sampling unit size. As a general rule the best size for the sampling unit is the smallest one possible (Elliott 1977). For a population dispersed in a random fashion any sample size is as efficient (gives an estimate of equal precision) as any other (Elliott 1977). Species may also be dispersed in a contagious or regular fashion. A regular distribution of organisms is unlikely, but a contagious one frequently occurs; for the latter type, a small sampling unit size is most efficient (Elliott 1977 and references therein). Brower and Zar (1977) illustrate some methods of testing for the three kinds of distributions.

If the biomass for all species at a site is of interest, one should use a sampling unit size that maximizes the number of different species included in the sample. As in the performance curve illustrated in Fig. 7–1, if one plots the total number of species (instead of biomass) against the cumulative sampling unit size, the point where the collector's curve (Pielou 1977) levels off is a rough estimate of the optimum sampling unit size (see Pielou 1977; Mueller-Dombois 1974). Another approach is to use a sampling unit size that minimizes the estimate of the variance (the square of the standard deviation) of the mean (Gonor and Kemp 1978).

2. Sample number. The number of samples to be taken can be determined on the basis of the desired degree of inclusiveness or precision of the data. A performance curve (Brower and Zar 1977) of the type illustrated in Fig. 7–1 plots cumulative biomass against the cumulative number of samples. Brower and Zar state that the number of

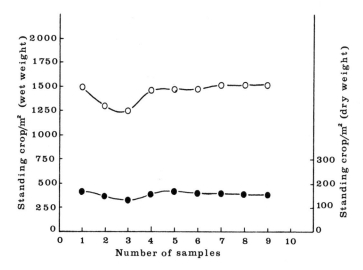

Fig. 7–1. Performance curve using data for *Sargassum muticum* from British Columbia. Wet weight (○); dry weight (●).

replicates is sufficiently large when the cumulative mean has become insensitive to the variations in the data.

If one can state the degree of precision desired in the data, then the number of samples can be determined by preliminary sampling (Brower and Zar 1977; Elliott 1977; Gonor and Kemp 1978; Sokal and Rohlf 1969). Sokal and Rohlf (1969) discuss the "true difference" method for calculating the number of samples required. Elliott (1977) defines an index of precision D used to obtain a sample size of given precision:

$$D = \text{standard error/arithmetic mean}$$

$$= 1/\bar{x}\sqrt{s^2/n} \tag{1}$$

In this equation, s is the standard deviation, n the number of sampling units in each random sample, and x the sample arithmetic mean. Then, if the acceptable error D is 20% of the mean, the number of sampling units required to obtain that precision is

$$n = s^2/D^2\bar{x}^2 = s^2/0.2^2\bar{x}^2 = 25s^2/\bar{x}^2 \tag{2}$$

Two important points must be kept in mind. First, the index of precision can be used only with random samples. Second, the number of sampling units required for a given degree of precision will vary from one time of year to another and from one site to another. If one were to adhere strictly to the admonition of equal sampling units and a given degree of precision, a sampling effort at any given place

Table 7–1. *Some examples of sampling unit size, number of samples, and quadrat placement methods in destructive sampling for standing crop information*

Algae examined	Quadrat size (m)	Quadrat number	Placement technique	Reference
Understory	0.25	6	Random	Calvin and Ellis (1978)
All	0.25	20	Random	Foreman (1977)
Kelp	0.5	2	Random (?)	Jupp and Drew (1974)
Kelp	1.0	1	Subjective	Kain (1977)
All	Varied	Varied	Random (?)	Lieberman et al. (1979)
All	0.25	Varied	Random (?)	Mann (1972)
Kelp	1.0	1	Subjective	Norton et al. (1977)
All	1.0	Varied	Subjective	Sakai (1977)
Gelidiales	0.03	20	Random (?)	Santilices (1978)

or time would have to include as many sampling units as were required for the most diverse site. This would clearly lead to difficulties if the requirement for more sampling units were not realized until late in the sampling program. Realistically, then, Equations 1 and 2 can be used as a guide to the minimum number of sampling units required, but some estimate may have to be made of the variance and when it is likely to be greatest during the sampling program.

Some common sampling unit sizes and sample numbers used in algal biomass estimates are given in Table 7–1.

C. Sample frequency

In order to determine changes in the standing stock over time, one must sample repeatedly. The frequency of such repeated sampling will be determined by the detail in which changes must be known. At the least, one must sample at times of minimum and maximum standing stock, times that will be decided by previous experience or data. Westlake (1969) suggested sampling at the peak of the standing stock, 2 wk before and 2 wk after, and then every 1–2 mo.

D. Sample collection

Methods of collecting the algae must be suited to the purpose of the research. Paint scrapers or small chisels will suffice in most cases.

The method of transport to the laboratory must also be considered. If biomass and species identification is important, then buckets, with the water changed frequently, will be adequate. For long trips it is advisable to wrap the seaweeds in damp newspaper and place the packages in plastic bags. It always helps to keep the bags cool. For chemical analyses of phycocolloids I have preserved the seaweeds in dry ice, and for long-term storage, freeze-drying will preseve many chemical properties.

I find chemical preservation, as with Formalin, obnoxious. Again, depending on the final use of the seaweeds, freezing in a home freezer may be sufficient. Some algae (*Rhodomela* and *Sargassum*) can be preserved well this way, but some of the delicate red algae (*Bonnemaisonia*) will be destroyed. For further details on collection, see Abbott and Tsuda (Chap. 4).

E. Sample analysis

1. Standing stock

a. Wet or dry weight. Wet weight is generally not a good way to express standing stock. The variation in the quantity of superficial water associated with the algae makes comparison of data difficult. One can standardize the measurements to some degree by blotting the algae dry or by spinning the material to remove water. For more details, see Brinkhuis (Chap. 22).

Dry weight is defined as the weight of the tissue remaining after all water has been removed. Often, it will also be necessary to remove salts by washing in fresh water before drying. Dry weight is usually obtained by oven-drying the material to constant weight in labeled containers; Gonor and Kemp (1978) suggest a temperature of 70°C, but see also Brinkhuis (Chap. 22). Standing stock is frequently reported in the literature as dry weight.

b. Ash-free dry weight. Ash-free dry weight is determined by oxidizing the organic material in the sample at high temperatures and then subtracting the weight of the remaining "ash" from the original dry weight. Ash-free dry weight thus is the weight of the organic portion of the sample. Gonor and Kemp (1978) suggest using a muffle furnace set at 500°C, for 12 to 24 h. Careful procedures must be followed to obtain an accurate weight of the oxidized sample; for details see Brinkhuis (Chap. 22).

2. Productivity estimates. One way of estimating productivity is to measure the difference in the weight of the standing stock obtained at different sampling dates (Bellamy et al. 1973). At best, such a procedure gives a minimum estimate of productivity, since losses due to grazing, exudation, and storms cannot be accurately assessed. For a discussion, see Brinkhuis (Chap. 22).

IV. Sample data

A. Sample location, size, and frequency

The following data were obtained by the author in a study of the standing stock of *Sargassum muticum* in the Strait of Georgia, British Columbia, Canada. They will serve to illustrate the procedures discussed in the preceding sections.

The study site was chosen on the basis of its accessibility under most weather conditions, general representation of the standing stock of *S. muticum* in the sites surveyed, and sufficient abundances of algae to tolerate repeated sampling.

The location of the samples was determined by eye and random number. A transect line was placed with permanent end points (anchor bolts hammered into holes drilled into the substratum); the position of the line was in the approximate middle of the community and parallel to the shore. Distances were marked on the line, and the location of the sampling units was determined by a random number table. A decision was made before the sampling that alternate numbers would refer to sampling sites above and below the transect line. This was thus stratified random sampling, since the transect line was placed subjectively but the sampling units were located in a random fashion. The size of the sampling unit was determined by harvesting four randomly placed quadrats, for each of six different quadrat sizes (Table 7–2). Before this study it had already been determined that the largest unit that could be harvested was 2500 cm^2; hence, no larger unit was attempted. As shown by Table 7–2, the variance of the mean (shown as its square root, the standard deviation) drops sharply at a sample unit size of 1600 cm^2. This fact, plus the chance that at other sample dates the variance might increase, led to the choice of a sampling unit size of 2500 cm^2. If sampling units of smaller size had been used, it would have been necessary to take more on any given sampling date for a given precision.

The number of sampling units per sampling date also had to be determined. Since the size of the sampling unit was now fixed at 50 × 50 cm, a series of such quadrats were randomly placed to determine the number required. Figure 7–1 and data in Table 7–3 show the performance curve method for both dry and wet weights; four to five samples are sufficient. When the index of precision D is used to calculate the number of sampling units (based on data in Table 7–3), for a precision of 15% ($D = 0.15$),

$$n = (\text{variance})/(D^2)(\text{mean})$$

$$= (923.4)/(0.023)(24{,}336) = 1.65. \qquad (3)$$

Table 7–2. *Example of data used to calculate sampling unit size: weights of Sargassum muticum in May 1975, British Columbia*

Sample size (cm^2)	Dry weight per m^2				Standard deviation
	Sample 1	Sample 2	Sample 3	Sample 4	
100	686	660	600	450	105.6
200	697	590	555	790	106.7
400	660	970	720	1510	387.3
800	426	746	771	595	159.0
1600	663	725	626	738	52.7
2500	669	781	603	689	73.5

Table 7–3. *Sample data for calculating the number of samples required: Sargassum muticum in May 1975, British Columbia*

Sample number	Wet weight ($g \cdot 2500\ cm^{-2}$)	Dry weight ($g \cdot 2500\ cm^{-2}$)	Cumulative average, wet wt ($g \cdot 2500\ cm^{-2}$)	Cumulative average, dry wt ($g \cdot 2500\ cm^{-2}$)
1	1500	169	1500	169
2	1125	125	1312	147
3	1300	112	1483	161
4	1950	211	1308	135
5	1525	181	1469	154
6	1500	169	1480	160
7	1650	151	1507	160
8	1500	135	1506	157
9	1500	150	1505	156

Similarly, for a precision of 10% ($D = 0.10$), n is equal to 3.79; in this case four quadrats, each 50 × 50 cm, would be required to attain precision at the 10% level. Again, given ideal circumstances, one might wish to use smaller quadrats and more of them in order to gain more replicates and greater degrees of freedom.

As mentioned, the variation in the area being sampled will change over time. In the study of *Sargassum muticum* standing stock, the data were obtained in May 1975; a calculation of the precision of the data in July of that same year showed that it had dropped to $D = 0.29$, or 29%. When six samples, each 50 × 50 cm, were taken, the precision was 18%; as a general guide, 20% is probably an acceptable accuracy for benthic samples (Elliott 1978). In this project, however, other factors limited the number of quadrats to four, and the lower precision had to be accepted. To obtain data on the temporal variation

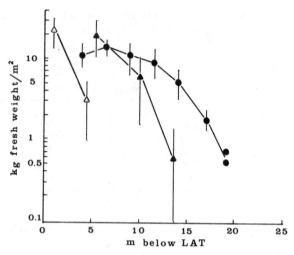

Fig. 7–2. Data on standing crop of *Laminaria hyperborea* at various depths and localities in Great Britain. Symbols indicate different sampling sites; LAT is lowest astronomical tide; vertical bars indicate confidence limits (*P* = .05). Redrawn from Kain (1977).

of the standing stock, the site was sampled approximately once per month.

B. *Standing stock data*

As already pointed out, standing stock information can be used in a variety of ways (see also Bellamy et al. 1973). The use of this information in productivity estimates, the determination of caloric content, and population statistics is detailed elsewhere in this book (see Littler and Arnold, Chap. 17; Arnold and Littler, Chap. 18; and Carefoot, Chap. 23). Figure 7–2 is an example of standing stock data from Kain (1977).

V. Critical evaluation

Destructive sampling for standing stock estimates entails some problems. The first problem is the destruction of the habitat that ensues. Although most research projects do not destroy large areas, the potential for site deterioration is great, and care must be taken that this does not occur. Also, since the same site cannot be studied more than once, variability due to the sampling of different plots may obscure some patterns.

Another problem is that the sampling program sometimes does not accurately estimate the standing stock. The techniques described in this chapter will help to ensure that the estimate is, in fact, accurate; however, practical limitations may force one to diverge from ideal theory. Gonor and Kemp (1978) point out another problem: the likelihood that repeated standing stock measurements in the same area will be biased by the effects of previous sampling. They point out that this is particularly true for species with long lives and low recruitment capacity. Although most marine algae produce copious spores and gametes, recruitment into cleared sites can be altered by the growth of other algae or changed herbivore densities. Similarly, such changes can alter the composition of sites adjacent to the cleared areas. Theoretically, one can avoid such effects by sampling over a very large area, but in practice one may have to contend with the consequences. In addition, if one samples over a very large area, the environmental differences that are then bound to occur will also confound comparison of the results.

Another problem may occur when one is using standing stock to estimate productivity. If different species are used in the estimate, then each species may attain its maximum standing stock at a different time (Teal 1980). As a result, the estimate of production will be less accurate.

VI. Alternative techniques

There are no complete alternatives to destructive sampling for obtaining standing stock measurements of benthic marine algae. Remote sensing, photographic sampling, and in situ size measurements all require a harvest at some time in order to establish a correlation between the nondestructive measurements made and biomass. However, one should carefully evaluate each of these nondestructive methods before deciding on a destructive sampling program; a combination of these two methods may give the required information and thus lessen the impact on a site.

VII. References

Bellamy, D. J., Wittick, A., John, D. M., and Jones, D. J. 1973. A method for the determination of seaweed production based on biomass estimates. In *A Guide to the Measurement of Primary Production under Some Special Conditions*, pp. 27–33. UNESCO, Paris.

Brower, J. E., and Zar, J. H. 1977. *Field and Laboratory Methods for Ecology*. Brown, Dubuque, Iowa. 194 pp.

Calvin, N. I., and Ellis, R. J. 1978. Quantitative and qualitative observations on *Laminaria dentigera* and other subtidal kelps of southern Kodiak Island, Alaska. *Mar. Biol.* 47, 331–36.

Elliott, J. M. 1977. *Some Methods for the Statistical Analysis of Samples of Benthic Invertebrates*, 2nd ed. Freshwater Biological Publ. no. 25, Freshwater Biological Association, Ambleside, England. 160 pp.

Foreman, R. E. 1977. Benthic community modification and recovery following intensive grazing by *Strongylocentrotus droebachiensis*. *Helgo. Meeres.* 30,
   468–84.

Gonor, J. J., and Kemp, P. F. 1978. *Procedures for Quantitative Ecological
   Assessments in Intertidal Environments.* U.S. Environmental Protection Agency
   no. 600/3-78-087, Corvallis, Ore. 103 pp.

Green, R. H. 1979. *Sampling Design and Statistical Methods for Environmental
   Biologists.* Wiley, New York. 257 pp.

Grieg-Smith, P. 1964. *Quantitative Plant Ecology.* Butterworths, London. 256
   pp.

Jupp, B. P., and Drew, E. A. 1974. Studies on the growth of *Laminaria
   hyperborea* (Gunn.) Fosl. 1. Biomass and productivity. *J. Exp. Mar. Biol. Ecol.*
   15, 185–96.

Kain, J. M. 1977. The biology of *Laminaria hyperborea*. 10. The effect of
   depth on some populations. *J. Mar. Biol. Assoc. U. K.* 57, 587–607.

Lewis, J. R. 1976. Long-term ecological surveillance: practical realities in the
   rocky littoral. *Oceanogr. Mar. Biol. Annu. Rev.* 14, 371–90.

Lieberman, M., John, D. M., and Lieberman, D. 1979. Ecology of the subtidal
   algae on seasonally devastated cobble substrates off Ghana. *Ecology* 60,
   1151–61.

Mann, K. H. 1972. Ecological energetics of the seaweed zone in a marine
   bay on the Atlantic coast of Canada. 1. Zonation and biomass of seaweeds.
   Mar. Biol. 12, 1–10.

Mueller-Dombois, D., and Ellenberg, H. 1974. *Aims and Methods of Vegetation
   Ecology.* Wiley, New York. 547 pp.

Norton, T. A., Hiscock, K., and Kitching, J. A. 1977. The ecology of Lough
   Ine. 10. The *Laminaria* forest at Carrigathorna. *J. Ecol.* 65, 919–41.

Pielou, E. C. 1977. *Mathematical Ecology.* Wiley, New York. 385 pp.

Poole, R. W. 1974. *An Introduction to Quantitative Ecology.* McGraw-Hill, New
   York. 532 pp.

Sakai, Y. 1977. Vegetation structure and standing crop of the marine algae
   in the *Laminaria* bed of Otaru City, Hokkaido, Japan. *Jap. J. Ecol.* 27, 45–
   51.

Santilices, B. 1978. Multiple interaction of factors in the distribution of some
   Hawaiian Gelidiales (Rhodophyta). *Pac. Sci.* 32, 119–47.

Sokal, R. R., and Rohlf, F. J. 1969. *Biometry.* Freeman, San Francisco. 776
   pp.

Tanner, C., Hawkes, M. W., Lebednik, P. A., and Duffield, E. 1977. A hand-
   operated suction sampler for the collection of subtidal organisms. *J. Fish.
   Res. Bd. Can.* 34, 1031–4.

Teal, J. M. 1980. Primary production of benthic and fringing plant communities. In Barnes, R. K., and Mann, K. H. (eds.), *Fundamentals of Aquatic
   Ecosystems.* Blackwell, Oxford. 229 pp.

Westlake, D. F. 1969. Macrophytes. In Vollenweider, R. A. (ed.), *Primary
   Production in Aquatic Environments*, pp. 103–7. International Biological Programme (IBP) Handbook no. 12, Blackwell, Oxford.

Wiegert, R. G. 1962. The selection of an optimum quadrat size for sampling
   the standing crop of grasses and forbs. *Ecology* 43, 125–9.

8: Nondestructive sampling

MARK M. LITTLER AND DIANE S. LITTLER

Department of Botany, National Museum of Natural History, Smithsonian Institution, Washington, D.C. 20560

CONTENTS

I. Introduction

Modern ecological sampling represents a quantitative discipline designed to produce statistically interpretable analyses of biotic distribution and abundance patterns within defined habitats. Such quantitative objectives are usually beyond the powers of resolution when general visual surveys (Kenchington 1978), arbitrary scales (see Scheer 1978), or other subjective approaches are employed. The goals and objectives must be carefully considered throughout a study, because these determine the analytical techniques used, which in turn influence the nature of the data collected.

Nondestructive sampling, by utilizing permanently marked sampling locations that can be precisely relocated and reassessed, provides a powerful tool for the quantification of natural and anthropogenic changes in macroalgal standing stocks. The ultimate objective, stated very simply, is to obtain sufficient understanding to explain why a given population or community exists in a particular pattern at a specific time and place. An adequately large number of samples must be taken for proper statistical treatment, and this requires that the method be rapid and simple to use.

One purpose of the photogrammetric technique, detailed in Sec. II, is to produce a permanent historic record of photosamples that depict the status of the biota at a given time. This method has been the standard required for large-scale rocky intertidal studies contracted by the U. S. Department of the Interior (e.g., Bureau of Land Management, U.S. National Park Service, and Minerals Management Service) for nearly a decade. Many of the sampling considerations are the same as those discussed by De Wreede in Chap. 7 (Sec. I.B.1) and will be cross-referenced where appropriate. Because faunistic components are often important in determining temporal or spatial patterns of marine macrophytes, it is worthwhile to include relevant animals. The techniques presented here, as well as those given by De Wreede (Chap. 7), are equally applicable to sessile animal populations.

[162]

II. Method: photogrammetric

The principal method of nondestructive recording of standing stocks is the photogrammetric technique of undisturbed sampling [first developed for seaweeds by Littler (1968, 1971)] that yields parallax-free samples. Such samples can be used to generate precisely detailed and highly reproducible quantitative information, that is, cover, density (number of individual organisms per unit of area), and frequency (percentage of sample plots in which a given species occurs). Since the first time that photographic sampling was alluded to as a possibility for ecological studies in the marine environment (Ernst 1957), it has been developed independently (e.g., Littler 1968, 1971, 1980a; Johnston et al. 1969; Connell 1970) and undergone diffuse application (Lundälv 1971; Vadas and Manzer 1971; Dayton et al. 1974; Laxton and Stablum 1974; Torlegård 1974; Wilson 1974; Paine 1974; Ott 1975; Drew 1977; Vance 1979; George 1980; Karlson 1980). The method now represents one of the most widespread and sophisticated techniques for permanently recording marine algal standing stocks.

Random sampling (determined by some mechanical means) is theoretically required to obtain unbiased results (Southwood 1966). Randomness is usually achieved by means of a line transect laid out along a predetermined compass bearing, a table of random numbers being used to determine the precise sampling locations. Although bias is eliminated by random sampling, it is not always desirable for intertidal work because of the marked patchiness typically present. An advantage of systematic assessment at fixed intervals within a given area is that the array of samples is relatively easy to achieve and relocate for repetitive study. In addition, this arrangement of plots may give a more accurate picture of macrophyte distributions and abundances because samples are spread over the entire area to be analyzed (Greig-Smith 1964; Poole 1974; Loya 1978). Uniform intervals cannot be used if the biota itself is distributed in a systematic pattern (coincident to that of the sample array), such as a linear arrangement within a narrow zone related to tidal height or along a rock fissure. Haphazard techniques, such as throwing a marker or dropping a quadrat without looking at the sample area, are not random and should be avoided because marginal areas tend to be undersampled and plots often become arrayed inadvertently at fixed distances, often with undetected bias. Therefore, some stratified random sampling pattern is usually preferable as a compromise in rocky intertidal community studies.

The general location of each study area can be determined by reference to aerial photographs and maps of the region. After extensive ground reconnaissance of a given area, the precise location of the upper end of each transect should be determined (optimally by consensus of several experienced marine biologists) along a biologically representative part of the shoreline. For intertidal work, several transect tapes (Leitz Symlon fiber glass tapes, Ben Meadows Co.), positioned at random within representative areas and as dictated by the steepness of the shoreline and topography, are laid perpendicular [by means of a sighting compass (Suunto Fast-Accuracy Compass, Ben Meadows Co.) or dumpy level (David White Meridian Construction Level, Surveyors Service Co.)] to the waterline at each study site from immediately above the high water level of intertidal organisms to just below the waterline at low tide, thus providing locations for a minimum of at least four samples per community type. To establish permanent sample locations, holes are drilled (Gas Hand Drill, McMaster-Carr Supply Co.) and eyebolts cemented into the substratum at the upper, middle, and lower ends of the transect lines; this ensures the precise replacement of the transects during subsequent studies. Because of the patchy nature of marine coastal habitats, it is sometimes desirable to sample each patch separately. This involves division of a heterogeneous habitat into homogeneous units, termed strata; each stratum can then be sampled randomly (i.e., stratified random sampling).

In southern California (Littler 1980a,b), sampling optimization analysis by the Poisson statistic (Wilson 1976) revealed that at least 30 samples (0.15 m^2) were required to assess adequately a typical rocky intertidal site. Therefore, ~40 rectangular quadrats were placed along transect lines at various intervals (depending on the steepness of the shoreline), providing permanent, stratified plots for sampling temporal and spatial distributions of organisms. A random numbers table is used to determine quadrat locations along a given line. To furnish statistically adequate replication, no fewer than four quadrats of each size should be taken in a given 0.3-m tidal interval, whenever possible. This is accomplished by adding quadrats to the immediate right and left sides (in some cases upper and lower sides) of quadrats known to be at tidal heights that are "undersampled." Quadrat locations are marked permanently with metal studs, stainless steel nails, marine epoxy (e.g., Sea Goin' Poxy Putty Multi-Purpose-1324, Permalite Plastics Corp., Pro-Line-Splash Zone Compound), or various bolts set in "hard-rock" cement (see also De Wreede, Chap. 7, and Foster et al., Chap. 10).

A large number of small quadrats is usually preferable to a smaller number of large plots per unit of equal area assessed (Green 1979), since greater sample numbers permit a better assessment of between-sample variability. The size of the plots should be commensurate with the size of the organisms themselves and their density (Kershaw 1973). The number of quadrats sampled must be sufficient to include the majority of species present in the study area (see also De Wreede, Chap. 7). This number is often determined by plotting a species–area curve consisting of the cumulative species total plotted against the number of samples taken. An adequate sample number should be well into the portion of the curve where additional samples are contributing few new taxa. There is extensive literature dealing with the problem of sample size, usually involving the concept of "minimal area" or species–area curves. Minimal area, however, is a controversial concept (see Poore 1964) and, in some cases, is of limited utility.

In quadrat assessments, the plot should be no smaller than the largest individual organism to be sampled. We have found a rectangular 0.15-m^2 quadrat to be adequate for sampling most intertidal macrophytes; however, 1.0-m^2 or larger quadrats are necessary for larger kelps (e.g., *Eisenia, Egregia,* and *Macrocystis*). Rectangular quadrats have an advantage over square or circular plots of equal area because they tend to incorporate a greater diversity of populations, it being less probable that a rectangle will fall completely within a given clump or patch of organisms. In this regard, a number of studies (Clapham 1932; Pechanec and Stewart 1940; Bormann 1953) have shown that elongate rectangular quadrats may furnish a more accurate analysis of the composition of a population or community than an equal number of square or circular plots having the identical area. This is especially true when the long axes of the quadrats are oriented parallel to the axes of environmental gradients within the habitat being studied. We also designed our labeled quadrats for photogrammetric sampling (30 × 50 cm) to coincide with the proportions of the 35-mm film used for recording the data. These dimensions permit framing of the sample at a comfortable distance without having to resort to special wide-angle lenses, excessively long quadrapods, or even balloon-suspended frames (Rützler 1978) for holding the camera.

Relative tidal heights for each quadrat are measured from a permanent reference point by means of a stadia rod (Fiber Glass leveling rod, Ben Meadows Co.) and a standard (20-power) surveyor's transit or dumpy level. The height of this reference point is determined in relation to mean lower low water by measuring, at six or more places along the shoreline on successive days, the midpoints between low and high wave peaks at the time of the predicted (U.S. Department of Commerce Tide Tables, National Ocean Survey) low tides. The repeatability of measurements checked on different site visits should be within 0.1 m.

Considerable care must be taken to minimize trampling and other forms of disturbance to the biotic communities under nondestructive study. Photogrammetric sampling is advantageous in this regard, since the field time at a given quadrat is minimal.

Physical descriptions of each study area (e.g., date, time, tidal stages, wave heights, air and water temperature, cloud cover, and salinity) should be recorded at the time of each visit. Many of the environmental considerations are the same as those outlined for monitoring by Foster et al. (Chap. 10), and the interested reader should refer to their treatment. Oceanographic literature and climatological data can be used, when available, to characterize further the relevant environmental features.

Data are obtained by photographing the numbered quadrats perpendicular to the substratum with two Nikonos cameras equipped with waterproof electronic flash units. We have used quadrapod mounts with framers to good advantage, particularly at SCUBA depths (Lewbel et al. 1981). In the case of stratified macrophyte communities, lower layers are photographed after upper canopies are moved aside until the organisms covering the primary substratum have been recorded. Affixed to the upper left corner of each quadrat is a gray plastic label that is marked with a wax pencil to identify permanently each of the photosamples. One camera contains 35-mm Kodachrome-64 slide film (Kodak) and the other contains Ektachrome color infrared (IR) slide film. Miniature tape recorders in waterproof housings (Wet · Tape, Sound-Wave Systems, Inc.) and plastic- (polyethylene-) coated paper (PolyPaper, Nalge Company) can be used for note taking on the contents of the photosamples. For every sample, a taxonomist records the taxa, counts the individuals (if possible), and, using a cross-wire grid, visually estimates the cover of each species in a detailed section-by-section format (quadrats are subdivided into 20 equal sections). The understory can be quantified at this time by the cross-wire point-quadrat method if overstory is extremely dense. The only organisms removed from the permanent undisturbed quadrats are small voucher samples taken for taxonomic purposes (Tsuda and Abbott, Chap. 4). However, this is done only when adequate voucher material is unavailable elsewhere.

It is worth noting that many studies stop at this level of quantification (by estimation in situ). We have found that such approximations usually cannot be repeated precisely (i.e., often exceeding ±25% for dominant organisms) because of parallax and variability among and within observers. Observer differences are influenced by various degrees of field distractions and stresses, which can be pronounced during heavy surf and nighttime low-tide conditions. The recorded in situ information is used in the laboratory for cryptic organisms, dense understory layering, and density counts and to minimize taxonomic and other problems encountered in the interpretation of photosamples.

In the laboratory, the developed pairs of transparencies are projected simultaneously (the IR below the color) onto two sheets (each 21×28 cm) of fine-grained white Bristol paper. The paper contains a grid pattern of dots at ~2.0-cm intervals on the side of the reflected light; this has been shown (Littler 1980a) to be an appropriate density (e.g., ~1.0 cm^{-2}) for consistently reproducible estimates of cover. Two replicate scorings are performed for each transparency (after movement between assessments) with a minimum of 150 dots per scoring. Sousa (1979) discusses the advantages of placement of dots at uniform fixed intervals versus mechanically randomized placement of dots. Briefly, uniform arrangements are easier to locate and faster to score, but can result in over- or underestimation of linearly arranged organisms. Random dot placements are confusing to score and can result in over- or underestimation of clumped organisms but are technically required for statistical analysis. As in quadrat placement, some stratified random arrangement usually is employed as a compromise. Red dots contrast best with the biological detail shown by the projected color transparencies; black dots are used in conjunction with the IR transparencies. The photographs are aligned and focused without regard to the field of dots to ensure unbiased assessments. The percent cover values are expressed as the number of "hits" for each species divided by the total number of dots contained within the quadrats. Reproducibility is consistently high and seldom varies more than ±5% for a given taxon. Species not abundant enough to be scored by the replicated grid of point intercepts are assigned an arbitrary cover value of 0.1%, or the field estimates are used.

For coarse forms, it is feasible to use a planimeter (Keuffel & Esser compensating polar planimeter, Forestry Suppliers, Inc.) to trace the circumference of individuals and obtain the projected area from photographs (Littler 1971). The Zeiss MOP-30 image analyzer (Carl Zeiss, Inc.) provides a similar, but more precise, faster, and relatively expensive alternative to the planimetric method.

The color IR transparencies are indispensible to the delineation of the various species of primary producers (e.g., blue-green algae are dominant forms that can be discerned reliably on wet dark substrata only by the use of color IR photography) and to the assessment of the status of their health. Each species fluoresces differently in the IR band, according to its chlorophyll content and health (the percentage of dead branches on an algal thallus can be seen more clearly in IR).

The method as described here does not allow for the quantification of microalgae, small flora, and infauna when they occur in low abundances. We realize that these may be metabolically very active, but their analysis requires special techniques and expertise that comprise problems outside the scope of this text. For this reason, the present discussion is restricted to macroepibiota that can be discerned in the field with the unaided eye. However, it is straightforward to quantify microbiota (e.g., turfs of filamentous algae) when it is present in high abundance, and residual microalgae from voucher sampling are identified and retained for future analysis (Tsuda and Abbott, Chap. 4). Also, it is possible to use macrophotography to assess microalgae on a finer scale (e.g., Vance 1979). Macrophotogrammetry works well on heavily grazed biotic reef systems when used in conjunction with microscopic assessments of extensive voucher samples.

III. Analyses of data

Information obtained by the photogrammetric method (undisturbed) and by the harvest method (disturbed; see De Wreede, Chap. 7) provides quantitative information on the distribution of standing stocks in relation to tidal height, depth, distance along a transect line, or some environmental gradient or among treatments and controls during experiments. These data are summed and averaged to interpret differences in populations and communities among sites or experimental treatments as well as seasonally within sites. Species cover, frequency, and density fluctuations are calculated as a function of various vertical or horizontal gradients throughout the intertidal zone. When photogrammetric measurements of cover are used in conjunction with disturbed assessments of biomass for very large sample sizes, it is possible to generate precise regressions for biomass on cover (e.g., Littler 1978). Subsequently, cover data alone can be interpolated to estimate biomass for most of the abundant taxa.

Diversity measurements have been used widely by those responsible for determining the effects of human stresses on biotic communities. Species diversity is often measured by indices that include components of both richness and equability. The problem with any single index is that both components are confounded. Many diversity indices also contain the underlying assumption that the ecological importance of a given species is proportional to its abundance. These problems can be reduced by using the standardly applied Shannon and Weaver (1949) index (H' incorporating both richness and evenness) along with separate indices for richness [counts of taxa, Margalef's (1968) D'] and evenness [Pielou's (1975) J']. Such indices are calculated for the cover data using natural logarithms and applied as supplementary information to quantify seasonal changes and to provide between-site comparisons of community structure. Poole (1974) has indicated that, regarding the Shannon–Weaver index, the base of the logarithms is very much open to choice; however, in most ecological cases, natural logarithms should be used. By simply multiplying H'_e diversity values by the factor 1.443, one can obtain H'_2 numbers.

To characterize seasonal and between-site groupings in an unbiased manner, the undisturbed cover data from each site visit for every macrophyte species are subjected to hierarchical cluster analyses (flexible sorting) by the Bray and Curtis (1957) percentage of distance statistic (Smith 1976). This produces a dendrogram of assemblages that are then interpreted according to their environmental affinities and used to map the prevalent yearly and biogeographical patterns for the various sites.

IV. Critical evaluation

Much of our knowledge of benthic marine organisms is based upon anecdotal observations, although some studies have employed "quasi-quantitative" methods. Among such methods are those in which diagrammatic sketches within sample units are made and subsequently used to obtain estimates of abundance (e.g., Manton 1935; Abe 1937), visual approximations of cover (e.g., Nicholson and Cimberg 1971; Widdowson 1971), and the utilization of grids (e.g., Caplan and Boolootian 1967) to estimate the abundance of organisms. Such in situ assessments are usually time-consuming and often physically exhausting, thereby severely limiting the number of samples that can be taken within the field time available (e.g., during a SCUBA dive or low-tide period). A significant problem associated with all of these visually based in situ techniques is that of parallax (due to movement of the observer and organisms relative to the sampling devices), which has been shown (Littler 1971) to be an unsatisfactorily large source of error when one is measuring the cover of space-occupying organisms.

The undisturbed sampling method, involving permanently marked sampling locations, provides a powerful tool for the quantification of subtle seasonal and yearly biological differences, and this is what we have emphasized. This procedure has the advantage of being rapid and simple to use, thus enabling the investigator to take a greater number of samples per unit of time. When used with color IR film, the technique permits the quantification of blue-green algae, the predominant cover organisms in most rocky intertidal habitats. In addition, a high degree of quality control is achieved because photosamples scored by various individuals can be reviewed by the total research staff, including senior taxonomic personnel, to ensure standardization and accuracy in the quantification process. The IR transparencies also emphasize unhealthy thalli with reduced chlorophyll contents that are often masked by accessory pigments; these would otherwise not be visible by color photography or to the unaided eye. In the case of complex multilayered communities, an assistant may be required to hold upper canopies to the side as the various strata are photographed. If this is not possible (see Foster et al., Chap. 10), the field estimates of understory biota must be incorporated. Another important feature of the technique is that permanent historical data sets (i.e., photosamples) are obtained which depict the status of the biota at a given point in time; these may be useful at a future date for purposes not originally intended. In addition, changes (e.g., due to human disturbance, seasonality, recruitment, and growth) can be demonstrated by direct comparisons of photosamples taken of the identical quadrats over different sampling periods.

V. Alternative (plotless) techniques

Plotless methods essentially reduce quadrat measurements to linear or point recordings of distances and represent economical means of sampling biotic communities while avoiding problems of plot size and shape. However, they are not as powerful for detecting subtle changes as are permanently marked fixed quadrats. Dimensionless points and line intercepts cannot be replaced precisely in the same area, and a variance of only a few millimeters often can change estimates of abundance significantly.

Continuous line-intercept, transect sampling has been discussed (Loya and Slobodkin 1971; Loya 1972) and found to be considerably more cost efficient than quadrat techniques, because a greater portion of the overall habitat can be covered per unit of effort and problems of bottom topography can be avoided. This is among the most traditional of methods used by plant ecologists for recording the frequency and coverage of vegetation. Although the line transect theoretically has no breadth, measurements of the coverages of individual species intercepted along its length are possible, since they are proportional to the areas covered by each species concerned. Adequate transect length is determined by using the species versus transect-length curve in the same manner that the species–area curve is used in quadrat assessments. A detailed analysis of the methods for selecting the optimal sampling unit is given in Cochran (1963) and other standard publications (Greig-Smith 1964; Williams 1964; Southwood 1966; Kershaw 1973).

Plotless and line-intercept methods are most appropriate for relatively sessile, discrete, widely spaced organisms that are easily mapped. For example, cover of macroalgae under kelp canopies has been assessed by a simple random point device (Foster 1982) consisting of a loosely strung knotted string, fastened to each end of a metal bar, that is pulled tight to determine various points. Plotless methods are more difficult to use for dense macrophyte communities, where it is often impossible to distinguish individual plants. However, where layering is extensive and dense, the cross-wire sighting method in situ is especially useful. A major disadvantage of point methods is the extreme variability of results, which necessitates the recording of an inordinate number of points before an adequate sample is obtained, as well as the vulnerability of the method to subjective bias (Loya 1978).

Detailed analyses and descriptions of plotless techniques are given by Cottam and Curtis (1956) and Loya (1978) and include (1) closest-individual method, (2) nearest-neighbor method, (3) random-pairs method, (4) point-centered quadrat method, and (5) wandering-quarter method. All of these involve measuring point to center of plant distances. The mean point-to-plant distance squared gives the mean area per plant or average overall coverage. Of these methods, the point-centered quadrat method (or point-quarter method) has been found (Cottam and Curtis 1956) to be the least variable for distance determinations and to provide more data on individual species per sampling point, while having the lowest susceptibility to subjective bias. The assumption when any plotless method is used is that individuals of all species are randomly dispersed (Cottam and Curtis 1956; Greig-Smith 1964). In most cases, when this assumption does not hold, there appears to be no significant error. However, in situations where marked deviation from random dispersion is obvious, preliminary comparisons with quadrat sampling should be made before plotless sampling is undertaken (Greig-Smith 1964). Equations for utilizing plotless data to obtain vegetational density, frequency, and importance values are provided and comprehensively discussed by Cox (1976).

VI. Acknowledgments

Work leading to the development of the photogrammetic method was supported by Atomic Energy Commission Contract AT-(04-3)-235 during the mid 1960s. Further refinements were made under (1) a grant by the Office of Water Research and Technology, USDI, under the Allotment Program of Public Law 88-379, as amended, and by the University of California, Water Resources Center, as a part of OWRT Project No. A-054-CAL and WRC Project W-491 and (2) Bureau of Land Management, USDI, Contract 08550-CT5-52.

VII. References

Abe, N. 1937. Ecological survey of Iwayama Bay, Palao. *Palao Trop. Biol. Station Stud.* 1, 217–324.

Bray, J. R., and Curtis, J. T. 1957. An ordination of the upland forest communities of southern Wisconsin. *Ecol. Monogr.* 27, 325–49.

Bormann, F. H. 1953. The statistical efficiency of sample plot size and shape in forest ecology. *Ecology* 34, 474–87.

Caplan, R. I., and Boolootian, R. A. 1967. Intertidal ecology of San Nicolas Island. In Philbrick, R. N. (ed.), *Proceedings of the Symposium on the Biology of the California Islands*, pp. 203–17. Santa Barbara Botanic Gardens, Santa Barbara, Calif.

Clapham, A. R. 1932. The form of the observational unit in quantitative ecology. *J. Ecol.* 20, 192–7.

Cochran, W. G. 1963. *Sampling Techniques*, 2nd ed. Wiley, New York. 413 pp.

Connell, J. H. 1970. A predator–prey system in the marine intertidal region. 1. *Balanus glandula* and several predatory species of *Thais*. *Ecol. Monogr.* 40, 49–78.

Cottam, G., and Curtis, J. T. 1956. The use of distance measures in phytosociological sampling. *Ecology* 37, 451–60.

Cox, G. W. 1976. *Laboratory Manual of General Ecology*, 3rd ed. Brown, Dubuque, Iowa, 232 pp.

Dayton, P. K., Robilliard, G. A., Paine, R. T., and Dayton, L. B. 1974. Biological accommodations in the benthic community at McMurdo Sound, Antarctica. *Ecol. Monogr.* 44, 105–28.

Drew, E. A. 1977. A photographic survey down the seaward reef-front of Aldabra Atoll. *Atoll. Res. Bull.* 193, 1–6.

Ernst, J. 1957. Quelques remarques sur la photographie sousmarine en couleurs des communautes algales. *Ecol. Alg. Mar.* 81, 273–6.

Foster, M. S. 1982. The regulation of macroalgal associations in kelp forests. in Srivastava, L. (ed.), *Synthetic and Degradative Processes in Marine Macrophytes*, pp. 185–205. DeGruyter, Berlin.

George, J. D. 1980. Photography as a marine biological research tool. In Price, J. H., Irvine, D. E. G., and Farnham, W. F. (eds.), *The Shore Environment*, vol. 1: *Methods*, pp. 45–115. Systematics Association Special Volume 17(a), Academic Press, London.

Green, R. H. 1979. *Sampling Design and Statistical Methods for Environmental Biologists*. Wiley, New York. 257 pp.

Greig-Smith, P. 1964. *Quantitative Plant Ecology*, 2nd ed. Butterworths, London. 242 pp.

Johnston, C. S., Morrison, I. A., and MacLachlan, K. 1969. A photographic method for recording the underwater distribution of marine benthic organisms. *J. Ecol.* 57, 453–9.

Karlson, R. H., 1980. Alternative competitive strategies in a periodically disturbed habitat. *Bull. Mar. Sci.* 30, 894–900.

Kenchington, R. A. 1978. Visual surveys of large areas of coral reefs. In Stoddart, D. R., and Johannes, R. E. (eds.), *Coral Reefs: Research Methods*, pp. 149–61. UNESCO, Paris.

Kershaw, K. A. 1973. *Quantitative and Dynamic Plant Ecology*, 2nd ed. American Elsevier, New York. 308 pp.

Laxton, J. H., and Stablum, W. J. 1974. Sample design for quantitative estimation of sedentary organisms of coral reefs. *Biol. J. Linn. Soc.* 6, 1–18.

Lewbel, G. S., Wolfson, A., Gerrodette, T., Lippincott, W. H., Wilson, J. L., and Littler, M. M. 1981. Shallow-water benthic communities on California's outer continental shelf. *Mar. Ecol. Progr. Ser.* 4, 159–68.

Littler, M. M. 1968. Development of reef-building crustose coralline algae measurement techniques. In *Western Society of Naturalists 49th Meeting*, Abstracts, p. 17.

Littler, M. M. 1971. Standing stock measurements of crustose coralline algae (Rhodophyta) and other saxicolous organisms. *J. Exp. Mar. Biol. Ecol.* 6, 91–9.

Littler, M. M. 1978. Arithmetic mean regression, product moment correlation coefficient equations for wet and dry weight as a function of percent cover for biota throughout the southern California bight during 1975–77. In Littler, M. M. (ed.), *Intertidal Study of the Southern California Bight*, pp. IV-1.1.B-1 to IV-1.1.B-23. Bureau of Land Management, U.S. Department of the Interior, Washington, D.C.

Littler, M. M. 1980a. Southern California rocky intertidal ecosystems: methods, community structure and variability. In Price, J. H., Irvine, D. E. G., and Farnham, W. F. (eds.), *The Shore Environment*, vol. 2: *Ecosystems*, pp. 565–608. Systematics Association Special Volume 17(b), Academic Press, London.

Littler, M. M. 1980b. Overview of the rocky intertidal systems of southern California. In Power, D. M. (ed.), *The California Islands: Proceedings of a Multidisciplinary Symposium*, pp. 265–301. Santa Barbara Museum of Natural History, Santa Barbara, Calif.

Loya, Y. 1972. Community structure and species diversity of hermatypic corals at Eilat, Red Sea. *Mar. Biol.* 13, 100–23.

Loya, Y. 1978. Plotless and transect methods. In Stoddart, D. R., and Johannes, R. E. (eds.), *Coral Reefs: Research Methods*, pp. 197–217. UNESCO, Paris.

Loya, Y., and Slobodkin, L. B. 1971. The coral reefs of Eilat (Gulf of Eilat, Red Sea). *Symp. Zool. Soc. Lond.* 28, 117–39.

Lundälv, T. 1971. Quantitative studies on rocky-bottom biocoenoses by underwater photogrammetry. *Thalassia Jugoslavica* 7, 201–8.

Manton, S. M. 1935. Ecological surveys of coral reefs. *Sci. Rep. Great Barrier Reef Exped.* 3, 273–312.

Margalef, R. 1968. *Perspectives in Ecological Theory.* University of Chicago Press, Chicago. 111 pp.

Nicholson, N. L., and Cimberg, R. L. 1971. The Santa Barbara oil spills of 1969; a post-spill survey of the rocky intertidal. In Straughan, D. (ed.), *Biological and Oceanographical Survey of the Santa Barbara Channel Oil Spills,* pp 325–401. Allan Hancock Foundation, University of Southern California, Los Angeles.

Ott, B. 1975. "Quantitative Analysis of Community Pattern and Structure on a Coral Reef Bank in Barbados, West Indies." Ph.D. dissertation, McGill University, Montreal. 156 pp.

Paine, R. T. 1974. Intertidal community structure: experimental studies on the relationship between a dominant competitor and its principal predator. *Oecologia* 15, 93–120.

Pechanec. J. F., and Stewart, G. 1940. Sagebrush-grass range sampling studies: size and structure of sampling unit. *J. Amer. Soc. Agron.* 32, 669–82.

Pielou, E. C. 1975. *Ecological Diversity.* Wiley, New York. 165 pp.

Poole, R. W. 1974. *An Introduction to Quantitative Ecology.* McGraw-Hill, New York. 532 pp.

Poore, M. E. D. 1964. Investigation in the plant community. *J. Ecol.* 52(suppl.), 213–26.

Rützler, K. 1978. Photogrammetry of reef environments by helium balloon. In Stoddart, D. R., and Johannes, R. E. (eds.), *Coral Reefs: Research Methods,* pp. 45–52. UNESCO, Paris.

Scheer, G. 1978. Application of phytosociologic methods. In Stoddart, D. R., and Johannes, R. E. (eds.), *Coral Reefs: Research Methods,* pp. 175–96. UNESCO, Paris.

Shannon, C. E., and Weaver, W. 1949. *The Mathematical Theory of Communication.* University of Illinois Press, Urbana. 117 pp.

Smith, R. W. 1976. "Numerical Analysis of Ecological Survey Data." Ph.D. dissertation, University of Southern California, Los Angeles.

Sousa, W. P. 1979. Disturbance in marine intertidal boulder fields: the nonequilibrium maintenance of species diversity. *Ecology* 60, 1225–39.

Southwood, T. R. E. 1966. *Ecological Methods with Particular Reference to the Study of Insect Populations.* Chapman & Hall, London. 391 pp.

Torlegård, A. K. I. 1974. Under-water analytical system. *Photogram. Eng.* 40, 287–93.

Vadas, R. L., and Manzer, F. E. 1971. The use of aerial color photography for studies on rocky intertidal benthic marine algae. In Anson, A. (ed.), *Proceedings of the 3rd Biennial Workshop on Aerial Color Photography in the Plant Sciences and Related Fields,* pp. 255–66. American Society of Photogrammetry, Falls Church, Va.

Vance, R. R. 1979. Effects of grazing by the sea urchin, *Centrostephanus coronatus,* on prey community composition. *Ecology* 60, 537–46.

Widdowson, T. B. 1971. Changes in the intertidal algal flora of the Los Angeles area since the survey by E. Yale Dawson in 1956–1959. *Bull. S. Calif. Acad. Sci.* 70, 2–16.

Williams, C. B. 1964. *Patterns in the Balance of Nature and Related Problems in Quantitative Ecology.* Academic Press, London. 324 pp.

Wilson, D. P. 1974. *Sabellaria* colonies at Duckpool, North Cornwall, with a note for May 1973. *J. Mar. Biol. Assoc. U.K.* 54, 393–436.

Wilson, J. L. 1976. Data synthesis. In *Southern California Baseline Study, Final Report*, vol. 3, rep. 5.2. U.S. Department of the Interior, Bureau of Land Management, Washington, D.C.

9: Remote sensing and mapping

T. BELSHER

Antenne COB (CNEXO), Station Biologique, 29211, Roscoff, France

L. LOUBERSAC AND G. BELBEOCH

COB (CNEXO), B.P. 337, 29273 Brest Cédex, France

CONTENTS

I. Introduction

On marine shorelines with tides of large amplitude, a considerable area may be emergent for several hours per day during the equinoctial spring tides. In those exposed intertidal zones, plant distributions are governed by a complex series of parameters that vary seasonally as well as annually. Hence, these intertidal algal assemblages are accessible for only a few hours each day on an irregular basis. The acquisition of information concerning such populations is a challenge to scientists, industrialists, and planners for whom these data are indispensable. Remote sensing is a technique well suited for gathering information about intertidal plants because it surmounts numerous problems. However, such information must be backed by knowledge acquired on the ground (ground truth).

A. Repetitive inventory of potential algal resources

1. Sparcity of data. Analysis of existing documents shows that they were produced in response to periodic demand for data on a particular well-defined sector of coastline. When one is dealing with populations in rapidly fluctuating environments, this information is soon outdated.

2. Benthic algae as bioindicators. Benthic algae may integrate the characteristics of the water mass in which they exist and thus constitute bioindicators of possible pollution (Belsher 1977). However, it is necessary to distinguish between the variations due to natural factors and those caused by acute or chronic pollution.

3. Economic interest. As examples, populations of *Ascophyllum nodosum* and *Fucus serratus* are of particular interest for their alginate contents and as potential animal food. In the future, other genera, such as *Gelidium, Gracilaria, and Delesseria,* may prove highly useful to the pharmaceutical industry. Therefore, it is necessary to know where and when they can be found in large quantities.

B. Automated qualitative and quantitative cartography of seaweeds

The purposes of automatic digital processing of high-resolution satellite data over several wavelengths are to provide thematic maps of seaweed distributions including data on surface cover, species or plant associations density information, temporal patterns, and health.

The evaluation of potential algal resources, and especially those of the intertidal flora, requires both qualitative and quantitative mapping. It is preferable that this information be obtained in an automated and repetitive way. It is imperative that ground-truth data be obtained by a consistent and accurate method. The primary advantage of this method is the capacity to survey rapidly and synoptically large regions or areas inaccessible from land and sea. The drawbacks linked to the coarseness of scale may be solved by future programs involving very high resolution satellites.

II. Equipment

A. Remote sensing by aerial photography

1. Photographic emulsions (films). Aerial photography is done by using several types of photographic emulsions simultaneously. Films are chosen according to their respective performances. The procedure makes it easier to compare and interpret photographs and facilitates the performance of good-quality cartography.

a. Black and white. Panchromatic, the classic black and white film, has a sensitivity that covers the visible field, between 400 and 700 nm (0.4–0.7 μm). The limit of resolution lies between 50 and 100 lines per millimeter. Infrared film is sensitive in the near infrared, from 700 to 900 nm (0.7–0.9 μm). Resolution limits range from 32 to 80 lines per millimeter.

b. Color. Natural color emulsion provides color information in the visible region. Resolution limits are from 40 to 80 lines per millimeter. False-color infrared film is sensitive to parts of both the visible and the near infrared (from 500 to 900 nm). The information is recorded as a color picture. Data from the near infrared appear primarily as oranges and reds; hence, the name "false color" is applied to that emulsion (resolution limit, 32–63 lines per millimeter).

2. Optical densitometry. The transformation of qualitative images to quantitative information requires measurements of the areas covered by the various identified populations. Since such populations may have overlapping and indistinct limits, a densitometer can be used to measure the extent of a population electronically with great rapidity and precision. By utilizing ground-truth information, it is possible to calibrate an average biomass to correspond with each color.

3. Airplanes

a. Light. A good price–quality ratio can be obtained by using a high-winged plane of the Piper type. A simple device that wedges the camera (Hasselblad is excellent for this operation) vertically along the back door improves the quality and usefulness of the photographs. However, the ideal position is a hatch in the floor of the plane, which can be used regardless of the type of plane (high or low wings). However, lack of stability (due to rolling and pitching) often makes it necessary to correct for vertical error (obliqueness) to obtain a record of sufficient cartographic value.

Oblique photogrammetric sampling by helicopter or light airplane can be done to obtain a rapid overview; a helicopter is preferred, as one can then obtain instant ground truth. A verbally recorded transcript of the observations during the flights must be included (Littler and Littler 1979).

b. Heavy. The use of stable heavy aircraft (B-17 for instance) with a long flight endurance and specially equipped camera (vacuum chamber, large-sized emulsions) permits the acquisition of very good quality photographs that are directly usable for cartographic work. However, the organization of airborne missions with this type of aircraft is complex and costly. In addition, the dependence on atmospheric conditions is great, and the immobilization of the aircraft and crew awaiting possible weather changes can make the cost of the mission still higher.

B. Remote sensing by satellite

Green plants, and seaweeds in particular, reflect light differently in various parts of the spectrum. For instance, reflection is particularly strong in the near-infrared region, whereas absorption is stronger in the red, and this information can be recorded on photographic emulsions using filters. However, the advent of satellites with increased radiometric precision has brought new techniques for recording the intensity of reflected light with imaging radiometers.

Scanning radiometers use a rotating mirror that collects electromagnetic energy reflected by the target and focuses it on a sensitive detector at the chosen wavelength. This detector transforms the electromagnetic energy into an electric current, the variations of which are recorded on magnetic tape. Scanning radiometers may be airborne or satellite borne (e.g., see Table 9–1).

Image processing. The necessary configuration consists of an interactive system of image processing and computer (Fig. 9–1). The software

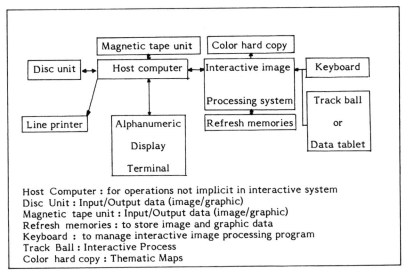

Fig. 9–1. Interactive system of image processing and computer.

includes digital image processing techniques. An example is GIPSY, developed by Belbeoch (1982) at COB (CNEXO), France.

III. Method

A. Acquisition of data: analog methodology

The mapping of intertidal zones requires consideration of the following points:

1. Flight plan. The flight plan is calculated according to the tidal cycle to utilize the same tidal height while maintaining a constant altitude in order to maintain the same scale. Photographs must be obtained as close to perpendicular as possible. They should be taken so as to obtain 60% overlap of pictures along a given flight axis and 40% overlap for a series of pictures (flight axis 1) over another flight axis (flight axis 2).

2. Simultaneous acquisition. This requires simultaneous photography using a diverse range of specialized emulsions (multispectral shooting).

a. Color. The color of an object results from the light transmitted or reflected by partial absorption and from light emitted due to the effect of incident radiation. Such information must be obtained in the field for the visible spectrum by means of color film. This method enables one to make a better distinction between land plants and

Table 9–1. Satellites used for remote sensing

Satellite	Equipment[a]	Characteristics

First generation
Landsat 1, 2, 3
	One multispectral scanner
	  4: 0.5–0.6 μm
	  5: 0.6–0.7 μm
	  6: 0.7–0.8 μm
	  7: 0.8–1.1 μm
	Resolution, 60 × 80 m (not useful for
	  seaweed observation)
	Three return-beam vidicon
	Visible and near infrared
		Orbit, heliosynchronous
		Altitude, 930 km
		Repetitivity, 18 days

Second generation
Landsat D (launch, 1982)
	One thematic mapper
	  1: 0.45–0.52 μm
	  2: 0.52–0.60 μm
	  3: 0.63–0.69 μm
	  4: 0.74–0.90 μm
	  5: 1.55–1.75 μm
	Resolution 30–40 m (relatively well
	  adapted to seaweed observation)
	  6: 10.4–12.6 μm
	Resolution, 120 m
		Equivalent to *Landsat 1, 2, 3*
		Repetitivity, 16 days

Spot I (predicted launch, early 1985)

Two instruments, high resolution, visible
Multispectral mode:
    XS1: 0.50–0.59 μm
    XS2: 0.61–0.68 μm
    XS3: 0.74–0.89 μm
Resolution, 20 m (shows good promise
    for seaweed observation)
Panchromatic mode:
    P: 0.51–0.73 μm
Resolution, 10 m

Orbit heliosynchronous
Altitude, 832 km
Repetitivity, 26 days with vertical views
    and 2.5 days with oblique views
    (latitude 45°)
Stereoscopy

[a] Numbers represent channel : wavelength.

intertidal plants. Distinctions within seaweed populations are difficult or even impossible to make, because differences generally appear in the black or dark brown range.

b. False-color infrared. Green plants, and particularly benthic algae, show up well on emulsions sensitive to the near infrared due to rediffusion of that portion of the spectrum. This rediffusion diminishes in accordance with the thickness of the covering water mass. It is also dependent on the health of the seaweeds. Under optimal conditions (shallow submersion, healthy plants), the algal populations appear on these emulsions in various colors and are easier to interpret than on an infrared black and white emulsion. A comparison of the results obtained with the two types of emulsions is particularly instructive. They enable one to determine the extent of populations and some geomorphic aspects as well (Belsher 1982).

3. Correlations between ground truth and data obtained by remote sensing. Excellent correlations must be obtained between firsthand samples of the populations and the data obtained by remote sensing. This is indispensable for interpretation and for establishing the credibility of the final documents. A reference sector is selected according to its representativeness, diversity, and accessibility. All surveys are carried out in a homogeneous manner after the calculation of a surface of sampling for every population concerned. The sampling areas are delimited by strips of orange cloth (this color being the most conspicuous from a plane). This enables one to calibrate scale, camera angle, and correspondence between the mark and the population in which it is set. Afterward, the variance of the results concerning the occupied surfaces can be calculated.

4. Sketch of photo interpretation. Zones of similar appearance are delimited on the photographs, which are marked and codified on the map of the sector. These homogeneous zones are called "isophenic" (territories of equal significance).

5. Preparation of ground truth. To prepare for the interpretation of the photos, a route is calculated so as to cross several series of isophenic zones in order to ensure their respective homogeneity. There are several sampling techniques, based on the aerial photographic coverage obtained, for determining the number of samples to be taken within each of the isophenic zones.

1. Systematic sampling technique: A transparent plastic sheet regularly squared according to a mesh adapted to the scale of the photo is placed on an isophenic zone. The sampling stations are determined by the points of intersection of the squaring.

2. Selective sampling technique: A number of stations are arbitrarily chosen regardless of the size of the isophenic zone.

3. Random sampling technique: A rhodoid plate the size of the photo is placed on the latter, a number of randomly distributed holes having been drilled into the plate. The holes indicate the location of the stations.

The principle for obtaining a minimal sampling area is simple: One determines the area of the survey required to obtain as efficiently as possible a sample representative of the population (see De Wreede, Chap. 7, and Littler and Littler, Chap. 8).

To obtain the qualitative minimal area, a species–area curve is constructed for every one of the main populations (see Boudouresque and Belsher 1979). Afterward all samples within the same population will have an identical area and will be easily compared.

Several authors (particularly Nedelec, 1979) advocate determining a quantitative minimal area through the area–Kulczynski similarity coefficient method.

6. Final photo interpretation. The photos having been qualitatively and quantitatively interpreted, one can plan a mosaic of photos covering the zone that have cartographic value.

B. Acquisition of data: numerical methodology

Most earth-observation satellites carry spectroradiometers. The spectral signatures obtained by these instruments vary according to the predominating species and their densities (Figs. 9–2 and 9–3). However, these signatures are susceptible to attenuation and distortion by meteorological conditions. Thus, two steps are indispensable for the interpretation of spectral signatures.

1. Spectroradiometric approach: ground truth. The radiometric channel used by the satellite should be tested on the ground (e.g., SPECTER spectroradiometer of the OTC/IO Department of the National French Center of Space Studies, CNES, and the Group for the Development of Remote Sensing, GDTA). Spectroradiometer samples are taken among the populations to be sampled, which enable one to obtain qualitative (the species represented, respective dominances, etc.) as well as quantitative data (density and biomass).

2. Spectroradiometric record by remote sensing. This radiometric channel was also tested (e.g., Daedalus spectroradiometer during Spot mock air campaigns managed by the CNES and the GDTA); our tests were carried out at an average altitude of 7200 m.

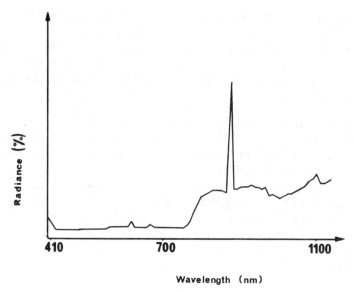

Fig. 9–2. Spectroradiometric record of *Ascophyllum nodosum*.

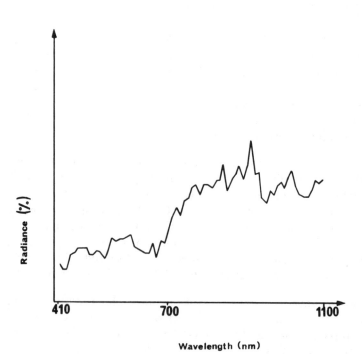

Fig. 9–3. Spectroradiometric record of *Fucus vesiculosus* and *F. serratus*.

C. Treatment of data

1. Statistical treatment. Phytosociological charts are generated, which bring out the analytical parameters (e.g., number of species, diversity index, and reproduction coefficient) and synthetic parameters [qualitative and quantitative dominances of the main systematic groups (Boudouresque 1971)] of each of the surveys. Appropriate coding of data facilitates computer treatment. Classic analyses of similarity bring out regroupings of surveys whose homogeneity is tested. Afterward, with homogeneity taken into account, it is possible to generalize from a reference sector to a wider stretch.

2. Densitometric analysis. This analysis utilizes a video system that reads the data. The signal produced is corrected (by a memory device) for heterogeneity of lighting. Then, the picture is split into 10 or more categories of grays in a linear or logarithmic progression situated between two extremes selected by the operator. A color is assigned to every shade of gray; after transposition into 10 colors, the picture is restored on a cathode screen. Equal-density areas are thus recognized, with as many equal-density categories being recognized as the technical device permits (usually up to 256).

It suffices to adapt a color to every equal density to obtain a document in colored equal densities, a treated picture of the original photograph. One can then retain by utilizing the color analyzer, only the desired color or colors and visualize areas of equal density. Areas of such sites of equal density are obtained with an integrator. The data obtained are then converted to absolute area measures (Rudelle and Demarq 1978).

3. Cartography. Maps are made from the mosaic of photos obtained after careful assembly of the corrected parts. Stereorestitution, through projection, is a technique well known to all cartographers and enables one to obtain a photographic document of cartographic value.

4. Automatic cartography using high-resolution satellite data. A typical methodology based on the numerical processing of high-resolution satellite data (software GIPSY, described by Belbeoch 1982) was applied to intertidal tropical and temperate zones and especially to seaweeds (Loubersac and Belbeoch, in press; Loubersac 1983a,b). The main results show that high-resolution satellite data (*Spot* and *Landsat* thematic mapper) may be very helpful for quick cartography and taking inventories of seaweed canopies in regions where human access is difficult. Furthermore, the methodology shows the potential to discriminate intertidal seaweed canopies as specific major domi-

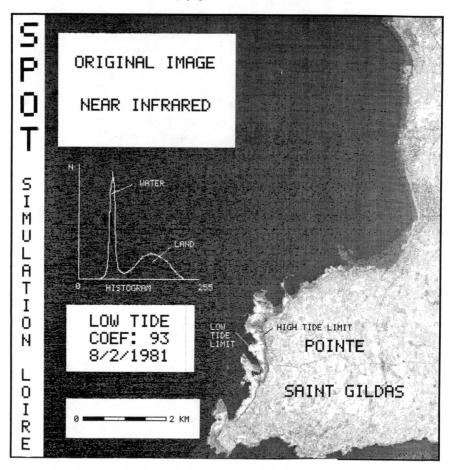

Fig. 9–4. *Spot* simulation, Loire (France): original image, near infrared.

nants and to provide surface information on subtidal areas in shallow water (Figs. 9–4 to 9–6).

IV. Sample data: qualitative and quantitative cartography

A. *Intertidal zone of Plogoff*

During an observational flight conducted on March 1979 at lowest low level, with a 112 tide coefficient, a photographic coverage was made in false-color infrared. Coverage extended from Audierne to the Pointe du Raz (France). This intertidal zone is generally narrow and difficult to approach. The goal in this case was to determine the biogeographical limits of specific populations. The "vertical photog-

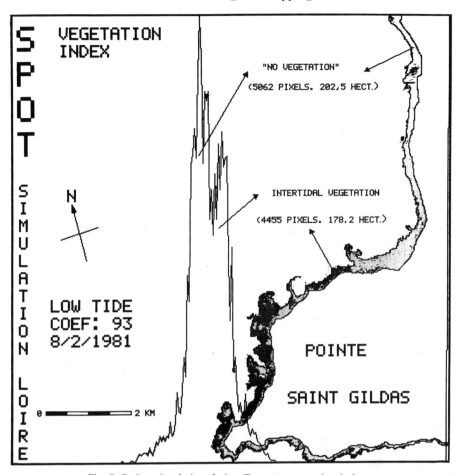

Fig. 9–5. *Spot* simulation, Loire (France): vegetation index.

raphy" criterion necessary for quantitative cartography was not observed.

One zone was selected for this example: from Sillon to the limit of Lervily (Fig. 9–7). The bottom consists of a flat substratum over which about 150 m is colonized by populations of *Fucus serratus* and *Chondrus crispus*, which are above populations of *Laminaria hyperborea* and *Laminaria saccharina*.

B. *Intertidal zone of Penly*

1. Qualitative cartography. One 10-km reference sector, from which the aerial cover of an intertidal zone will be interpreted, is given as an example (Fig. 9–8). After comparison and analysis of the data

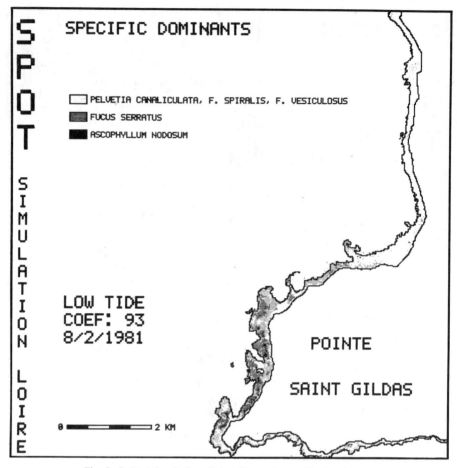

Fig. 9–6. *Spot* simulation, Loire (France): specific dominants.

from the color and infrared documents, three populations were mapped:

1. Ulvales, mainly on a mass of fallen rock
2. Fucales, essentially *Fucus serratus*
3. Heterogeneous populations, with a dominance of endolithic algae

2. Quantitative cartography

a. Surface. The surface of the intertidal reference zone is 7.3 ha, and the biota covers 2.68 ha; the distribution (according to species) is also measured with optical densitometry. For the two most important populations, this sampling yielded an average biomass per square meter, and consequently an assessment of the existing stock could

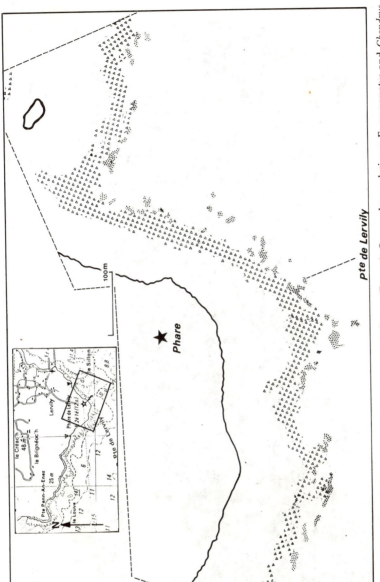

Fig. 9–7. Zone A, intertidal zone of Lervily, south Brittany (France). Seaweed populations: *Fucus serratus* and *Chondrus crispus* (triangles); Laminariales (stippling).

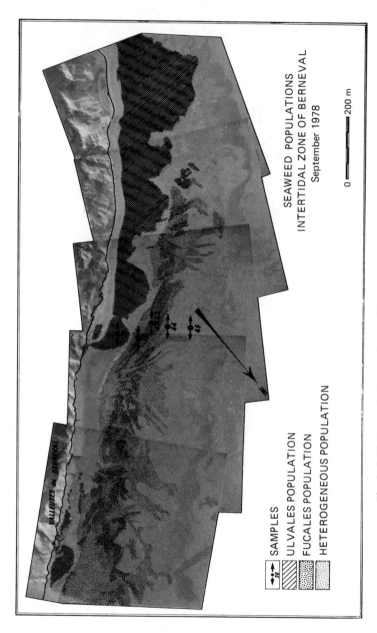

SAMPLES

ULVALES POPULATION

FUCALES POPULATION

HETEROGENEOUS POPULATION

SEAWEED POPULATIONS
INTERTIDAL ZONE OF BERNEVAL
September 1978

0 ⊢━━━━━ 200 m

VALLEUSES de BERNEVAL

Fig. 9–8. Valleuses de Berneval, near Dieppe (France): seaweed populations.

Table 9–2. *Reference and sampling sector of intertidal algal populations,*
1979

Parameter	Population		
	Ulvales	Fucales	Heterogeneous
Surface area (m^2)	9,192	13,965	3,669
Average biomass (g·m^{-2})	39.61	972.42	
Assessment of stock (metric tons)	0.365	13.58	

be proposed (Table 9–2). For the whole of this sector, the available
stocks in 1979 for the Ulvales and Fucales were, respectively, 14
metric tons and 310 tons, and the total stock of algal material was
estimated to be 327 tons.

   b. Assessment of the stock and annual fluctuation. An aerial survey
similar to the first one was made 1 year later. Sampling and treatment
methodology, identical with that just described, were applied to the
reference sector. Repetition of data acquisition and treatment strategy
from year to year permits one to calculate the fluctuations of the
intertidal seaweed cover and thereby of the stock itself. The timing
of the surveys can, of course, be adapted to the abundance cycles of
the algal populations during the year. Thus, between 1979 and 1980
in the reference sector, the area occupied by the *Fucus serratus*
population increased by 8%, but its biomass underwent an approximately three- to fourfold decrease.

   In effect, the variations of the seaweed cover do not necessarily
imply variations of the biomass in the same direction. The emergent
seaweeds, whether young or old, more or less branched, show up on
the infrared emulsion; also, they appear, disappear, and change
from year to year and even from season to season. The responsible
factors are not known. The amplitude of the natural variations of
seaweed populations is not yet well known, and many years of
methodical and experimental study will be necessary to yield an
accurate appraisal. Accordingly, it is necessary to choose the reference
zone and sampling strategy with the greatest possible scientific
accuracy as well as to obtain representative biological ground truth.

C. Automatic and thematic maps

High-resolution satellites provide data from which thematic maps
can be made. These maps supply both qualitative and quantitative
information on seaweeds.

1. Qualitative data. Identification of dominants, even to the species
level, is possible through high-resolution satellite data processing.

This discrimination necessitates a good knowledge concerning the variations of the vegetative state of the different species with time. Qualitative density information is acquired through the use of a density or biomass index.

2. *Quantitative data.* Automatic surface information is provided through the use of a vegetation index. Also, seaweed succession on intertidal surfaces can be determined over time (hour and coefficient of tide). The quantitative assessment of density requires the establishment of a correlation between ground-truth data and digital level ratios.

V. Critical evaluation

A. Analog method

1. *Accuracy.* The classic photographic techniques remain for the moment unmatched in accuracy.

2. *Repeatability.* This is closely connected with meteorological conditions and linked to the best tide coefficients. These conditions often make exact replication unpredictable.

3. *Time.* The assembly of the photographs leading to a mosaic with cartographic value is the work of a specialist and requires a great deal of time and meticulous work.

B. Numerical method

1. *Resolution.* The resolution of *Landsat 1, 2, 3* is not well suited, except on fairly large seaweed zones, to the technique of automated cartography. It is likely that the high resolution of the second-generation earth-observation satellite data (Spot, Landsat TM) could be useful for intertidal seaweed observation and show future potential for assessing surface canopies of subtidal kelps in shallow waters. The mixing of the very high resolution of a panchromatic channel (10 m) with multispectral channels (20 m) would provide very accurate information. *Landsat 4* data are not yet available, but it is likely that they could be useful for seaweed observations.

2. *Accuracy.* Satellite data cannot take the place of aerial photography for high resolution. A comparison between precision of surfaces estimated from aerial photography and *Spot* simulated data show the latter to be inferior by 5%.

3. *Repetitivity.* The high potential repetitivity of *Spot* (2.5) would be an advantage for the study of vegetation changes over short time intervals.

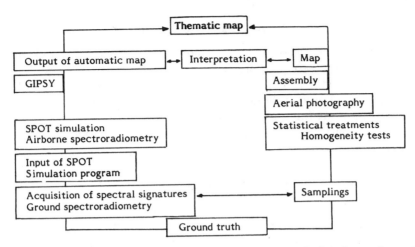

Fig. 9–9. Schematic drawing of a comparison of the two methodologies: analog and numerical.

4. Coverage (scale). Spot data can provide 1/25,000 maps. *Landsat D* will probably allow map scales between 1/50,000 and 1/100,000 maps. *Landsat 1, 2, 3* data allow mapping at a scale of 1/200,000.

5. Time comparison. A comparison, for example, of surface estimation shows that the automatic digital processing of numeric satellite data provides a time gain factor of 25 when compared with classical photo interpretation.

6. Accessibility. One of the great advantages of digital image processing of satellite data is the capability of extrapolation of results from an accessible zone to a nonaccessible zone.

7. Cost. A comparison of cost between aerial photography and high-resolution satellites data acquisition (*Landsat D* and *Spot*) on the basis of prices announced in January 1983 shows that satellite data are 10 times cheaper to acquire than aerial photographs (per square kilometer).

C. Comparison of the two methodologies

Because of their accuracy, the data obtained by the classic photographic techniques enable one to improve the interpretation of satellite information. The two methods, analog and numerical, are therefore complementary. Our present process can be symbolized as shown in Fig. 9–9.

VI. Alternative techniques

The use of an airborne laser at low altitude, which causes the excitation of the accessory pigments (i.e., fluorescence of chlorophyll) has been used by Topinka and Korjeff (in press) for subtidal populations. These authors indicate that this technique is one of the most promising for defining, both qualitatively and quantitatively, the main characteristics of the subtidal phytobenthic populations of large coastal areas.

VII. Conclusions

Because of their morphological characteristics, marine macrophytes are an excellent group of organisms for analysis by remote sensing. The most promising results, given the present state of the technique, are obtained from the intertidal habitat, with the subtidal zone proving more difficult to assess. However, in the next few years, with the use of earth-observation satellites with a high-resolution capacity and with active sensors, mainly synthetic aperture radar, there will be a significant breakthrough in capability. The fluctuations of algal populations can now be determined rapidly and easily and will lay the groundwork for predictive patterns. The rational economic management of exploited and potentially exploitable species will ensue, facilitated by thematic maps derived from global data obtained in an efficiently automated way. Finally, continuous remote sensing monitoring on a global scale should lead to a better understanding of the functional ecology of coastal marine ecosystems.

VIII. Acknowledgment

The data concerning the sectors of Plogoff and Penly came from studies ordered and financed by Electricity of France (EDF) as part of a preliminary environmental impact study on the effluent originating from the proposed installation of thermonuclear power stations on the coastline.

IX. References

Belbeoch, G. 1982. *GIPSY: Logiciel général de traitement d'images numériques.* Rapport CNEXO (SEO), Brest, France. 250 pp.

Belsher, T. 1977. "Analyse de répercussions de pollutions urbaines sur le macrophytobenthes de Méditeranée." Ph.D. dissertation, Aix-Marseille II, France. 287 pp.

Belsher, T. 1982. Télédétection et cartographie du phytobenthos intertidal. *Penn ar Bed,* 108–9; 65–73.

Boudouresque, C. F. 1971. Méthodes d'étude qualitative et quantitative du benthos (en particulier du phytobenthos). *Téthys* 3, 79–104.

Boudouresque, C. F., and Belsher, T. 1979. Le peuplement algal du port de Port Vendres: recherches sur l'aire minimale qualitative. *Cah. Biol. Mar. Roscoff* 3, 259–69.

Littler, M. M., and Littler, D. S. 1979. *Rocky Intertidal Island Survey*, vol. 2: Report 5.0 in Southern California Intertidal Survey Year III, pp. 11-5.0-1 to II-5.0-5. Bureau of Land Management, Los Angeles.

Loubersac, L. 1983a. Remote sensing of temperate and tropical intertidal zones using SPOT simulated data. In *Seventeenth International Symposium on Remote Sensing of Environment*, Ann Arbor, Mich., May 9–13, 1983.

Loubersac, L. 1983b. Applications des données satellitaires haute résolution à l'observation du milieu littoral: le cas des simulations SPOT dans le cadre de la veille écologique des côtes Bretonnes. In *Proceedings of an EARSeL/ ESA Symposium on Remote Sensing Applications for Environmental Studies*, Brussels, Belgium, April 26–29, 1983 (ESA SP-188 July 1983), 79–85.

Loubersac, L., and Belbeoch, G. (in press). Application des données simulées SPOT à la télédétection du milieu intertidal. In *Symposium International de la SIP, Toulouse*, vol. 2,

Nedelec, H. 1979. *Etude Structurale et Problèmes d'Echantillonnage dans une Phytocénose Portuaire*. Mem. D.E.A., Univ. P. et M. Curie, Paris. 38 pp.

Rudelle, J., and Demarcq, Y. 1978. Techniques de pointe à l'Université de Picardie: ARISTIDE, nouveau né dans l'équipement scientifique de la région. *Picardie Inf.*, 31, 27–9.

Topinka, J. A., and Korjeff, W. A. (in press). The characterisation of marine macroalgae by fluorescence signatures.

10: Subtidal techniques

MICHAEL S. FOSTER

Moss Landing Marine Laboratories, P.O. Box 223, Moss Landing, California 95039

THOMAS A. DEAN AND LARRY E. DEYSHER

Kelp Ecology Project, 531 Encinitas Blvd., Encinitas, California 92024

CONTENTS

I. Introduction

The shallow, rocky subtidal zone (0–30 m deep) has more seaweed species and greater macroalgal biomass than any other marine habitat. However, humans are adapted for life on land, not under water; experimental analyses of the intertidal environment are now common, whereas subtidal research is still largely descriptive. The subtidal ecologist must often work under conditions of low visibility and against surge and current forces while relatively weightless. As Barilotti (1980: 15) aptly stated, "Studying subtidal algae with SCUBA is analagous to studying chaparral in a heavy rain, with 200 mph winds that change direction every several seconds, by crawling around in a raincoat, observing through a stovepipe, and breathing through a hose." In addition, cold and depth decompression limit working time, and cumbersome equipment and communication via primitive hand signals limit the efficient use of the time available. Pressure, buoyancy, water motion, corrosive salt water, and fouling pose special problems for equipment use and maintenance. To work efficiently, one must overcome these difficulties or at least reduce them. There have been considerable improvements in underwater equipment and techniques since Kitching et al. (1934) first descended with a diving helmet and bathing suit into *Laminaria* beds for 15-min observations. Numerous investigators now regularly use SCUBA for collecting, and a variety of methods are available for in situ descriptive and experimental work with subtidal macroalgae.

In this chapter, we focus on methods unique to research in the subtidal zone using SCUBA. The difficulty with these methods, and perhaps one reason that few standard techniques have been developed in subtidal research, is that the same question might be investigated in entirely different ways in a tropical lagoon and along a wave-exposed Antarctic coast. The methods we describe were generally developed for use in temperate areas, but most could be modified for other conditions. We encourage further thought about all of the methods described here, since most can probably be improved, and new and better ones developed.

[200]

II. The diver

To work efficiently, research diving requires that the diving itself be automatic. The diver must be well trained and comfortable under rigorous field conditions. Poor training, inexperience, health problems, and improper or malfunctioning equipment will at least inhibit performance and can be very dangerous. Details of training, basic equipment, and diving procedures can be found in the excellent text for research divers by Miller (1979). In addition to the work itself, one should also be aware that transport to the research location can affect diver performance, and every effort should be made to ensure that all divers arrive rested, warm, and alert. As much planning as possible should be done in the relative comfort of a shore laboratory, because decision making is difficult on moving ships, especially after one has spent some time under water.

Perhaps most important to diver performance is suitable thermal protection. Wet suits are most widely used and come in a variety of thicknesses and styles. However, cold is still a problem with the best of these at water temperatures below 15°C (Miller 1979), especially during long bottom times and periods of reduced activity. Variable-volume dry suits made of wet-suit material are warmer and have become widely available.

Miller (1979) reviews more complex diving equipment not discussed in this chapter, including surface air supplies, electronic communication, and hot-water diving suits. This equipment is most efficient for work in a limited area, and the latter can greatly increase bottom time in very cold water. SCUBA is used by the majority of research divers because it is less expensive than other methods and allows more mobility. Special equipment necessary for deep diving and the use of underwater habitats and lockout submersibles that greatly increase bottom time with only a single decompression are reviewed by High et al. (1973) and Earle (Chap. 11).

III. Basic techniques

A. *General considerations*

Basic subtidal research techniques applicable to a variety of organisms and habitats are described by Finnish IBP-PM Group (1969), Holme and McIntyre (1971), Woods and Lythgoe (1971), Kinne and Burcheim (1973), Drew et al. (1976), and Miller (1979). We shall discuss some of these as they apply to macroalgal studies, as well as others from the literature and our personal experience. Before adopting any technique, one should carefully evaluate its relevance to the questions being asked as well as its applicability to the habitat or alga being investigated.

B. Underwater tools

A variety of devices can be used to record information under water. Small amounts of information can be recorded on small, roughened plastic slates with a pencil attached to the slate on a long piece of surgical tubing so it will not float away. Foster (1976) gives plans for a slate with a frosted Mylar writing surface that can be attached to the arm, leaving both of the diver's hands free when he or she is not writing. A larger version is described by Ogden (1977). Plastic can be silk-screened to make formatted data sheets, or clear Mylar sheets can be used over a formatted slate. The legibility of a diver's writing is usually poor, especially when he or she is wearing gloves in cold water. Formatting before the dive reduces the amount of underwater writing and also helps to ensure that all required information is collected during the dive. Information can also be recorded with pencil on underwater paper (Nalgene PolyPaper, Markson Scientific) attached to a slate with clips. This paper can be formatted in a copy machine, but care must be taken to avoid damage to the machine from overheating (see technical details supplied with paper). Underwater tape recorders can also be used to record information but require a modified regulator mouthpiece or a full-face mask to make intelligible recordings. A complete recording unit can be purchased (Sound Wave Systems) or constructed from component parts and an underwater housing (Ikelite Underwater Systems). Before purchasing or constructing a recorder, one should evaluate its reliability and ease of use under anticipated field conditions.

Underwater objects can be magnified with the relatively simple underwater hand lens described by Pratt (1973). Neushul (1972) describes a portable underwater microscope. However, it is expensive and difficult to use in all but very calm water.

Careful marking of subtidal research areas is essential, particularly if visibility is low. There are numerous methods and devices for doing this, most of which are similar to those described for intertidal zone use by De Wreede (Chap. 7). Fence or swing anchors are simple to install in soft bottoms and can be used to anchor buoys or attach equipment (Foster 1980). If the substratum is soft rock or coral, stakes (iron bars, large nails, etc.) can be driven into the bottom. For hard substrata, we commonly use case-hardened nails (concrete nails) driven through short pieces of colored plastic bicycle handle bar tape. Fouling organisms can be easily removed from the plastic and it is durable. If cracks are present, pitons or large, square, galvanized boat nails provide a strong marker and anchor. Nails and pitons that last for a number of years can be driven into the substratum with 3-lb or greater short-handled sledgehammers (the force of regular hammers is significantly reduced by water drag). Markers and fasteners can also be driven into the substratum with a hammer-in or explosive underwater tool (Ramset Fastening Systems). One must exercise extreme care when using the latter, especially on hard basalt or granite substrata. An alternative to forcing a fastener is to drill a hole into the bottom. Epoxy (Brolite Z spar epoxy putty, no. A-788, Koppers Co., is especially good under water), concrete, plastic screw anchors driven into the hole, or self-anchoring expansion bolts (available from Ramset and in most hardware stores) can be used for attachment.

Holes can be made by hand with a hammer and star drill or hammer tools used with masonry anchors (available from Ramset) or, preferably, with an air-driven impact hammer, drill, or hammer drill (Black and Decker Manufacturing Co. and various other manufacturers). All of the air tools operate at ~100 psi (69 Pa) and can be run from the first stage of most single-hose SCUBA regulators attached to a SCUBA tank. Star drills can be welded to the ends of standard impact hammer tools, and masonry bits are used with the drills. Air drills require downward force, so the diver or tool must be heavily weighted. All air-driven tools should be flushed with oil or similar lubricant as soon as possible after use and stored in diesel fuel; they are not made for underwater use and rust rapidly. The tools should be cleaned of excess oil, etc., before use to avoid contamination of field sites. Similar hydraulic drills and hammers (Stanly Hydraulic Tools) are quiet and more powerful and do not release clouds of bubbles that obscure visibility. They are expensive and require surface support but should be considered if extensive bottom manipulations are necessary. Other types of tools and equipment developed for commerical diving can be useful for subtidal research. These are frequently advertised in the journals *Sea Technology* and *Journal of the Marine Technology Society*, and in commercial diving supply catalogs (e.g., M and E Marine Supply).

Fiberglass measuring tapes (15–100 m long; Forestry Suppliers) are excellent for underwater use, as are collapsible aluminum quadrats. The latter do not float and can be folded for transport and unfolded around large plants and other obstacles in the sampling area. Small equipment and large specimens are easy to carry in canvas diving bags with nylon mesh bottoms available in most diving stores. Objects tend to snag in bags made entirely of nylon mesh, and small objects are easily lost.

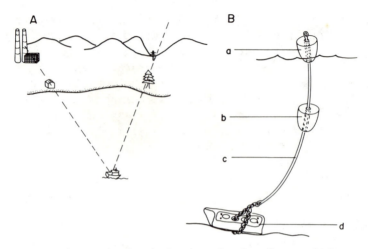

Fig. 10–1. (A) Diagram showing site location using shore lineups. (B) Buoy system. Surface (a) and subsurface (b) torpedo-shaped buoys reduce fouling by drift plants and prevent line from abrading on the bottom at low tide. Buoy line (c) has upper eye for boat mooring and lower eye for attachment to chain around engine block anchor (d).

C. Site location

Relocating a permanent subtidal research site can be both time-consuming and frustrating. We recommend a combination of triangulation or line-of-sight positioning from shore landmarks, surface buoys, and underwater markers. If the site is close to shore, lineups provide the simplest and most accurate shore landmark system. With this system, pairs of objects on shore line up to define a direction. Two pairs in different directions define a point over the site, and the system is accurate to ±5 m (much greater than compass bearings alone) if the objects in a pair are widely separated and the angle between pairs is about 90° (Fig. 10–1A). Compass bearings should be taken to roughly locate the lineup pairs, and a sketch must be made for future location. To locate the position of a study site on a nautical chart, bearings on three objects on the chart near the horizon can be taken with a sextant. Objects not near the horizon can be used to relocate a position, but not on a chart (the farther above the horizon the objects are, the greater will be the error toward shore). Electronic positioning is an alternative and necessary if the shore is distant or obscured by fog. Electronic systems using shore transponders (e.g., Motorola Mini-Ranger, Motorola Inc.) are accurate to ±1 m. Although expensive, they can be rented and are excellent if numerous sites must be relocated.

Surface buoys are excellent for relocation *if* they stay in place. Their composition and anchoring will vary according to site. D. Coon (personal communication) reports good success with taut-moored spar buoys. The system shown in Fig. 10–1B has worked well for us in large swells. The subsurface buoy keeps the line from abrading on the bottom at low tide, and the torpedo shape reduces entanglement by drift kelp. In areas where there is a great deal of boat traffic, lines should be weighted about 3 m below the surface to keep buoyant line from floating to the surface and entangling boat propellers during low tide. Large buoys and chain or steel cable must be used if theft is a problem.

Sites can be marked under water with nails or other materials, as described earlier. It is advisable to place a line 10–30 m long on the bottom parallel to shore close to the site but not close enough to cause abrasion of study areas. If all shore and surface relocation systems fail, divers can usually position themselves seaward but in the vicinity of the site and then swim toward shore until the line is located. Leaded nylon line (available in most marine supply stores) is good for this purpose as well as for marking transects and, since it does not float, reduces the danger of diver entanglement. The line should be laid and attached in sections so only a portion will be lost if hooked by an anchor. Underwater sonar beacons or subsurface acoustic release buoys can also be placed in the site. The former can be relocated by means of ship or diver-held directional receivers (EFCOM Communication Systems), and the latter brought to the surface with a sonar signal from a ship (Helle Engineering Inc.).

One should be aware that some metals, particularly copper and brass, are toxic. We are unaware of any evidence that small amounts of materials made from these metals are toxic in the subtidal zone (other than surface effects). However, the potential exists and, along with possible corrosion problems (particularly when different metals are used in close proximity), should be carefully considered before use.

D. Collecting

General collecting procedures applicable under water are discussed by Tsuda and Abbott (Chap. 4). Mesh bags are better than plastic for large specimens, because water currents will not flush out previously collected plants. Small specimens are easily carried away and difficult to grasp under water; their collection can be aided with a hand-operated suction sampler described by Tanner et al. (1977). If a suction sampler is not available, small plants must be placed in plastic bags or small vials to prevent loss. If vials are used, they must be perforated to equalize pressure and preferably have attached lids.

E. Descriptive surveys

If visibility permits, large-scale surveys of the more common large algae can be accomplished from the surface with a view box or by towing a diver (see Tsuda and Abbott, Chap. 4). Methods for determining sample size and location when a defined area is being described are discussed by De Wreede (Chap. 7), Foster and Sousa (Chap. 13), and Littler and Littler (Chap. 8); they depend to a large extent on the questions being addressed. Plants must be collected for biomass estimates, and this can be difficult in the subtidal zone because plants dislodged by scraping may be dispersed by surge. This problem can be eliminated by the use of an enclosed "box quadrat" and suction sampler (Jones 1971) or a highly portable airlift sampler operated entirely under water (Chess 1978). Both devices can lift samples to a support ship or into a mesh bag that can be changed by a diver.

If nondestructive methods are adequate or required, then counts or percentage of cover estimates within quadrats or along transects are appropriate. If vegetation layering is minimal, cover can be estimated from photographs of known areas (Littler and Littler, Chap. 8). With layering, photographs miss understory plants and thus lead to an underestimation of true cover. Moreover, overstory plants cannot be moved aside for successive photographs as they can be in the intertidal zone. Some form of a point-quadrat sampling device is most useful under these circumstances (see Foster and Sousa, Chap. 13). With all of these nondestructive techniques, we have found a combination of writing slate and small vial holder (for collecting vouchers of unknown species in the field) to be particularly useful (Fig. 10–2).

F. Relocating plants

In situ studies of growth, reproduction, and survivorship often require repeated relocation of the same plants. The great range of size and form in the algae necessitates a variety of tagging methods, but most have been developed for kelps. Like site relocation, plant relocation is time-consuming. We prefer double tags to enhance relocation, one on the bottom next to the plant (concrete nails with numbered plastic tags, markers cemented to the substratum, etc.) and one on the plant. Double marking increases the chances of finding a plant if one tag is lost and is necessary in survivorship studies to make certain that the plant and not just the plant tag is lost. Plastic-coated wire and electrical cable ties, tie wraps, or clamps

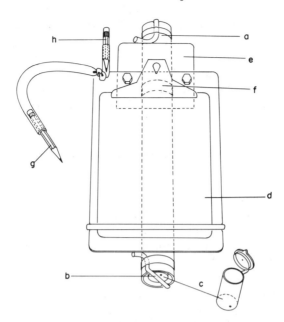

Fig. 10–2. General-purpose writing slate with collecting vials. (a) PVC tube; (b) surgical tubing to hold collecting vials in tube (over each end); (c) snap-top plastic collecting vial with small holes for pressure equalization; (d) aluminum clipboard with clip at top and surgical tubing at bottom to hold underwater paper; (e) plastic backing bolted to clipboard and attached to tube with stainless steel hose clamp (f); (g) pencil held in surgical tubing; (h) spare pencil. The PVC tube can vary in size and length to accommodate different sizes and numbers of vials. Lead weights are attached to hose clamp for negative buoyancy.

(available in a variety of sizes from electrical supply stores) fitted with appropriate numbered markers (preferably smooth plastic for long life, to inhibit fouling, and for easy removal of organisms that do foul) are excellent for marking individual kelps with robust stipes but may abrade the stipes if water motion is high. Rubber bands could also be used to avoid abrasion but might not last as long as plastic. Any of these materials can also be attached around haptera if present. One can map the position of large plants such as *Macrocystis* effectively by noting their perpendicular distance from a permanent line on the bottom marked at 5-m intervals. Particular patterns of small holes can be punched in blades for identification as well as growth studies (Mann 1972). Nicholson (1970) marked *Nereocystis luetkeana* with ink tattoos; this method should work with most large blades. Small plants are more difficult to tag without damage. Johansen and Austin (1970) tagged articulated coralline fronds with plastic-coated wire and plastic labels, and Barilotti and Silverthorne

(1972) tagged *Gelidium robustum* fronds with numbered plastic bird bands. Some tag colors may attract fishes, and tag attraction should be tested before extensive use. To our knowledge, no one has successfully marked small, filamentous species so that they can be relocated in the field (for laboratory–field techniques, see Sec. V.D). At present, these must be relocated using distance and/or angles from substratum markers.

IV. Physical–chemical measurements

A. General considerations

Physical and chemical measurements are collected by the subtidal phycologist in order to support biological data (Wheeler and Neushul 1981). The variate to be measured and the temporal and spatial scale over which measurements are made can vary widely depending on the question at hand. For example, several instantaneous estimates of light attenuation may be sufficient to determine relative amounts of radiant energy at two locations, whereas absolute values of irradiance taken continuously over a period of several weeks may be needed to relate algal growth to irradiance. Therefore, it is important to determine first the appropriate variate to be measured and the corresponding temporal and spatial scales. Unfortunately, the ideal measurements are often difficult to obtain because instrumentation is not available or is costly.

We do not intend to provide an exhaustive list of equipment suppliers here. A more complete listing of instrument manufacturers is given in the "Guide to Scientific Instruments" published yearly as a supplement to *Science* (American Association for the Advancement of Science, Washington, D.C.) and in "Buyers Guide/Directory" published yearly by *Sea Technology* (Compass Publications).

B. Light

Lüning (1981) reviewed light relative to seaweed biology, and in Chap. 2 Ramus thoroughly discusses light, light measurement, and light-measuring devices. We only briefly discuss a few techniques and devices particularly relevant to subtidal research.

There are several ways of measuring light in submarine environments. Among the simplest is to determine the relative levels of attenuation of light by using a Secchi disk (30 cm in diameter, white), lowered from a boat to determine vertical transmission of surface light. Several empirical relationships have been developed to calculate attenuation coefficients (the rate at which irradiance declines with depth) from Secchi disk readings so that Secchi disk data can be used to estimate relative irradiance changes with depth (Holmes 1970; Weinberg 1976; Nyquist 1979; Walker 1980).

A more accurate way of measuring radiant energy available to benthic algae is by using a quantum irradiance meter (Jerlov 1968). This instrument measures the total number of quanta over a given range of wavelengths and is generally appropriate for use in algal studies since photosynthesis is a quantum response (Tyler 1973). Quantum sensors generally record light in the photosynthetically active range of 350 or 400 to 700 nm. In some instances it may be necessary to use wavelength sensors restricted to a more narrow band. For example, Lüning and Dring (1972) and Lüning and Neushul (1978) showed that gametogenesis in Laminariales was mediated by blue light. Therefore, it is most appropriate to measure the number of quanta available in blue wavelengths to study this process. In addition to sensing the appropriate wavelengths, the light-collecting surface over the sensor should collect light from the same directions as the algae do in the field (see Ramus, Chap. 2). Collecting surfaces must be frequently cleaned of fouling organisms and debris if they are left in situ for long periods.

Instrumentation for measuring underwater irradiance can be purchased (e.g., Li-Cor; Biospherical) or built according to published specifications (Jerlov and Nygard 1969; Booth 1976; Lüning and Dring 1979). Several other instruments (Woodward and Yaqub 1979; Fitter et al. 1980) designed for terrestrial environments can be modified for underwater use.

C. Temperature

Temperature is important to benthic algae because it can affect survival, photosynthesis and respiratory rates, phenology or reproductive timing, and potential reproductive output. As with irradiance, one must first determine the spatial and temporal scales over which to measure and whether mean, median, minimum, or maximum temperature over a given period is the most important measure. Once these have been decided, a number of devices exist to measure temperature in situ. These range from handheld thermometers to thermistors with digital readouts. Many of these instruments are designed for underwater use and are commercially available (e.g., Ryan Instruments Co.) or are easily assembled (e.g., Woodward and Yaqub 1979). A monolithic temperature transducer (National Semiconductor Corp.) has been used successfully in instruments developed and used by ECOsystems Management Co. for subtidal temperature monitoring.

D. *Water motion*

Wave surge directly affects benthic algae by removing parts of plants or dislodging the entire plant from the substratum (Rosenthal et al. 1974; Foster 1982a) and by abrading plants against nearby objects. In addition, surge can affect the behavioral patterns of grazers (Lissner 1980) and influence physical parameters such as sediment resuspension. Currents can have similar effects and, like surge, also affect dispersal and nutrient availability. Denny (Chap. 1) discusses water motion in the intertidal zone. Many of the methods he describes are applicable to subtidal habitats, and additional subtidal methods are discussed here.

Total water motion has been determined in subtidal environments by measuring the dissolution rate of artificial materials over time (see Denny, Chap. 1, for details). However, data obtained by this method may be difficult to interpret, because the rate of weight loss is a function of other parameters (e.g., suspended solids) and may be a measure of both water motion and scour (Craik 1980). R. Day and M. S. Foster (unpublished) have used plaster hemispheres glued to plywood frames to measure substratum abrasion by *Ecklonia radiata*. Boards with hemispheres were fastened under *E. radiata* plants and weight loss compared with control boards in clearings affected only by water-motion-suspended solids.

Subtidal water motion can also be measured by either Savonious rotor or electromagnetic current meters. The former suffer from susceptibility to fouling and relatively slow response time. Response time is usually not critical in subtidal water motion measurements, because water motion "events" last longer than in the intertidal zone. However, instrument response times should be evaluated to ensure that they are appropriate for the water motion of interest (see Denny, Chap. 1, for further details). Electromagnetic devices are more suitable; they have extremely fast response times and no moving parts, and an additional advantage is that the data can be resolved to indicate water movement due to waves or currents. These instruments are generally deployed to detect only horizontal water movement. They are available from Marsh–McBirney and Bendix Corporation.

Wave height and period can be estimated by eye on a routine basis (Gerard 1976) for a relative measure of water motion from waves. In shallow water, where most seaweeds are found, surge velocity on the bottom can be calculated from various wave parameters. Appropriate equations can be found in Neumann and Pierson (1966) and most other oceanography texts. In addition to some of the simple methods discussed by Denny (Chap. 1), Charters et al. (1969) and

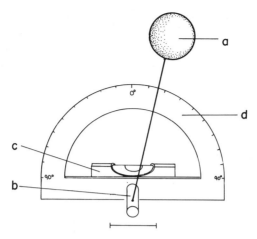

Fig. 10–3. Diver-held device for measuring surge. (a) Buoyant ball attached to string; (b) plastic rod; (c) line level; (d) large plastic protractor to measure deflection of string. If sphere is a Ping-Pong ball, $s = 0.85(\tan \theta)^{1/2}$ (s is water speed in meters per second, and θ is the angle of string deflection). Ping-Pong balls implode at a depth of ~10 m, and different materials (wood, etc.) and ball diameters require different constants. Barilotti (1980) found that turbulence around the ball reduces accuracy at speeds of >1 m·s^{-1}. Device developed by A. Charters, M. Foster, and D. Barilotti. Scale = 10 cm.

Barilotti (1980) used diver-held instruments for instantaneous measurements of wave surge, the former a tensiometer with a spherical sensor and the latter a simple device for measuring deflection of a tethered sphere (Fig. 10–3).

Pressure-sensing devices and wave riders can also be used to measure waves. The former can record both wave height and period, and the data can be used to determine orbital velocities. A variety of such instruments are made commercially (e.g., Marsh–McBirney and the Danish Hydraulic Institute). Wave rider instruments (Louis Adamo, Inc.) are spherical buoys deployed at the surface and attached to a mooring. Real-time data are sent by radio signal to a remote shore station.

Currents can be measured by either Eulerian or Lagrangian techniques. The Lagrangian method follows the course of a particular water mass over time (e.g., using drogues). This method has been used to determine movements of vegetation drifting on the bottom with tagged plants, mushroom-shaped bottom drifters (Kahl Scientific Instruments), and perforated Ping-Pong balls (Josselyn et al. 1983). Local surface drift of floating kelp has been measured using marked oranges (J. Yaninek, personal communication). Fixed (Neushul et al. 1976; Josselyn et al. 1983) and buoyed collecting devices (Zieman et al. 1979) can be used to determine drift composition and biomass. Drift plants may be important agents of long-range dispersal (Deysher and Norton 1982), and further investigations of drift dispersal are certainly needed. Eulerian methods record net movement (generally horizontal) across a stationary point and are more appropriate for measuring water motion past an attached alga. Motion is measured by electromagnetic meters (as indicated previously) or by several types of mechanical devices that utilize various propeller arrangements. Among the most successfully used instruments in nearshore ocean environments are vector-measuring current meters as developed by Weller (1978). These filter out water motion due to waves and thus record only the net water flow due to currents (down to 2 cm $\cdot$ s^{-1}). They are available through Deep Ocean Systems and EG&G.

E. Sediments

Sedimentation can bury and abrade benthic algae, and sediments suspended in the water column influence the amount of light reaching the bottom. The relative amount of sediment abrasion has been measured by observing the weight loss of a plaster ball tethered to the bottom (Craik 1980). However, as mentioned earlier, a method such as this may measure both water motion and abrasion. It may also be unreasonable to assume that the rates at which plant materials and plaster are abraded are equal or even proportional to one another.

Measuring burial by sediments is easily accomplished with larger algae by observing sediment depth on a ruled stake. However, sedimentation may be most important to small algae or to small life stages. For example, Devinny and Volse (1978) showed that gametophytes of *Macrocystis pyrifera* were killed when buried by a thin film of sediment. Net accumulation of sediment on small areas of rocks or artificial substrata can be estimated by vacuuming the surface and weighing the sediment. However, in most benthic environments, substrata are probably buried and unburied by thin films of sediments on a daily or more frequent basis, and documenting important burial episodes may be difficult.

The most widely used measure of sedimentation is the amount of sediment that accumulates in sediment tubes or "traps." These traps measure the vertical flux of suspended sediments and may give rough estimates of potential burial and scour. Gardner (1980a,b) reviewed sediment trap technology and showed that tubular traps with a height-to-diameter ratio of between 2 and 3 gave the best representation of vertical sediment flux.

F. Nutrients

Both macronutrients (nitrogen and phosphorus) and micronutrients (iron, manganese, etc.) are essential for the growth and maintenance of algae. The availability of nutrients to benthic algae is generally determined by collecting water samples and analyzing them according to standard techniques (see Wheeler, Chap. 3). Collection is done with pumps or bottle systems operated from a boat or by diver-held bottles or pumps. Diver collection ensures precise sample location.

Sampling for nutrients is generally carried out on a weekly (or less frequent) basis. However, in many marine environments, the availability of nutrients is determined by oceanographic events such as internal waves and tides that occur on smaller time scales. Therefore, relatively infrequent water sampling may provide a poor estimate of nutrient availability. Unfortunately, there are no methods for determining nutrient concentration on either an integrated or a continuous basis. Johnson and Petty (1982) developed a system for phosphate analysis that can sample at a rate of over 90 times per minute. It can also be used for nitrates and ammonia (M. Neushul, personal communication) and may be adaptable for underwater use.

In some cases, it may be possible to estimate nutrients on the basis of continuous temperature profiles. Nutrient concentrations and temperature correlations can be developed, and concentrations can be predicted on the basis of temperature (Jackson 1977; Gerard 1982). These techniques are especially applicable when changes in nitrogen and phosphorus are associated with upwelling.

G. Substrata

The substratum composition of a given area can have profound effects on the distribution and abundance of algae. Most species of benthic algae are restricted to hard substrata, so the relative abundance of soft and hard bottom can be of great importance. Moreover, the amount of soft sediment in an area may indirectly affect plants because sediments are a source of suspended particles that can reduce light transmission and cause abrasion or scouring. The types of hard substrata present may also be important. Small cobbles, for example, are subject to higher turnover rates than larger boulders or bedrock, and algae attached to smaller cobbles may have higher mortality (Sousa 1979). Where continuous hard substrata exist, rock type (hardness) affects algal mortality during storms (Foster 1982a), and substratum rugosity (Foster 1975b) and angle may also be important. Different slopes of hard substrata can offer different advantages to algae or their competitors and predators.

The distribution of substrata over large underwater areas may best be determined with side-scanning sonar (Tucker and Stubbs 1961), as commonly done by geologists and marine surveyors (e.g., Stride 1961). On a smaller scale, the percentages of sand, cobble, and other material can be determined with methods similar to those described in Sec. III.E on surveys (e.g., percentage of cover estimates and line transects). Other substratum characteristics (slope, relief, surface area, rugosity, etc.) can be estimated visually or, preferably, quantified. Luckhurst and Luckhurst (1978) provided an estimate of rugosity by laying a length of chain over some distance and comparing the actual surface distance with the linear distance. The slope of a particular surface can be measured with an inclinometer, with a compass and protractor device as used in terrestrial studies (Fovargue and Perino 1979), or with a protractor, string, and small float.

H. Data-logging devices

Many of the instruments described for the measurement of irradiance, temperature, currents, and waves produce a large amount of data in a relatively short time. Various means are available for recording and storing these data. In some cases, telemetry may be used to send data to a shore station (e.g., Harger 1979; Lüning and Dring 1979), or data can be stored in situ for later retrieval. In situ methods include storing a printout from a strip chart recorder (Ryan Instruments) or printed paper tape (Li-Cor). The reading from a mechanical counter may also be recorded on movie film (General Oceanics). All of these methods require transfer of the data to a computer, a labor-intensive and time-consuming process. This problem has been partially resolved by the use of magnetic tape recorders (e.g., Datel Intersil). The tapes can be retrieved and data transferred to computer-compatible tapes that can be input directly into the computer. Information stored by some electronic data loggers (A.D. Data Systems) can be input directly into the computer.

We have successfully used data loggers to record temperature and irradiance data on an erasable programmable read-only memory (EPROM) integrated circuit (designed and manufactured by ECO-systems Management). These have proved to be more reliable than electromechanical data loggers and provide even simpler data transfer to a computer than can be provided by cassette tapes.

V. In situ experiments

A. General considerations

The emphasis in ecological studies during the past decade has shifted from descriptive investigations, in which the effects of various environmental factors are inferred from observed correlations, to manipulative experiments, in which important environmental variables can be selectively controlled. The control of selected variables in these field experiments is usually not as precise as in laboratory experiments due to the multitude of factors that must be considered and the often unpredictable nature of the marine environment. However, these experiments, with their innovative and increasingly precise control of environmental factors, have substantially advanced our knowledge of subtidal algal ecology.

B. *Productivity*

In situ productivity of subtidal macrophytes has been measured from oxygen evolution and uptake in large chambers (Hatcher 1977) by the use of plastic bags to incubate parts of plants in ^{14}C (Towle and Pearse 1973) and with standard light–dark bottle techniques (Healey 1972; Ramus et al. 1976; Hoffman and Dawes 1980; Smith 1981). Heine (1983) successfully adapted intertidal productivity methods described by Littler and Murray (1974) for use in the subtidal zone. Light and dark bottles were incubated in situ with water motion provided by magnetic stirrers driven with compressed air from a SCUBA tank.

Growth measurements can be converted to biomass changes and used as a measure of productivity. This has been done for kelps (Mann 1972; Gerard 1976; Mann and Kirkman 1981) and could be done for smaller plants where in situ growth can be measured by comparing successive photographs of the same frond (Johansen and Austin 1970; Barilotti and Silverthorne 1972). For further discussion of productivity methods, see Littler and Arnold (Chap. 17).

C. *"Seeding"*

The direct seeding of spores or other propagules offers great potential in both autecological and community studies investigating the factors that influence the distribution and density of algal populations. Paine (1979) inoculated intertidal quadrats with *Postelsia palmaeformis* zoospores released from mature sporophytes. These sporophytes were removed from a nearby population and held at the experimental site by stainless steel mesh bolted to the substratum. Reed (1981) inoculated subtidal areas with red algal spores by nailing mesh bags full of reproductive blades to the substratum. This technique is applicable to almost all macroalgae.

North (1976) used both zoospores and developing embryos to seed *Macrocystis pyrifera* in subtidal areas. Concentrated zoospore solutions were released directly on the bottom and under large plastic sheets spread over the bottom. The latter were designed to minimize drift of the zoospores away from the inoculation site. In the more successful embryo seedings, gametophytes were first cultured on plastic sheets in running seawater. Newly fertilized embryos were then removed from the plastic sheets and concentrated, and the concentrated suspension piped by a hose from a boat to the bottom.

Deysher and Norton (1982) inoculated intertidal and subtidal quadrats with embryos of *Sargassum muticum* (Yendo) Fensh. This species offers a number of advantages for this type of inoculation experiment; individual plants contain both male and female receptacles, ensuring that eggs will be fertilized immediately after release from the receptacle, and release of eggs is synchronous within a population, ensuring an adequate supply of embryos for experiments. The inoculation procedure consisted of removing fronds with new embryos still attached and holding these fronds in contact with the substratum for 24 h. The fronds were held on the substratum under pieces of monofilament nylon netting attached to nails driven into sandstone.

D. Outplanting

Techniques for culturing various algal life history stages in the laboratory and then outplanting them to the field are the basis for a number of large aquaculture industries, for example, nori (*Porphyra*) and kombu (*Laminaria*). These techniques have been extensively covered in the literature (Tseng 1981), and only a brief overview is presented here.

Laboratory outplant techniques are especially useful for microscopic algal life history stages (e.g., *Laminaria* gametophytes) that are presently impossible to manipulate entirely in the field. Hsiao and Druehl (1973) outplanted gametophytes of *Laminaria saccharina* grown on glass slides. The slides were periodically returned to the laboratory for determination of gametophyte survival and sporophyte recruitment. An interesting technique in this study was the staining of the outplanted gametophytes with a fluorescent dye (Cole 1964), which made it possible to discriminate between outplanted and naturally settled gametophytes. The technique, however, requires the use of a fluorescence microscope.

L. E. Deysher and T. A. Dean (unpublished data) have developed techniques for outplanting gametophytes of the giant kelp *Macrocystis pyrifera*. In these studies, nylon rope substrata inoculated with *Macrocystis* zoospores are outplanted on plates at approximately monthly intervals to a number of field stations (Fig. 10–4A). To ensure that substrata both within and between outplants have an equal density of gametophytes, a standard zoospore concentration and volume of inoculation solution are used. In addition, gametophyte densities on

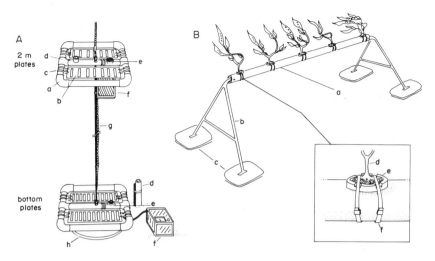

Fig. 10–4. Station setups used for *Macrocystis pyrifera* gametophyte outplants (A) and juvenile sporophyte transplants (B). Letters on the gametophyte station (A) indicate (a) floating PVC frame; (b) plexiglass plates holding pieces of nylon rope containing gametophytes (ropes attached with cable ties); (c) surgical tubing with stainless steel hooks used to attach plates to frame; (d) sediment tube; (e) irradiance sensor; (f) irradiance–temperature integrator and logger; (g) stainless steel and safety cable; (h) iron anchor plate. Letters on juvenile sporophyte station (B) indicate (a) 3-m length of PVC pipe; (b) steel frame; (c) 20-kg iron anchor plates; (d) juvenile *Macrocystis* stipe and holdfast; (e) PVC collar; (f) cable ties. In both types of stations, a reinforcing bar can be driven into the substratum through holes in the anchor plates to increase anchor strength.

the outplant ropes are determined before outplant by means of a dipping-cone microscope. (The microscope and an adaptation of it for underwater use are discussed by Neushul 1972.) The outplanted substrata are returned to the laboratory after 6 wk for determination of sporophyte recruitment from the outplanted gametophytes. Recruitment of sporophytes from the naturally settled gametophytes is monitored by pairing uninoculated substrata with each inoculated plate (see Chapman, Chap. 12, for details regarding demographic studies).

Laboratory outplant techniques have also been used for many years to study the juvenile sporophytes of a number of Laminariales (Sundene 1961; Tseng 1981). In these studies, the gametophyte generation is cultured in the laboratory until a large population of sporophytes is produced. Culture techniques have varied among all of these studies. L. E. Deysher, T. A. Dean, and D. C. Barilotti (unpublished data) have developed a simple, low-cost method that requires only a constant-temperature room maintained for *Macrocystis pyrifera*, at 15°C. This can be easily done with an ordinary room air conditioner. The substrata inoculated with gametophytes are the same as those used for the gametophyte outplants described earlier. The gametophyte inoculation density, however, is reduced by a factor of 10 because high gametophyte survival in the laboratory produces too dense a population of sporophytes.

Substrata are cultured in 10 liters of Erdschreiber's medium (McLachlan 1973) that has been filtered to remove particles down to 0.2 μm to avoid contamination by diatoms and other algae. All equipment is sterilized with 0.1% bleach solution and washed with 10% HNO_3. The culture vessels consist of rectangular polyethylene trash cans aerated with an aquarium airstone. The inoculated substrata are held off the bottom approximately midway in the culture medium, by means of polyvinyl chloride (PVC) racks. This ensures good circulation of the medium, which is important for sporophyte development. Light is provided by daylight fluorescent bulbs held ~45 cm above the substrata and providing a quantum irradiance of 45 $\mu E \cdot m^{-2} \cdot s^{-1}$. Under this constant irradiance, sporophytes are produced in 9 to 12 d and outplanted at a size of 0.3 to 0.5 mm.

In contrast to the spores and gametophytes of the Laminariales, the large size of fertilized Fucales eggs (~150 μm) makes them quite easy to handle in the laboratory. Pollock (1969) settled eggs of *Fucus distichus* onto corrugated plastic strips in the laboratory and outplanted them at various field locations. L. E. Deysher (unpublished data) inoculated asbestos panels with fertilized eggs of *Sargassum muticum* and outplanted them to intertidal and subtidal locations. The inoculation was carried out as follows. The asbestos plates were placed on the bottom of a 200-liter aquarium and *Sargassum* fronds with embryos were floated on the surface. Embryos were distributed evenly on the plates as a result of falling from the receptacles through the large volume of water.

E. Transplanting

The transplanting of established plants (or pieces of plants) from one location to another in the field is the most common method of conducting field experiments. Transplanting makes it possible to determine the effects of various environmental factors by comparing the responses of the transplanted individuals with those of the control population. Transplant controls, plants moved with the transplants and then returned to the original site, must also be used to determine the effects of transplantation itself. Transplanting also allows various densities and assemblages of plants to be manipulated.

Techniques for securing transplanted individuals in new environments have been developed for species ranging from small red to

the largest brown algae. Barilotti and Silverthorne (1972) transplanted individual *Gelidium robustum* by chipping small pieces of rock
with attached plants from the substratum, cementing the pieces into
holding structures, and then bolting these to the substratum at the
transplant site. Plants on small rocks can also be attached with
underwater epoxy (Foster 1982b). Other methods of transplanting
small algae include the use of plastic clips (Barilotti 1980) and placing
plants between strands of nylon line (Brinkhuis and Jones 1976).

A number of techniques appropriate for other large algae have
been developed for transplanting the giant kelp *Macrocystis pyrifera*.
One of the most successful methods of attaching small plants directly
to the bottom was developed to transplant juveniles (~1–3 m long)
into subtidal areas dominated by *Pterygophora californica* (R. McPeak,
personal communication). In this method, the *Pterygophora* stipes are
cut ~3 cm above the bottom, leaving a small stub to which the
haptera of the *Macrocystis* plant are tied with rubber bands. In areas
where natural attachment sites such as *Pterygophora* stipes are not
available, the plants can be tied to spikes driven into the substratum.

To assess juvenile *Macrocystis* growth in different habitats while
eliminating bottom effects (sea urchin grazing, burial, etc.), T. A.
Dean (unpublished data) placed plants on PVC pipes held ~0.5 m
off the bottom by means of iron legs bolted onto iron plate anchors
(Fig. 10–4B). The plants were held on the 8 cm diameter PVC pipe
with electrical cable ties. In another study of *Ecklonia radiata* growth,
in which bottom effects were not of interest, H. Kirkman (personal
communcation) attached adult plants to suspended ropes with small
nylon line wrapped around the holdfast and rope. When attached
this way, the plants hang down or move with currents, but this does
not affect growth. *Ecklonia radiata* and other kelps are damaged if
attached by the stipes (H. Kirkman, personal communication). Holdfasts can also be secured to boulders with pieces of inner tube or
sewn with small-diameter nylon rope onto heavy lines anchored to
the bottom. Gordon and De Wreede (1978) successfully transplanted
Egregia menziesii by first attaching plants to boards with pieces of
inner tube and then attaching the boards to the bottom.

Plants can also be provided with their own anchors for transplantation. D. C. Barilotti (personal communication) moves plants to
unstable bottoms by the use of bags of cement. The cement bags are
enclosed in burlap bags to provide reinforcement and then placed
on the bottom. Holdfasts are attached by driving iron stakes through
them into the bag immediately after the bags are placed on the
bottom. The stakes are then held firmly in place when the cement
hardens. J. Woessner, B. Harger, and M. Neushul (personal communication) established a stand of 700 adult *Macrocystis* plants on

sand by placing individual holdfasts in nylon mesh bags filled with ~20 kg of gravel. They removed plants to be placed in the bags from natural stands by cutting the holdfast with a pruning saw. This was done so that about half of the holdfast remained. After the cut holdfast was placed in the bag, the opening was loosely closed around the base of the primary stipe or below the prostrate branching system with a drawstring or electrical cable tie. Hapteral growth eventually bound the whole system together, and the haptera grew out through the mesh, ultimately covering the entire nylon bag anchor.

The success of these methods is dependent largely on the amount of water motion at the transplant site, and field tests should be done first to avoid the loss of all transplants during the first storm. This is particularly true of large, nonfloating plants with stipes (e.g., *Laminaria*) that, if attached in an upright position, can be forced over onto the substratum if not securely anchored.

F. Abiotic manipulations

The effects of such physical factors as light, temperature, and nutrients can be best studied under carefully controlled laboratory conditions. However, in some cases, either because the factor of interest cannot be suitably controlled in the laboratory or the plant cannot be adapted to laboratory culture, controlled field experiments must be used. The most common experimental design is to select field sites in which only the factor of interest varies among sites. In some cases suitable sites can be found naturally, but in most experiments some type of manipulation must be performed to alter the factor of interest at the experimental site.

Monitoring the controlled physical factors is important either to ensure that these variables are equal among sites or to quantify the difference in these factors so the difference can be used in the analysis of the experiment.

1. Light.
The most common method of altering irradiance in the field has been to remove the canopy of overstory algae from an experimental site while leaving the canopy intact at the control site (Norton and Burrows 1969; Dayton 1975; Pearse and Hines 1979; Reed and Foster 1984). Light can also be increased by transplanting plants to a shallower depth. However, this may alter other environmental variables in addition to irradiance and must be used with care. An ingenious method of increasing irradiance is the use of a mirror under individual fronds of *Gelidium robustum* (Barilotti 1980). The mirror reflects light back onto the undersides of fronds held by small clips, thereby increasing the light available for photosynthesis. Irradiance was decreased in experimental plots of eelgrass, *Zostera marina*, by the construction of 1-m^2 shades over the substratum (Backman

and Barilotti 1976). This study was conducted in a shallow coastal lagoon where currents and wave surge were negligible. In high-energy environments, however, such shades may be difficult to keep in place.

2. Water motion. Most studies on the effects of water motion have dealt with differences in algal morphology in calm versus exposed locations (Sundene 1964; Norton 1969; Gerard and Mann 1979). In these studies, either outplanted or transplanted individuals were placed in sites with widely different current regimes and compared after a period of growth. As discussed earlier, care must be taken to ensure that only water motion varies among the experimental sites.

A unique method of altering the current regime at a single experimental site was designed by Jones (1959). In this experiment a floating raft was constructed with narrow sluices that could be oriented into the current. Some sluices could be closed at both ends, reducing water motion to any algae transplanted into them. Adequate water circulation into the closed sluices was ensured by holes drilled into the bottom. This design helped alleviate problems associated with any possible covariance of factors among sites.

3. Temperature. Field experiments on temperature effects are difficult because it is difficult to obtain different temperature regimes without changes in other physical parameters. For example, decreases in temperature with depth are usually associated with decreases in irradiance and increases in nutrients. Temperatures are usually quite uniform over large areas at the same depth, and local differences are often accompanied by differences in turbulence and nutrients.

L. E. Deysher (unpublished data) manipulated temperature by transplanting small *Sargassum muticum* from San Diego to Bamfield, British Columbia. The plants (less than 1 cm long) used in this experiment had been recruited onto small asbestos plates attached to cement blocks at a depth of -1.5 m (mean lower low water) in Mission Bay, San Diego. The asbestos plates were packed in styrofoam chests, air-freighted to Vancouver, British Columbia, and then reattached to cement blocks in front of the Bamfield Marine Laboratory. Water clarity, turbulence, and depth were quite similar between the San Diego and Vancouver Island sites. However, temperatures in Mission Bay ranged from 10 to 24°C, whereas those at Bamfield ranged from 5 to 14°C. Photoperiod was also different between the two sites, but this factor did not appear to play a significant role in the large differences in growth observed.

4. Nutrients. The effects of nutrients on a wide variety of algal species have been experimentally evaluated under field conditions. The scale of these experiments has ranged from effects on gametogenesis in

Macrocystis pyrifera to effects on productivity of entire beds of the same species (North et al. 1982). The primary problem with field nutrient experiments is maintaining an elevated concentration of nutrients for the duration of the experiment, since fertilizer is constantly removed from the experimental site by both advection and diffusion. The most common method for maintaining elevated nutrient concentration has been to put fertilizer into porous containers (Chapman and Craigie 1977; Tseng 1981).

Elevated nitrogen concentrations in the study of *Macrocystis* gametophytes (described earlier) were maintained by the use of Osmocote fertilizer (Sierra Chemical Co.). Osmocote is a blend of ammonium nitrate and phosphate salts in pellets coated with a semipermeable polymer film. The polymer film controls the rate of release of fertilizer, which is dependent on polymer thickness and temperature. The fertilizer was placed in nylon-mesh covered trays directly below nylon rope substrata inoculated with *Macrocystis* gametophytes, and the fertilizer trays were changed weekly during the course of the 6-wk experiment. Water samples for the determination of nutrient concentrations were drawn by vacuum pressure through perforated polyethylene tubes that alternated with the nylon rope substrata on the outplant plate. The vacuum was supplied by glass bottles that could be sealed at the surface and then opened by means of a stopcock after being connected to the collection system. The water samples were immediately brought back to the boat, filtered, and frozen on dry ice.

The large-scale fertilization of a southern California *Macrocystis* bed (North et al. 1982) was conducted by dispersing ammonium sulfate and ammonium phosphate crystals from a crop-dusting helicopter. Preliminary studies had shown that ammonium sulfate crystals weighing 0.7 mg or less dissolved completely in seawater in less than 30 s while falling through a vertical distance of less than 1 m. In addition, dye dispersal studies showed that the *Macrocystis* canopy, where 50% or more of the kelp biomass is found in the upper 1 m of the water column, acted as a barrier to currents, thereby increasing the residence time of nutrient-rich water. Water samples indicated that after the distribution of 1 ton of fertilizer over the 0.15-km^2 kelp canopy, ammonium concentrations were initially elevated 50–90 μm over ambient in the canopy and remained higher than ambient concentrations for at least 2 h.

Fertilization of commercial-scale *Laminaria* cultivation in China has been practiced since the late 1950s (Tseng 1981). Early methods consisted of placing porous clay bottles filled with fertilizer throughout the *Laminaria* beds. However, this method was costly, labor intensive, and not very effective in distributing the fertilizer to the plants. More recently, distribution has been accomplished by simple spraying of the fertilizer over beds.

Gerard (1982) used a novel transplant method in a study of internal nitrogen reserves in *Macrocystis*. She moved an adult plant from an inshore kelp forest with a relatively high ambient concentration of nutrients to an offshore area with a low ambient concentration of nutrients.

5. Sedimentation. Sediment instability, burial, and scour can cause high mortality of *Macrocystis* microscopic stages in the laboratory (Devinny and Volse 1978) and, based on descriptive studies and natural experiments, can significantly influence algal distribution and abundance in the field (Foster 1975a; Daly and Mathieson 1977; Taylor and Littler 1982; Littler et al. 1983). However, these effects are difficult to quantify, and to date no field experiments have been conducted in subtidal habitats. This area appears to be ripe for innovative experimentation.

G. Biotic manipulations

1. Plant–plant interactions. Most subtidal field experiments on plant–plant interactions, either intraspecific or interspecific, have involved removing selected individuals from experimental areas. This removal has changed the light regime by either reducing overstory algal canopies (Dayton 1975; Kain 1975; Pearse and Hines 1979; Harkin 1981; Reed and Foster 1984) or opening substrata to recruitment by other algal species (Kain 1975; Pearse and Hines 1979; Deysher and Norton 1982; Reed and Foster 1984). These and other methods for competition experiments are reviewed by Denley and Dayton (Chap. 25).

2. Grazing. Experimental evaluations of grazing have generally been conducted by completely excluding or removing grazers from an area and comparing the resulting algal population with that of an unmanipulated control site. Methods of doing this are discussed by Vadas (Chap. 26). A few methods particularly applicable to subtidal habitats are discussed here.

Devices for excluding subtidal grazers have ranged from small cages enclosing areas of less than 1 m² (Foster 1975b) to elaborate fences enclosing areas of greater than 100 m² (Pace 1976). D. Laur (personal communication), using modifications of a basic cage developed by R. Van Wagenen (personal communication) to include sea urchins but exclude sea otters, developed a set of subtidal cage types that are relatively easy to install, can withstand considerable water motion, and can be used to exclude different organisms

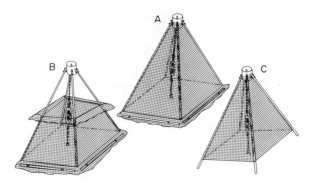

Fig. 10–5. Cages designed to exclude large fish and invertebrates. The frames are made from PVC tubing and anchored by a central chain and turnbuckle to a single eyebolt in the substratum. One leg is adjustable (small tubing inside large). Screen is made from plastic or nylon mesh and attached to frame with tie wraps. (A) Total exclusion; (B) large invertebrate exclusion–fish access (mesh upper lip helps prevent entry of large, crawling organisms such as sea stars); (C) fish exclusion–large invertebrate access (mesh extends to within 5 to 10 cm of substratum). Bottom of mesh in A and B held to substratum with PVC bars anchored with bolts to the bottom. Light controls (for North Temperate areas) consist of C-type cages with mesh on south and east sides only. Cage size depends on the experiment, and other controls may be necessary (see Foster and Sousa, Chap. 13, and Vadas, Chap. 26). Cage design courtesy D. Laur and R. Van Wagenen.

selectively (Fig. 10–5). Cages or other structures used to manipulate organisms may also alter other factors, and well-designed controls are necessary (see Foster and Sousa, Chap. 13, and Vadas, Chap. 26).

A refinement of the total exclusion cage is a series of enclosures containing various densities of the grazer of interest (Ebert 1977). This design allows a more quantitative estimate of the effect of the grazer to be made. However, the behavior of many grazers precludes this type of design, and confinement may alter normal behavior.

3. Dispersal. The effective dispersal distance of reproductive propagules is obviously important for the distribution and density of algal populations. This parameter was determined for *Sargassum muticum* with an experiment on a newly constructed artificial reef (Deysher and Norton 1982). A small population of fertile plants was transplanted to the center of the reef, and the resulting recruitment of juveniles was mapped at periodic intervals. Because the reef was more than 5 km from any natural population of *S. muticum*, the pattern of recruitment observed on the reef could be ascribed totally to dispersal of embryos from the transplanted population. If one can discount growth from preexisting propagules or dispersal from drift plants, dispersal can also be estimated by mapping juveniles around natural or artificially isolated plants or stands (Anderson and North 1965; Guzman del Proo and De la Campa de Guzman 1969; see also Chapman, Chap. 12).

H. Underwater platforms

In situations where measurements, transplants, or experiments may be impossible to perform on the natural bottom, underwater platforms for holding plants, artificial substrata, and physical–chemical measuring devices may be an appropriate substitute, and entire artificial reefs may be used if available. Some of these platforms are discussed in Secs. V.D and V.E (outplanting and transplanting), and others are reviewed by Foster (1980) and Foster and Sousa (Chap. 13).

VI. Perspective

As this chapter suggests, we now have many of the tools and procedures necessary for efficient underwater work and experimentation. Other methods applicable to in situ studies of subtidal vegetation are discussed by others in this volume. Their use should provide more precise, quantitative descriptions of the subtidal environment and speed the transition from descriptive, correlative studies to experimental investigations of subtidal seaweed ecology.

VII. Acknowledgments

We thank D. C. Barilotti, V. Breda, J. Carter, D. Coon, M. Denny, B. Harger, D. Laur, M. Littler, D. Littler, and M. Neushul for their suggestions and editorial comments.

VIII. References

Anderson, E. K., and North, W. J. 1965. *In situ* studies of spore production and dispersal in the giant kelp, *Macrocyctis*. In Young, E. G., and McLauchaln (eds.), *Proceedings of the Fifth International Seaweed Symposium*, pp. 73–86. Pergamon Press, Oxford.

Backman, T., and Barilotti, D. C. 1976. Irradiance reduction: effects on standing crops of the eelgrass, *Zostera marina*, in a coastal lagoon. *Mar. Biol.* 24, 33–40.

Barilotti, D. C. 1980. Genetic considerations and experimental design of outplanting studies. In Abbott, I. A., Foster, M. S., and Eklund, L. F. (eds.), *Pacific Seaweed Aquaculture*, pp. 10–18. California Sea Grant College Program, La Jolla.

Barilotti, D. C., and Silverthorne, W. 1972. A resource management study of *Gelidium robustum*. In Nisizawa, K. (ed.), *Proceedings of the Seventh International Seaweed Symposium*, pp 255–61. University of Tokyo Press, Tokyo.

Booth, C. R. 1976. The design and evaluation of a measurement system for photosynthetically active quantum scalar irradiance. *Limnol. Oceanogr.* 25, 496.

Brinkhuis, B. H., and Jones, R. F. 1976. The ecology of temperate salt-marsh fucoids 1. Occurrence and distribution of *Ascophyllum nodosum* ecads. *Mar. Biol.* 34, 337–48.

Chapman, A. R. O., amd Craigie, J. S. 1977. Seasonal growth in *Laminaria longicruris*: relations with dissolved inorganic nutrients and internal reserves of nitrogen. *Mar. Biol.* 40, 197–205.

Charters, A. C., Neushul, M., and Barilotti, D. C. 1969. The functional morphology of *Eisenia arborea*. In Margalef, R. (ed.), *Proceedings of the Sixth International Seaweed Symposium*, pp. 89–105. Direccion General de Pesca Maritima, Madrid.

Chess, J. R. 1978. An airlift sampling device for *in situ* collecting of biota from rocky substrata. *Mar. Technol. Soc. J.* 12, 20–3.

Cole, K. 1964. Induced fluorescence in gametophytes of some Laminariales. *Can. J. Bot.* 42, 1173–81.

Craik, G. J. 1980. Simple method for measuring the relative scouring of intertidal areas. *Mar. Biol.* 59, 257–60.

Daly, M. A., and Mathieson, A. C. 1977. The effects of sand movement on intertidal seaweeds and selected invertebrates at Bound Rock, New Hampshire, U.S.A. *Mar. Biol.* 43, 45–55.

Dayton, P. K. 1975. Experimental studies of algal canopy interactions in a sea otter-dominated kelp community at Amchitka Island, Alaska. *Fish. Bull.* 73, 230–7.

Devinny, J. S., and Volse, L. A. 1978. The effects of sediments on the development of *Macrocystis pyrifera* gametophytes. *Mar. Biol.* 48, 343–8.

Deysher, L. E., and Norton, T. A. 1982. Dispersal and colonization in *Sargassum muticum* (Yendo) Fensholt. *J. Exp. Mar. Biol. Ecol.* 56, 179–95.

Drew, E. A., Lythgoe, J. N., and Woods, J. D. (eds.). 1976. *Underwater Research*. Academic Press, New York. 430 pp.

Ebert, T. A. 1977. An experimental analysis of sea urchin dynamics and community interactions on a rock jetty. *J. Exp. Mar. Biol. Ecol.* 27, 1–22.

Finnish IBP-PM Group. 1969. Quantitative sampling equipment for the littoral benthos. *Int. Rev. Hydrobiol.* 54, 185–93.

Fitter, D. J., Knapp, P. H., and Wilson, J. 1980. Stand structure and light penetration. 4. A sensor for measuring photosynthetically active radiation. *J. Appl. Ecol.* 17, 183–93.

Foster, M. S. 1975a. Algal succession in a *Macrocystis pyrifera* forest. *Mar. Biol.* 32, 313–29.

Foster, M. S. 1975b. Regulation of algal community development in a *Macrocystis pyrifera* forest. *Mar. Biol.* 32, 331–42.

Foster, M. S. 1976. A mini-slate for recording information underwater. *Underwater Natur.* 9, 14–15.

Foster, M. S. 1980. The use of substratum manipulations in field studies of seaweed colonization and growth. In Abbott, I. A., Foster, M. S., and Eklund, L. F. (eds.), *Pacific Seaweed Aquaculture*, pp. 23–31. California Sea Grant Program, La Jolla.

Foster, M. S. 1982a. The regulation of macroalgal asociations in kelp forests. In Srivastava, L. (ed.), *Synthetic and Degradative Processes in Marine Macrophytes*, pp. 185–205. De Gruyter, Berlin.

Foster, M. S. 1982b. Factors controlling the intertidal zonation of *Iridaea flaccida* (Rhodophyta). *J. Phycol.* 18, 285–94.

Fovargue, A., and Perino, J. 1979. A simple device for measuring microtopography in ecological studies. *Ohio J. Sci.* 79, 130–2.

Gardner, W. 1980a. Sediment trap dynamics and calibration: a laboratory evaluation. *J. Mar. Res.* 38, 17–39.

Gardner, W. 1980b. Field assessment of sediment traps. *J. Mar. Res.* 38, 41–52.

Gerard, V. 1976. "Some Aspects of Material Dynamics and Energy Flow in a Kelp Forest in Monterey Bay, Ca." Ph.D. dissertation, University of California, Santa Cruz. 173 pp.

Gerard, V. 1982. Growth and utilization of internal nitrogen reserves by the giant kelp, *Macrocystis pyrifera*, in a low nitrogen environment. *Mar. Biol.* 66, 27–35.

Gerard, V., and Mann, K. H. 1979. Growth and production of *Laminaria longicruris* (Phaeophyta) populations exposed to different intensities of water movement. *J. Phycol.* 15, 33–41.

Gordon, D. K., and De Wreede, R. E. 1978. Factors influencing the distribution of *Egregia menziesii* (Turner) Areschoug (Phaeophyta, Laminariales) in British Columbia, Canada. *Can. J. Bot.* 56, 1198–1205.

Guzman del Proo, S. A., and De la Campa de Guzman, S. 1969. Investigaciones sobre *Gelidium cartilagineum* en la costa occidental de Baja California, Mexico. In Margalef, R. (ed.), *Proceedings of the Sixth International Seaweed Symposium*, pp. 179–86. Direccion General de Pesca Maritima, Madrid.

Harger, B. W. 1979. "Coastal Oceanography and Hard Substrate Ecology in a California Kelp Forest." Ph.D. dissertation, University of California, Santa Barbara. 427 pp.

Harkin, E. 1981. Fluctuations in epiphyte biomass following *Laminaria hyperborea* canopy removal. In Levring, T. (ed.), *Proceedings of the Tenth International Seaweed Symposium*, pp. 303–8. De Gruyter, Berlin.

Hatcher, B. G. 1977. An apparatus for measuring photosynthesis and respiration of intact large marine algae and comparison of results with those from experiments with tissue segments. *Mar. Biol.* 43, 381–5.

Healey, F. P. 1972. Photosynthesis and respiration of some Arctic seaweeds. *Phycologia* 11, 267–71.

Heine, J. N. 1983. Seasonal productivity of two red algae in a central California kelp forest. *J. Phycol.* 19, 146–52.

High, W. L., Ellis, I. E., Schroeder, W. W., and Loverich, G. 1973. Evaluation of the undersea habitats—Tektite II, Hydro-Lab, and Edalhab—for scientific saturation diving programs. *Helgo. Meeres.* 24, 16–44.

Hoffman, W. E., and Dawes, C. J. 1980. Photosynthetic rates and primary production by two Florida benthic red algal species from a salt marsh and a mangrove community. *Bull. Mar. Sci.* 30, 358–64.

Holme, N. A., and McIntyre, A. D. (eds.). 1971. *Methods for the Study of Marine Benthos.* Blackwell, Oxford. 334 pp.

Holmes, R. W. 1970. The Secchi disk in turbid coastal waters. *Limnol. Oceanogr.* 15, 688–94.

Hsiao, S. I., and Druehl, L. D. 1973. Environmental control of gametogenesis in *Laminaria saccharina.* IV. *In situ* development of gametophytes and young sporophytes. *J. Phycol.* 9, 160–4.

Jackson, G. A. 1977. Nutrients and production of giant kelp, *Macrocystis pyrifera*, off southern California. *Limnol. Oceanogr.* 22, 979–95.

Jerlov, N. G. 1968. *Optical Oceanography.* Elsevier, Amsterdam. 194 pp.

Jerlov, N. G., and Nygard, K. 1969. *A Quanta and Energy Meter for Photosynthetic Studies.* University of Copenhagen Institute of Physical Oceanography Report 10, Copenhagen. 19 pp.

Johansen, H. W., and Austin, L. F. 1970. Growth rates in the articulated coralline *Calliarthron* (Rhodophyta). *Can. J. Bot.* 48, 125–32.

Johnson, S. K., and Petty, R. L. 1982. Determination of phosphate in sea water by flow insertion analysis with injection of reagent. *Anal. Chem.* 54, 1135–7.

Jones, N. S. 1971. Diving. In Holme, N. A., and McIntyre, A. D. (eds.), *Methods for the Study of Marine Benthos*, pp. 71–9. Blackwell, Oxford.

Jones, W. E. 1959. Experiments on some effects of certain environmental factors on *Gracilaria verrucosa* (Hudson) Papenfuss. *J. Mar. Biol. Assoc. U.K.* 38, 153–67.

Josselyn, M. N., Cailliet, G. M., Niesen, T. M., Cowen, R., Hurley, A. C., Connor, J., and Hawes, S. 1983. Composition, export, and faunal utilization of drift vegetation in the Salt River submarine canyon. *Est. Coast. Shelf Sci.* 17, 447–65.

Kain, J. 1975. Algal recolonization of some cleared subtidal areas. *J. Ecol.* 63, 739–65.

Kinne, O., and Burcheim, H. P. (eds.). 1973. Man in the sea: *in situ* studies of life in oceans and coastal waters. *Helgo. Meeres.* 24, 1–535.

Kitching, J. A., Macan, T. T., and Gilson, H. C. 1934. Studies in sublittoral ecology. 1. A submarine gully in Wembury Bay, South Devon. *J. Mar. Biol. Assoc. U.K.* 19, 677–705.

Lissner, A. L. 1980. Some effects of turbulence on the activity of the sea urchin *Centrostephanus coronatus* Verill. *J. Exp. Mar. Biol. Ecol.* 48, 185–93.

Littler, M. M., Martz, D. R., and Littler, D. S. 1983. Effects of recurrent sand deposition on rocky intertidal organisms: importance of substrate heterogeneity in a fluctuating environment. *Mar. Ecol. Progr. Ser.* 11, 129–39.

Littler, M. M., and Murray, S. N. 1974. The primary productivity of marine macrophytes from a rocky intertidal community. *Mar. Biol.* 27, 131–5.

Luckhurst, B. E., and Luckhurst, K., 1978. Analysis of the influence of substrate variables on coral reef fish communities. *Mar. Biol.* 49, 317–23.

Lüning, K. 1981. Light. In Lobban, C. S., and Wynne, M. J. (eds.), *The Biology of Seaweeds*, pp. 326–55. University of California Press, Berkeley.

Lüning, K., and Dring, M. J. 1972. Reproduction induced by blue light in female gametophytes of *Laminaria saccharina*. *Planta (Berl.)* 104, 252–6.

Lüning, K., and Dring, M. J. 1979. Continuous underwater light measurement near Helgoland (North Sea) and its significance for characteristic light limits in the sublittoral region. *Helgo. Meeres.* 32, 403–24.

Lüning, K., and Neushul, M. 1978. Light and temperature demands for growth and reproduction of laminarian gametophytes in southern and central California. *Mar. Biol.* 45, 297–310.

Mann, K. H. 1972. Ecological energetics of the seaweed zone in a marine bay on the Atlantic coast of Canada. 2. Productivity of the seaweeds. *Mar. Biol.* 14, 199–209.

Mann, K. H., and Kirkman, H. 1981. Biomass method for measuring productivity of *Ecklonia radiata*, with the potential for adaption to other large brown algae. *Aust. J. Mar. Freshw. Res.* 32, 297–304.

McLachlan, J. 1973. Growth media: marine. In Stein, J. (ed.), *Handbook of Phycological Methods: Culture Methods and Growth Measurements*, pp. 25–51. Cambridge University Press, Cambridge.

Miller, J. W. (ed.). 1979. *NOAA Diving Manual*, 2nd ed. National Oceanic and Atmospheric Administration, U.S. Government Printing Office, Washington, D.C. (unpaginated).

Neumann, G., and Pierson, W. J. 1966. *Physical Oceanography*. Prentice-Hall, Englewood Cliffs, N.J. 545 pp.

Neushul, M. 1972. Underwater microscopy with an enclosed incident-light dipping-cone microscope. *J. Microscop.* 95, 421–4.

Neushul, M., Foster, M. S., Coon, D. A., Woessner, J. W., and Harger, B. W. 1976. An *in situ* study of recruitment, growth, and survival of subtidal marine algae: techniques and preliminary results. *J. Phycol.* 12, 397–408.

Nicholson, N. L. 1970. Field studies on the giant kelp, *Nereocystis*. *J. Phycol.* 6, 177–82.

North, W. J. 1976. Aquacultural techniques for creating and restoring beds of giant kelp, *Macrocystis* spp. *J. Fish. Res. Bd. Can.* 33, 1015–23.

North, W., Gerard, V., and Kuwabara, J. 1982. Farming *Macrocystis* at coastal and oceanic sites. In Srivastava, L. (ed.), *Synthetic and Degradative Processes in Marine Macrophytes*, pp. 247–64. De Gruyter, Berlin.

Norton, T. A. 1969. Growth, form, and environment in *Saccorhiza polyschides*. *J. Mar. Biol. Assoc. U.K.* 49, 1025–45.

Norton, T. A., and Burrows, E. M. 1969. Studies on marine algae of the British Isles. 7. *Saccorhiza polyschides* (Light f.) Batt. *Br. Phycol. J.* 4, 19–53.

Nyquist, G. 1979. Relationships between Secchi disk transparency, irradiance attenuation, and beam transmittance in a fjord system. *Mar. Sci. Commun.* 5, 333–59.

Ogden, J. C. 1977. A scroll apparatus for the recording of notes and observations underwater. *Mar. Technol. Soc. J.* 11, 13–4.

Pace, D. R. 1976. "Environmental Control of Red Sea Urchin (*Strongylocentrotus franciscanus*) Vertical Distribution in Barkely Sound, British Columbia." Ph.D. dissertation, Simon Fraser University, Vancouver. 98 pp.

Paine, R. T. 1979. Disaster, catastrophe, and local persistence of the sea palm, *Postelsia palmaeformis*. *Science* 205, 685–7.

Pearse, J. S., and Hines, A. H. 1979. Expansion of a central California kelp forest following the mass mortality of sea urchins. *Mar. Biol.* 51, 83–91.

Pollock, E. G. 1969. Interzonal transplantation of embryos and mature plants of *Fucus*. In Margalef, R. (ed.), *Proceedings of the Sixth International Seaweed Symposium*, pp. 345–56. Direccion General de Pesca Maritima, Madrid.

Pratt, W. 1973. Macro snooping. *Skin Diver*, September, pp. 38–9.

Ramus, J., Beale, S. I., and Mauzerall, D. 1976. Correlation of changes in pigment content with photosynthetic capacity of seaweeds as a function of water depth. *Mar. Biol.* 37, 231–8.

Reed, D. C. 1981. "The Effects of Competition for Light and Space on the Perennial Algal Assemblage in a Giant Kelp (*Macrocystis pyrifera*) Forest." M.A. thesis, California State University, San Francisco. 56 pp.

Reed, D. C., and Foster, M. S. 1984. The effects of canopy shading on algal recruitment and growth in a giant kelp forest. *Ecology* 65, 937–48.

Rosenthal, R. J., Clarke, W. D., and Dayton, P. K. 1974. Ecology and natural history of a stand of giant kelp, *Macrocystis pyrifera*, off Del Mar, California. *Fish. Bull.* 72, 670–84.

Smith, W. D. 1981. Photosynthesis and productivity of benthic macroalgae on the North Carolina continental shelf. *Bot. Mar.* 24, 279–84.

Sousa, W. P. 1979. Experimental investigations of disturbance and ecological succession in a rocky intertidal algal community. *Ecol. Monogr.* 49, 227–54.

Stride, A. H. 1961. Mapping the sea floor with sound. *New Scientist* 10, 304–6.

Sundene, O. 1961. Growth in the sea of *Laminaria digitata* sporophytes from culture. *Norw. J. Bot.* 9, 5–24.

Sundene, O. 1964. The ecology of *Laminaria digitata* in Norway in view of transplant experiments. *Nytt. Mag. Bot.* 11, 83–107.

Tanner, C., Hawkes, M. W., and Lebednik, P. A. 1977. A hand-operated suction sampler for the collection of subtidal organisms. *J. Fish. Res. Bd. Can.* 34, 1031–4.

Taylor, P. R., and Littler, M. M. 1982. The roles of compensatory mortality, physical disturbance, and substrate retention in the development and organization of a sand-influenced, rocky-intertidal community. *Ecology* 63, 135–46.

Towle, D. W., and Pearse, J. S. 1973. Production of the giant kelp, *Macrocystis*, estimated by *in situ* incorporation of ^{14}C in polyethylene bags. *Limnol. Oceanogr.* 18, 155–9.

Tseng, C. K. 1981. Commercial cultivation. In Lobban, C. S., and Wynne, M. J. (eds.), *The Biology of Seaweeds*, pp. 680–725. University of California Press, Berkeley.

Tucker, M. J., and Stubbs, A. R. 1961. Narrow-beam echo-ranger for fishery and geological investigations. *Br. J. Appl. Phys.* 12, 103–10.

Tyler, J. E. 1973. Lux vs quanta. *Limnol. Oceanogr.* 18, 810.

Walker, T. 1980. A correction to the Poole and Atkins Secchi disk/light attenuation formula. *J. Mar. Biol. Assoc. U.K.* 60, 769–71.

Weinberg, S. 1976. Submarine daylight and ecology. *Mar. Biol.* 37, 291–304.

Weller, R. A. 1978. "Observations of Horizontal Velocity in the Upper Ocean Mode with a New Vector Measuring Current Meter." Ph.D. dissertation, University of California, San Diego. 169 pp.

Wheeler, W. N., and Neushul, M. 1981. The aquatic environment. In Lange, O. L. Nobel, P. S., Osmond, C. B., and Ziegler, H. (eds.) *Physiological Plant Ecology: Responses to the Physical Environment, Encyclopedia of Plant Physiology, New Series*, vol. 12A, pp. 229–47. Springer-Verlag.

Woods, J. D., and Lythgoe, N. (eds.). 1971. *Underwater Science.* Oxford University Press, London. 330 pp.

Woodward, R. I., and Yaqub, M. 1979. Integrator and sensors for measuring photosynthetically active radiation and temperature in the field. *Appl. Ecol.* 16, 545–52.

Zieman, J. C., Thayer, G. W., Robblee, M. B., and Zieman, R. T. 1979. Production and export of sea grasses from a tropical bay. In Livingston, R. J. (ed.), *Ecological Processes in Coastal and Marine Systems*, pp. 21–33. Plenum Press, New York.

11: Equipment for conducting research in deep waters

SYLVIA A. EARLE

California Academy of Sciences, Golden Gate Park, San Francisco, California 94118

CONTENTS

I. Introduction

Methods of observing and collecting subtidal marine plants historically have included the use of various nets, dredges, and hooks. Such equipment made it possible for researchers such as William Randolph Taylor (1928) to obtain samples of plants growing near Dry Tortugas, Florida, in depths below 100 m. Humm (1954) reported the occurrence of the green alga *Anadyomene menziesii* in the Gulf of Mexico from trawl samples that established the deep-water habitat (100 + m) of this unusual plant. Adey and MacIntyre (1973) noted that rhodoliths occur abundantly in samples taken on the continental shelf to 200-m depth.

Although useful for acquiring qualitative information, the difficulties encountered with remote sampling techniques can be compared to the problems one might experience in trying to study the characteristics of a forest by flying in an aircraft, blindly dragging a net, hoping to snare something significant from the "depths" below. Data concerning arrangements, abundances, substrata, associations with animals, and numerous other factors are difficult if not impossible to assess using nets and dredges. Until recently, however, botanists who were curious about plant life in deep water had to be content with fragmentary samples fortuitously obtained.

The advent of self-contained underwater breathing apparatus (SCUBA) in the 1950s greatly increases the ability of scientists to gain direct access to the sea in depths to ~60 m. Although often regarded as a recreational device rather than a serious tool for scientific research, SCUBA has become established and accepted as a valuable asset for marine research. With SCUBA, the sea is accessible to scientists for routine work, although many special precautions must be taken (Foster et al., Chap. 10).

Compared with the ease of studying terrestrial or intertidal environments, botanists are restricted in their access to subtidal habitats. If standard "snorkle diving" techniques were applied to a terrestrial situation, excursions into a forest or field would be limited in terms

[234]

of distance and duration by the length of time one could hold one's breath.

By the use of conventional SCUBA methods, excursion time could be increased to as much as 1 h, but the distance one could go into the forest would have physiological as well as temporal constraints. Applying the general rule of no-decompression diving to a terrestrial hillside, one could spend no more than 20 min at a distance 30 m from the starting point. An excursion 60 m away would be limited to not more than 5 min and would be coupled with the dizzying narcotic effects of breathing nitrogen under pressure. Beyond 60 m, one could observe the plants beyond only distantly without being able to travel to where they were growing.

Considering such limitations, it is no wonder that knowledge of plant life in the sea is still so rudimentary; rather, the wonder is that so much has been discovered.

II. Saturation diving

The development of saturation diving in the 1960s and 1970s greatly increased the time that scientists could remain at the underwater study site of their choice and, to some extent, increased the depth range as well. The first civilian use of an underwater saturation diving facility, during the Tektite Program, took place in 1969 and 1970. Scientists and engineers remained underwater for 10 to 60 d, living in a four-room facility at 15 m depth and making excursions into the surrounding area as much as a quarter of a mile away. This approach has a terrestrial counterpart in field studies in which investigators set up camp at locations for several days or weeks.

Participants in the Tektite Program were required to have sound scientific credentials, and their proposed research projects were subjected to peer review. They also had to be certified SCUBA divers and to be in reasonably good health. There were no other special requirements or qualifications. Selection was based primarily on the merits of scientific research proposals and scientific credentials rather than diving expertise.

Two research projects dealt specifically with assessing plant life during the Tektite Program (Earle 1972; Mathieson et al. 1975), and in both instances extensive use was made of rebreather units as well as compressed air. Rebreathers, like the life support units worn by astronauts, continuously recycle air, chemically removing CO_2 and replacing O_2 as it is consumed. Excursions with a single standard SCUBA tank (72 ft^3) rarely exceed 1 h, but the rebreathers used during the Tektite Program provided enough air for ~12 h. In

Fig. 11–1. *Hydrolab,* a saturation diving facility capable of housing four people in depths of 15 to 20 m for a week. Presently, *Hydrolab* is operated at the West Indies Laboratory in St. Croix, U.S. Virgin Islands, by the National Oceanic and Atmospheric Administration. Photograph by S. Earle.

practice, excursions were made by scientists for 2 to 6 h at a time in depths to 26 m. Some scientists spent as many as 12 h per day in the water in depths below 15 m.

During 2 wk of saturation, the equivalent of more than 2 mo of diving time using conventional SCUBA methods was possible (Earle 1972). Although a considerable amount of follow-up analysis of data was required, the ability to concentrate effort within a short time and to make ecological and behavioral observations through 24-h cycles proved to be valuable. Time considerations aside, the perspective gained as a resident, as opposed to an in-and-out visitor, contributed to the opportunity to observe thoroughly and better understand the ecosystems in the area.

More than 50 underwater habitats have been developed and used in various parts of the world since the mid-1960s (Miller 1975) and are widely used by commercial divers in the offshore oil industry.

With mixtures of exotic gases (helium and argon), some dives are made in depths beyond 500 m.

The system most often used for scientific research is *Hydrolab* (Fig. 11–1), a saturation diving facility designed and built by Perry Oceanographics, Inc., launched in 1968, and operated almost continuously since then. More than 300 scientists have experienced the use of saturation diving techniques in the *Hydrolab*, and numerous in situ studies have been conducted on the ecology, distribution, morphology, and physiology of marine plants by scientists using *Hydrolab* as a subsea base of operations. Examples of botanical studies that have been conducted from *Hydrolab* include research reported by Hurley et al. (1981) and Josselyn et al. (1983). The system is presently operated by the West Indies Laboratory at St. Croix, U.S. Virgin Islands, with support from the National Oceanic and Atmospheric Administration (NOAA). Other NOAA-supported facilities and contacts for acquiring them, as of 1985, are listed in Table 11–1.

In 1975 the use of *Hydrolab* was coupled with an unusual submersible, the *Johnson-Sea-Link* (Fig. 11–2), which was used as a "taxi" to transport *Hydrolab* scientists to and from distant locations. The submersible is equipped with a transparent acrylic sphere that houses a pilot and one observer, who remain at one atmosphere (surface pressure) throughout the duration of an excursion. The spherical shape functions as a lens under water, and objects appear to be more distant than they actually are. However, a trained observer quickly learns to compensate for this. An aft compartment with small portholes can be pressurized to allow a diver to swim outside the submersible for work, then return, using the submarine as a portable decompression chamber.

Several missions were designed to take advantage of the unique combination of a saturation laboratory and transportation via submersible. One project involved the use of the *Johnson-Sea-Link* to travel over the edge of a vertical drop-off for lockout dives and observations to obtain data on the distribution of plants relative to depth and light. During the first such excursion (45 min in 76-m depth), a new genus of green algae was discovered growing on a sheltered ledge. Later described as *Johnson-sea-linkia profunda* (Eiseman and Earle 1983), the plant is one of a sizable number of species that occurs only in deep water, more than 60 m, where the light level is no more than 1% of surface intensity.

Without access to such situations, it is not likely that many of these deep-water species would be known at all, and studies concerning their ecology would be virtually impossible to execute.

A new habitat is under development with support from the NOAA

Table 11–1. *Undersea research facilities sponsored by the National Oceanic and Atmospheric Administration as of 1985*

I. Southern California

Program description

By 1986, NOAA's National Underwater Research Program at the University of Southern California will include a saturation habitat system designed to accommodate 6 people for missions lasting up to 14 days with dives to 40-m depth. As an interim measure, scientists are offered an operational program of SCUBA, surface-supplied umbilical diving, and a portable underwater way station to gain working access to a variety of environments, habitats, and biological communities. Included are midwater, sand, rocky substrates, as well as kelp beds.

Contact

Program Science Director
National Undersea Research Program at the University of Southern California
Catalina Marine Science Center
P.O. Box 398
Avalon, California 90704

II. U.S. East Coast and Gulf of Mexico

Program description

In cooperation with the University of North Carolina and an advisory group, the South-eastern Consortium for Undersea Research, NOAA supports a 21.5-m ship, *R/V Seahawk*, equipped to support SCUBA, surface-supplied air and mixed gas, and wet-bell diving operations to a depth of 80 m. A remotely operated vehicle (ROV) may be available to complement diving operations.

Contact

Program Science Director
National Undersea Research Program at the University of North Carolina (Wilmington)
601 South College Road
Wilmington, North Carolina 28403

III. Caribbean Sea

Program description

In cooperation with West Indies Laboratory of Fairleigh Dickinson University, NOAA supports the operation of *Hydrolab*, an undersea habitat. *Hydrolab* accommodates 4 people for 7-day missions. It is located in 16-m depth in the Salt Rivercanyon No. 5 mi offshore. Excursions from the habitat are done on special SCUBA gear supplied by the *Hydrolab* program. Dives are made to depths of 40 m and sometimes 50 m with special approval.

Contact

Program Science Director
National Undersea Research Program at Fairleigh Dickinson University
West Indies Laboratory
Teague Bay, Christian Sted
St. Croix, U.S. Virgin Island 00820

Table 11–1. (*cont*)

IV. New England coast

Program description

Specific facilities, to be determined by research requests, will be operated on a "lease-trial" basis by NOAA in cooperation with the University of Connecticut. Included may be submersibles, ROV systems, and 1-atm suits.

Contact

Program Science Director
National Underwater Research Program at the University of Connecticut
Avery Point Campus
Groton, Connecticut 06340

V. Hawaii and other Pacific locations

Program description

NOAA, in cooperation with the University of Hawaii's Undersea Research Laboratory, operates the two-person submersible *Makalii* (depth capability 400 m) and the observation/camera-supplied ROV *Snoopy* (depth capability 400 m). The equipment is made available through NOAA grants for appropriate scientific research proposals that are accepted.

Contact

Program Science Director
National Underwater Research Program at the University of Hawaii
2540 Dole St., Hol-401
Honolulu, Hawaii 96822

Note: Proposals from qualified scientists are solicited and the facilities described here are provided via grants for approved projects.

(Table 11–1) for operation at the University of Southern California's marine station at Santa Catalina Island. It is expected to be available for scientific projects by 1985.

III. Manned submersibles

Technology is now available that gives scientists unprecedented working access under water. Two *Johnson-Sea-Link* submersibles, designed by Edwin A. Link and developed and operated by the Harbor Branch Foundation at Fort Pierce, Florida, are being used for a wide range of scientific research including many significant new discoveries about the ecology and distribution of deep-water plants (Eiseman 1978, 1979).

For example, while using the *Johnson-Sea-Link*, Dr. Mark Littler and colleagues (Littler et al. 1985) determined a new maximum depth for benthic plants growing in the sea. It had been thought that plants could not survive below ~100 m because of the low light intensity, until dredged material and then direct observations proved this estimate to be conservative. In 1968 divers using the predecessor

Fig. 11–2. *Johnson-Sea-Link* submersible, operated by the Harbor Branch Foundation, Fort Pierce, Florida. Photograph by S. Earle.

of the *Johnson-Sea-Link*, the Link-designed *Deep Diver*, obtained samples of *Lobophora*, *Caulerpa*, *Microdictyon*, and two filamentous red algae alive in 210-m depth in the Bahamas (Bold and Wynne 1978). Dr. Littler observed two species of crustose Rhodophyta growing attached in the Bahamas to 268 m while conducting continuous transect studies with a television camera (Littler et al. 1985). A complete videotape recording was made along a transect extending from the limit of light penetration (520 m) to the top of a submerged mound in 76-m depth. To obtain accurate, complete documentation of the arrangement and abundance of conspicuous species, a grid was superimposed on the video screen, and appropriate counts, measurements, and permanent records were made at random intervals. Voucher specimens, as well as material for fish preference and primary productivity experiments, were obtained to complement the photographic record by the use of a manipulator arm outside the submersible that was operated with a toggle switch control from within the *Johnson-Sea-Link*.

Several other small research submersibles are available for scientific study in the United States and elsewhere. Among them are the *Makalii*, a two-person system operated by the University of Hawaii in depths to 400 m and the U.S. Navy submersible *Alvin* operated by the Woods Hole Oceanographic Institution and capable of working to nearly 4000 m. The NOAA contributes to the support of these vehicles as a part of the National Underwater Research Program (Table 11–1), and entertains proposals from qualified scientists who are interested in using such facilities.

The cost of operating two- to four-person submersibles ($7000 to $25,000 per day with support vessel) has discouraged their use except on a limited basis. Moreover, the manipulative working capability is very restricted.

Visibility in submersibles is limited by the size of the small portholes (i.e., *Alvin* and the *Makalii*) or by the presence of necessary internal and external equipment (*Johnson-Sea-Link*). Despite such drawbacks, submersibles are proving to be invaluable as means of providing direct observation and working capability, particularly in depths below the range of divers. Age, physical health, and prior underwater experience are not limiting factors in the use of submersibles by scientists. Traveling as a passenger under water is comparable to riding in an automobile, and the safety record is considerably better for submersibles.

IV. One-person atmospheric diving systems

The high cost of operating submersibles that require not only a dedicated trained pilot and support team but also a large support ship, their cumbersome nature, and the difficulty of accomplishing effective work with conventional manipulator systems have led to the development of a new generation of equipment that, in effect, combines the advantages of a submersible (1 atm, no decompression, no physiological restraints, extended time, and extended depth) with the advantages of an unrestrained diver (mobility, dexterity, low cost, transportability, and simplicity of operation).

The first such system was designed by an Englishman, Joseph Peress, in the early 1930s and was used for successful salvage work on the sunken vessel *Lusitania* in depths to 133 m. Called *Jim* (after the first person willing to use it), the system (Fig. 11–3) has been reconfigured and used for work in the offshore oil industry in depths to 650 m.

Fifteen *Jim* units are now being operated worldwide. In 1979, a series of dives was made to evaluate the potential of the system for scientific use (Earle 1980a,b). The maximum depth attained was 417

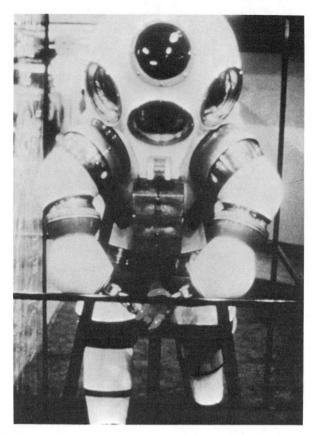

Fig. 11–3. *Jim*, a tethered one-person 1-atm diving system capable of operating in a depth of 650 m. Photograph by Phil Nuytten.

m, 10 km offshore from Makapuu Point, Oahu, Hawaii. No attached plants were observed, but drifting blades of *Ulva lactuca* were common in depths below 300 m.

Advantages of the one-person 1-atm diving system for scientific research proved to be simplicity of operation and, most importantly, the fact that the person conducting the research was also operating the system. *Jim* is not only a submersible that one can wear, but a system that responds to almost instinctive directions from the operator. Disadvantages include the fact that it is somewhat bulky and requires muscle power for all movement of the arms, legs, and manipulators.

A motor-driven version of the one-person 1-atm diving system was designed in 1976 by another Englishman, Graham Hawkes. Called *Wasp* (Figs. 11–4 and 11–5), this "flying" diving system has been

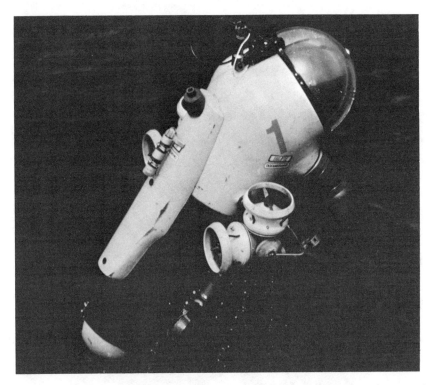

Fig. 11–4. *Wasp*, a tethered motor-powered one-person diving system or "personal submersible" used for research and industrial applications. Photograph by G. Hawkes.

used for two scientific projects, although most of the 18 units in service are engaged by the offshore oil industry.

Dr. Bruce Robison and Dr. Alice Alldredge of the University of California at Santa Barbara conducted a National Science Foundation–sponsored project in 1982 that involved eight scientists who trained to use *Wasp* in depths to nearly 650 m for behavioral studies of zooplankton in the Santa Barbara Channel (Robison 1983).

Dr. Joseph MacInnis led an archeological expedition in the spring of 1983 that included the scientific use of a *Wasp* to explore and recover material, including biological samples, from a sunken ship, the *Breadalbane*, under the ice in the Canadian arctic (MacInnis 1983).

Jim and *Wasp* require muscle power to operate pincherlike claws and the articulated metal "sleeves" that encase the operator's arms. In 1978, Graham Hawkes incorporated the use of internally operated mechanical manipulators with a one-person microsubmersible, a system called *Mantis* (Fig. 11–6). Although widely used for off-shore industrial applications (25 units are presently in service), *Mantis* has

Fig. 11–5. *Wasp,* showing mode of access and articulated metal sleeves. Photograph by S. Earle.

not yet been used for scientific research. The system has been evaluated for its potential by several scientists (Earle 1983a). Its small size, ease of operation, low cost, and high degree of maneuverability will very likely result in the use of *Mantis* for various research applications.

One-person systems can be leased from various diving operation companies for about 20% of the cost of *Alvin* (Robison 1984). *Wasp* and *Jim* units are available through Oceaneering, International and Can-Dive, Inc. *Mantis* can be leased from various companies, including International Underwater Contractors, Cal-Dive International, and, in the United Kingdom, H.M.B. Subworks.

Fig. 11–6. *Mantis*, a tethered microsubmersible for one person using two mechanical manipulators to perform work in depths to 700 m. Photograph by G. Hawkes.

A new system, *Deep Rover* (Fig. 11–7), has been designed by Hawkes (Earle 1981; Hawkes 1983a,b,c; Rawlins and Hawkes 1983). *Deep Rover* features a transparent acrylic sphere that can accommodate one or two operators in depths to a maximum of 1000 m. Totally autonomous, with power and life support sufficient to last for a week if necessary, *Deep Rover* combines many of the advantages of a large submersible (space, time, and comfort) with the advantages of a lightweight, low-cost, self-operated atmospheric diving suit.

Critically important to scientists accustomed to working in a laboratory, or even as a free-swimming diver, are adequate dexterity and sensory feedback from manipulators. Just as astronauts cannot allow their own hands and arms to be exposed to the hostile environments encountered in space, scientists working in deep water (unless using specialized saturation systems) must use pressurized sleeves (as in *Jim* and *Wasp*) or adopt mechanical substitutes for arms and hands (as in *Johnson-Sea-Link, Alvin, Mantis,* and *Deep Rover*).

V. Remotely operated systems

More than 400 remotely operated vehicles are being used for industrial purposes subsea (Earle 1983b), and several are being operated by scientific researchers for general survey and other subsea tasks. Many are equipped with video cameras that transmit images to a surface operator, who can record the terrain below in a manner similar to that used by Littler in the *Johnson-Sea-Link*. Instead of

Fig. 11–7. *Deep Rover,* a microsubmersible launched in 1984 for one or two operators. The system is equipped with two sensory manipulators and can be operated either tethered or autonomously in depths to 1000 m. Photograph by John Bouvier.

traveling under water, the operator can stay topside, commanding the machines below to gather information. Systems equipped with appropriate manipulators, cameras, and sampling containers have already been used to perform at least some of the tasks that previously required the physical presence of a scientist subsea.

Geologists and geophysicists have used deep-tow systems extensively for the past 20 years. Largely developed by the navy with operational support from the National Science Foundation, deep-tow vehicles have proved to be invaluable for gathering information about the nature of the deep sea.

An example of a remotely operated system is the towed vehicle *Angus,* a system that is equipped with cameras to record subsea terrain. Operated by the Woods Hole Oceanographic Institution, *Angus* "discovered" the now well-known communities of giant tube-worms and associated life near the Galápagos Islands (Grassle 1982) that later were explored and studied by scientists in *Alvin.*

At the Harbor Branch Foundation, a remotely operated vehicle, *Cord,* is deployed from a surface vessel, where an operator views and records video images and governs the vehicle's actions. The U.S.

Navy employs numerous kinds of remotely operated systems. *Cord* is an example of one that has been used successfully for ecological survey work (Salizar 1970).

New manipulator systems make it possible to sense touch, motion, force, and sound with dexterity sufficient for an operator to sample selectively even delicate, fragile plants and secure them in containers for transport to the surface (Hawkes 1983a,b,c). It is possible that such systems may, in the future, be used not only with manned submersibles, but also with remotely operated vehicles.

VI. Conclusions

Just as the exploration of microscopic and submicroscopic realms and access to the inner workings of living cells had to await the development of appropriate technology, so has it been necessary to await the advent of SCUBA, habitats, submersibles, video systems, manipulators, and other equipment to understand even the most fundamental facts about life in the sea. We still know very little about the taxonomy and composition of deep-water algal communities and next to nothing about depth distributions, abundances, population biology, and community dynamics. Areas of investigation concerning productivity in blue light, effects of pressure on metabolism, heterotrophy, competition, herbivory, recruitment, natality, and mortality phenomena remain untouched for all algae restricted to the ocean's deep waters. How do plants survive under arctic ice through 6 mo of darkness? What are the implications of plant life in the deep sea to human life on shore? With available technology and developments underway, at least some of these questions can soon be addressed. Until we begin to understand processes at these levels, it will remain impossible to make definitive statements concerning the role of algae or other members of the food web in the biological dynamics of this vast and largely neglected oceanic realm.

VII. References

Adey, W. H., and MacIntyre, I. G. 1973. Crustose coralline algae; a reevaluation in the geological sciences. *Geol. Soc. Amer. Bull.* 84, 833–904.

Bold, H., and Wynne, M. 1978. *Introduction to the Algae.* Prentice Hall, Englewood Cliffs, N.J. 706 pp.

Earle, S. A. 1972. The influence of herbivores on the marine plants of Great Lameshur Bay, with an annotated list of species. In Collette, B. B., and Earle, S. A. (eds.), *Results of the Tektite Project: Ecology of Coral Reef Fishes*, pp. 17–44. Science Bulletin 14, Natural History Museum of Los Angeles County, Los Angeles.

Earle, S. 1980a. A walk in the deep. *Nat. Geogr.* 157, 624–31.

Earle, S. 1980b. *Exploring the Deep Frontier*. National Geographic Society, Washington, D.C. 296 pp.

Earle, S. 1981. The descent of man. *Science '81*. 2, 44–52.

Earle, S. 1983a. Application of one man atmospheric diving systems for research and exploration. *Mar. Technol. Soc. J.* 17, 29–39.

Earle, S. 1983b. Will robots replace man in the sea? *Sea Technol.* 24, 69.

Eiseman, N. 1978. Observations of the marine algae occurring from 30 to 100 m. depths off the east coast of Florida. *J. Phycol.* 14, 25 (abstract).

Eiseman, N. 1979. Marine algae of the east Florida continental shelf 1. Some new records of Rhodophyta, including *Scinaia incrassata*, sp. nov. (Nemalionales: Chaetangiaceae). *Phycologia* 18, 355–61.

Eiseman, N., and Earle, S. 1983. *Johnson-sea-linkia profunda*, a new genus and species of deep water chlorophyta from the Bahama Islands. *Phycologia* 22, 1–6.

Grassle, J. F. 1982. The biology of hydrothermal vents: a short summary of recent findings. *Mar. Technol. Soc. J.* 16(33), 8.

Hawkes, G. 1983a. The future of atmospheric diving systems and associated manipulator technology, with special reference to a new submersible, *Deep Rover*. *Mar. Technol. Soc. J.* 17, 51–60.

Hawkes, G. 1983b. Advances in diver alternative systems. In *Proceedings of the Society of Petroleum Engineers*, paper no. 11741, pp. 699–706.

Hawkes, G. 1983c. Advanced manipulator concepts and applications. In *Conference Proceedings ROV '83*, pp. 72–81. The Marine Technology Society, San Diego Section.

Humm, H. J. 1954. Rediscovery of *Anadyomene menziesii*, a deep-water green alga from the Gulf of Mexico. *Bull. Mar. Sci.* 6, 346–8.

Hurley, A., Cailliet, G., Josselyn M., Niesen, T., Cowen, R., Hawes, S. and Conner, J. 1981. *The Sources, Dispersal, and Utilization of Benthic Drifting Plants in the Salt River Canyon*. Missions 80-2 and 80-7, final report, National Underseas Laboratory Program, NOAA, Washington, D.C. 47 pp.

Josselyn, M., Calliet, G., Niesen, T., Cowen, R., Hurley, A., Conner, J., and Hawes, S. 1983. Composition, export and faunal utilization of drift vegetation in the Salt River Submarine Canyon. *Est. Coast. Shelf Sci.* 17, 447–65.

Littler, M. M., Littler, D. S., Blair, S., and Norris, J. N. 1985. Deepest known plant life discovered on an uncharted seamount. *Science* 227, 57–59.

MacInnis, J. 1983. Exploring a 140-year-old ship under arctic ice. *Nat. Geogr* 164, 104a–d.

Mathieson, A., Fralick, R. A., Burns, R., and Flahive, W. 1975. Phycological studies during Tektite II, at St. John, U.S.V.I. In Earle, S., and Lavenberg, R. (eds.), *Results of the Tektite Program: Coral Reef Invertebrates and Plants*, pp. 77–103. Science Bulletin 20, Natural History Museum of Los Angeles County, Los Angeles.

Miller, J. W. (ed.). 1975. *The NOAA Diving Manual: Diving for Science and Technology* (National Oceanic and Atmospheric Administration). U.S. Government Printing Office, Washington, D.C. 550 pp.

Rawlins, J., and Hawkes, G. 1983. One man atmospheric diving systems: safety considerations. *Mar. Technol. Soc. J.* 17, 40–50.

Robison, B. 1983. Midwater biological research with the Wasp ADS. *Mar. Technol. Soc. J.* 17, 21–7.

Robison, B. 1984. Submersibles as tools for midwater biology. Special Session on New Developments in Ocean Science Instrumentation from a User's Perspective. Abstract, AGO/ASLO Meeting, New Orleans, January 1984.

Salizar, M. H. 1970. Photoazis in the deep-sea urchin *Allocentrotus fragilis* (Jackson). *J. Exp. Mar. Biol. Ecol.* 5, 254–64.

Taylor, W. R. 1928. *The Marine Algae of Florida with Special Reference to the Dry Tortugas.* Publ. 379, Carnegie Institution, Washington, D.C. 219 pp.

12: Demography

A. R. O. CHAPMAN

*Department of Biology, Dalhousie University, Halifax, Nova Scotia,
Canada B3H 4J1*

CONTENTS

[251]

I. Objective

The objective of a demographic study is to determine the effects of population parameters on the density or size of a single species population. The four primary population parameters that affect density are natality, mortality, immigration, and emigration. Most seaweeds are attached to a surface so that immigration and emigration are of only minor importance. Secondary population parameters that are frequently studied include age class distribution, sex ratios, and ploidy ratios. Seaweed populations are relatively easy to manipulate in the field, and this makes it possible to measure the significance of various processes on population structure.

II. Equipment

A. Density determinations

1. Quadrats. Quadrat size is matched to plant size (see Sec. III.A.1). Quadrats of 100 cm^2 and 0.25 m^2 should be made from 5-mm-diameter stainless steel wire bent into shape. Quadrats of 5 m^2 used in kelp forests are made from 2.5-cm angle iron or angle aluminum bolted or welded together.

2. Transect line. A transect line is needed for stratified random sampling (see Sec. III.A.2). A 30-m graduated fiberglass tape (Forestry Suppliers Inc.) weighted with clip-on lead shot is used.

B. Tagging and plant-locating equipment
1. Tags
   a. Surveyor's flagging (Forestry Suppliers Inc.)
   b. Polyester hair ribbon (5 mm)
   c. Nylon monofilament tags (Fig. 12–1)
   d. Cinch-up plastic tags (FT-4, Floy Tag Co.)
2. Plant location by position
   a. Strung frame quadrat: 2.5-cm bolted aluminum stock, 40 × 40 cm, strung with twine at 1-cm intervals

[253]

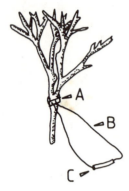

Fig. 12–1. Nylon monofilament tag (developed by J. Pringle and G. Sharp) on a red algal specimen. A, Latex rubber plug through which monofilament passes; used to cinch up tag. B, Monofilament. C, Numbered, hollow "spaghetti" label.

   b. Epoxy for marking quadrat positions (Sea Goin' Poxy, Permalite Plastics Corp).

   c. Pantograph (Forestry Suppliers Inc).

   d. Nikonos camera, 35-mm lens with close-up tube and wire frame (Ehrenreich Photo Optical Inc)

   e. Box frame for photogrammetry: 1.25-cm copper tubing with soldered copper elbows; frame 50 cm on a side with 2-cm alternating black- and white-banded markings (Johnston et al. 1969)

C. Cultivation equipment

All details can be found in Chapman (1973) and McLachlan (1973).

D. Fecundity measurements

  1. Microscopes and accessories

    a. Compound microscope

    b. Dissecting microscope with ×400 magnification

    c. Sedgewick Rafter chamber (Clay Adams)

    d. Hemocytometer

  2. Freezing microtome (American Optical)

  3. Zeiss Ibas image analyzer (Carl Zeiss)

  4. Gyratory shaker (New Brunswick Scientific Corp.)

  5. Top-loading balance

  6. Vacuum filtration apparatus (Gelman)

E. Ploidy determinations

  1. Resorcinal–acetal reagents and equipment described by Craigie and Leigh (1978)

  2. Equipment for examination of chromosomes (see Godward 1966 for details)

III. METHODS

A. Measurement of density

The study of density regulation by population parameters is central to any demographic analysis. Therefore, the measurement of density is clearly an important starting point. Density is the number of individuals per unit area. Normally, one can estimate density by counting the number of individuals in quadrats placed within a study area. The choices of quadrat size, quadrat number, and quadrat position are important in estimating density (see De Wreede, Chap. 7).

1. Quadrat size and shape. Quadrat size is normally determined by the size of the plants being studied. When there are thousands of individuals per square meter, as in *Chordaria flagelliformis* populations, a quadrat size of 100 cm² is suggested (Rice and Chapman 1982). When plants are large and number perhaps one or two per square meter, a quadrat size of 4.5 m² is useful. For intermediate sizes and densities, 0.25-m² quadrats are commonly used. The shapes and sizes of quadrats influence the variance of the density estimates when plants are clumped. Very large quadrats reduce variances because they include low- and high-density patches in the same count (Kershaw 1964).

2. Quadrat number and position. There is no general rule prescribing the number of quadrats required for density estimates. The best way to obtain an acceptable estimate is to take 5 samples and obtain a mean density. The number of quadrat samples is increased to 10 and a new mean calculated. When the mean values remain approximately constant with increasing sample number, an appropriate estimate of density has been achieved.

Quadrats are best arrayed in a stratified random fashion. A transect line is placed along a shore gradient and quadrats placed randomly at chosen levels on the transect. Thus, for instance, quadrats would be located at random distances laterally between the 3- and 4-m marks on the transect line. Subsequently, random samples might be made between the 6- and 8-m marks, and so on. Random distances from the transect line are determined from random number tables (not haphazardly).

3. Identifying individuals. Most seaweeds have a modular construction by which the products of one zygote or spore can differentiate into many individuals. In practice, it is often impossible to discriminate between the products of separate zygotes and spores. In these cases, it is best to consider the population biology of the modular components (ramets) rather than the genets (Harper 1977). For example, Cousens (1981) studied the demography of upright fronds of *Ascophyllum nodosum* regardless of zygotic origins.

B. Measurement of recruitment rates

1. Microscopic stage bank. In many seaweeds there is a huge bank of unseen benthic microscopic stages. This bank is analogous to the seed bank of vascular plants (Harper 1977). In both cases, the number of individuals vastly exceeds the number of visible adult plants. The members of the microscopic stage bank in seaweeds may be alternate ploidy stages in the life history or simply stunted, microscopic forms of the adult plants. There are no resistant resting stages in seaweeds, so that the analogy to a seed bank is rather tenuous in this regard.

Estimating the size of the bank of microscopic stages is not easy but has been attempted for two *Laminaria* species by placing pieces of substratum from the kelp forest floor under sprays of running seawater (A. Chapman, unpublished data). Epiphyte competition is minimal under seawater sprays. Growth of microscopic stages made counting with a dissecting microscope possible within 6 wk.

The size of the microscopic stage bank at any one time is a static population parameter. A greater appreciation of the dynamics of the bank can be obtained by the introduction of sterilized pieces of substratum into the seaweed bed at regular intervals through the reproductive season. These can be retrieved on a regular basis for counting of microscopic stages in the manner outlined earlier.

C. Measurement of age- and size-specific mortality

Mortality estimates are central to any demographic study. To measure mortality, it is necessary to follow the history of known individuals until their deaths. It is therefore essential to identify individuals either through tagging or mapping.

Mortality (like other demographic rates) is measured in relation to age and/or size. Age is usually difficult to determine in a nondestructive manner, and it is best to follow the fate of a cohort of individuals recruited at the same time. The usual starting point is when plants first become visible to the unaided eye. If population parameters are to be related to the size of plants, then measurement of size must be nondestructive.

1. Tagging. Tagging seaweeds is not at all straightforward because it may influence death rates (see Foster et al., Chap. 10). Ideal subjects such as *Laminaria digitata* in semisheltered locations can be tagged with surveyor's flagging secured around the stipe. Polyethylene flagging should be used; vinyl flagging quickly breaks up. Although tags may survive in the sea for 12 to 15 mo, they should be replaced at shorter intervals as they show signs of deterioration.

In exposed and semiexposed locations, flagging quickly disintegrates. In these cases, polyester fabric ribbon is recommended. Ribbon comes in a variety of widths that can be matched to the size of the seaweed being studied. The ribbon and polyethylene tags can be numbered with an alcohol-base felt pen. These types of tags have persisted in Nova Scotia kelp beds for more than 2 years.

Tags for *Chondrus crispus* have been made from nylon monofilament and latex rubber by J. Pringle and G. Sharp (personal communication). A numbered "spaghetti" label (from Floy Tag Co.) is threaded onto the nylon monofilament (Fig. 12–1).

Because seaweeds vary so much in morphology and texture, it is difficult to recommend universal tagging systems. A wide variety of tags is manufactured by the Floy Tag Co. (see Sec. II.F), and it may be possible to find suitable types from among that company's selection.

2. Mapping. Some seaweeds simply cannot be tagged without being damaged. In these cases, it is necessary to identify individuals from their position on the substratum. This procedure has been used by Chapman and Goudey (1983) to map the positions of individual *Leathesia difformis* plants. The quadrat described in Sec. II.B is positioned in the desired location so that plants to be studied are enclosed within its bounds. Dabs of freshly mixed epoxy (Sec. II.B) are applied to the rock under each corner of the quadrat. The quadrat is then pushed into the epoxy and withdrawn, leaving imprints of the corners. The quadrat can be easily relocated at a later date. Plants are identified from their coordinate positions in the strung grid of the quadrat. Obviously, the smaller the plant, the smaller the grid spacings should be.

Harper (1977: 572, fig. 19/12b) has described the use of a pantograph to map the positions of buttercup plants in fields. A pantograph is a draftsman's instrument consisting of jointed rods fitted together so that they form an extendable arm joining two drawing pens. The instrument is commonly used to transfer precision plans from one surface to another, with facility for reduction or enlargment. It can be used to record the positions of individual plants on a vegetation map. The method is probably impractical on rough substratum. Forestry Suppliers Inc. can supply a precision instrument.

Photography provides a rapid means of mapping plant positions. Two-dimensional images have been used to map the positions of early stages of *Chondrus crispus* before tagging (D. Bhattacharya, unpublished data). A Nikonos with a 35-mm lens, close-up tube, and wire frame (Sec. II.B) is suitable for this kind of work.

Ordinary two-dimensional images of three-dimensional plants are sometimes difficult to interpret. Photogrammetric techniques provide three-dimensional images and have been used by Svane and Lundläv (1981) to study mortality in sessile ascidians. The technique involves photographing the same field twice from different positions on a metal frame. The two images are then viewed simultaneously in a stereoscope to give a three-dimensional image.

The simplest, but perhaps most effective photogrammetric technique for positioning and relocating seaweeds was described by Johnston et al. (1969). The apparatus is described in Sec. II.B. The box frame is placed in position, and photographs are taken from the following angles: (1) nearly vertical downview and (2) three views at an angle of depression of ~45°. The position of an individual plant can be worked out by plotting a scale plan of the metal frame.

D. Measurement of age- and size-specific fecundity and fertility

Fecundity is the total number of eggs or spores produced by an organism. Fertility is the number of *viable* eggs or spores. Total potential fecundity can be estimated by counting eggs or spores in sporangia and oogonia before they are released. Scagel (1961) made early estimates of individual plant fecundity in *Nereocystis leutkeana* and *Rhodymenia pertusa* in this way. In kelps, estimates are made from sorus sections cut parallel to the surface with a freezing microtome (Sec. II.D). Microscope counts within a known area of sorus can then be made. One then determines the area of sorus by tracing its outline with a wax pencil onto sheets of transparent polyethylene. The area of the tracings can be determined with an image analyzer (Sec. II.D).

The fecundity of British fucoids was measured by counting the following parameters on each plant: the number of receptacles, the number of conceptacles per receptacle, and the numbers of oogonia and antheridia per conceptacle (Vernet and Harper 1980). The number of gametangia per conceptacle was determined microscopically after the receptacles were softened for 15 min in 1 M sodium carbonate.

Counting eggs or spores before release is often not possible. In such cases, it is best to release the reproductive bodies into a known volume of water in a conical flask on a shaker (Sec. II.D). Shaking is necessary to retard attachment. Water samples are taken and reproductive body density counted in a 1.0-ml Sedgewick Rafter chamber (Sec. II.D). After a first attempt at spore or egg release, the sporangia or oogonia should be checked microscopically for complete discharge. If discharge is incomplete, the process should be repeated. Spore release in *Chondrus crispus* has been successfully quantified in this way (D. Bhattacharya, personal communication).

An estimate of viability of spores and/or eggs is required for measuring fertility. A viability estimate is obtained by culturing reproductive structures and observing whether germination occurs. Methods for cultivation are described by Chapman (1973).

E. Measurement of reproductive effort

The term "reproductive effort" is often confused with "fecundity." Reproductive effort is a dimensionless number that is the proportion of energy at a given age that is allocated to reproduction. In plants, reproductive effort is normally measured as the proportive of total plant weight (dry) that is allocated to reproduction disseminules. Alternatively, reproductive effort can be defined in terms of the total biomass expenditure involved in reproduction. This broader definition includes sterile structures in reproductive organs, as well as spores and eggs.

Vernet and Harper (1980) obtained counts of the eggs and sperms produced in British fucoids (see Sec. III.D). These numbers were converted to dry weights in the following way. The average volumes of eggs and sperms were determined microscopically. Vernet and Harper (1980) assumed a density of 1.0 and a dry weight–fresh weight ratio of 0.22 to calculate the weight of each gamete type.

G. Rosenberg (personal communication) estimated the weights of spores of *Laminaria longicruris* in the following way. Spores were released into suspension after drying of sorus material (Chapman 1973). The spores were filtered under slight vacuum through a coarse (25-μm) filter and trapped on preweighed fiberglass filters (Whatman GF/C). The filters were then reweighed after drying at 60°C for 3 d. The spore density in suspension was counted on a hemocytometer (Guillard, 1973) so that spore weights on the filters could be converted to individual spore weights.

F. Measurement of reproductive dispersal

Quantifying reproductive dispersal in seaweeds is a difficult task. The basic objective is to find the relationship between the number of disseminules collected and the distance from the point of parental release. This relationship has been studied only rarely in seaweeds (Anderson and North 1966; Dayton 1973; Deysher and Norton 1981) because it is difficult, in a more or less continuous stand, to distinguish the disseminules of one individual from those of others of the same species. All of the methods used to date depend on the discovery (and subsequent counting) of recruits beyond the boundaries of stands of reproductive adults. Anderson and North (1966) moved plants of *Macrocystis pyrifera* to a barren area made devoid of fleshy seaweeds by sea urchin grazing. Settlement of sporelings around kelp transplants (after the removal of sea urchins) gave a quantitative estimate of reproductive dispersal. This method clearly has limited application.

None of the methods used to date discriminates between the dispersal of spores and gametes and the dispersal of reproductively mature plants (or plant fragments) that have been dislodged.

G. Age class structures, sex ratios, and ploidy ratios

Age class structures, sex ratios, and ploidy ratios are population attributes studied at single points in time. The parameters are therefore static, and great care is necessary if they are to be used in interpreting population dynamics.

1. Age class distribution. Random samples of individuals are aged and the age class frequency determined. In some species that have been studied carefully [e.g., *Laminaria hyperborea* (Kain 1963)] this is a relatively straightforward procedure. Kain recommends the following procedure for aging *L. hyperborea*. A longitudinal cut is made from just above the holdfast region to the basal end of the stipe, passing through one of the rows of haptera, if possible. One then determines age by counting annual growth lines. During the growing season of *L. hyperborea*, there is considerable variation in the time of rapid secondary growth. Plants of the same age may appear to have different numbers of growth rings. However, the pale smooth texture of new secondary growth allows this to be taken into account.

Determination of age by procedures other than counting growth rings is fraught with problems. For example, *Ascophyllum nodosum* forms one vesicle on each axis in each year (Baardseth 1968). Theoretically, age can be determined by counting vesicles. However, Cousens (1981) showed that most axes are broken. Finding a "minimum" age by counting vesicles upward from the holdfast along the route that gives the highest total is not acceptable.

2. Sex ratios. The composition of many dioecious species populations deviates markedly from the expected male/female ratio of 1. For example, males of *Chondrus crispus* appear to be rarer than females (Hanic 1973). When studying the sex ratios of a dioecious species, one should sample the population in random quadrats (Sec. II.A). The identification of males and females is part of classical phycology, and the reader is referred to Fritsch (1935, 1945) and Kylin (1956) for methods.

3. Ploidy ratios. Many seaweeds exist in independent haploid and diploid phases. In this respect, they differ from the exclusively diploid higher plants and animals most frequently studied in demographic investigations. Ploidy status can be determined from vegetative morphology when there are dissimilar components of the life history or from reproductive morphology when plants are reproductively mature (for descriptions of ploidy-specific reproductive structures see Fritsch 1935, 1945; Kylin 1956; Bold and Wynne 1978). In a population study, ploidy status should be determined in immature as well as mature plants. When diploids and haploids are vegetatively identical, this presents difficulties. In the family Gigartinaceae (Florideophyceae) diploids and haploids can be distinguished by the resorcinol–acetal test (for details, see Craigie and Leigh 1978). In some cases, there is no alternative to making chromosome preparations for direct microscopic examination of ploidy status. Descriptions of procedures for visualising the exceptionally small chromosomes of seaweeds are given in Godward (1966).

H. Intraspecific interaction

Members of the same species may interact competitively, and the study of this interaction in wild populations is an important part of population biology. In plant populations, interactions are often studied by measuring the effects of plant density on mortality. These effects can be quantified by observing variations in the parameter of interest (e.g., mortality) under naturally occurring variations in density. This approach was used by Black (1974) to show how densities of larger individuals of the kelp *Egregia laevigata* influence the mortalities of new recruits of the same species (see Denley and Dayton, Chap. 25, for details). An alternative approach is to manipulate plant densities and to observe the effects. Increasing densities in natural stands is usually not possible. A decrease in natural stand density is usually obtained by cutting away visible plants. Ideally, a range of densities and control plots should be set up.

Crowding effects of conspecific neighbors are not always simply the result of increasing densities. In the case of cohorts of annuals that may recruit over short time spans, increasing size of individual plants contributes considerably to crowding. Density and plant size components of crowding are expressed in Deevey's coefficient of crowding $C_c = 2\pi r^2 N^2$, where r is the radius of an individual and N is density (Southwood 1966).

IV. Sample data

A. Age-specific mortality schedules

Data on mortality in populations are best expressed in a life table budget. Table 12–1 is a sample table for the kelp *Pelagophycus porra*

Table 12–1. *Life table for Pelagophycus porra*

Age (months)	l_x	d_x	$1000q_x$	L_x	T_x	e_x
0–1	1000	157	157	921.5	6832.0	6.8
1–2	843	143	170	771.5	5910.5	7.0
2–3	700	71	101	664.5	5139.0	7.3
3–4	629	28	44	614.5	4474.5	7.1
4–5	600	43	72	578.5	3860.0	6.4
5–6	557	43	77	535.5	3281.5	5.9
6–7	514	43	84	492.5	2746.0	5.3
7–8	471	43	91	450.0	2253.5	4.8
8–9	429	28	65	414.5	1083.5	4.2
9–10	400	57	143	371.5	1389.0	3.5
10–11	343	129	376	278.5	1017.5	3.0
11–12	214	71	332	178.5	739.0	3.5
12–13	143	29	203	128.5	560.5	3.9
13–14	114	57	500	85.5	432.0	3.8
14–15	57	14	246	50.0	346.5	6.1
15–16	43	14	326	36.0	296.0	6.9
16–17	29	0	0	29.0	260.5	9.0
17–18	29	0	0	29.0	231.5	8.0
18–19	29	0	0	29.0	202.5	7.0
19–20	29	0	0	29.0	173.5	6.0
20–21	29	0	0	29.0	144.5	5.0
21–22	29	0	0	29.0	115.5	4.0
22–23	29	0	0	29.0	86.5	3.0
23–24	29	0	0	29.0	57.5	2.0
24–25	29	14	483.0	21.5	28.5	1.0
25–26	14	14	1000.0	7.0	7.0	0.5
26–27	0	0	—	—	—	—

Note: The parameters are as follows: l_x, number of survivors at beginning of age interval x; d_x, number of plants dying in age interval x; $1000q_x$, number of plants dying per 1000 alive at beginning of age x; L_x, number of plants alive betwen age x and age $x + 1$; T_x, sum of months of life remaining to those age x; e_x, average life expectancy of those age x.
Source: After Coyer and Zaugg-Haglund (1982).

(data from Coyer and Zaugg-Haglund 1982). The raw data in the column headed l_x (number of individuals alive at the beginning of age interval x) are converted to give age-specific mortality q_x and age-specific life expectancy e_x in the manner shown (Table 12–1).

The construction of the e_x column in a life table depends on the censusing of numbers of cohort survivors at equally spaced intervals through time. This condition may not be obtained when wind and surf conditions prevent censusing on chosen days. In this case, the survivorship data for the selected time intervals must be calculated

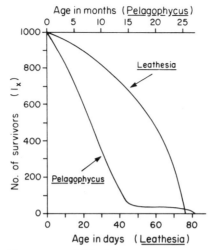

Fig. 12–2. Survivorship curves exhibited by a summer annual seaweed (*Leathesia difformis*) and a perennial seaweed (*Pelagophycus porra*). After Chapman and Goudey (1983) and Coyer and Zaugg-Haglund (1982).

from an equation that describes the survivorship curve (1_x versus x). Chapman and Goudey (1983) used this method to calculate age-specific life expectancy in *Leathesia difformis*. They found that the survivorship curve could be described by the Gompertz equation,

$$N = a \exp(-be^{kt})$$

where N is the number of survivors, t is time, and a and k are constants. The method is described by Batschelet (1971).

B. Survivorship curve

A survivorship curve is obtained by plotting the number of cohort survivors against age. The shape of the curve is used to compare life history strategies of populations. Figure 12–2 shows survivorship curves of *Pelagophycus porra* and *Leathesia difformis*. Clearly, relative mortality of the youngest stages differs markedly in the two species.

It is difficult to compare the survivorship curves of species with different longevities. A solution to this problem lies in the use of the Weibull function to describe the curves (Pinder et al. 1978). Cousens (1981) used the Weibull curve function to analyze survivorship in *Ascophyllum nodosum*.

C. Depletion curve

The term "depletion curve" is used to describe the rate of loss of individuals from a population of unknown age structure (Harper

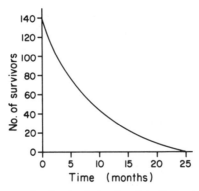

Fig. 12–3. Depletion curve for *Laminaria longicruris* individuals in mixed age and size classes.

Table 12–2. *Fecundity of two Laminaria species (10 plants among a sample of 102 analyzed, 24 August 1981)*

Plant no.	Species	Sorus area (cm^2)	No. of sporangia	No. of spores
1	*L. longicruris*	978	2.73×10^8	8.75×10^9
2	*L. longicruris*	0	0	0
3	*L. longicruris*	0	0	0
4	*L. longicruris*	817	2.28×10^8	7.3×10^9
5	*L. digitata*	0	0	0
6	*L. digitata*	344	7.16×10^7	2.3×10^9
7	*L. digitata*	295	6.13×10^7	1.96×10^9
8	*L. digitata*	89	1.86×10^7	5.9×10^8
9	*L. digitata*	0	0	0
10	*L. digitata*	0	0	0

Note: Sporangial densities: *L. longicruris,,* $2.79 \times 10^5 \pm 1.88 \times 10^4$ cm^{-2}; *L. digitata*, $2.08 \times 10^5 \pm 1.11 \times 10^4$ cm^{-2}. Thirty-two spores per sporangium assumed for both species.

1977). A depletion curve for *Laminaria longicruris* is shown in Fig. 12–3.

D. Fecundity of two species of Laminaria

The spore productions of 10 individuals of *Laminaria longicruris* and *L. digitata* are shown in Table 12–2. The 10 individuals are from a sample of 102 plants analyzed in August 1981. The fecundity is obviously of astronomical dimensions. At a single point in time, a single kelp individual may bear over 8 billion spores (A. Chapman, unpublished data).

Table 12–3. *Recruitment of juveniles of Egregia laevigata after removal of adult plants and/or other sessile organisms*

Sessile organisms	Old kelp	
	Present	Absent
Present	175 ± 31.4	534.4 ± 104
Absent	81.6 ± 39.4	376.4 ± 57

Note: Juvenile densities are given as the number of organisms per square meter ± standard error.
Source: Black (1974).

E. Reproductive effort in some fucoids

Cousens (1981) calculated the reproductive effort of *Ascophyllum nodosum* at a variety of tidal heights at Polly Cove in Nova Scotia. At this site, reproductive effort varied between 61 and 66%. Cousens's calculations were based on measurements of total receptacle weight, including gametes and sterile tissue. In contrast, Vernet and Harper (1980) calculated the reproductive effort of a variety of British fucoids on the basis of the proportion of total body weight allocated to gametes alone. Among all of the species tested only 0.1–0.4% of biomass was allocated to gametes.

F. Bank of microscopic stages in Laminaria longicruris and L. digitata

Ceramic bricks were left in a mixed kelp bed of two *Laminaria* species for the period 6 October 1981 to 5 November 1981. After this period, the settled microscopic stages of both kelps were grown to visible size and counted (Sec. III.B.1). Juveniles of the two species were identified by the benzidine test (Kain 1971). The mean densities of *L. longicruris* and *L. digitata* were $511{,}121 \text{ m}^{-2}$ and $270{,}040 \text{ m}^{-2}$, respectively. Since the mean densities of macroscopic *L. longicruris* and *L. digitata* in the same bed were only 1.2 and 3.2 m^{-2}, respectively, it can be seen that there must be massive mortality among benthic juvenile stages.

G. Intraspecific competition in Egregia laevigata

Black (1974) carried out experimental clearings of large individuals of the kelp *Egregia laevigata* to test their effects on juvenile recruitment. Experimental clearings were done where other sessile organisms were either present or absent (removed manually). It is obvious (Table 12–3) that there was a significant increase in juvenile recruitment when adult kelp was absent.

V. Problem areas

A. *Estimating recruitment of microscopic stages*

Trapping and estimating the settlement of plants from the rain of disseminules are presently the weakest links in any seaweed demographic study. The recruitment data of Anderson and North (1966), Dayton (1973), and Deysher and Norton (1981) suggest that dispersal is limited, at least in some species. Therefore, if adult distribution is patchy, the rain of spores and eggs is likely to be patchy. Furthermore, the disseminules may be substratum selective so that the trapping surfaces may be either more or less attractive than other available substrata.

Since trapped juvenile stages are microscopic, they are usually unidentifiable. They must therefore be grown to some identifiable stage before counting. However, Kain (1975) has shown that, when trapped microscopic stages are grown in static culture, the proportions of species changes over time. In addition, the reproductive potential of the microscopic stages varies with degree of nutrient enrichment and illumination provided to the cultures.

B. *Tag loss and tag-induced mortality*

Mortality among macroscopic plants is usually estimated by the rate of loss of tagged individuals from the population. If tags are lost before the deaths of plants, an increased mortality estimate will result. At least some of the plants in a study should be tagged twice (see Foster et al., Chap. 10); a rate of loss of tags from these plants can be calculated independently of plant death.

Plant tags often increase mortality by mechanical injury or by increasing form drag. Tag-induced losses and injuries should be estimated from a comparison between tagged individuals and those identified by mapping alone.

C. *Integrating spore and egg production over time*

The methods described in Sec. III.D for measuring fecundity refer to plants sampled at single points in time. In practice, it is difficult to integrate production over time, because new gametangia and sporangia are being formed while older reproductive structures are being shed. Some attempt should be made to study the dynamics of this process. In fucoids the loss of mature receptacles and growth of new ones can be followed on individually marked branches. Similarly, the rates of erosion and the addition of new sorus material might be measured in kelps (albeit with great difficulty).

VI. References

Anderson, E. K., and North, W. J. 1966. *In situ* studies of spore production and dispersal in the giant kelp *Macrocystis*. In Young, E. G., and McLachlan, J. L. (eds.), *Proceedings of the Fifth International Seaweed Symposium*, pp. 73–86. Pergamon Press, Oxford.

Baardseth, E. 1968. *Synopsis of Biological Data on Ascophyllum nodosum (Linnaeus) Le Jolis. Food and Agriculture Organization Fisheries Synopsis*, no. 38, Rome (pagination variable).

Batschelet, E. 1971. *Introduction to Mathematics for Life Scientists*. Springer-Verlag, Berlin, 493 pp.

Black, R. 1974. Some biological interactions affecting intertidal populations of the kelp *Egregia laevigata. Mar. Biol.* 28, 189–98.

Bold, H. C., and Wynne, M. J. 1978. *Introduction to the Algae: Structure and Reproduction*. Prentice-Hall, Englewood Cliffs, N.J. 706 pp.

Chapman, A. R. O. 1973. Methods for macroscopic algae. In Stein, J. R. (ed.), *Handbook of Phycological Methods: Culture Methods and Growth Measurements*, pp. 87–104. Cambridge University Press, Cambridge.

Chapman, A. R. O., and Goudey, C. L. 1983. Demographic study of the macrothallus of *Leathesia difformis* (Phaeophyta) in Nova Scotia. *Can. J. Bot.* 61, 319–23.

Cousens, R. 1981. "The Population Biology of *Ascophyllum nodosum* (L.) Le Jolis." Ph.D. dissertation, Dalhousie University, Halifax, Nova Scotia. 273 pp.

Coyer, J. A., and Zaugg-Haglund, A. C. 1982. A demographic study of the elk kelp *Pelagophycus porra* (Laminariales, Lessoniaceae), with notes on *Pelogophycus* × *Macrocystis* hybrids. *Phycologia* 21, 399–407.

Craigie, J. S., and Leigh, C. 1978. Carrageenans and agars. In Hellebust, J. A., and Craigie, J. S. (eds.), *Handbook of Phycological Methods: Physiological and Biochemical Methods*, pp. 109–31. Cambridge University Press, Cambridge.

Dayton, P. K. 1973. Dispersion, dispersal and persistence of an annual intertidal alga, *Postelsia palmaeformis* Ruprecht. *Ecology* 54, 433–8.

Deysher, L., and Norton, T. A. 1981. Dispersal and colonization of *Sargassum muticum* (Yendo) Fensholt. *J. Exp. Mar. Biol. Ecol.* 56, 179–95.

Fritsch, F. E. 1935. *The Structure and Reproduction of the Algae*, vol. 1. Cambridge University Press, Cambridge. 791 pp.

Fritsch, F. E. 1945. *The Structure and Reproduction of the Algae*, vol. 2. Cambridge University Press, Cambridge. 939 pp.

Godward, M. B. E. 1966. *The Chromosomes of the Algae*. Arnold, London. 212 pp.

Guillard, R. L. 1973. Division rates. In Stein, J. R. (ed.), *Handbook of Phycological Methods: Culture Methods and Growth Measurements*, pp. 289–311. Cambridge University Press, Cambridge.

Hanic, L. A. 1973. Cytology and genetics of *Chondrus crispus*. In Harvey, M. J. (ed.) *The Biology of Chondrus crispus*, pp. 34–52. Proceedings of the Nova Scotian Institute of Science, vol. 27 (suppl.), Halifax.

Harper, J. L. 1977. *Population Biology of Plants*. Academic Press, London. 892 pp.

Johnston, C. S., Morrison, I. A., and MacLachlan, K. 1969. A photographic method for recording the underwater distribution of marine benthic organisms. *J. Ecol.* 57, 453–9.

Kain, J. M. 1963. Aspects of the biology of *Laminaria hyperborea*. 2. Age, weight and length. *J. Mar. Biol. Assoc. U.K.* 43, 129–51.

Kain, J. M. 1971. *Synopsis of Biological Data on Laminaria hyperborea*. *Food and Agriculture Organization Fisheries Synopsis* no. 87, Rome. 66 pp.

Kain, J. M. 1975. Algal recolonization of some cleared subtidal areas. *J. Ecol.* 63, 739–65.

Kershaw, K. A. 1964. *Quantitative and Dynamic Ecology*. Arnold, London, 183 pp.

Kylin, H. 1956. *Die Gattungen der Rhodophyceen*. CWK Gleerups Förlag, Lund, Sweden. 673 pp.

McLachlan, J. 1973. Growth media: marine. In Stein, J. R. (ed.), *Handbook of Phycological Methods: Culture Methods and Growth Measurements*, pp. 25–51. Cambridge University Press, Cambridge.

Pinder, J. E., Wiener, J. G., and Smith, M. H. 1978. The Weibull distribution: a new method of summarizing survivorship data. *Ecology* 59, 175–9.

Rice, E. L., and Chapman, A. R. O. 1982. Net productivity of two cohorts of *Chordaria flagelliformis* (Phaeophyta) in Nova Scotia, Canada. *Mar. Biol.* 71, 107–11.

Scagel, R. F. 1961. Marine plant resources of British Columbia. *Bull. Fish. Res. Bd. Can.* 127, 1–39.

Southwood, T. R. E. 1966. *Ecological Methods*. Chapman & Hall, London. 391 pp.

Svane, I., and Lundälv, T. 1981. Reproductive patterns and population dynamics of *Ascidia mentula* O. F. Müll. on the Swedish west coast. *J. Exp. Mar. Biol. Ecol.* 50, 163–82.

Vernet, P., and Harper, J. L. 1980. The cost of sex in seaweeds. *J. Linn. Soc. Biol.* 13, 129–38.

13: Succession

MICHAEL S. FOSTER

Moss Landing Marine Laboratories, P.O. Box 223, Moss Landing, California 95039

WAYNE P. SOUSA

Department of Zoology, University of California, Berkeley, California 94720

CONTENTS

I. Introduction

The study of succession has been of major importance to community ecologists since Clements (1916) proposed the facilitation model (early successional species facilitate the establishment of later ones) of the successional process as an integral part of his view that climax communities are "superorganisms" (see also Odum 1969). Accumulated evidence now suggests that this model is incorrect for most marine communities on hard substrata (Lee 1966; Foster 1975a; Connell and Slatyer 1977; Sousa 1979a), and there is little evidence that communities, like organisms, are highly integrated systems (Horn 1974, 1976; Connell and Slatyer 1977). Concurrent with changes in the succession paradigm have come changes in our view of climax communities. Rather than being in some fairly stable equilibrium, entire communities or portions of them are now thought to be in various states of recovery from disturbance (Connell and Slatyer 1977; Connell 1978; Sousa 1979b). Climax is rarely, if ever, achieved, and successional processes are probably an integral part of most communities. Succession studies have thus become central to the study of community structure.

A variety of methods is possible and necessary for investigating succession in the context just discussed. Different communities may require different research approaches, and these approaches will also depend on the hypotheses being tested. The use of particular methods must be based on a careful consideration of research objectives and thorough experimental evaluation to ensure that they will answer the questions asked. Our objective is to discuss and evaluate methods that have been used for the investigation of macroalgal succession on hard substrata, and we hope that this will provide a basis for the development of new methods.

Because methods are determined partly by the process being investigated, it is important to define the process. We define "succession" in the broadest possible sense, and without any assumption of cause, as changes observed in a community following a disturbance. A community is a group of species occurring in some defined area, and a disturbance is any event that generates space for colonization.

[270]

If the reseach objective is to understand natural community changes and their cause, then this definition requires knowing community composition before disturbance and what the disturbance was (or at least that the community was altered).

Methods for studying succession have also been used to answer questions about algal distribution (Hruby 1975; Lubchenco 1980; Schonbeck and Norton 1980; Foster 1982), survivorship (Neushul et al. 1976), dispersal (Anderson and North 1966; Dayton 1973; Paine 1979; Chapman 1981; Deysher and Norton 1982), reproduction and seasonality (Northcraft 1948; Lee 1966; Foster 1975a; Kain 1975; Neushul et al. 1976; Emerson and Zedler 1978; Hruby and Norton 1979; Hawkins 1981), and the effects of pollution on algal assemblages (Murray and Littler 1978; Rastetter and Cooke 1979). We consider these methods basic to a variety of field studies and suggest that only with their more widespread use will we gain a better understanding of algal population and community dynamics.

II. Artificial substrata

A. *General considerations*

Artificial substrata placed directly in the sea have generally been used to simulate a disturbance that removes all organisms. Such a disturbance may be rare in nature except as a result of rock fracture, lava flows, landslides, and so on, which expose previously unoccupied rock. These substrata thus may represent a relatively artificial disturbance and have been used primarily in subtidal habitats where conditions make manipulation and sampling of natural substrata difficult. Artificial substrata can be "seeded" with spores or adults of particular species to represent other levels of disturbance that do not remove all organisms (Foster 1980), but this has been attempted in only one algal succession study (Harger et al. 1981). If successional processes in the natural community are of interest, one must also question whether the artificial substrata are good analogs of the natural bottom.

Plates and blocks of various kinds have the great advantages of ease in handling and sampling in both the laboratory and field and, unlike immovable natural substrata, can be used in nondestructive laboratory analyses of microscopic organisms. Foster (1980) and Cairns (1982) review details of construction and placement of various substratum types. The system developed by Neushul et al. (1976) exemplifies how such substrata can be incorporated into a field–laboratory experimental system that allows a description of succession from microscopic stages to reproductively mature adults (Fig. 13–1).

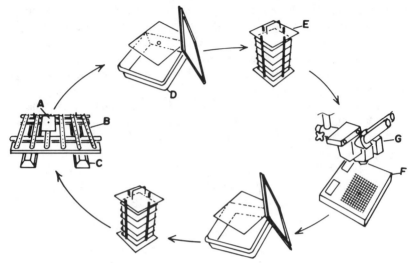

Fig. 13–1. System for nondestructive sampling of artificial substrata. Clockwise from left: subtidal platform for holding 20 × 20 cm plexiglass plates (A) fastened to a PVC frame (B) with stainless steel bolts. Frame is anchored to support platform of concrete parking-lot bumpers (C). The platform can also be used to mount light and temperature recorders, sediment traps, etc. Plates are unbolted from the frame by divers, placed in individual plastic trays with snap-on lids (D), and transported to the laboratory in a carrying rack (E). Sampling includes microscopic examination (G) under a grid placed over the plates while they are held in seawater-filled trays (F) cooled with ice packs. Higher resolution can be obtained with a dipping-cone microscope. Data can be read into a tape recorder or entered directly into a computer. After sampling, the procedure is reversed, the plates being returned to the platform. Redrawn from Neushul et al. (1976).

These substrata are also relatively easy to manipulate (see Sec. II.B.) and provide standardized replicates.

B. Composition

Although artificial substrata have many advantages, difficulties arise when natural succession in a particular community is of interest and these substrata are used as a substitute for clearing areas on the bottom. Artificial substrata may differ from natural surfaces in color and reflectivity, chemical characteristics, surface roughness and macrotopography, and porosity. The last-named affects water-holding capacity and is particularly important in the intertidal zone (Nienhuis 1969; Den Hartog 1972). In fouling studies, some of these factors have been examined, particularly chemical composition [Woods Hole Oceanographic Institution (WHOI) 1952; Long 1974]. A number of studies have shown that surface texture and topography can affect algal settlement and species composition (Ogata 1953; Johansen and

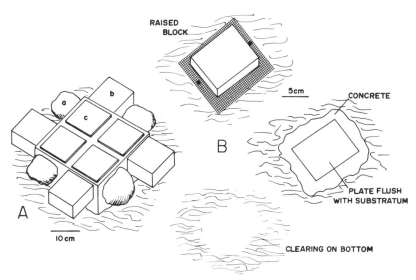

Fig. 13–2. (A) Substratum module used to test the effects of substratum type on succession: (a) piece of dead coral; (b) block cut from dead coral; (c) PVC plate (all epoxied to a large concrete block). (B) Experiment to test the effects of mounting substrata above the bottom: raised block fastened to bottom on stainless steel mesh; flush plate cemented into excavation; control clearing to test for substratum type and concrete effects. Frame A courtesy of M. Hixon; B, courtesy of S. Kennelly).

Austin, 1970; Foster 1975b; Luther 1976; Harlin and Lindbergh 1977; Norton and Fetter 1981), and surface characteristics of artificial substrata must therefore approximate those of the natural rock in the area. Alteration of surfaces has been attempted by adding particles (Harlin and Lindbergh 1977) or objects (Foster 1975b) of known size or molding plates from natural surfaces (Risk 1973). Although natural rock can be mounted on frames or cut into slabs, the cut surface is usually very smooth, and newly exposed rock may inhibit settlement, as discussed in Sec. III.

Numerous studies have evaluated the influence of substratum type on fouling (see review in WHOI 1952). Neushul et al. (1976) found similar subtidal communities on concrete and plexiglass, whereas Day (1977) reported differences in algal abundance on plastic versus limestone. Hixon and Brostoff (1981, 1982) compared succession on natural substrata (pieces of dead coral) with that on artificial substrata [polyvinyl chloride (PVC) and cut blocks of coral with flat surfaces]. Using the design illustrated in Fig. 13–2A, M. A. Hixon (unpublished data) found that the number of algal species was similar on the flat coral blocks and PVC plates but lower on both of these than on uncut dead coral, especially when fish grazing was intense. He suggested that more species occurred on the natural substrata because their greater surface heterogeneity provided more microhabitats and refugia from grazing. Brock (1979) also found that a heterogeneous surface enhanced the species richness and biomass on terra cotta tiles exposed to grazing by parrotfishes in a laboratory microcosm.

C. Size

The question(s) being addressed by a study usually dictate the size(s) of substrata used. For example, one may want to simulate the average size of cleared patches generated by a specific kind of disturbance such as the free space produced in areas of kelp where holdfasts are torn loose from the bottom during storms. If succession after such a disturbance is the process of interest, substratum size should approximate the attachment area of a holdfast. (Questions regarding the dynamics of extant algal assemblages are best addressed using natural substrata, but in situations in which this is not logistically feasible, artificial substrata may be meaningfully substituted if potential artifacts attendant on their use are carefully evaluated.) Other questions may require that the substratum be large enough to accommodate the largest possible individual of a particular species or to ensure that all species in the natural community are represented. A rigorous estimate of the latter size can be obtained by first using test substrata of various sizes to construct species–area curves (or abundance–area curves or both, depending on the questions being asked) and then using these to determine the minimum area on which all species coexist. Greig-Smith (1983) presents a thorough discussion of species–area determination that is applicable to succession studies.

Substratum size can also introduce artifacts. Community composition is altered within 1 to 2 cm of the edges of plates (Foster 1975b; Day 1977; Borowitzka et al. 1978), an effect at least partly due to alterations in water flow. Variation caused by edge effects can be reduced with larger plates and/or sampling only in the center. Water flow may also be altered by attachment structures, but this has not been investigated.

D. Position

Whether placed on platforms (Fager 1971, Foster 1975a, Neushul et al. 1976), on suspended racks (Coe and Allen 1937), or directly on the bottom (Foster 1975b; Robles and Cubit 1981), the location of substrata may alter larval and spore availability, growth and development, and access by mobile animals that can affect succession. Little is known about these effects, since investigators rarely compare succession on artificial substrata with that on cleared natural surfaces in the surrounding community. Foster (1975a) found no short-term qualitative differences between communities established on concrete blocks placed on raised platforms (~30 cm above the bottom) and communities in cleared areas on the bottom. In contrast, D. C. Barilotti (unpublished data) and T. A. Dean and L. Deysher (unpublished data) have found significant differences in the recruitment and growth of juvenile *Macrocystis pyrifera* sporophytes on artificial substrata placed on the bottom and 1–2 m above the bottom, suggesting that substratum position (changes in environmental variables over very small distances) can affect succession. Some species of algae, particularly those that grow as turfs and colonize open space primarily by vegetative propagation, will often be underrepresented on isolated artificial substrata (Sousa 1979a).

Less mobile grazers such as limpets and sea urchins may have difficulty maneuvering over angular surfaces, whereas more mobile grazers such as fishes, coiled snails, and amphipods will not have this problem. On the contrary, complex installations may create refugia for such species, causing the local effects of their grazing to be exaggerated (Robles and Cubit 1981). Kennelly (1983) made a very careful study of the effects of substratum placement on early (2 wk) succession (Fig. 13–2B). Algal cover was not significantly different between cleared areas on the bottom and blocks mounted flush with the bottom. However, algal cover on the tops of 5-cm-thick blocks attached to the bottom on wire mesh was significantly higher than the former treatments. A variety of evidence suggested that the raised surfaces were less accessible to small grazers such as amphipods.

III. Natural substrata

If the object of a study is to determine the patterns or mechanisms of change within an extant community at a particular site, natural substrata are preferable to artificial substrata.

A. Methods and degrees of clearing

The initiation of succession on natural substrata requires the removal of attached organisms. Depending on the question being addressed, the removal may be partial or complete. Rarely do disturbances of natural algal assemblages remove all plant tissues from the rock surface. The impact of large storm waves may remove algal blades or portions thereof, leaving stipes and holdfasts to regenerate. More extreme physical and biological disturbances, such as those caused by the impact of logs, sand or ice scouring, or grazing by dense populations of sea urchins or gastropods, will remove most or all algal stipes and even portions of holdfasts, but rarely, unless very chronic, will they remove entire plants. The revegetation of a disturbed site results from a combination of regeneration of surviving plants and recruitment of new plants from settling spores or zygotes.

This distinction between primary and secondary succession is too rarely made in studies of marine algal succession. In future studies it will be important to do so, because the life histories of attached macroalgae should be viewed in the context of the range of disturbance intensities to which they are normally exposed. Often it will be useful to mimic experimentally the natural levels of damage incurred by a particular algal assemblage. In a number of studies, the erect portions of plants have been removed, leaving the holdfasts and understory undamaged (Printz 1959; Dayton 1975a,b; Lubchenco 1980; Foster 1982; Reed and Foster, 1984).

Only in a few instances is it possible for the investigator to impose at will a disturbance identical to the natural one. Sousa (1980) studied the influence of different intensities of disturbance on the algal assemblages of intertidal boulder fields. The most common form of disturbance to which these assemblages are subjected is that caused when boulders are overturned by wave action. By overturning sets of boulders for different lengths of time, then righting them to allow colonization and regrowth, it was possible to impose experimentally an essentially natural disturbance in a range of intensities.

Some studies require that a natural rock surface be cleared of all organisms and sterilized. These include (1) studies in which the documentation of patterns of primary succession per se are of interest, (2) studies in which the effects of prior occupants of the substratum on subsequent colonization are tested and (3) studies with the aim of monitoring temporal patterns of recruitment from planktonic propagules by making sterilized surfaces available for colonization periodically throughout the year. The standard approach to sterilizing natural rock surfaces has been to remove large organisms with a knife or paint scraper and then vigorously abrade the surface with a wire brush. In the intertidal, the plot may then be burned with a weed burner, a propane torch, a mixture of white gas and diesel fuel (Dayton 1971), or alcohol (Murray and Littler 1978). Repeated burnings are often required to kill the basal systems of tenacious species (Dayton 1971). In place of burning, some workers have applied toxic chemicals such as 5% Formalin (Castenholz 1961; Emerson and Zedler 1978), formaldehyde (Wilson 1925), alcohol (Castenholz 1961), and phosphoric acid (Underwood 1980; Underwood and Jernakoff 1981). Steam cleaning is also possible. Several workers (Padilla 1981; Gaines 1982) have found that applications of concentrated sodium hydroxide in the form of commercial lye or aerosol oven cleaner, in combination with scraping and brushing, effectively remove encrusting algae or the basal systems of erect forms. An advantage of this method is that lye is readily soluble in seawater, so that probably little residue remains on the rock surface after a few tidal inundations, in contrast to Formalin or petroleum-based burning techniques. Infrared photographs can be used to test the effectiveness of all these techniques (Murray and Littler 1978).

More drastic methods of creating clean rock surfaces include sandblasting (Lubchenco and Menge 1978; Lubchenco 1980), removal of surface rock layers with a hammer and chisel (Northcraft 1948; Castenholz 1961; Littler and Doty 1975), or exposing virgin rock surfaces with dynamite (Dayton 1971; Kawashima 1972). Of these methods, sandblasting has the least effect on the surface texture, chemistry, current flow, and so on. Dayton (1971) suggested that a certain amount of leaching or weathering of a newly exposed rock surface is required before recruitment can occur. Reed and Foster (1984) found that areas cleared to bare rock with a pneumatic hammer in a subtidal *Pterygophora californica* bed were not colonized as rapidly or as heavily by brown algae as were nearby areas where only the overstory was removed. When feasible, sterlization by burning or by the application of water-soluble chemicals seems preferable to methods that expose virgin rock surfaces.

B. Areal extent of clearing

Just as the intensity of disturbances should be considered in studies of algal succession, so should their areal extent. In most habitats, the spatial scale of natural disturbances ranges from a few square centimeters to several square meters. The sizes of plots employed in experimental studies of algal succession have spanned a similar range; however, in only a few studies (Sousa 1979a, 1984; Suchanek 1979) have the effects of plot size on patterns of colonization and temporal species replacements been explicitly examined. If grazers find refuge in areas outside of the cleared area, for example, in neighboring mussel beds (Suchanek 1978, 1979; Paine and Levin 1981; Sousa 1984), and forage only a fixed distance from the refuge, the centers of large clearings will be ungrazed, whereas the entirety of small clearings will be heavily grazed. Similarly, plants that colonize cleared areas by growing vegetatively inward from the edge and/or whose spores are dispersed only a short distance from the parent plant will fill small clearings more rapidly than large plots (Sousa 1979a) simply because of their greater edge-to-area ratio. The recruitment and growth of plants in small clearings are also more likely to be influenced (e.g., through shading, whiplash, competition for nutrients, etc.) by the surrounding adult plants than they are in large clearings. It is often instructive to study succession in experimental clearings of several sizes within the observed range.

C. Frequency of clearing

Natural stands of macroalgae are disturbed at different frequencies depending on their location, the season, and the stability of the substratum to which they are attached. In two studies of rocky intertidal communities, the frequency of natural physical disturbances has been explicitly measured (Sousa 1979a; Paine and Levin 1981). Although a number of studies have examined the influence of season on patterns of algal succession (Northcraft 1948; Lee 1966; Foster 1975a; Kain 1975; Paine 1977; Emerson and Zedler 1978; Sousa 1979a; Lubchenco and Cubit 1980; Hawkins 1981), there have been only two experimental investigations of how the frequency of disturbance affects algal community composition. Sousa (1979b) stabilized small boulders by gluing them to stationary lengths of redwood, thereby eliminating disturbance caused by storm waves. He compared the pattern of algal colonization on these with that on control boulders of similar size that were subjected to natural frequencies and intensities of disturbance. Emerson and Zedler (1978) compared patterns of colonization in a mat of coralline alglae under conditions of (1) no experimental disturbance, (2) experimental clearing once per year (and in each season), and (3) biweekly clearing.

IV. Manipulations

A. General considerations

There is growing evidence that a variety of interacting physical and biological factors play important roles in the process of species replacement during succession. Despite the usefulness of some laboratory studies, it is usually preferable to conduct such investigations in the natural environment (Connell 1974). In the simplest of field experiments, one factor per treatment is manipulated while others are allowed to vary naturally. The effects of such a manipulation should be compared against contemporaneous changes measured in nearby control areas. Connell (1974, 1975) discusses the advantages of this procedure over so-called natural experiments, in which the effects of a natural change in the environment are simply observed.

B. Abiotic environment

As discussed in Sec. II.B, surface characteristics of both artificial and natural substrata can be manipulated in a variety of ways. However, these manipulations may affect grazer access as well as spore settlement and growth by altering microhabitats that may be refuges from grazing.

Nutrients can be varied by the placement of containers of slow-release fertilizer (e.g., Osmocote, Sierra Chemical Co.) in the vicinity of plots or under modified artificial substrata (see Foster et al., Chap. 10, for details). In both experiments, actual nutrient increase must be measured in the field at the plate surfaces because dilution is rapid. Light can be decreased by the placement of shades over the bottom (Backman and Barilotti 1976) and increased by the removal of overstory canopies (Pearse and Hines 1979; Reed and Foster 1984) or by the use of mirrors (Barilotti 1980).

Sediment was manipulated by Neushul et al. (1976), but the time between removals may have been too long to produce a significant alteration. Sediment has highly significant effects on young stages of *Macrocystis pyrifera* in the laboratory (Devinny and Volse 1978) and warrants further attempts at field manipulation, with fences or cages (see Sec. IV,C.1.). Similarly, barriers could be used to reduce water velocity, but these may also increase sedimentation. At present, in situ investigations of water velocity are limited to comparisons of different natural areas where other factors may also vary. Submergence or at least water coverage can be increased in the intertidal zone by the construction of artificial tide pools that drain over study areas (Hatton 1938; Frank 1965; Dayton 1970).

Ecology is replete with examples of factor interaction, especially in studies of terrestrial plant growth (Harper 1977). The single-factor experiments already discussed are naive in the sense that such factors as light, nutrients, and temperature may vary simultaneously and interact in nature. Barilotti (1980) briefly describes the use of factorial design for in situ marine algal experiments and points out the difficulty of performing the large experiments necessary to determine factor interaction. However, single factors are a beginning, and as can be seen from this short discussion, much experimental work with abiotic factors remains to be done.

C. Biotic environment

1. Effects of consumer species. Numerous studies have demonstrated that grazing can influence algal assemblages, and predators can do so indirectly by reducing the density of grazers and/or reducing the abundance of sessile invertebrates that may compete with algae for space. These animals can obviously affect succession. Methods for studying grazer effects are reviewed by Vadas (Chap. 26), and most are applicable to consumer species in general.

As discussed by Vadas (Chap. 26), fences and cages have been used extensively to exclude or include grazers. However, like artificial substrata and their attachment structures, these devices may introduce artifacts that alter factors other than grazing. Robles and Cubit

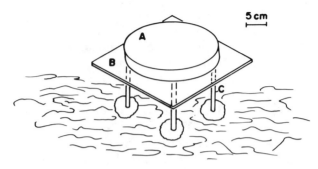

Fig. 13–3. Structure for excluding crabs and limpets from an artificial substratum. (A) Concrete substratum; (B) plastic baffle to exclude crabs; (C) iron frame that raises substratum to exclude limpets. Limpets were further excluded with copper paint around base of frame (see text). Drawn from Robles (1982).

(1981) and Robles (1982) used a particularly innovative experimental design that employed neither cages nor fences to examine the effects of a variety of herbivores, including crabs, limpets, and dipteran larvae on the successional dynamics of a high intertidal algal assemblage. To document patterns of algal colonization on concrete plates subjected to various regimes of grazing, crabs and limpets were excluded from some plates by raising the plates 8 cm off the substratum on iron frames. The frames were fitted with plexiglass baffles that further prevented crabs from reaching the plates. Limpets were excluded from some plates with copper paint (59 ml Shining Armor copper paint, no. 606, Illinois Bronze Powder and Paint Co., plus 29 g of copper powder, Luco copper lining bronze, Leo Uhlfelder Co.). Finally, dipteran larvae were excluded with repeated hand removals by means of forceps or with applications of insecticide (Malathion 50, Chevron Chemical Co.). Controls for the latter manipulations were, respectively, sham removals in which the algae were disturbed with forceps but no larvae were removed and repeated applications of the insecticide solvent minus the active ingredient. The structures employed in this study are illustrated in Fig. 13–3. In the final analysis, the successful measurement of the impact of consumers on algal successions depends on the cleverness of the investigator in designing experiments and establishing appropriate controls.

2. Competition. The role of competitive interactions in succession can be studied with selective removals or additions of putative competitors. Methods and experimental design considerations for studying algal competition in general are discussed by Denley and Dayton (Chap. 25) and are applicable to succession studies. When investigating the mechanisms by which established sessile organisms influence the recruitment, growth, and survival of algae, one might desire to test whether the observed effects result from biological properties of the established species in question (e.g., allelopathy) or are due simply to its physical structure. Such tests could employ sterile artificial materials that mimic the structure of the resident species. Such materials have been used successfully in several studies of assemblages of sessile invertebrates (Suchanek 1979; Russ 1980; Dean 1981). Tropical encrusting coralline algae commonly settle on dead *Acropora palmata*, and this substratum has been simulated with PVC pipe (Adey and Vassar 1975). Harlin (1973) used fibrous polypropylene strips as a mimic of seagrass blades in her pioneering study of epiphyte–host relationships. Custom-made plastic models of giant kelp (*Macrocystis pyrifera*) blades have been used in studies of sessile invertebrates that foul living kelp blades (S. Schroeter, J. Dixon, and J. Kastendiek, unpublished data). Such artificial plants might be used creatively to study the effects of physical versus biological properties of algae on the outcome of competitive or epiphytic interactions.

V. Measures of algal abundance

Like the studies themselves, methods of determining the results may vary depending on the algae of interest (e.g., turf species versus large kelps), the questions being asked, the research approach (e.g., artificial versus natural substrata), and the facilities and time available for sampling. Many terrestrial ecological techniques are applicable to marine habitats, and most are reviewed by Greig-Smith (1983). One usually has to decide whether sampling will be destructive or nondestructive, how often samples should be taken, whether to sample the entire area under study or take subsamples, whether to measure biomass, cover, or individuals (or some combination of these), and, finally, how to make these measurements.

De Wreede, in Chap. 7, discusses destructive sampling, and Littler and Littler, in Chap. 8, discuss nondestructive sampling. Succession studies often require numerous samples in time. Hixon and Brostoff (1981) had to use over 1000 replicates to sample destructively their experiment on the effects of grazing on succession. Destructive sampling may reduce time spent in the field, since measurements can be done in the laboratory. Artificial substrata have the advantage of allowing nondestructive laboratory analysis (Foster 1975a; Neushul et al. 1976). Sampling schedules should reflect the questions being investigated and organisms examined. Variable schedules may be preferred because change is often most rapid early in succession. This might make weekly sampling for the first few months followed by monthly sampling most appropriate.

Cover is most often used as a measure of abundance in successional studies, because it is related most directly to utilization of space and light resources and is often the easiest abundance parameter to measure. The most complete measures of cover include canopies and the substratum and may require using some form of the point-quadrat technique (Foster 1975a; Greig-Smith 1983).

Methods for determining biomass are discussed by De Wreede (Chap. 7). Both cover and biomass can be used in diversity and similarity calculations (Lyons 1981), as can "individuals" or branches. As discussed by Harper (1977), parts of a plant (ramets) may be better indicators of ecological effects than individuals (genets). Depending on the variance, one measure of abundance can often be used to estimate others by correlation techniques. Community analyses may require the use of different techniques and abundance measures for different subassemblages (e.g., crust and turf) and strata (e.g., cover on the substratum and overstory cover).

In addition to abundance, a measure of similarity between various successional assemblages may be desirable. A number of indices are available, and these vary in their sensitivity to changes in number of species and relative abundance (see reviews by Horn 1966; Hurlbert 1978; Cailliet and Berry 1978; Green 1980; Greig-Smith 1983).

Changes in diversity during succession have been of great interest to community ecologists (for review, see Pielou 1975), and a variety of diversity measures can be derived from species composition and abundance data. However, there is considerable debate over both the biological meaning of some diversity indices and their sensitivity to changes in community composition (Hurlbert 1971; May 1975). The use of a particular index should be thoroughly evaluated and, if possible, compared with other indices derived from the same data (Hixon and Brostoff 1982).

VI. Experimental design and sampling schedules

By adopting a multifactorial design, those investigating marine algal succession will be able to evaluate simultaneously the influence of a variety of disturbance characteristics on subsequent changes in the algal assemblage. The patterns observed in such studies will generate a number of hypotheses that can, in turn, be tested with more detailed experiments. This general approach is outlined in Table 13–1. Conceivably, this approach is also applicable, albeit on a smaller scale, to epiphytic (e.g., on seagrass or kelp plants) or epizoic (e.g., on mollusk shells) assemblages.

The design of experiments called for in step 2 of Table 13–1 will vary with the specific system under study. Treatments conducted in a particular season should be established on the same date and close

Table 13–1. *Proposed approach to the investigation of successional patterns and mechanisms of species replacements*

Step 1

Observe the natural regime of disturbance to which an algal assemblage is subjected; make quantitative measurements if possible

Step 2

On the basis of these observations, design and conduct a multifactorial experiment to reveal the patterns of succession that occur under a variety of realistic regimes of disturbance that differ in
1. Intensity
2. Areal extent
3. Frequency of occurrence
4. Season of occurrence
5. Various combinations thereof

Step 3

Formulate and test specific hypotheses concerning mechanisms of successional species replacements (e.g., Connell and Slatyer 1977; Sousa 1979a); this may involve studies of
1. Interspecific competition
2. The impact of grazing
3. The tolerance of species to physical stress

to one another to ensure that the environmental conditions and the availability of propagules are as similar as possible across treatments and replicates. Treatments should be assigned to plots or artificial substrata in such a way as to avoid possible environmental differences other than those experimentally imposed. This usually involves random allocation (Hurlbert 1984). Repeated measurements of abundance within a particular set of plots pose serious problems for statistical comparison of treatment effects, because the values obtained on successive dates are not statistically independent. In such cases, the use of analysis of variance with time of sampling as one factor is inappropriate (Underwood 1981; Hurlbert 1984). Instead, treatment effects should be compared on one or a few sampling dates only. Adjustment of the "experimentwise" error rate (Sokal and Rohlf 1981: 241) is required if tests for treatment effects are conducted on data from several sampling dates. Alternatively, statistical procedures for repeated measures could be applied (Winer 1971).

Problems associated with nonindependent samples in time can be eliminated if enough replicates of each treatment can be established at the outset, such that a different randomly chosen subset of plots can be sampled on each date. This would also allow for destructive sampling if, for example, estimates of biomass were desired. Rarely, when natural substrata are used, will such a design be practical because of the limited space available for the establishment of experimental plots in natural habitats; however, when artificial or small discrete natural substrata are employed, such a design is feasible (e.g., Hixon and Brostoff 1981). One kind of information that is lost in such designs is that concerning temporal patterns of change within specific replicates.

A wide variety of parametric and nonparametric methods of statistical analysis are available. Their use and misuse are discussed in numerous readable texts (e.g., Siegel 1956; Steel and Torrie 1960; Snedecor and Cochran 1967; Sokal and Rohlf 1981). See Underwood (1981) for an excellent review of methods of analysis of variance in experimental marine biology and ecology, and Hurlbert (1984) for a thorough discussion of experimental design.

It is impossible to state the number of replicates required for a particular experiment without an a priori estimate of the variability within treatments. If such estimates are available, from preliminary trials, for example, one can solve iteratively (Sokal and Rohlf 1981: 263) for the sample size needed to be certain, at some chosen percentage of time, of detecting a true difference between treatments at an α level of significance (see also De Wreede, Chap. 7). We have not found a single study of marine algal succession in which sample size was based on such a calculation. Most commonly, the number of replicates employed reflects other constraints, such as the amount of time available for sampling and space available for establishing experimental plots. Obviously, the more replication the better.

VII. Summary

There now exists a considerable body of descriptive information on successional patterns in assemblages of marine macroalgae. However, the mechanisms that cause temporal change in the composition of these assemblages are not as well documented. We also know little about the sources of variability in successional sequences. In our view, controlled field experimentation using either artificial or natural substrata is the most powerful method of investigating these aspects of successional dynamics. This approach entails (1) the formulation of falsifiable a priori hypotheses based on careful observations and (2) the design of controlled field experiments with a specific statistical analysis in mind.

Successful experiments are often difficult to conduct under field conditions, and one is bound to encounter frustrating pitfalls – for example, ineffective treatments or controls and artifactual changes in the biotic or abiotic environment caused by the experimental apparatus (for a review of such problems, see Dayton and Oliver

1980). Difficulties such as these do not negate the validity of the experimental approach. Instead, they challenge our imaginations as scientists to develop more satisfactory experimental designs. We hope that this chapter will serve the same purpose.

VIII. Acknowledgments

We thank Mark and Diane Littler and the Board of Advisors for their very helpful suggestions and editing.

IX. References

Adey, W. H., and Vassar, J. M. 1975. Colonization, succession and growth rates of tropical crustose coralline algae (Rhodophyta, Cryptonemiales). *Phycologia* 14, 55–69.

Anderson, E. K., and North W. J. 1966. *In situ* studies of spore production and dispersal in the giant kelp, *Macrocystis*. In Young, E. G., and McLachlan, J. L. (eds.), *Proceedings of the Fifth International Seaweed Symposium*, pp. 73–86. Pergamon Press, Oxford.

Backman, T., and Barilotti, D. C. 1976. Irradiance reduction: effects on standing crops of the eelgrass, *Zostera marina*, in a coastal lagoon. *Mar. Biol.* 24, 33–40.

Barilotti, D. C. 1980. Genetic considerations and experimental design of outplanting studies. In Abbott, I. A., Foster, M. S., and Ecklund, L. F. (eds.), *Pacific Seaweed Aquaculture.* pp. 10–18. California Sea Grant Program, La Jolla.

Borowitzka, M. A., Larkum, A. W., and Borowitzka, L. J. 1978. A preliminary study of algal turf communities of a shallow coral reef lagoon using an artificial substratum. *Aquat. Bot.* 5, 365–81.

Brock, R. E. 1979. An experimental study on the effects of grazing by parrot fishes and role of refuges in benthic community structure. *Mar. Biol.* 51, 381–8.

Cailliet, G. M., and Berry, J. P. 1978. Comparison of food array overlap measures useful in fish feeding habit analysis. In Lipovsky, S. J., and Simenstad, C. A. (eds.), *Proceedings of the Second Pacific Northwest Technical Workshop on Fish Food Habit Studies*, pp. 67–79. Washington Sea Grant Program, Seattle.

Cairns, J. (ed.). 1982. *Artificial Substrata.* Ann Arbor Science Publishers, Ann Arbor, Mich. 279 pp.

Castenholz, R. W. 1961. The effect of grazing on marine littoral diatom populations. *Ecology* 42, 783–94.

Chapman, A. R. O. 1981. Stability of sea urchin dominated barren grounds following destructive grazing of kelp in St. Margaret's Bay, eastern Canada. *Mar. Biol.* 62, 307–11.

Clements, F. E. 1916. *Plant Succession.* Publ. 242, Carnegie Institute, Washington, D.C. 512 pp.

Coe, W. R., and Allen, W. E. 1937. Growth of sedentary marine organisms on experimental blocks and plates for nine successive years at the pier of the Scripps Institution of Oceanography. *Scripps Inst. Oceanogr. Bull. Tech. Ser.* 4, 101–36.

Connell, J. H. 1974. Ecology: field experiments in marine ecology. In Mariscal, R. N. (ed.), *Experimental Marine Biology*, pp. 21–54. Academic Press, New York.

Connell, J. H. 1975. Some mechanisms producing structure in natural communities: a model and evidence from field experiments. In Cody, M. L. and Diamond, J. (eds.), *Ecology and Evolution of Communities*, pp. 460–90. Belknap Press, Cambridge, Mass.

Connell, J. H. 1978. Diversity in tropical rain forests and coral reefs. *Science* 199, 1302–10.

Connell, J. H., and Slatyer, R. O. 1977. Mechanisms of succession in natural communities and their role in community stability and organization. *Amer. Natur.* 111, 1119–44.

Day, R. W. 1977. Two contrasting effects of predation on species richness in coral reef habitats. *Mar. Biol.* 44, 1–5.

Dayton, P. K. 1970. "Competition, Predation and Community Structure: The Allocation and Subsequent Utilization of Space in a Rocky Intertidal Community." Ph.D. dissertation, University of Washington, Seattle. 174 pp.

Dayton, P. K. 1971. Competition, disturbance, and community organization: the provision and subsequent utilization of space in a rocky intertidal community. *Ecol. Monogr.* 41, 351–89.

Dayton, P. K. 1973. Dispersion, dispersal, and persistence of the annual intertidal alga, *Postelsia palmaeformis* Ruprecht. *Ecology* 54, 433–8.

Dayton, P. K. 1975a. Experimental evaluation of ecological dominance in a rocky intertidal algal community. *Ecol. Monogr.* 45, 137–59.

Dayton, P. K. 1975b. Experimental studies of algal canopy interactions in a sea otter-dominated kelp community at Amchitka Island, Alaska. *Fish Bull.* 73, 230–7.

Dayton, P. K., and Oliver, J. S. 1980. An evaluation of experimental analyses of population and community patterns in benthic marine environments. In Tenore, K. R., and Coull, B. C. (eds.), *Marine Benthic Dynamics*, pp. 93–120. Belle W. Baruch Library of Marine Science Publ. 11, University of South Carolina Press, Columbia.

Dean, T. A. 1981. Structural aspects of sessile invertebrates as organizing forces in an estuarine fouling community. *J. Exp. Mar. Biol. Ecol.* 53, 163–80.

Den Hartog, C. 1972. Substratum: multicellular plants. In Kinne, O. (ed.), *Marine Ecology*, vol. 1, pp. 1277–89. Wiley-Interscience, New York.

Devinny, J. S., and Volse, L. A. 1978. Effects of sediments on the development of *Macrocystis pyrifera* gametophytes. *Mar. Biol.* 48, 343–8.

Deysher, L., and Norton, T. A. 1982. Dispersal and colonization in *Sargassum muticum* (Yendo) Fensholt. *J. Exp. Mar. Biol. Ecol.* 56, 179–95.

Emerson, S. E., and Zedler, J. B. 1978. Recolonization of intertidal algae: an experimental study. *Mar. Biol.* 44, 315–24.

Fager, E. W. 1971. Pattern in the development of a marine community. *Limnol. Oceanogr.* 16, 241–53.

Foster, M. S. 1975a. Algal succession in a *Macrocystis pyrifera* forest. *Mar. Biol.* 32, 313–29.

Foster, M. S. 1975b. Regulation of algal community development in a *Macrocystis pyrifera* forest. *Mar. Biol.* 32, 331–42.

Foster, M. S. 1980. The use of substratum manipulations in field studies of seaweed colonization and growth. In Abbott, I. A., Foster, M. S., and Eklund, L. F. (eds.), *Pacific Seaweed Aquaculture*, pp. 23–31. California Sea Grant College Program, La Jolla.

Foster, M. S. 1982. Factors controlling the intertidal zonation of *Iridaea flaccida* (Rhodophyta). *J. Phycol.* 18, 285–94.

Frank, P. W. 1965. The biodemography of an intertidal snail population. *Ecology* 46, 831–44.

Gaines, S. D. 1982. "The Role of Consumer Guild Structure in Community Organization: Tests in Temperate and Tropical Intertidal Communities." Ph.D. dissertation, Oregon State University, Corvallis. 131 pp.

Green, R. H. 1980. Multivariate approaches in ecology: the assessment of ecologic similarity. *Annu. Rev. Ecol. Syst,* 11, 1–14.

Greig-Smith, P. 1983. *Quantitative Plant Ecology,* 3rd ed. University of California Press, Berkeley. 359 pp.

Harger, B. W., Coon, D. A., Foster, M. S., Neushul, M., and Woessner, J. 1981. Settlement, growth and reproduction of benthic algae within *Macrocystis* forests in California. In Fogg, G. E., and Jones W. E. (eds.), *Proceedings of the Eighth International Seaweed Symposium,* pp. 342–52. The Marine Science Laboratories, University College of North Wales, Manai Bridge.

Harlin, M. M. 1973. "Obligate" algal epiphyte: *Smithora naiadum* grown on a synthetic substrate. *J. Phycol.* 9, 230–2.

Harlin, M. M., and Lindbergh, J. M. 1977. Selection of substrata by seaweeds: optimal surface relief. *Mar. Biol.* 40, 33–40.

Harper, J. L. 1977. *Population Biology of Plants.* Academic Press, New York 892 pp.

Hatton, H. 1938. Essais de bionomie explicative sur quelques especes intercotidales d'algues et d'animaux. *Ann. Inst. Oceanogr. Monaco* 17, 241–348.

Hawkins, S. J. 1981. The influence of season and barnacles on the algal colonization of *Patella vulgata* exclusion areas. *J. Mar. Biol. Assoc. U.K.* 61, 1–15.

Hixon, M. A., and Brostoff, W. N. 1981. Fish grazing and community structure of Hawaiian reef algae. In Gomez, E. D., Birkeland, C. E., Buddemeier, R. W., Johannes, R. E., Marsh, J. A., Jr., and Tsuda, R. T. (eds.), *Proceedings of the Fourth International Coral Reef Symposium,* pp. 507–14. Marine Sciences Center, University of the Phillipines, Quezon City, Philippines.

Hixon, M. A., and Brostoff, W. N. 1982. Differential fish grazing and benthic community structure on Hawaiian reefs. In Cailliet, G. M., and Simenstad,

C. A. (eds.), *Proceedings of the Third Pacific Technical Workshop on Fish Food Habits Studies*, pp. 249–57. Washington Sea Grant Program, Seattle.

Horn, H. S. 1966. Measurement of "overlap" in comparative ecological studies. *Amer. Natur.* 100, 419–24.

Horn, H. S. 1974. The ecology of secondary succession. *Annu. Rev. Ecol. Syst.* 5, 25–37.

Horn, H. S. 1976. Succession. In May, R. M. (ed.), *Theoretical Ecology*, pp. 187–294. Saunders, Philadelphia.

Hruby, T. 1975. Observations of algal zonation resulting from competition. *Est. Coast. Mar. Sci.* 4, 231–3.

Hruby, T., and Norton, T. A. 1979. Algal colonization on rocky shores in the Firth of Clyde. *J. Ecol.* 67, 65–77.

Hurlbert, S. H. 1971. The non-concept of species diversity: a critique and alternative parameters. *Ecology* 52, 577–86.

Hurlbert, S. H. 1978. The measurement of niche overlap and some relatives. *Ecology* 59, 67–77.

Hurlbert, S. H. (1984). Pseudoreplication and the design of ecological field experiments. *Ecol. Monogr.* 54, 187–211

Johansen, H. W., and Austin, L. F. 1970. Growth rates in the articulated coralline *Calliarthron*. *Can. J. Bot.* 48, 125–32.

Kain, J. M. 1975. Algal recolonization of some cleared subtidal areas. *J. Ecol.* 63, 739–65.

Kawashima, S. 1972. A study of life history of *Laminaria angustata* Kjellm. var. *longissima* Miyabe by means of concrete block. In Abbott, I. A., and Kurogi, M. (eds.), *Contributions to the Systematics of Benthic Marine Algae of the North Pacific*, pp. 93–109. Japanese Society of Phycology, Kobe.

Kennelly, S. J. 1983. An experimental approach to the study of factors affecting algal colonization in a sublittoral kelp forest. *J. Exp. Mar. Biol. Ecol.* 68, 257–76.

Lee, R. K. 1966. Development of marine benthic algal communities on Vancouver Island, British Columbia. In Taylor, R. L., and Ludwig, R. A. (eds.), *The Evolution of Canada's Flora*, pp. 100–20. University of Toronto Press, Toronto.

Littler, M. M., and Doty, M. S. 1975. Ecological components structuring the seaward edges of tropical Pacific reefs: the distribution, communities and productivity of *Porolithon*. *J. Ecol.* 63, 117–29.

Long, E. R. 1974. Marine fouling studies off Oahu, Hawaii. *Veliger* 17, 23–36.

Lubchenco, J. 1980. Algal zonation in the New England rocky intertidal community: an experimental analysis. *Ecology* 61, 333–44.

Lubchenco, J., and Cubit, J. 1980. Heteromorphic life histories of certain marine algae as adaptations to variations in herbivory. *Ecology* 61, 676–87.

Lubchenco, J., and Menge, B. A. 1978. Community development and persistence in a low rocky intertidal zone. *Ecol. Monogr.* 48, 67–94.

Luther, G. 1976. Bewuchsuntersuchungen auf natursteinsubstratem in Gezeitenbereich des Nurdsylter Wattenmeeres: algen. *Helgo. Meeres.* 28, 318–51.

Lyons, N. I. 1981. Comparing diversity indices based on counts weighted by biomass or other importance values. *Amer. Natur.* 118, 438–42.

May, R. M. 1975. Patterns of species abundance and diversity. In Cody, M. L., and Diamond, J. M. (eds.), *Ecology and Evolution of Communities*, pp. 81–120. Belknap Press, Cambridge, Mass.

Murray, S., and Littler, M. M. 1978. Patterns of algal succession in a perturbated marine intertidal community. *J. Phycol.* 14, 506–12.

Neushul, M., Foster, M. S., Coon, D. A., Woessner, J. W., and Harger, B. W. 1976. An *in situ* study of recruitment, growth, and survival of subtidal marine algae: techniques and preliminary results. *J. Phycol.* 12, 397–408.

Nienhuis, P. H. 1969. The significance of the substratum for intertidal algal growth on the artificial rocky shore of the Netherlands. *Int. Rev. Hydrobiol.* 54, 207–15.

Northcraft, R. D. 1948. Marine algal colonization on the Monterey Peninsula, California. *Amer. J. Bot.* 35, 396–404.

Norton, T. A., and Fetter, R. 1981. The settlement of *Sargassum muticum* in stationary and flowing water. *J. Mar. Biol. Assoc. U.K.* 61, 929–40.

Odum, E. P. 1969. The strategy of ecosystem development. *Science* 142, 262–70.

Ogata, E. 1953. Some experiments on the settling of spores of red algae. *Bull. Soc. Plant Ecol. Tokyo* 3, 128–34.

Padilla, D. 1981. "Selective Agents Influencing the Morphology of Coralline Algae." Master's thesis, Oregon State University, Corvallis. 80 pp.

Paine, R. T. 1977. Controlled manipulations in the marine intertidal zone and their contributions to ecological theory. In *Changing Scenes in Natural Sciences, 1776–1976*, pp. 245–70. Special Publ. 12, Academy of Natural Science, Philadelphia.

Paine, R. T. 1979. Disaster, catastrophe and local persistence of the sea palm *Postelsia palmaeformis*. *Science* 205, 685–7.

Paine, R. T., and Levin, S. A. 1981. Intertidal landscapes: disturbance and the dynamics of pattern. *Ecol. Monogr.* 51, 145–78.

Pearse, J. S., and Hines, A. H. 1979. Expansion of a central California kelp forest following the mass mortality of sea urchins. *Mar. Biol.* 51, 83–91.

Pielou, E. C. 1975. *Ecological Diversity*. Wiley, New York. 165 pp.

Printz, H. 1959. Investigations of the failure of recuperation and repopulation in cropped *Ascophyllum* areas. *Avhandlinger utigitt av Det Norske Videnskaps-Akademi i Oslo I* 3, 1–15.

Rastetter, E. B., and Cooke, W. J. 1979. Responses of marine fouling communities to sewage abatement in Kaneohe Bay, Oahu, Hawaii. *Mar. Biol.* 53, 271–80.

Reed, D. C., and Foster, M. S. (1984). The effects of canopy shading on algal recruitment and growth in a giant kelp forest. *Ecology* 65, 937–48.

Risk, M. J. 1973. Settling plates of cold-cure acrylic plastic replicated from natural surfaces. *Limnol. Oceanogr.* 18, 801–2.

Robles, C. D. 1982. Disturbance and predation in an assemblage of herbivorous diptera and algae on rocky shores. *Oecologia* 54, 23–31.

Robles, C. D., and Cubit, J. 1981. Influence of biotic factors in an upper intertidal community: dipteran larvae grazing on algae. *Ecology* 62, 1536–47.

Russ, G. R. 1980. Effects of predation by fishes, competition, and structural complexity of the substratum on the establishment of a marine epifaunal community. *J. Exp. Mar. Biol. Ecol.* 42, 55–69.

Schonbeck, M. W., and Norton, T. A. 1980. Factors controlling the lower limits of fucoid algae on the shore. *J. Exp. Mar. Biol. Ecol.* 43, 131–50.

Siegel, S. 1956. *Nonparametric Statistics.* McGraw-Hill, New York. 312 pp.

Snedecor, G. W., and Cochran, W. G. 1967. *Statistical Methods,* 6th ed. Iowa State University Press, Ames. 593 pp.

Sokal, R. R., and Rohlf, F. J. 1981. *Biometry,* 2nd ed. Freeman, San Francisco. 859 pp.

Sousa, W. P. 1979a. Experimental investigations of disturbance and ecological succession in a rocky intertidal algal community. *Ecol. Monogr.* 49, 227–54.

Sousa, W. P. 1979b. Disturbance in marine intertidal boulder fields: the nonequilibrium maintenance of species diversity. *Ecology* 60, 1225–39.

Sousa, W. P. 1980. The responses of a community to disturbance: the importance of successional age and species' life histories. *Oecologia* 45, 72–81.

Sousa, W. P. (1984). Intertidal mosaics: patch size, propagule availability, and spatially variable patterns of succession. *Ecology* 65, 1918–35.

Steel, R. G., and Torrie, J. H. 1960. *Principles and Procedures of Statistics.* McGraw-Hill, New York. 481 pp.

Suchanek, T. H. 1978. The ecology of *Mytilus edulis* L. in exposed rocky intertidal communities. *J. Exp. Mar. Biol. Ecol.* 31, 105–20.

Suchanek, T. H. 1979. "The *Mytilus californianus* Community: Studies on the Composition, Structure, Organization, and Dynamics of a Mussel Bed." Ph.D. dissertation, University of Washington, Seattle. 286 pp.

Underwood, A. J. 1980. The effects of grazing by gastropods and physical factors on the upper limits of distribution of intertidal macroalgae. *Oecologia* 46, 201–13.

Underwood, A. J. 1981. Techniques of analysis of variance in experimental marine biology and ecology. *Oceanogr. Mar. Biol. Annu. Rev.* 19, 513–605.

Underwood, A. J., and Jernakoff, P. 1981. Effects of interactions between algae and grazing gastropods on the structure of a low-shore intertidal algal community. *Oecologia* 48, 221–33.

Woods Hole Oceanographic Institution. 1952. *Marine Fouling and Its Prevention.* United States Naval Institute, Annapolis, Md. 388 pp.

Wilson, O. T. 1925. Some experimental observations of marine algal successions. *Ecology* 6, 303–11.

Winer, B. J. 1971. *Statistical Principles in Experimental Design.* 2nd ed. McGraw-Hill, New York. 907 pp.

14: Biomechanics

M. A. R. KOEHL

Department of Zoology, University of California, Berkeley, California 94720

STEPHEN A. WAINWRIGHT

Department of Zoology, Duke University, Durham, North Carolina 27706

CONTENTS

I. Introduction

A. Use of biomechanics in ecological research

A major goal of ecology is to discover and to explain quantitatively the mechanisms responsible for the distribution and abundance of organisms in time and space. Both physical and biological factors determine the distribution of macroalgae. An important physical factor in the distribution and form of seaweeds is the degree to which the plants are exposed to wave action and water currents. Moving water can transport nutrients, wastes, and propagules to and from seaweeds, but it can also rip plants from the substratum.

Which features of an alga enable it to withstand moving water? The theory and practice of fluid and solid mechanical engineering allow us to study quantitatively how algal gross morphology affects the magnitude of the flow-induced forces plants must sustain in particular habitats, as well as the distribution of mechanical stresses within the thalli (stress is a force per cross-sectional area of tissue bearing the force). Mechanics also enable us to investigate how algal microscopic anatomy affects the deformation and possible breakage of plants in response to those stresses. A spoonful of mechanics can provide a rich, multilevel understanding of how seaweed structure contributes to survival in mechanically rigorous habitats.

Disturbance can have important effects on the diversity of macroalgal communities. Moving water that rips organisms off the substratum is a major agent of disturbance in many rocky shore communities. Using a biomechanical approach, one can assess the susceptibility of various members of such communities to physical disturbance.

Biomechanics is a tool, not a subject that is an end in itself. The tool allows ecologists to use simple physical principles and engineering practices to study the role of mechanical factors, such as wave action, in limiting the distribution of particular macroalgae. In this chapter we shall focus on some simple techniques for studying the mechanics of biological materials but shall first mention a few leads into the literature for those interested in the fluid mechanics of algae or in

[292]

the effects of gross morphology on stress distributions within the bodies of organisms bearing forces.

B. *Determining mechanical stresses in algae in nature*

A number of techniques have been developed to work out forces on and stresses in organisms in nature; we shall mention a few, citing references that include more procedural details.

The main mechanical loads that macroalgae bear are probably due to moving water. Ways of measuring water-flow-induced forces [drag force, for plants in steady currents; acceleration reaction force and drag, for plants in waves] on organisms in the field are described by Koehl (1977a) and Denny (1982). Alternative ways of assessing flow-induced forces involve measuring water velocities and accelerations encountered by the organisms (e.g., Koehl 1977a,d; LaBarbera and Vogel 1976; Tunnicliffe 1980) and then calculating drag and acceleration reaction on the organisms for such flow conditions (e.g., Koehl 1977a; Vogel 1981). Alternatively, one can measure drag forces on the organisms at those velocities (e.g., Charters et al. 1969; Koehl 1977a) in a flow tank (e.g., Vogel and LaBarbera 1978; Charters and Anderson 1980).

When a benthic alga is subjected to a load, such as the force imposed on it by moving water, it can be deformed in a number of ways (Fig. 14–1). Note that, when an alga is pulled (Fig. 14–1B) or bent (Fig. 14–1C) by a load, some or all of the tissue in its thallus is stretched and thus experiences tensile stress. The magnitude and location of such stresses in stiff organisms subjected to flow-induced forces or other mechanical loads can be directly measured by means of strain gauges (William T. Bean; Measurements Group, Vishay) glued to the organisms in the field (e.g., Tunnicliffe 1980) or laboratory (e.g., Vosburgh 1977). Even if an organism is too deformable or lubricious, as most algae are, to have strain gauges glued to it, estimates of the magnitude of the stresses within the tissues can be calculated if the force on the plant is known. Descriptions of the types of stresses produced in structures of various shapes subjected to loads applied in different ways can be found in standard engineering texts (e.g., Roark and Young 1975; Faupel and Fisher 1981), as can formulas for calculating the magnitudes of those stresses. The biologist can find in Alexander (1968) or Wainwright et al. (1976) expressions for stresses in structures of certain regular shapes, such as cylinders, loaded in particular uncomplicated ways. For algae undergoing large deformations in flowing water, one should choose the stress equations derived for large deformations rather than the standard small-deformation formulas (e.g., see Charters et al. 1969).

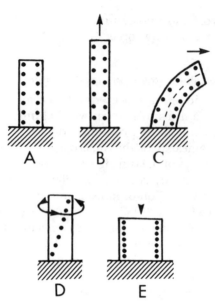

Fig. 14–1. Some ways in which a sessile organism can be deformed by a force. The arrows indicate the direction in which the organism is being deformed, the spots mark positions on the organism, and the hatched blocks represent the substratum. (A) Undeformed. (B) Pulled. (C) Bent. (Note that tissue on one side of the organism is stretched and on the other side is compressed. The dashed line indicates the neutral axis where tissue is neither stretched nor compressed.) (D) Twisted. (E) Compressed.

Analyses of flow-induced forces on, and stress distributions in, algae not only will enable one to design biologically relevant mechanical tests, but should also reveal ways in which the gross morphology of the plants affects their mechanical performance (e.g., see Charters et al. 1969; Neushul 1972; Koehl 1977a,b; Wainwright and Koehl 1976; Koehl and Wainwright 1977).

C. Deciding which mechanical tests to use

Mechanical tests can be used to measure basic mechanical properties of algal tisues (such as stiffness, resilience, strength, or toughness), to indicate the sorts of molecular mechanisms likely to be responsible for those properties, or to ascertain the response of algae to mechanical loading situations they would experience under various conditions in nature.

Which test(s) is chosen depends on the questions being asked. If one is interested in the deformability of an alga, the relevant tissues should be subjected to force–extension tests (see Sec. II) or the relevant structure to bending tests (see Sec. II.E). If the capacity of the tissues to recover their original shape after bearing a load is of interest, cyclic force–extension tests (Sec. III) or creep-recovery tests (Sec. IV) are appropriate. If one must know the breakability of the tissues, force–extension tests should be conducted until the specimens fracture (Sec. II). If one wants to study molecular adaptations of an alga to a mechanical habitat, long-term creep tests (Sec. IV) will point the way.

If the questions of concern are, for example, How do species A and B respond mechanically to wave action? Tidal currents? Rasping herbivores? then one should design algal loading regimes that simulate the situations of interest (Sec. V). In such simulations, the type (tensile, compressive, or shear; see Fig. 14–1) and magnitude of the stress applied to a specimen should mimic those in nature (see Sec. I.B). Furthermore, the rate at which the specimen is deformed during the experiment, as well as the duration of the deformation, should be the same as in the real-world situation. Such simulations should be specifically designed once these relevant variables have been measured or calculated for the algae in nature.

The types of specimens subjected to mechanical testing also depend on the questions asked. For example, if the behavior of whole stipes or blades is of interest, the relevant tests should be conducted on the entire structure. However, if the goal is to determine which tissues are the main load-bearing components of the structure (e.g., If a herbivore removes this tissue, what difference does it make mechanically?) or to work out the mechanical design of the alga (e.g., What morphological features of this alga render it so flexible?), the relevant tests should be conducted on isolated tissues from the plant (e.g., tissue just from the center and the periphery of a stipe or just from the proximal and distal ends of a stipe) and the results correlated with the microscopic structure of those tissues (Sec. I.D) and with their location in the plant. For an example of working out the mechanical design of an alga, see Koehl and Wainwright (1977).

Engineers have developed a large battery of mechanical tests to study the performance of materials. We describe here only a few basic types of tests that are relatively easy to perform and that yield a large amount of information relevant to ecological questions. For readers interested in conducting other types of tests, in more extensive discussions of molecular interpretation of test results, or in mathematical descriptions of mechanical behavior, we suggest consulting books on polymer mechanics, such as those of Nielsen (1965), Ferry (1970), or Aklonis et al. (1972). More information about the mechanics of biological materials can be found in Wainwright et al. (1976) or Vincent (1982).

Values should be expressed in SI units. For definitions of these units, as well as useful conversion tables, we recommend Mechtly (1973).

D. Importance of morphology in seaweed mechanics

The mechanical behavior of an alga depends on its gross and microscopic morphology. It is instructive for an ecologist to measure morphological correlates of the functional (in this case, mechanical) properties of algae and the physical environmental factors they encounter for several reasons. First, plant species are recognized by their morphology. Second, every function of an organism is permitted, controlled, and limited by its structure at lower levels of organization. Therefore, there is a high degree of interconnection among ecological, physiological, and developmental features of an organism via morphological parameters. Finally, morphological parameters can usually be readily measured with a high degree of accuracy and thus can be a sound basis for any analysis of a complex system.

A few mechanically important morphological parameters can be measured in a matter of seconds on fresh specimens that take less than 10 min to prepare. External linear measurements made with vernier calipers to $\pm$ 0.1 mm will allow accurate estimates to be made of the cross-sectional area of the blade, stipe, or holdfast of an alga. Alternatively, sections can be photographed with a Polaroid camera, attached to a microscope if necessary. The profiles in the photograph can be cut out and weighed and their weight compared with similarly prepared photographs of square millimeters.

Noncircular cross sections of an algal thallus prompt one to ask whether the wide or narrow side is aligned parallel to prevailing flow directions. This alignment is important in the consideration of bending of a seaweed thallus. When a structural element bends, the convex side is in tension, the concave side is in compression (Fig. 14–1C), and a plane somewhere between the two surfaces is called the "neutral axis" because there is no stress there. The resistance to bending of a structural element is called its "flexural stiffness" EI, where E is the stiffness of the tissue (see elastic modulus described in Sec. II) and I is the second moment of area of the cross section. Second moment of area measures the distribution of material around the neutral axis of the section and is given by $\int y^2 \, dA$, where dA is an increment of area that is distance y from the neutral axis. For regular shapes, there are handy formulas for I that require the measurement of a radius and a face or two of the section. For a solid circular section, I is $\pi r^4/4$, where r is the radius, and for an oval shape is $\pi ab^3/4$, where a is the radius of the neutral axis and b is the radius perpendicular to a (Wainwright et al. 1976). Note that a small increase in the width of an alga in the direction of bending can lead to a large increase in its flexural stiffness.

A frozen section 10–20 μm thick of a tissue or organ can be made with a cryostat (an ordinary rotary microtome housed in a cooler that maintains a temperature around −20°C). Using the Polaroid photoprofile technique, one can determine the area percentage and then calculate the volume percentage of any tissue or cell wall component. By studying both cross and longitudinal sections, one can determine the orientations in an alga of continuous cell walls, which are often load-bearing structures.

By viewing these frozen sections between crossed Polaroid filters on a microscope, one can see the distribution of high concentrations of macromolecules that have a preferred orientation; such regions will shine brightly. With a first-order red interference compensator (available from Edmund Scientific Co.), one can readily determine the direction of this preferred orientation by determining the sign (+ or −) of birefringence by the simple methods described in Bennet (1961), Chamot and Mason (1958), and Wahlstrom (1960). One expects high strength and stiffness in places and directions where large molecules have preferred orientation. The location of these regions corresponds to the distribution of stress-resisting tissue components.

II. Force–extension tests

A. *Purposes*

Force–extension tests permit one to measure the stiffness, strength, and toughness of a piece of material. In such a test, a specimen is pulled until it breaks; the specimen's extension and the force with which it pulls back at each extension are measured throughout the test. So that results from different tests can be compared, extension is expressed as "extension ratio" λ, where

$$\lambda = L/L_o \tag{1}$$

Here, L is the extended length of the specimen and L_o is the original unstretched length of the specimen between the grips of the device used to pull it. Note that λ is a dimensionless number. Similarly, force is expressed as "stress" σ, where

$$\sigma = F/A \tag{2}$$

Here, F is the force (in newtons) with which the specimen pulls back against being pulled, and A is the cross-sectional area of the specimen (in square meters). Results of force–extension tests are plotted in Fig. 14–2. For examples of stress–extension ratio curves for kelp

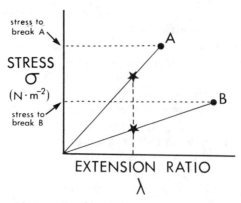

Fig. 14–2. Stress (σ)–extension ratio (λ) curves for pieces of tissue pulled until they broke. Note that, at a given extension, specimen A pulled back harder than specimen B (compare stars on the graphs). The slope (i.e., the elastic modulus, E) of line A is greater than that of line B; specimen A was stiffer than specimen B. Also note that the stress at which specimen A broke was greater than that at which specimen B broke (compare circles on the graph); specimen A was stronger than specimen B.

tissues, see Koehl and Wainwright (1977), or for various other plant and animal tissues, see Wainwright et al. (1976) or Vincent (1982).

One may encounter "stress–strain" curves in the literature. Some authors define strain ϵ in terms of the original length of the specimen (L_o),

$$\epsilon = \Delta L / L_o \tag{3}$$

where ΔL is an increment in length. Other authors define strain in terms of the length of a specimen at a given instant; such "true strain" (ϵ_T) is given by

$$\epsilon_T = \int_{L_0}^{L} dL/L = \ln L/L_o \tag{4}$$

where dL is a small increment in length, and L is the actual length of the specimen just before that increment is added. If one chooses to plot a σ–λ or a σ–ϵ curve, one calculates σ using the original cross-sectional area of the specimen. If one plots a σ_T–ϵ_T curve, one calculates "true stress" σ_T using the actual instantaneous cross-sectional area of the specimen (see the discussion of Poisson's ratio in Wainwright et al. 1976). Obviously, σ–λ curves are the easiest to do, which is why we have chosen to describe them here. However, before comparing results with those in the literature, one should make sure that all the curves being compared were calculated in the same way.

The "elastic modulus" E of a material is a measure of its stiffness,

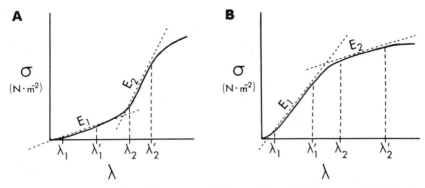

Fig. 14–3. Examples of nonlinear stress (σ)–extension ratio (λ) curves (solid lines). The slopes used to calculate elastic moduli (E_1 and E_2) for different portions of the curves (for extension ratios of λ_1 to λ_1' and for λ_2 to λ_2', respectively) are indicated by dashed lines.

(i.e., its resistance to deformation). Compare stress–extension curves A and B in Fig. 14–2. Note that, at a given extension, specimen A pulls back harder than specimen B; specimen A resists deformation more than specimen B. The slope of curve A is steeper than that of curve B. The elastic modulus is the slope of a σ–λ curve; the higher the E, the stiffer is the material. The E for a particular part of a σ–λ curve can be calculated using the formula

$$E = \Delta\sigma/\Delta\lambda \tag{5}$$

where $\Delta\sigma$ is the increment in stress accompanying a given increment in extension ratio $\Delta\lambda$. Since $\Delta\lambda$ is dimensionless, the units of E are the same as those for σ: newtons per square meter. Some materials (such as those in Fig. 14–2) have essentially one elastic modulus. Other materials are stiffer at some extensions than at others, (Fig. 14–3). In such cases, one elastic modulus can be calculated for a particular range of extension ratios, and another for a different range of extension ratios.

Many biological materials are stiffer when pulled quickly than when pulled slowly. Therefore, it is important to define the extension rate λ at which a test is conducted,

$$\dot{\lambda} = \Delta\lambda/\Delta t \tag{6}$$

where $\Delta\lambda$ is the increment in extension ratio made during a time interval, Δt. Since $\Delta\lambda$ is dimensionless, the units of extension rate are reciprocal seconds. Obviously, for stiffness measurements to be biologically relevant, tests should be conducted at $\dot{\lambda}$'s similar to those experienced by the tissues in the field, and E's should be calculated for realistic $\dot{\lambda}$'s, as discussed later.

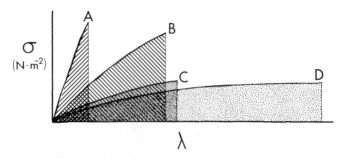

Fig. 14–4. Examples of stress (σ)–extension ratio (λ) curves for specimens of four different tissues pulled until they broke. The area under each curve (hatched or stippled) represents the work (joules) per volume (cubic meters) to break the specimen; the larger this area, the tougher is the specimen.

The strength of a material is defined as the stress required to break it (σ_{BRK}), given by

$$\sigma_{BRK} = F_{BRK}/A \qquad (7)$$

where F_{BRK} is the force required to break the specimen and A is the cross-sectional area. Either the original cross-sectional area of the specimen or the instantaneous cross-sectional area may be used. We recommend the former because of ease of measurement. Which A has been used should be specified when σ_{BRK} is reported, or compared with σ_{BRK} in the literature. The greater the σ_{BRK} of a tissue, the stronger it is. Specimen A in Fig. 14–2 is stronger than specimen B.

Glass is stronger than leather, but it is generally more difficult to break a boot than a bottle because leather is tough whereas glass is brittle. The tougher a material, the greater the amount of work required to break it. The area under a σ–λ curve of a specimen pulled until it breaks is a measure of the work required to break the specimen. So that toughness values can be compared for different specimens, toughness should be expressed as work (in joules) per volume (in cubic meters) of tissue absorbing that work. Sometimes the toughness of a material appears to depend on the length of the specimen tested. Therefore, it is important to make toughness measurements on specimens of the same length if comparisons are to be made. Sometimes toughness is expressed in joules per square meter for specimens of a defined length rather than as joules per cubic meter.

In Fig. 14–4 specimens B and D are tougher than specimen A or C. By studying the relationship of strength (σ_{BRK}), stiffness (E), extensibility (λ_{BRK}), and toughness in Fig. 14–4, one will see that there are several different ways that a specimen can be tough. One easy way of measuring toughness is to cut out and weigh the paper area under a σ–λ curve; toughness w would then be calculated as

$$w = c \text{ (work per volume}/s) \tag{8}$$

where c is the weight of the cutout area under the σ–λ curve, and s is the weight of a standard area of paper representing a known work (expressed in joules) per volume (expressed in cubic meters) for the scales used on the σ–λ graph. [Note that the units of stress are newtons per square meter and those of extension ratio are meters per meter; thus, the units of the area under such a curve are (newtons) (meters) per cubic meter, which equals joules per cubic meter.] Another easy way to measure toughness is to plot the σ–λ curve on very fine graph paper, to count the number of squares under the curve, and to multiply by the appropriate conversion factor to get joules per cubic meter. [If there is no need to calculate σ and λ, one can divide the area under the force (in newtons) versus extension (in meters) curve by the volume of the specimen between the grips (in cubic meters) to determine the toughness.]

B. Equipment

A number of instruments are available for conducting force–extension tests; we mention only a few examples that we have used. Mechanical testing instruments run the full range from huge machines equipped with electronic recording devices to small field-portable machines not requiring electricity. All of these devices consist of a pair of grips that hold opposite ends of a strip of material; these are pulled apart, the material thereby being stretched. The machines have some sort of mechanical or electronic means of measuring force and extension as the specimen is stretched. The instruments can generally be set up to bend, compress, or shear specimens, as necessary.

Generally, the large machines, such as Instron universal testing machines, are the best to use, if possible. Problems due to machine friction and vibration are minimized, a wide range of specimen types can be accommodated, specimens can be pulled over a wide range of steady rates, and a continuous record of force versus extension can be made for the duration of a test. Furthermore, Instrons are versatile and can be programmed to apply complex cyclic loading regimes. Instrons are usually available in mechanical engineering departments of universities.

At the opposite end of the spectrum of mechanical testing machines are small, portable, hand-operated devices, such as the Ametek LCTM, CTM, or RM universal testing machines. These small, inexpensive machines can be used in the field but have several disadvantages: Only a small range of specimen types can be tested (e.g., strong algae cannot be broken in such small machines, the force transducer may not be sensitive enough for flimsy algae, or the extension scale may not be sensitive enough for very stiff algae); machine friction and vibration can bias test results for delicate specimens; the range and steadiness of extension rates depend on the investigator's skill at turning cranks; and no time record is kept of force or extension (values are read on mechanical scales or dials). One way to eliminate the last problem is to film the dials during the course of a test.

A compromise between the Instron and Ametek machines is the Hounsfield tensometer (TecQuipment Inc.) This bench-top testing machine can be hand-operated and permits force and extension to be recorded mechanically but can also be set up to extend specimens and to record force and extension electronically.

C. Method

1. Preparation of specimens. Fresh algae should be used for mechanical testing. We have found no significant difference between the mechanical properties of freshly collected algal stipes, stipes from plants kept in running seawater for a day or two, and damp stipes kept chilled in coolers for a day or two while being shipped. Similar determinations should be made for algae if they will not be tested immediately after collection. If an intact stipe or frond is to be tested, use a razor blade to cut a length of the structure equal to the L_0 to be used plus the length on either end required for gripping. A specimen should never be wider than the grips. If a particular type of tissue is to be used, cut strip of it out of the alga; take care that the strip is of uniform width and thickness. Specimens should be cut just before testing and should be kept in seawater of the appropriate temperature. Any specimens with surface nicks or flaws should be discarded (unless, of course, the study includes the mechanical effects of particular types of flaws, such as grazer marks or epiphytes). The cross-sectional area of each specimen should be measured, as described earlier.

2. Gripping specimens. A range of grips are available, or can be machined, to hold specimens of different shapes in a testing machine. It is a challenge to clamp slippery algae so that they do not slide out of the grips when pulled. Unfortunately, if algal tissue is gripped too tightly, it can be damaged. We have found that, if mucilage is wiped off the parts of the specimen to be clamped and if rough paper toweling is glued (with cyanoacrylate contact cement) to the surfaces of the tissue that would otherwise come into contact with the grips, slippage and damage can be minimized. Alternatively, the grips can be padded with neoprene rubber and the ends of the specimen wrapped with dry paper toweling. Of course, specimens should be observed during the course of a test so that if grip problems occur, the test results can be discarded. Once a specimen is mounted in a testing machine, its length between the grips (L_o) should be measured with calipers.

3. Maintenance of specimens. If the duration of a test is only a few seconds, a specimen can be tested in air. If a test will last longer, the specimen should be bathed in seawater kept at the appropriate temperature (for some examples, see Gosline 1971 or Koehl 1977b,c).

D. Critical evaluation

It is important to watch a specimen carefully during a force–extension test and to discard results subject to the following artifacts. If a specimen slips in the grips, it will appear to be tougher and less stiff than it actually is. If a specimen breaks in or at a grip, the values for strength and toughness are likely to be too low. Furthermore, a nick in a specimen can cause fracture to occur at a lower stress and extension than it would in an unflawed specimen.

If the results of force–extension tests are used to interpret the functional morphology or ecological distribution of algae, the tests should be set up in biologically relevant ways: Is a tensile (i.e., pulling) force–extension test appropriate for the alga? When an alga is subjected to a load (such as the drag force imposed on it by moving water), it can be deformed in a number of ways (Fig. 14–1). Tensile tests are appropriate for plants that are bent in nature as well as for those that are pulled. One should observe the manner in which the algae under study are deformed in the field. If a tensile test is inappropriate, one can refer to one of the books cited at the end of Sec. I.C for advice on tests using other modes of deformation.

Are the main load-bearing tissues of the algae being tested? Are the values reported for stiffness E taken at a range of stresses likely to be encountered by the algae in nature? Do the algae break at stresses much greater than those they encounter in the field? To answer these questions, one should measure or calculate estimates of the loads the algae bear in nature and then calculate the magnitude of the stresses in their tissues when they bear such forces (see Sec. I.B).

Specimens should be pulled at rates similar to the extension rates they experience in nature. Such rates, for very deformable algae, can be determined from movies of plants in the field on which bright strings or spots have been affixed at known distances apart (another method for estimating extension rates is described by Koehl 1977b). If values for stiffness, strength, or toughness for different tissues are to be compared, all the tissues should be pulled at the same λ, or all should be pulled at the λ's that are biologically appropriate for each.

E. Alternative techniques

An intact alga in the field can be pulled or bent with a spring scale and the distance apart of marked spots on the plant can be measured. For a description of how to determine E from bending a stipe, see White (1974: ch. 5). Similarly, an intact plant can be collected and hung overhead; weights can be added to a basket suspended from the free end of the alga, and the distance between spots measured. Both these techniques offer several advantages. Any artifacts introduced by disrupting the structural integrity of an alga by chopping it into pieces to be gripped in a testing machine can be avoided; the places in an alga where deformation and failure occur can be observed; and no fancy equipment is required to perform the force–extension tests, so that they can be conducted inexpensively and at remote field sites without electricity. The disadvantages of these techniques are that no continuous record of force and extension is obtained, and no good control or record of λ can be achieved. Another difficulty of the second type of test is that algae often continue to extend after a weight is added; hence, λ increases with time and is therefore difficult to measure in some repeatable and interpretable way. Of course, desiccation due to extended periods of atmospheric exposure must be avoided.

III. Cyclic force–extension tests

A. Purposes

A piece of algal tissue can be pulled and then returned to its original length by means of a testing device such as an Instron or a tensometer. Such cyclic tests permit the resilience of the tissue to be measured: The area under a force–extension curve represents the work required to stretch the specimen (hatched area in Fig. 14–5A); the area under the return curve (stippled area in Fig. 14–5A) represents the work stored in the tissue as "strain energy" that is used for elastic recoil. The greater the ratio of the area under the return curve to the area under the force–extension curve, the more resilient is the specimen. The specimen in Fig. 14–5A is more resilient than that in Fig. 14–5B. The more resilient an alga, the more likely it is to bounce back to resting shape before the next wave hits.

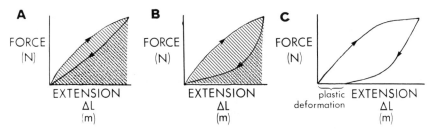

Fig. 14–5. Examples of force–extension (ΔL) curves for specimens pulled to a given extension and then returned at the same rate to their original length. Arrows pointing up and to the right indicate the curve for a specimen being pulled; arrows pointing down and to the left indicate the curve for a specimen being returned. The hatched area represents the work (joules) required to pull the specimen to that extension, and the stippled area represents the work stored elastically in the specimen and used for elastic recoil. Specimen A was more resilient than specimen B. Specimen C underwent plastic deformation and hung slack before it was returned to its original unstretched length.

Many materials undergo permanent ("plastic") deformation when pulled to extensions lower than those that break them. If a piece of an alga is subjected to a cyclic test and the return curve reaches zero stress before λ is back to 1 (i.e., force reaches zero before the grips of the testing machine have returned to their original distance apart of L_o), the specimen has undergone "plastic deformation" (Fig. 14–5C). Some tissues can recover from plastic deformation with time; a repertoire of repeated cyclic tests with different intervals of "rest" between tests can reveal such behavior. More importantly, cyclic tests can be designed to mimic biologically relevant extension regimes (such as an alga might encounter in the back-and-forth flow associated with waves) so that the resilience and plasticity of tissues can be measured and compared.

For a discussion of the use of cyclic force–extension tests to identify stress-softening behavior and the biological significance of such behavior, see Vincent (1982) or Koehl (1982).

B. Equipment, method, and critical evaluation

Cyclic tests should be done only on a testing machine, such as the Instron, the hysteresis of which is trivial compared with that of the specimen. The alga should be prepared, gripped, and maintained in the testing machine, as described earlier. The grips should be moved apart to a distance less than that which would break the specimen and then immediately moved back together at the same rate to their original position; the force and extension should be measured throughout this procedure. If more than one cycle of pulling is done, the time between successive cycles should be measured. The areas under the force–extension and return curves should be measured, as already described.

C. Alternative techniques

Resilience can also be measured by free-vibration tests (see Nielson 1965) or by forced-vibration tests (see Nielsen 1965; Gosline 1971). Plastic flow can be measured by creep-recovery tests, which are described in the next section.

IV. Creep tests

A. Purposes

Many deformable biological materials exhibit time-dependent mechanical behavior (i.e., their response to a load depends not only on the magnitude of the force, but also on the rate, duration, and history of force application). One sort of mechanical test that can be used to describe the time-dependent behavior of a tissue is a creep test. In this test a constant *stress* (force/area) is applied to a specimen and its extension with time is measured. The extensions of the specimen after measured time intervals are used to calculate the compliance D, a measure of extensibility of the specimen,

$$D(t) = \frac{\Delta L(t)/L_o}{\sigma} \tag{9}$$

where $\Delta L(t)$ is the increase in length of the specimen at time t, L_o is the specimen's original length, and σ is the stress applied to the tissue. An example of how the compliance of tissues on different time scales is related to the habitat and behavior of organisms is given in Koehl (1977c).

Creep tests also can be used to indicate the sorts of macromolecular mechanisms that are likely to be responsible for the mechanical behavior of a tissue. Many pliable biological tissues are composed of fibers (e.g., cellulose and collagen) dispersed in an amorphous matrix of highly hydrated polymers (e.g., alginic acid and glycosaminoglycans). When such a tissue stretches in response to a load, the fibers tend to become aligned with the stress axis and to slide relative to each other as the polymer molecules in the matrix rearrange. The mechanical behavior of the matrix of such tissues has been likened to that of other polymeric materials (e.g., Gosline 1971; Wainwright et al. 1976).

Figure 14–6 shows examples of compliance $D(t)$ versus the log of time for different types of polymeric materials subjected to creep tests. Polymer solution A is more dilute than B; the molecules in A

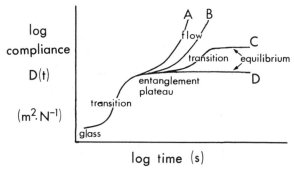

Fig. 14–6. Examples of plots of the log of compliance $D(t)$ versus the log of time for different types of polymeric materials subjected to creep tests. For an explanation, see Sec. IV.A. Graph based on information given in Ferry (1970), Aklonis et al. (1972), and Wainwright et al. (1976).

and B are *not* covalently cross-linked to each other. Solution C is a lightly cross-linked polymer network, and D is a heavily cross-linked polymer network. When the materials stretch (i.e., when their compliance increases), the polymer molecules rearrange in response to the stress applied to the tissue. Such rearrangements occur when the molecules change their configurations as they undergo random thermal motion. Therefore, the higher the temperature, the more rapidly the events described in the following paragraphs take place.

At extremely short times after a stress has been applied, when the molecules have not yet rearranged, the material has a low compliance and is described as a "glass." Then, as the polymer molecules begin to straighten out in response to the stress, the compliance of the material rises (this is the first "transition" shown in Fig. 14–6). At some point, further stretching of the material becomes hindered by "molecular entanglements (physical tangles or other noncovalent attachments between molecules, such as hydrogen bonds.); the compliance does not continue to rise with time and the material shows an "entanglement plateau." The more concentrated the polymers in the material, or the more groups they contain that are capable of temporary attachment to other molecules, the longer is the entanglement plateau. Eventually, the molecules slip past each other and the compliance continues to rise ("flow" in A and B; "transition" in C). If the polymers are covalently bound to each other to form a network, these junctions between molecules prevent the material from extending indefinitely. Such polymer networks show an "equilibrium" plateau, where the compliance no longer rises with time. If the molecules in a network are very heavily cross-linked to each other, the transition between the entanglement and equilibrium plateaus may not be seen (as in D).

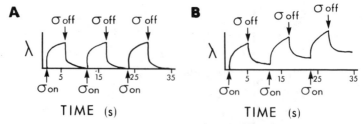

Fig. 14–7. Examples of results of cyclic creep-recovery tests designed to simulate wave-induced stresses in the tissues of two different marine organisms. Extension ratio λ is plotted versus time. Arrows pointing up indicate when stress σ was applied, and arrows pointing down indicate when the stress was removed. Tissue A bounced back to its resting length before "the next wave hit," whereas tissue B did not, and thus became stretched out as more and more "waves" pulled on it.

Thus, the results of creep tests can indicate which sorts of chemical and ultrastructural differences among species are likely to be responsible for differences in their mechanical behavior and thus point the way for further chemical or ultrastructural investigations. For examples of how creep tests or stress-relaxation tests have been used to point out mechanical adaptations or organisms at the molecular level of organization, see Gosline (1971) or Koehl (1977c).

Creep tests can also be designed to simulate the stress regime an alga might encounter in nature. After a specimen has been allowed to creep for a period of time, the stress can be removed so that the specimen is free to return to its original configuration. By means of such a creep-recovery test one can assess the ability of an alga to bounce back to resting shape. Results of creep-recovery tests designed to simulate wave-induced stresses on the tissues of two organisms are presented in Fig. 14–7.

B. Equipment and methods

A device for conducting creep tests is illustrated in Fig. 14–8. If one must work in the field with no electricity, the position of the tip of a long pointer, extending from one of the arms of the creep machine, can be traced onto a sheet of paper at timed intervals in order to calculate ΔL,

$$\Delta L(t) = x(t)c/d \tag{10}$$

where $x(t)$ is the distance the pointer tip moved by time t, c the length of the lever arm between the knife edge and the chain of the specimen grip, and d the distance between the knife edge and the tip of the pointer (Fig. 14–8). A different simple design for a creep machine is given by Vogel and Papanicolaou (1983).

A specimen of algal tissue should be prepared and gripped as

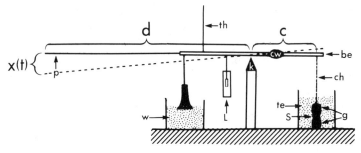

Fig. 14–8. Diagram of a device for conducting creep tests. The beam (be) sits on a knife edge (k) and is balanced (using the counterweight, cw) like a seesaw. The weight is then hung from the beam, and when the thread (th) is burned, the weight causes the tissue specimen (S) to stretch. The specimen is held between two grips (g), one of which is attached to the table and one of which is suspended by a light chain (ch) from the beam. The specimen is in a temperature-controlled bath of seawater (te). As the specimen stretches, its cross-sectional area decreases; since stress is force per cross-sectional area, the stress in the specimen would increase as it stretches. However, as the tissue stretches, the weight sinks into the water (w) and is bouyed up, thereby reducing the force on the specimen. The shape of the weight is such that the stress in the specimen is kept constant (for references on how to calculate the shape of the weight, see Koehl 1977c). Extension of the specimen can be measured electronically using a linear variable differential transformer (L) (available from Schaevitz Engineering) or visually using a pointer (p) (see Equation 10 in the text). The dashed line indicates the position of the beam and pointer at time t after the thread was burned.

described earlier. The specimen should be immersed in a bath of seawater held at the correct temperature; if the test is to run for a long time, an antibiotic should be added to the water to retard bacterial growth and decay of the tissue. Controls should be run on algal tissues to which no stress has been applied to measure how much elongation of the tissue is due to growth; controls should also be run in which the same weight applied to the alga is applied to a piece of very stiff material (such as metal or glass), so that any stretching of the creep machine can be measured.

C. Critical evaluation

As for other mechanical tests, one should ascertain whether the test is biologically relevant. Is the stress applied within the range encountered by the tissue in nature? Do the durations of stress application and recovery correspond to the durations of loading and unloading in the field? Are the load-bearing portions of the alga being tested? Is tension the appropriate mode of deformation?

Two other factors should also be considered when creep or creep-recovery test results are analyzed. One is that the tissue may have grown or degenerated during a long-term test. It is important not only to use the water baths and to run the controls already mentioned, but also to inspect the condition of the specimen carefully to make sure that it has not decayed. The other factor is that the shape of the hyperbolic weight (Fig. 14–8) or spiral pulley (Vogel and Papanicolaou 1983) used to apply the constant stress is calculated assuming that the specimens remain at a constant volume throughout the test. Therefore, the dimensions of the specimen at the beginning and end of the test should be carefully measured to make sure that no significant shrinking or swelling has occurred.

D. Alternative techniques

The creep test can be replaced by the stress-relaxation test, which can be performed using an Instron or tensometer. In such a test (described in Nielsen 1965; Gosline 1971; Wainwright et al. 1976) the specimen is "instantaneously" stretched to a given extension and held at that length; the stress with which it pulls on the grips is measured with time.

V. Design of mechanical tests

An advantage of using standard engineering tests is that results can be compared with a wide range of other results in the literature. However, a specific biomechanical question can sometimes be better answered by means of a new procedure designed for that question. For example, a question of interest to algal ecologists might be, How resistant are different types of seaweeds to rasping by gastropod radulae? Tissue hardness could be measured as an index of abrasion resistance (procedure described in Currey 1976). However, a more relevant measure could be made with a custom-built device that would permit a radula to be dragged across algal tissue at a defined rate and in a standardized fashion similar to the manner in which a gastropod rasps algae (Padilla 1982). A measure by which the "rasp resistance" of different algae could be compared should be defined (e.g., the minimum normal force that must be applied to the device to remove tissue when the radula is dragged across the alga or the cross-sectional area of the scrape made when a standard normal force is applied to the device as it is moved across the tissue). Similarly, a "gouge resistance" measure could be devised to compare the susceptibility of different algae to sea urchin teeth, or a "scour resistance" measure could be designed to compare the vulnerability of various tissues to abrasion by sand of defined grain sizes, and so on. The main points to keep in mind for all such custom-designed mechanical tests are that (1) the load should be applied in the same direction and at the same rate as in nature, (2) the procedure for load application should be standardized and repeatable, (3) the force used should be measurable [the specimen should be loaded with a known weight, or by a force transducer, e.g., Schaevitz Engineering; see Vogel (1981) for hints on how to build force transducers], and (4) the amount of deformation or damage produced should be measurable.

VI. Conclusions

The response of algae to mechanical loads, such as those imposed by water currents or crashing waves, is one important parameter affecting their distribution, their abundance, and the age structure of their populations. Thallus shape determines the distribution and magnitude of the stresses in an alga subjected to a load such as a flow-induced force. The parameters E, σ_{BRK}, and w of the tissues of an alga (all of which can be easily measured) determine the response of the plant to those stresses. Therefore, both gross and microscopic morphologies have an important influence on the way an alga is deformed and whether it will break in flowing water; mechanically important aspects of gross and microscopic structure can be easily measured. Mechanical tests can be designed to mimic stress conditions in the field, and similar tests can also be used to indicate molecular mechanisms responsible for the mechanical behavior of algal tissues. Thus, biomechanical analyses of algae can reveal how aspects of their structure are related to their distribution and behavior in the field.

VII. References

Aklonis, J. J., MacKnight, W. J., and Shen, M. 1972. *Introduction to Polymer Viscoelasticity*. Wiley, New York. 249 pp.

Alexander, R. 1968. *Animal Mechanics*. University of Washington Press, Seattle. 346 pp.

Bennet, H. S. 1961. Polarized light. In Jones, R. M. (ed.) *McClung's Handbook of Microscopical Technique*, pp. 591–677. Hafner, New York.

Chamot, E. M., and Mason, C. W. 1958. *Handbook of Chemical Microscopy*, 3rd ed. Wiley, New York. 502 pp.

Charters, A. C., and Anderson, S. M. 1980. Low-velocity water tunnel for biological research. *J. Hydrodynam.* 14, 3–4.

Charters, A. C., Neushul, M., and Barilotti, D. C. 1969. The functional morphology of *Eisenia arborea*. In Margalef, R. (ed.), *Proceedings of the Sixth International Seaweed Symposium*, pp. 89–105. Subsecretaria de la Marina Mercante, Madrid.

Currey, J. D. 1976. Further studies on the mechanical properties of mollusc shell material. *J. Zool. Lond.* 180, 445–53.

Denny, M. W. 1982. Forces on intertidal organisms due to breaking ocean waves: design and application of a telemetry system. *Limnol. Oceanogr.* 27, 178–83.

Faupel, J. H., and Fisher, F. E. 1981. *Engineering Design*, 2nd ed. Wiley, New York. 980 pp.

Ferry, J. D. 1970. *Viscoelastic Properties of Polymers*, 2nd ed. Wiley, New York. 671 pp.

Gosline, J. M. 1971. Connective tissue mechanics of *Metridium senile*: 2. Viscoelastic properties and macromolecular model. *J. Exp. Biol.* 55, 775–95.

Koehl, M. A. R. 1977a. Effects of sea anemones on the flow forces they encounter. *J. Exp. Biol.* 69, 87–105.

Koehl, M. A. R. 1977b. Mechanical organization or cantilever-like sessile organisms: sea anemones. *J. Exp. Biol.* 69, 127–42.

Koehl, M. A. R. 1977c. Mechanical diversity of connective tissue of the body wall of sea anemones. *J. Exp. Biol.* 69, 107–25.

Koehl, M. A. R. 1977d. Water flow and the morphology of zoanthid colonies. In *Proceedings of the Third International Coral Reef Symposium*, vol. 1: *Biology*, pp. 437–44. Rosentiel School of Marine and Atmospheric Science, University of Miami, Miami.

Koehl, M. A. R. 1982. Mechanical design of spicule-reinforced connective tissue: stiffness. *J. Exp. Biol.* 98, 239–67.

Koehl, M. A. R., and Wainwright, S. A. 1977. Mechanical design of a giant kelp. *Limnol. Oceanogr.* 22, 1067–71.

LaBarbera, M., and Vogel, S. 1976. An inexpensive thermistor flowmeter for aquatic biology. *Limnol. Oceanogr.* 21, 750–6.

Mechtly, E. A. 1973. *The International System of Units: Physical Constants and Conversion Factors*. NASA, Washington, D.C. 21 pp.

Neushul, M. 1972. Functional interpretation of benthic algal morphology. In Abbott, I. A., and Kurogi, M. (eds.), *Contributions to the Systematics of Benthic Marine Algae of the North Pacific*, pp. 47–74. Japanese Society of Phycology, Kobe.

Nielsen, L. E. 1965. *Mechanical Properties of Polymers*. Reinhold, New York. 274 pp.

Padilla, D. K. 1982. Limpet radulae as tools for removing tissue from algae with different morphologies. *Amer. Zool.* 22, 968.

Roark, R. J., and Young, W. C. 1975. *Formulas for Stress and Strain*, 5th ed., McGraw-Hill, New York. 624 pp.

Tunnicliffe, V. 1980. "Biological and Physical Processes Affecting the Survival of a Stony Coral, *Acropora cervicornis*." Ph.D. dissertation, Yale University, New Haven, Conn. 316 pp.

Vincent, J. V. F. 1982. *Structural Biomaterials*. Wiley, New York. 206 pp.

Vogel, S. 1981. *Life in Moving Fluids: The Physical Biology of Flow*. Willard Grant Press, Boston. 352 pp.

Vogel, S., and LaBarbera, M. 1978. Simple flow tanks for research and teaching. *Bioscience* 28, 638–43.

Vogel, S., and Papanicolaou, M. N. 1983. A constant stress creep testing machine. *J. Biomech.* 16, 153–6.

Vosburgh, F. 1977. The response to drag of the reef coral *Acropora reticulata*. In *Proceedings of the Third International Coral Reef Symposium*, vol. 1: *Biology*, pp. 477–82. Rosentiel School of Marine and Atmospheric Science, University of Miami, Miami.

Wahlstrom, E. E. 1960. *Optical Crystallography*, 3rd ed. Wiley, New York. 356 pp.

Wainwright, S. A., Biggs, W. D., Currey, J. D., and Gosline, J. M. 1976. *Mechanical Design in Organisms*. Wiley, New York. 423 pp.

Wainwright, S. A., and Koehl, M. A. R. 1976. The nature of flow and the reaction of benthic cnidaria to it. In Mackie, G.O. (ed.), *Coelenterate Ecology and Behavior*, pp. 5–21. Plenum, New York.

White, D. C. S. 1974. *Biological Physics*. Chapman & Hall, London. 293 pp.

15: Biogeographical analyses

LOUIS D. DRUEHL AND ROBERT G. FOOTTIT

Department of Biological Sciences, Simon Fraser University, Burnaby, British Columbia, Canada V5A 1S6

CONTENTS

I. Introduction

Biogeography is concerned primarily with describing biotic distributions and understanding why particular distributional patterns exist. The scales involved may cover several degrees of latitude or a few kilometers through an estuarine gradient. In the case of marine macrophytes, this objective has been approached by considering (1) correlations between seaweed distributions and environmental conditions, (2) taxonomic affinities between geographically separated floras, and (3) phylogenetic relationships within any given taxon over its distibutional range.

Testing for correlations between the geographical range of a seaweed taxon and the spectrum of environmental conditions encountered throughout the range has been perhaps the most common approach in studying biogeography (Druehl 1981a). Both the collection of pertinent environmental data and experimental testing of hypotheses are relatively straightforward.

For larger-scale studies of seaweed floras or, for example, disjunct distributional patterns that do not correlate with environmental conditions, an analysis of taxonomic affinities among species and/or groups throughout the range of their distribution may elucidate historical or evolutionary explanations for the observed distribution. With these approaches in mind, we shall introduce various methods for the acquisition and analysis of biogeographical data.

This chapter deals with methodology and does not attempt to review the field of marine biogeography. Topics of interest to marine plant biogeographers (e.g., island biogeography) are reviewed by Pielou (1979). Druehl (1981a)) summarizes the problems and advances in marine macrophyte biogeography.

II. Computer access

There are many computer packages that contain the univariate and multivariate statistical procedures needed in the analysis of marine macrophyte biogeography. Most large computing centers have implemented one or most of the commercially available, nationally

[316]

distributed computer packages, such as BMPD (Dixon 1981) and SPSS (Nie et al. 1975; Hull and Nie 1981). Both of these packages provide a wide range of univariate and multivariate statistical programs with excellent introductory write-ups and examples.

A number of packages have direct application to quantitative systematics, such as NT-SYS (Rohlf et al. 1974), which is a widely used library of programs related to many aspects of numerical taxonomy, and GEOVA (Mallis 1978), which is a series of computer programs for geographical variation analysis and numerical taxonomy. CLUSTAN (Wishart 1978) is one of the most comprehensive packages for performing cluster analysis. The SAS (SAS Institute, Inc. 1979) and SAS/GRAPH (SAS Institute 1980) packages provide programs for information storage and retrieval, statistical analysis, and graphics and for the flexible interaction of all three functions.

In addition, many textbooks on univariate and multivariate statistics contain FORTRAN programs that, with the appropriate job control language, can be implemented at most computer centers. For example, Harris (1975) provides programs that bridge gaps in programs available in some of the most widely used statistical packages. He also provides some useful comments on the optimal use of canned computer programs and discusses the relative merits of the more commonly used computer programs.

III. Correlations between seaweed distributions and environmental parameters

A. Introduction

If one starts from the premise that an observed biotic distributional pattern reflects extant environmental conditions, one can graphically model those conditions and test for a correlation between the environment and plant distribution. The results of this approach are visual and numerical statements of the probability that the observed distribution correlates with some range of environmental factor(s). These correlations can then be experimentally tested. Through such procedures, one may approach a cause-and-effect understanding of a biogeographical distribution. In reality, however, one can achieve only prima facie evidence supporting the role of the environment in determining seaweed distribution.

B. Temperature and salinity as environmental variables

A substantial pool of environmental data, both temporal and spatial, is required to test for correlations with observed macrophyte distributions. For some potentially important environmental parameters (e.g., nutrient concentration and submarine irradiance), there are

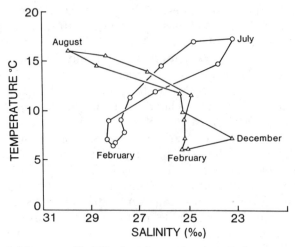

Fig. 15–1. Mean monthly *T/S* values for two locations having distinct floras.

few available data, but there does exist a sizable pool of temperature and salinity data. This information can be obtained from most oceanographic institutes, many marine stations, and the National Oceanographic Data Center.

We advocate the correlation of seawater temperatures and salinities with seaweed distributions, partially because of the availability of pertinent data, but more importantly because there is increasing evidence that these two parameters, when considered together, are significant in delimiting patterns of seaweed distribution (for discussion, see Druehl 1981a,b). We do not consider seawater temperature *T* and salinity *S* to be distinct environmental variables, but as one variable (*T/S*), which we refer to as *T/S* space.

Data on *T/S* can be graphically compared among locations having different floristics. Mean annual values can be compared, but this approach may be deceptive and mask real differences among locations (Fig. 15–1). Mean monthly values for those months that have extreme values of temperature and salinity can be compared; it is during such periods that plants most likely encounter the most stressful or beneficial conditions. Mean monthly values for all months of the year can be compared, which would allow one to follow the progression of *T/S* events as they are presented to the plant. Finally, one can compare the extreme events, as contrasted with mean *T/S* events, to which the plants are subjected.

These examples of *T/S* plotting provide a visual suggestion of correlations of this environmental parameter and seaweed distributions. To test statistically for significant *T/S* differences, one has two options. Graphically, one can surround mean *T/S* values with confi-

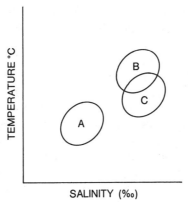

Fig. 15–2. Ninty-five percent confidence regions surrounding mean *T/S* values for three locations.

dence regions that form an ellipse (Sokal and Rohlf 1981: 594). In Fig. 15–2, situations A and B clearly occupy distinct *T/S* spaces. However, the statistical relationship between B and C indicates some overlap between the two true *T/S* population means.

To define the statistical relationship between two or more true population means positioned in close proximity, it is necessary to test using a multivariate analysis of variance (MANOVA). Analysis of variance (ANOVA) tests of such factorial designs are inadequate, since we are assessing the factor *T/S* as a single variable and not the dependent factors of *T* and *S*. A description of MANOVA and a FORTRAN program for its employment are given by Cooley and Lohnes (1971). Other MANOVA programs are available (BMDP, Dixon 1981; SPSS, Nie et al. 1975).

C. Other environmental parameters

The approach described for *T/S* can be applied to any number of environmental variables, provided that a significant data base exists and there is biological evidence suggesting synergistic interactions among the variables.

Another approach is the use of a canonical correlation analysis to describe the degree to which a distribution can be related to a variety of environmental parameters. Canonical correlations describe linear combinations of the environmental variables that maximize the correlation between environmental parameters and biotic distributions. Sullivan (1982) has used this approach in describing diatom distributions in relation to several environmental variables. This method is described by Nie et al. (1975). Packaged programs are CANCORR (subpackage of SPSS) and BMPD.

The foregoing approaches result in correlations between floristic patterns and environmental regimes. No evidence is provided by these methods to infer that the environmental differences are indeed responsible for the observed floristic differences. The purpose of using such tests is to distinguish differences among environments and to assist the researcher in narrowing down the number of variables that might be significantly delimiting seaweed distributions. Once potentially important variables have been identified, their actual effects on the plants can be tested though experimental studies, such as transplanting or laboratory manipulation (e.g., Druehl 1981b). The value of this approach is that it provides the researcher with a strong predictive tool in explaining differential seaweed distributions.

D. Biological interactions

Biological interactions (e.g., grazing and shading) are generally accepted as important determinants of the vertical distributions of marine macrophytes. The concept of biological interactions influencing large-scale horizontal distributions of seaweeds has been suggested (Gaines and Lubchenco 1982). Aspects indicative of biological interaction can be quantified and assessed as any environmental variable.

IV. Taxonomic affinities

A. Introduction

Since data bases often consist of series of morphological, ecological, or other biological measurements, the taxonomic affinities of marine macrophytes can be analyzed quantitatively. Numerical techniques provide an objective approach to the examination of such data in terms of patterns. Crovello (1970) provided a comprehensive review of the analysis of character variation in ecology and systematics. He treated this as a multistage decision process involving data accumulation, data analysis, data summarization, and evaluation of the results. In a later review, Crovello (1981) extended this approach to the study of quantitative biogeography, particularly historical biogeography, and again he provided a very useful outline of the stages involved in the analysis of biogeographical problems.

B. Univariate analysis

Univariate analysis of descriptive statistics compares the central tendencies and variation of samples. It gives one an initial impression of the characters and populations or taxa under study. Once the initial data set has been obtained, univariate statistics can be employed in the variable selection process. Summarization and comparison of variables and samples by the calculation of descriptive statistics, such as the mean, standard error, and coefficient of variation, can aid in the selection or deletion of variables for future analysis. Characters that are either invariant or too variable can be deleted, time and effort thereby being saved. The precision and accuracy of the measurement procedures can be tested at this stage also. Among the most useful general biostatistical textbooks are those of Sokal and Rohlf (1981) and Zar (1974). Crovello (1970) surveyed the procedures and problems of variable selection and the analysis and summarization of data.

C. Multivariate analysis

The geographical variation of an organism is believed to be the result of a multidimensional process arising from the adaptation of numerous features to many interdependent environmental and biological factors, the relationships of which change over space and time (Gould and Johnston 1972). For this reason, multivariate statistical techniques are most useful for the analysis of observed variability of characters. These procedures make possible the simultaneous analysis of the variation and covariation of a large number of characters from many individuals and populations or taxa. They provide a means of describing and summarizing patterns of variation and delineating groups of samples that are similar in these recognized patterns of variation. General reviews of the application of multivariate statistics to studies of biological variability are available (Blackith and Reyment 1971; Clifford and Stephenson 1975, Pimentel 1979; Sneath and Sokal 1973). Useful surveys of the application of multivariate statistical techniques to systematics are provided by Dunn and Everitt (1982) and Neff and Marcus (1980). Gould and Johnston (1972) and Thorpe (1976) review the use of multivariate statistical procedures in studies of geographical variation.

The taxonomic affinities of populations and taxa can be estimated by the calculation of similarity (or dissimilarity) measures. The type and complexity of the similarity coefficient used will depend on the type and number of characters measured. Sneath and Sokal (1973) discuss the relative merits of a variety of measures of taxonomic resemblance. The Mahalanobis generalized distance D^2 (Mahalanobis 1936), a multivariate measure of statistical distance between groups, is widely used in systematic studies and is considered superior to other measures of resemblance in that it takes into account the correlation between the characters and the covariation within groups.

Taxonomic affinities can be represented most usefully when both cluster analysis of a similarity measure and an ordination technique are used, a procedure that allows one technique to compensate for some of the disadvantages of the other. The relative merits of the use of cluster analysis and ordination techniques are discussed by Sneath and Sokal (1973). Cluster analysis, the clustering of populations of taxa into more inclusive groups, is useful in that it may produce partitions of populations or taxa that have a taxonomically meaningful order and that, as a result, supply useful summaries of taxonomic groups. For example, Murray and Littler (1981) used a cluster analysis and principal coordinates ordination to study the distribution patterns of the intertidal macrophyte floras of southern California in relation to temperature and the circulation of surface water. Ordination techniques do not impose a hierarchical structure on the data, and in cases where the variation is continuous, they may give a more meaningful taxonomic representation. Ordination techniques are particularly valuable for providing greater understanding of taxonomic affinities, because trends in variability can be associated with the attributes that cause them. For example, Lawson (1978) was able to obtain meaningful results in a study of the distribution of seaweed floras in the tropical and subtropical Atlantic Ocean using ordination by the reciprocal averaging method of Hill (1973). This method has an advantage in that it produces a simultaneous ordination of species and samples.

Two of the most commonly used multivariate statistical techniques in quantitative taxonomic studies are principal components analysis and multiple discriminant analysis, both of which can reduce a large amount of data to the most important factors or trends present. Principal components analysis is used to find relationships among variables and among individuals or samples assuming no a priori division of the individuals or samples into separate groups. Multiple discriminant analysis, which includes discriminant function analysis and generalized distance analysis, linearly combines discriminatory variables to produce a discriminant function that maximizes the distinction between two or more a priori groups. The Mahalanobis's generalized distance, mentioned earlier, can be derived from discriminant functions. Discriminant functions can also be used as an objective basis for the allocation of single specimens ("unknowns") into one of a number of a priori groups, thus serving as a check on the discriminatory power of the variables employed to calculate the original discriminant functions. For an example of the use of multiple discriminant analysis, see Widdowson's (1971) treatment of the brown alga *Alaria*.

D. Graphic approaches to species distribution

The distributions of taxa are often represented by maps and other graphics that provide useful summaries of the data. Crovello (1970) surveyed many of the types of biogeographical maps and graphics approaches. The development of high-resolution computer graphics packages such as SAS/GRAPH has greatly advanced this field. For example, the SYMAP, SYMVUE graphics packages (Laboratory for Computer Graphics) provide two-dimensional contour maps and three-dimensional graphic summaries of species distributions based on presence/absence data of species and enable one to make correlations with such parameters as geographical position and climatic and physiographic differences. Examples of the use of this approach are given in Crovello and Keller (1981).

Pielou (1979) provided information on a number of potentially useful techniques for determining the actual geographical ranges of related species in situations where inadequate sampling could result in a poor understanding of the degree of continuity in the range of a species. For example, from the comparison of the degree of overlap of seaweed species in the Atlantic and Pacific coasts of North and South America, Pielou (1978) speculated on the possible type of speciation process that might have produced the observed patterns. She found that there was a greater amount of overlap of the ranges of congeneric species in the Atlantic Ocean than in the Pacific Ocean. Pielou (1977) also described a method of representing seaweed distributions along shorelines as line segments and using these to analyze the latitudinal ranges of the plants.

Infraspecific geographical variation can also be mapped. The classic approach is to use isophene, or contour, maps that simply present the data. Other approaches, such as trend surface analysis (Sneath and Sokal 1973), are being used more frequently. Trend surface analysis condenses a large amount of data by representing the characters under study by a polynomial model and plotting the locations for a given character value on a contour map along with a goodness of fit criterion. Spatial classification methods, such as those of Gabriel and Sokal (1969), can be used to divide an area occupied by a species into geographically connected subareas and to divide a population exhibiting variation into homogeneous subpopulations that are geographically close, based on data consisting of sample locations on a map.

E. Numerical cladistics

The numerical cladistic approach has made many contributions to numerical biogeography. This involves the analysis of cladistic relationships among taxa in relation to designated geographical units, which may give a view of historical relationships of taxa under study. The most prominent approach has been that of vicariance biogeography (Platnick and Nelson 1978); however, Crovello (1981) has developed a more multifaceted approach that utilizes the taxon distribution data of biogeography and methods other than cladistics to determine taxon phylogenies. Methods in numerical cladistics are now being computerized; one example is a series of programs developed by J. Felsenstein (Department of Genetics, University of Washington, Seattle) for inferring phylogenies; programs for the construction of taxon cladograms are supplied.

V. References

Blackith, R. E., and Reyment, R. A. 1971. *Multivariate Morphometrics*. Academic Press, London. 412 pp.

Clifford, H. T., and Stephenson, W. 1975. *An Introduction to Numerical Classification*. Academic Press, London. 229 pp.

Cooley, W. W., and Lohnes, P. R. 1971. *Multivariate Data Analysis*. Wiley, New York. 364 pp.

Crovello, T. J. 1970. Analysis of character variation in ecology and systematics. *Annu. Rev. Ecol. Syst.* 1, 55–98.

Crovello, T. J. 1981. Quantitative biogeography: an overview. *Taxon* 30, 563–75.

Crovello, T. J., and Keller, C. 1981. The Indiana Biological Survey and rare plant data: an unending synthesis. In Morse E., and Henifin, M. S. (eds.), *Rare Plant Conservation: Geographic Data Organization*, pp. 133–47. New York Botanical Garden, New York.

Dixon, W. J. (ed.). 1981. *BMDP Statistical Software*. University of California Press, Berkeley. 726 pp.

Druehl, L. D. 1981a. Geographical distributions. In Lobban, C. S., and Wynne, M. J. (eds.), *The Biology of Seaweeds*, pp. 306–25. University of California Press, Berkeley.

Druehl, L. D. 1981b. The distribution of Laminariales in the north Pacific with reference to environmental influences. In Scudder, G. G. E., and Reveal, J. L. (eds.), *Evolution Today: Proceedings of the Second International Congress of Systematics and Evolutionary Biology*, pp. 55–67. Hunt Institute for Botanical Documentation, Carnegie–Mellon University, Pittsburgh. 486 pp.

Dunn, G., and Everitt, B. S. 1982. *An Introduction to Mathematical Taxonomy*. Cambridge University Press, New York. 152 pp.

Gabriel, K. R., and Sokal, R. R. 1969. A new statistical approach to geographic variation analysis. *Syst. Zool.* 18, 259–78.

Gaines, S. D. and Lubchenco, J. 1982. A unified approach to marine plant–herbivore interactions. 2. Biogeography. *Annu. Rev. Ecol. Syst.* 13, 111–38.

Gould, S. J., and Johnston, R. F. 1972. Geographic variation. *Annu. Rev. Ecol. Syst.* 3, 457–98.

Harris, R. J. 1975. *A Primer of Multivariate Statistics*. Academic Press, London. 332 pp.

Hill, M. O. 1973. Reciprocal averaging: an eigenvector method of ordination. *J. Ecol.* 61, 237–49.

Hull, C. H., and Nie, N. H. (eds.). 1981. *SPSS Update 7–9: New Procedures and Facilities for Releases 7–9.* McGraw-Hill, New York. 402 pp.

Lawson, G. W. 1978. The distribution of seaweed floras in the tropical and subtropical Atlantic Ocean: a quantitative approach. *Bot. J. Linn. Soc.* 76, 177–93.

Mahalanobis, P. C. 1936. On the generalized distance in statistics. *Proc. Nat. Inst. Sci. India* 2, 49–55.

Mallis, D. M. 1978. *GEOVAR: A Library of Computer Programs for Geographic Variation Analysis.* State University of New York at Stony Brook, New York. 49 pp.

Murray, S. N., and Littler, M. M. 1981. Biogeographical analysis of intertidal macrophyte floras of southern California. *J. Biogeogr.* 8, 339–51.

Neff, N. A., and Marcus, L. F. 1980. *A Survey of Multivariate Methods for Systematics.* Privately published, L. F. Marcus, Department of Invertebrates, American Museum of Natural History, New York. 243 pp.

Nie, N. H., Hull, C. H., Jenkins, J. G., Steinbrenner, K., and Bent, D. H. 1975. *SPSS: Statistical Package for the Social Sciences*, 2nd ed. McGraw-Hill, New York. 675 pp.

Pielou, E. C. 1977. The latitudinal spans of seaweed species and their patterns of overlap. *J. Biogeogr.* 4, 299–311.

Pielou, E. C. 1978. Latitudinal overlap of seaweed species: evidence for quasi-sympatric speciation. *J. Biogeogr.* 5, 227–38.

Pielou, E. C. 1979. *Biogeography.* Wiley-Interscience, New York. 351 pp.

Pimentel, R. A. 1979. *Morphometrics: The Multivariate Analysis of Biological Data.* Kendall/Hunt, Dubuque, Iowa. 276 pp.

Platnick, N. I., and Nelson, G. 1978. A method of analysis for historical biogeography. *Syst. Zool.* 27, 1–16.

Rohlf, F. J., Kishpaugh, J., and Kirk, D. 1974. *NT-SYS: Numerical Taxonomy System of Multivariate Statistical Programs.* Technical Report of the State University of New York at Stony Brook, New York. 91 pp.

SAS Institute Inc. 1979. *SAS User's Guide.* Raleigh, N.C. 494 pp.

SAS Institute Inc., 1980. *SAS/GRAPH User's Guide.* Raleigh, N.C. 66 pp.

Sneath, P. H. A., and Sokal, R. R. 1973. *Numerical Taxonomy.* Freeman, San Fancisco. 573 pp.

Sokal, R. R., and Rohlf, F. J. 1981. *Biometry*, 2nd ed. Freeman, San Francisco. 859 pp.

Sullivan, M. J. 1982. Distribution of edaphic diatoms in a Mississippi salt marsh: a canonical correlation analysis. *J. Phycol.* 18, 130–3.

Thorpe, R. S. 1976. Biometric analysis of geographic variation and racial affinities. *Biol. Rev.* 51, 407–52.

Widdowson, T. B. 1971. A statistical analysis of variation in the brown alga *Alaria. Syesis* 4, 125–43.

Wishart, D. 1978. *Cluston User Manual*, 3rd ed. Inter-University/Research Councils Series, Program Library Unit, Edinburgh.

Zar, J. H. 1974. *Biostatistical Analysis.* Prentice-Hall, Englewood Cliffs, N.J. 620 pp.

Section III

Ecological energetics

16: Respirometry and manometry

CLINTON J. DAWES

Department of Biology, University of South Florida, Tampa, Florida 33620

CONTENTS

[329]

I. Introduction and objectives

By means of manometric procedures, one measures the uptake or release of a gas in a closed vessel via changes in barometric liquid levels. Manometric techniques follow the simple gas law

$$PV = RT \tag{1}$$

where P is pressure, T temperature, V volume, and R a constant. Temperature is usually held constant. If both temperature and pressure are held constant, then the change in volume will reflect the amount of gas consumed or released. There are three types of manometers (Dixon 1952; Umbreit et al. 1972):

1. The constant-volume manometer: If the volume V is held constant, the change in pressure is a measure of the amount of gas consumed or released. This procedure is also called the Warburg method.

2. The constant-pressure manometer: By holding the pressure P constant, one can measure changes in volume to determine the amount of gas consumed or released. A Gilson respirometer makes use of this method.

3. The differential manometer: If both pressure and volume are allowed to change in two connected flasks, one can measure the amount of gas produced or released. The Barcroft method employs this procedure.

The objectives of this chapter are to present basic, usable manometric procedures and to assess their applicability to the measurement of macroalgal productivity and the abiotic factors that influence these rates. Detailed calculations and specific procedures for individual equipment as well as the theoretical basis for the procedures can be found elsewhere (Dixon 1952; Umbreit et al. 1972).

II. Equipment and types of manometers

A. Constant-volume manometer

The Warburg method utilizes a type of constant-volume manometer that was introduced by Barcroft and Haldane (1902) and Brodie

[330]

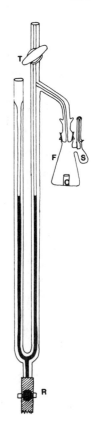

Fig. 16–1. The Warburg constant-volume respirometer measures changes in pressure through adjustment of manometric fluid in a closed system. The reaction flask (F) has a side arm with a gas vent (S) and a center well (C). The manometer can be vented to the atmosphere by a stopcock (T) and the manometer fluid level controlled by a screw clamp at the reservoir (R).

(1910) and extensively modified by Warburg (1962). Figure 16–1 is a diagram of a modern Warburg manometer, a constant-volume respirometer. The scale of the fluid-filled manometer is graduated in millimeters, with the smallest markings at 0.5 mm. The manometer tube has an open and closed (stopcock) end, and the liquid is adjusted to a given position (e.g., 250-mm level) before the pressure is recorded. Adjustment of the screw clamp (R) at the reservoir will alter the level of the fluid in the manometer. The unit has a detachable reaction flask (F) with one or more side arms that serve as entry ports or as vents. The reaction flask is held in a water bath of constant temperature and is shaken between readings to ensure even temperature and adequate mixing. Lights may be used under the bath (with a clear glass bottom) or on top to support photosynthesis (oxygen production measured). Respiration is measured in total darkness.

After an initial adjustment of the manometric fluid to some level (e.g., 250 mm, stopcock left open) on the reaction flask side, the stopcock is closed and the level of the fluid on the open side of the manometer is recorded (e.g., 249 mm). After 10 min, the fluid rises, for example, in the closed arm (respiration resulting in the consumption of oxygen) and falls in the open arm. By adjustment of the screw clamp (pressure clamp), the level of the fluid in the closed side is brought back to the original 250-mm reading, the gas in the flask being held constant but the level in the open arm side dropping (e.g., to 220 mm). The amount of gas consumed can then be calculated if the gas volume of the flask (V_g), the volume of fluid in the flask (V_f), the temperature of the system, the gas being consumed or exchanged, and the density of the fluid in the manometer are known. Equation 2 expresses this:

$$X = \Delta H K \tag{2}$$

Here, X is the amount of gas exchanged, H the change in the reading (millimeters) of the open arm of the mamometer, and K the flask constant, which is shown in Equation 3.

$$K = \frac{273/(V_g T + V_f \alpha)}{P_0} \tag{3}$$

In Equation 3, α is the solubility of the gas (expressed as milliliters gas per milliliter fluid) at a partial pressure of 1 atm, P_0 is standard pressure (760 mm Hg), and T is the absolute temperature (Kelvins) of the water bath (273°C + temperature in degrees Celsius).

It should be noted that the solubility of a gas is influenced by temperature. The solubility of oxygen, for example, decreases as temperature rises. Thus, α is usually taken from tables, and temperature is a critical factor that must be kept constant in a Warburg experiment.

Also, the value P_0 represents the initial atmospheric pressure in Equation 3 that is used to solve for K. Since the pressure in the room and temperature of the water bath can change during an experiment, one corrects for this by the use of a blank Warburg reaction flask. Any change in the blank manometer is used to correct the reading in the experimental manometer. A decrease in pressure (rising manometric fluid at closed end) would then be added to the known rise in pressure (photosynthesis) or subtracted from a decrease in pressure (respiration) of a reaction flask containing an algal sample.

B. Constant-pressure manometer

The constant-pressure respirometer is also termed a volumeter because changes in volume are measured. Two examples of this type of manometer are the Gilmont manometer and the Gilson differential respirometer (Fig. 16–2). In the Gilson volumeter, a typical manometric reaction flask and a digital micrometer determine changes in

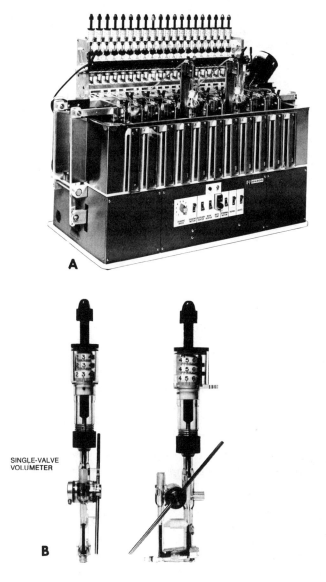

SINGLE-VALVE
VOLUMETER

Fig. 16–2. A Gilson respirometer, model GRP 20 (A), contains 20 volumeters, each (B) with a reaction flask, manometer, and digital readout. Gas volume changes are measured by movement of a piston when the digital mechanism is turned. All the volumeters are connected to a large reference flask, thus eliminating most barometric and temperature corrections. Photographs courtesy of Gilson Medical Electronics Inc.

gas volume (expressed as microliters of oxygen uptake or release). The main feature of the Gilson volumeter is that a number of plastic manometers (8–20) with reference flasks can be arranged in a water bath. The open ends of the reference arms are all connected to a common compensation vessel. The single compensation vessel allows for simultaneous evacuation and gassing of all reaction vessels and eliminates the tedious calculations for barometric changes as well as minimizes temperature errors. Gas changes are determined via a digital readout that measures the inward or outward movement of a plunger to maintain the original volume level of the manometric fluid at the closed end.

A correction for water vapor is required because, as the micrometer (digital) plunger is inserted, water vapor in each reaction flask system is condensed, and if the plunger is backed out, water is vaporized. Thus, the total water vapor changes and the following formula should be used:

$$\frac{273(P - P_{\mathrm{w}}) \, \Delta V_{\mathrm{g}}}{760T} \tag{4}$$

In this equation P is barometric pressure in millimeters of mercury, P_{w} is vapor pressure of water at temperature T (from standard tables), T is temperature (absolute), and ΔV_{g} is microliters of gas change measured.

Because most Gilson and Gilmont respirometers are constructed of plastic and Tygon tubing (except for glass reaction flasks), there is a problem of diffusion if the gas (oxygen or carbon dixoide) concentration in the flask is higher than that in the surrounding atmosphere. This can be controlled by the use of blank flasks with the gas only or the use of all-glass volumeters that have their own compensation flask.

C. Differential manometer

The Barcroft differential manometer (Fig. 16–3) consists of two flasks, each attached to different ends of the manometer so that the entire system can be closed off. The two halves are identical, and the flasks are submerged in a constant-temperature bath. One important use for this type of manometer is in the study of photosynthesis when carbon dioxide uptake and oxygen evolution are to be measured simultaneously. One flask contains the plant and is the reaction flask, whereas the other serves as the compensation vessel. Because any change in temperature or atmospheric pressure will equally affect the pressure and volume of the identical flasks, there will be no modification in the height of the manometric fluid and so

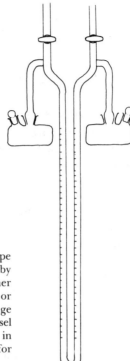

Fig. 16–3. A differential respirometer of the Barcroft type is a closed system in which the two flasks are connected by a manometer. One flask contains the alga, whereas the other flask functions as a compensation vessel. Consumption or release of oxygen in the reaction flask results in a change in fluid position in the manometer. The compensation vessel corrects for any atmospheric and temperature changes in the system, reducing the chance of error and the need for various mathematical corrections.

no special "blank" manometers are required. Indeed, the differential respirometer will outperform most Warburg or volumetric respirometers. The newer, single-valve Gilson respirometers, each with their own reference flask, can also be considered to be of the Barcroft type.

III. Methods and examples

This section is a brief review of a variety of procedures used to carry out manometric measurements. Specific methods can be found with the instructions accompanying various instruments.

A. *Carbon dioxide balance*

The Henderson–Hasselbalch equation (Equation 5; for derivation, see Umbright et al. 1972: 20–22) demonstrates that, at pH 7, there must be carbon dioxide in the atmosphere:

$$pH = pK' + \frac{\log[HCO_3]}{\log[CO_2]} \tag{5}$$

When atmospheric carbon dioxide concentration approaches zero, the factor (log of bicarbonate over carbon dioxide) becomes larger and pH increases, or if pH is held constant, bicarbonate concentration goes to zero. It is essentially impossible to keep the level of carbonic acid low because the reaction of carbonic acid with tissue buffers or seawater will increase the bicarbonate level. Thus, the simplest method is to add carbon dioxide to the gas phase, adjusting the partial carbon dioxide to obtain pH values of 6 to 8. The maintenance of a constant carbon dioxide level in the reaction flask is of considerable importance to phycologists attempting to follow photosynthesis.

A number of seawater buffers are described in the literature for growth medium buffers (see McLachlan 1973), although most are not useful in manometric studies. Ogata (1966) demonstrated that the carbon dioxide–absorbing capacity of Tris buffer [tris(hydroxymethyl)aminomethane] results in enhancement of photosynthetic activity. Dromgoole (1978) found that, at high light intensities, photosynthesis in species of the intertidal fucoid *Carpophyllum* was stimulated with standard bicarbonate-buffered artificial seawater. The artificial seawater medium contained a bicarbonate buffer mixture of 0.1 M $NaHCO_3$ and 0.1 M Na_2SO_3 typical of many culture media (McLachan 1973). Dromgoole further reported that the photosynthetic activities of *Carpophyllum* spp. were limited by the level of free carbon dioxide as well as fluctuations in pH. *Carpophyllum* occurs in the infralittoral fringe on the warm temperate coasts of New Zealand (Stephenson and Stephenson 1972); thus, the species may utilize free carbon dioxide directly in a manner similar to that proposed for other intertidal algae such as the red alga *Bostrychia* (Dawes et al. 1978).

The carbon dioxide balance procedure proposed in this chapter does not require the addition of bicarbonate buffer but rather relies on the bicarbonate buffering system of natural seawater and a maintenance of a given carbon dioxide atmosphere in the reaction flasks. Provided that the same level of atmospheric carbon dioxide is maintained in all experiments, photosynthetic rates should not increase due to fluctuations in carbon dioxide levels.

Although there are a number of methods for the maintenance of carbon dioxide in the reaction flask, perhaps the Pardee method (Pardee 1949) is one of the simplest. The formula to obtain a 1% carbon dioxide atmosphere is as follows:

6 ml	Diethanolamine (clear, odorless or discard)
15 mg	Thiourea
3 g	Potassium bicarbonate

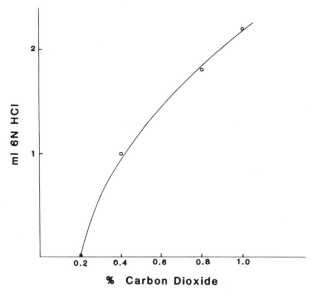

Fig. 16–4. Relationship between the amount of 6 N HCl added to the Pardee solution (Pardee 1949) and the resulting level of carbon dioxide maintained in the reaction flask atmosphere. After Pardee (1949).

2.2 ml   6 N hydrochloric acid (concentrated)
6.8 ml   Water (to bring to 15 ml)

The amount of hydrochloric acid, the difference being made up with water, will determine the level of carbon dioxide in the atmosphere (Fig. 16–4). The ingredients must be added in the order listed; otherwise, a precipitate will form. The solution is held in a stoppered flask overnight and used the following day. About 0.4 ml is placed in the side arm and 0.2 ml in the center well of the reaction flask.

In respiration studies, either oxygen consumption or carbon dioxide production can be measured. By removing the carbon dioxide in the reaction flask, one can determine a drop in pressure due to the consumption of oxygen. Control respiratory experiments without an external source of oxygen should be carried out in the flask; the result is usually a linear response by the macroalga over at least a 2-h period. A simple procedure is to add 0.2 ml of 1% potassium hydroxide in the center well of the reaction flask with a fan-folded paper wick. This method will retain a 0.5% carbon dioxide atmosphere, preventing a buildup in the reaction flask and a subsequent pH change in the medium. An alternative method is to use the Pardee carbon dioxide source that will maintain a 1% carbon dioxide atmosphere.

B. *Cleaning of glassware*

Umbreit et al. (1972) recommended two methods for the cleaning of reaction flasks. The first procedure uses potassium dichromate and sulfuric acid (63 g $KCrO_4$, 35 ml water, and concentrated sulfuric acid added to 1 liter) in which the reaction flasks are soaked for 24 h. Alternatively, the reaction flasks can be heated in nitric acid and sulfuric acid (1:1 solution of concentrated acids) for 30 to 60 min. In both cases, after acid cleaning, the flasks must be soaked in hot detergent and finally in distilled water. Before the acid treatment, the reaction flasks can be soaked in a solvent such as petroleum ether, hexane, or gasoline to remove the grease used in sealing the joints. If temperatures above 30°C are not to be used, we have found that petroleum jelly is an excellent seal and will wash off easily in hot tap water. Thus, the glassware can be cleaned in hot detergent followed by a number of rinses in hot deionized water. If stopcock grease or vacuum grease is used, stronger solvents are required. Because of the safety problems with gasoline, we use a 1:1 solution of xylene and 95% ethanol before washing in hot detergent and water. Excess grease should be removed from the portals of the reaction flasks with cotton swabs, and all air passages should be inspected and cleaned with pipe cleaners to remove possible water or grease plugs.

C. *Calibration of manometric equipment*

Accurate volume determination of reaction flasks is a detailed and specific procedure, and usually instructions are given by the manufacturer. In many cases, the volume is given for the equipment. If not, reaction flask volumes can be determined by filling to overflowing either with water or mercury, replacing the unit on a dry manometric joint, and sealing all fittings. After careful removal, the remaining liquid is poured into a graduated cylinder and the volume measured.

D. *Standard control determinations*

Photosynthetic or respiratory studies of macroalgae can yield erroneous results if too much plant material is used (biomass), because shading results. In some instances, a supplementary carbon dioxide source is not feasible, and thus if the run is too long, the rate of photosynthesis may decline with the removal of available carbon dioxide. It is therefore recommended that a linearity run be carried out. A graded series, by wet weight (e.g., 0.1, 0.5, 1.0, 2.0 g), of the alga should be run at a standard temperature and light intensity for 2 to 4 h with readings every 15 min. The resulting graph (Fig. 16–5) is an example of selection of a biomass (curve A, Fig. 16–5) to be

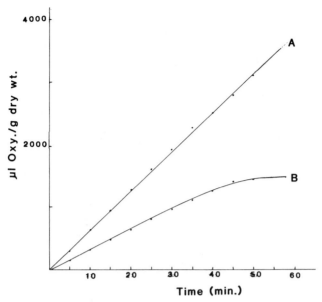

Fig. 16–5. Biomass–longevity graph showing a linear photosynthetic response over 60 min for the red alga *Hypnea musciformis* with 1 g wet wt (A). The photosynthetic response was lower (due to shading) and not linear (due to depletion of carbon dioxide) over the same period when 1.5 g wet wt of the alga was used (B).

used as well as the length of a manometric experiment for a particular alga.

E. Expression of rates

Macroalgal photosynthesis has been expressed in terms of dry weight (parts per million oxygen per gram dry weight per hour), wet weight, thallus area, grams of plant material per square meters, and photosynthetic pigments (microliters oxygen per milligram chlorophyll *a* per hour). It has been recommended (Dawes 1981) that all rates of oxygen evolution in photosynthesis be expressed as units of oxygen produced per milligram of chlorophyll *a* per unit time (minutes or hours).

No choice is completely satisfactory, and in many cases the choice of expression of photosynthetic rates will depend on the alga. Ramus (1978) has shown that light levels at which green algae saturate are, at least in part, dependent on seaweed morphology and anatomy. The optically transparent *Ulva lactuca* saturated at high light levels, and absorbance by the thallus was dependent on pigment content. The optically opaque *Codium fragile* saturated at low light, and the thallus absorbance was independent of pigment concentration. Thus, in one alga, pigment was related to photosynthetic rates, whereas in the other, this was not so. Differences between the alga's ecological background (habitat, nutrient sources, and shading) and chemical levels (number of active photosynthetic antennae; see Ramus et al. 1976) can influence the expression of photosynthesis via variation in the number of active chlorophyll *a* centers.

However, direct comparison of photosynthetic activity of filamentous and fleshy macroalgae based on dry weight is also a problem. A fleshy alga will have a limited amount of photosynthetically active cells (epidermis and outer cortex) when compared with an equal dry weight mass of a filamentous species (all cells photosynthetically active). As Ramus and co-workers (1976) noted, the use of biomass will yield one view of photosynthetic activity, whereas the use of chlorophyll *a* will yield another. Perhaps the usefulness of each unit should first be compared in a study of photosynthetic rates.

Respiration rates are invariably presented in terms of grams dry weight of the plant, and thus considerable variation in rates can occur when fleshy, thick-walled plants are compared with filamentous, thin-walled plants. A more accurate expression of respiratory rates would perhaps be one of ribulose bisphosphatase (RuBP) activity or simply milligrams of soluble protein. Although much of the soluble protein would be nonrespiratory RuBP, the large differences (40–80% by dry weight) in cell wall carbohydrate that is contained in macroalgae would be eliminated. The relatively simple Lowry procedure (Lowry et al. 1951), a colorimetric method, has been recommended for macroalgae (Dawes 1981).

F. Enzymes and particulate matter

Manometric methods are sufficiently accurate to measure enzymatic reactions such as those involved with oxygen and carbon dioxide uptake or evolution. Numerous procedures are presented by Umbreit et al. (1972), and only one example involving a macroalga (*Bostrychia binderi*) and the enzyme carbonic anhydrase will be mentioned here.

In any enzymatic assay, a most crucial step is extraction. This is especially important with carbonic anhydrase. Everson (1970) demonstrated that disruption of the chloroplasts themselves was necessary to measure total activity. Carbonic anhydrase activity in macroalgae has been found to be similar to that of land plants, and an important factor is the need for careful extraction (Graham and Smillie 1976). After comparison of procedures, Graham and Smillie (1976) utilized an extraction medium consisting of a Tris–borate buffer including dithiothreitol for protection of sulfhydryl groups and reduced sites; EDTA; Polyclar AT for "protection" from phenols (particulary in brown seaweeds); bovine serum albumin; and Triton X-100, a detergent for the lysis of chloroplasts. After about 10–20 g of algal material were washed in the appropriate medium, the macroalga was frozen in liquid nitrogen and ground in clean sand at 0°C by use of the extraction medium of Graham and Smillie (1976). The slurry was then filtered through cheesecloth, checked with a compound microscope to ensure that the chloroplasts were broken, and centrifuged at 3000 rpm, and the supernatant was used in Gilson reaction flasks. Specific activity is measured as the number of units (enzyme) per milligram protein or as the number of units per milligram chlorophyll *a*. Both protein and chlorophyll levels were determined on subsamples taken after grinding, but before filtration with cheesecloth.

Manometric determination of enzymatic activity should be checked with other procedures. For example, a colorimetric method (Graham and Smillie 1976) and an electrometric method (Bowes 1969) can be used to measure carbonic anhydrase levels. The colorimetric methods depend on measuring the time required for a pH indicator to change color (usually bromphenol blue, which goes from blue to yellow as the pH decreases). The electrometric methods are based on continual recordings of pH changes and the time required for a given change to occur.

G. Use of inhibitors

Manometric procedures allow for the use of inhibitors, and especially common are those that interfere with photosynthetic and respiratory activities of the macroalga (see Umbreit et al. 1972). For example, a number of compounds inhibit photorespiration and can be used to demonstrate that process in macroalgae (Fig. 16–6). Because the process of photorespiration utilizes oxygen, inhibition should show a higher rate of apparent photosynthesis if oxygen evolution is measured.

As an example, clean, young branches of *Hypnea musciformis* were placed in filtered (0.45 μm) seawater containing various concentrations (1.4–28.4 mM) of the inhibitor α-hydroxy-2-pyridinemethanesulfonic acid (α-hpms) adjusted to pH 8.4. After 15 min, the branches of *H. musciformis* were transferred to the reaction flasks of a Gilson respirometer (volumeter), and photosynthetic rates were measured at 20°C and 100 μE (photosynthetically active radiation, PAR). Levels of microliters oxygen were monitored every 15 min for 2 h. Each reaction flask held 0.6 ml of Pardee's carbon dioxide solution to maintain a 1% carbon dioxide atmosphere. As seen in Fig. 16–6, plants exposed to the inhibitor (curves A, B, C) showed increased oxygen production over those that were not inhibited (curve E,

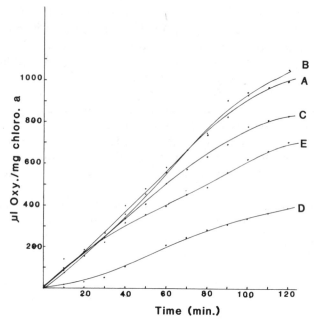

Fig. 16–6. The inhibitor α-hydroxy-2-pyridinemethanesulfonic acid prevents photo-respiration, a process that utilizes oxygen. Thus, when it is used in respirometric studies the amount of oxygen produced during photosynthesis increases if photores-piration is present. Photosynthetic rates expressed as microliters oxygen produced per milligram chlorophyll *a* rose when the inhibitor was used in lower concentrations: 1.42 m*M* (A), 7.42 m*M* (B), 14.28 m*M* (C) when compared with plants not exposed to the inhibitor (E). When the inhibitor was at a higher concentration (28.42 m*M*, D), there was a suppression of photosynthesis.

controls) with only the highest concentration of α-hpms causing a decrease in response (curve D, toxic effect at high concentrations) below that of the controls (Fig. 16–6).

H. Example of studying physical factors

A 5-d experiment is briefly outlined here and includes steps to be taken in the preparation of macroalgae and the testing of photosyn-thesis and respiration under various light, temperature, and salinity conditions. Only general procedures are given for the use of a Gilson respirometer, because details for each type of instrument vary. It should be noted that holding macroalgae in the laboratory over 5 d can result in some acclimation and thus, whenever possible, fresh collections should be made for each day.

Day 1: collection and preparation of plants. Healthy plants are sorted and washed in ambient seawater, and the salinity and temperature of the collection site noted. If photosynthesis is to be expressed in terms of chlorophyll *a*, a subsample should be prepared for chlorophyll *a* extraction (expression of photosynthetic rates) and another to be used for determination of soluble protein (expression of respiratory rates) each day of the study. To allow for acclimation to salinity, about 0.5 g wet weight of cleaned algae using entire plants (if possible – otherwise cut sections) should be placed in 125-ml Erlenmeyer flasks with 100 ml of filtered seawater of the various salinities. The 3-d acclimation proposed here is for macroalgae that are exposed to long-term salinity changes (e.g., estuarine forms). However, seaweeds of open coasts, where salinity variations are of a short term and mild nature, need not be held for more than 3 to 4 h before manometric study, and the salinity range can be reduced. The flasks can be shaken gently, aerated, and held in a growth chamber at ambient temperature and moderate light levels (60 $\mu E \cdot m^{-2} \cdot s^{-1}$). Algal material to be used in the temperature and light runs should be held in larger containers with adequate aeration. The carbon dioxide source (Pardee's) should be prepared for day 2.

Day 2: temperature series and control run. A biomass and linearity run should first be done using ambient temperature and light. Usually 10 ml of filtered seawater is sufficient for each 20-ml reaction flask. The macroalgal fragment will remain submerged if a shaker is used. Because of the possibility of diel changes in the photosynthetic rate of macroalgae (Hoffman and Dawes 1980), studies should be carried out at the same time of day for each experiment. After this, a temperature series can be carried out with 5°C steps from 5 to 30°C (six steps). It is best to begin at the lowest temperature, using ice to cool the bath rapidly, and then proceed to higher temperatures by removal of some water and replacement with hot water. Using 1–2 g of healthy plant material (as determined from the biomass run), one adds the carbon dioxide source to the side arm and center well and initiates the first temperature run. Usually a run should last ~30 min with readings every 10 min based on the linearity run. Photosynthetic rates (lights on) can be measured after a 30-min equilibration period, and with the lights off, respiration rates can be determined. After the temperature runs are made (either with same material or algal replacement between each run), the algal material is placed in weighing pans and dried overnight. The equipment is cleaned and set up for the next day's runs.

Day 3: light series. The procedure is similar to that of day 2, utilizing healthy macroalgal segments and a light series. To avoid changes in light quality caused by a rheostat when controlling light levels, it is recommended that bulb removal or neutral density screening be

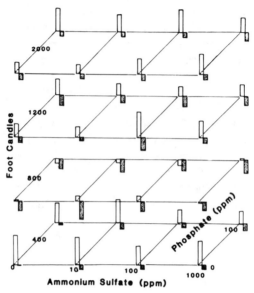

Fig. 16–7. The red alga *Hypnea musciformis* was grown for 2 wk at 20°C under four levels of light (400, 800, 1200, 2000 fc), with two levels of sodium phosphate (0 = 0.03 ppm, 100 = 3.0 ppm) and four levels of ammonium sulfate (0 = 0.05 ppm, 10 = 0.5 ppm, 100 = 5.0 ppm, 1000 = 50 ppm). Photosynthetic responses (above lines, clear bars) ranged from 100 to 4000 μl oxygen produced per gram dry wt each hour. Respiratory responses (below lines, shaded bars) ranged from 10 to 1600 μl oxygen consumed per gram dry wt each hour, as shown for the 20°C portion of the respirometric study. The graph was compared with those obtained for plants grown at 24 and 28°C to include seasonal response information.

used. In addition to the standard flood lamps of a respirometer, lights having a more balanced spectral quality of sunlight such as Gro-lux lamps can be placed overhead. Light quality and intensity should be measured at the reaction flask level in the water bath. The optimum temperature, determined on day 2, should be used for obtaining photosynthetic rates under various light levels. Light histories of the macroalgae should also be taken into consideration when various populations are compared.

Day 4: salinity series. This series involves macroalgae that were held in seawater of various salinities on day 1 and optimum temperature and light conditions as determined on days 2 and 3. Both photosynthetic and respiratory rates are determined.

Day 5: final steps. The dry weights are determined for each of the previous day's experiments, and the equipment cleaned. The procedures just outlined will enable one to investigate individual effects of such physical and chemical factors as desiccation, pH, nutrients, or the use of inhibitors with macroalgae (Dawes et al. 1976a,b). Desiccation can be maintained by holding the alga on a screen under the light system and controlled by weighing the alga during the dry period to determine the water loss over the drying period. It is also possible to carry out a complete random block experiment and a factorial experiment in which the effects of multiple factors can be studied. Thus, a study of the combined effects of light, temperature, and nutrients with the red alga *Hypnea musciformis* involving four levels of light and nitrate and two levels of phosphate at different temperatures required 64 distinct combinations with five replicates for each, as well as four control flasks for background information (Fig. 16–7; see also Dawes et al. 1976b).

IV. Critical evaluation

The use of manometry in studies of macroalgal physiology and responses to environmental parameters has improved phycologists' interpretation of the distribution of macroalgae (geographical and depth) and their tolerance to physical (light and temperature) and biological factors (reproductive state, shading, and pigment levels). Manometry, especially in multichannel respirometry, can utilize a large number of replicates and thus take into consideration the problem of variation among individuals. The measuring procedures are rapid, usually requiring only 2–3 h for a single run. Manometry is accurate and the results become rapidly available for interpretation, so that the data can be applied to the next series of experiments or maricultural studies. Furthermore, manometric results can be compared with results obtained by other procedures that measure oxygen uptake or evolution such as the oxygen probe, as well as ^{14}C uptake using BOD (biological oxygen demand) bottles (Hoffman and Dawes 1980; Littler, 1979). However, such comparisons are most useful with populations of the same macroalgal species.

For all of its advantages, manometery has not been extensively used in studies on algal physiology, perhaps because of the high cost of a modern respirometer (about $8000) and because scientists were discouraged by the complicated calculations the older Warburg constant-volume equipment required for each reaction flask. In situ measurements are considered preferable to laboratory studies, although the lack of control over environmental factors is usually a problem. The available light intensity and quality of artifical illumination in respirometry is poor, and the addition of sunlight-quality lighting systems is recommended.

Because of the requirement for small portions in the reaction flasks, most macroalgae must be cut into smaller segments. The result in many seaweeds is an increase in respiration rates from the wounded region (wound respiration). However, if the macroalga is cut and held for 6 to 12 h, usually the level of wound respiration is low to undetectable.

Another problem in manometric studies on macroalgae is the selection of plant material. Because of the small specimen size, as well as the natural variation in physiological activity, one must use a large number of replicates (e.g., 10–20). If the macroalga is large or has morphological differentiation (e.g., *Sargassum*), the study should include various portions of the plant (e.g., young and old portions and various organs such as blades, holdfasts, floats, and stipes) so that total plant productivity can be calculated. For the same reason, populations from different habitats (e.g., shaded vs. exposed and high vs. low intertidal sites) should be compared if a reasonable estimate of productivity or tolerance ranges is to be made for a population. This will require a large number of samplings.

Possibly the strongest criticism against manometric procedures is that the short-term measurements of photosynthetic and respiratory rates may not be useful in projecting long-term optimal growth responses to a set of environmental factors. Tolerances to high temperature can be misinterpreted. For example, the highest rates of photosynthesis and relatively low rates of respiration may occur at a higher than "normal" temperature over a short-term manometric experiment. When the same material that was held at the higher temperature is monitored at an intermediate temperature, the photosynthetic rate may be much lower than that previously determined. This is probably due to enzymatic damage at the higher temperature (Dawes 1981). Similarly, effects will be noted in plant responses to high and low light- and salinity-tolerance regimens when compared with culturing and growth responses on a growth gradient table. Thus, tolerance ranges derived from photosynthetic or respiratory rates must be viewed with caution.

The problem of diel rhythms in photosynthetic and respiratory rates has already been mentioned (Hoffman and Dawes 1980) and reiterated here as a warning that it must be considered in the performance of studies throughout the day or at different times during the day series. Despite the aforementioned problems, manometry has proved to be especially useful for identifying the general tolerances and peak areas of response of a macroalga for maricultural or growth experiments (Dawes et al. 1976a,b). Within a few days, respiratory and photosynthetic responses can be assessed and the desired temperature, light, and salinity regimes can be established. Manometric procedures are particularly useful for the monitoring of macroalgae held in growth gradient tables to determine how acclimated the specimens are to various physical factors. An exciting area in ecologial studies of macroalgae is that of enzymatic responses, especially in the evaluation of isozyme responses to temperature. The functional importance of isozymatic changes at different seasons in a population or among populations can thus be compared, and a closer approach to the genetic adaptations of macroalgal distribution should be possible.

V. References

Barcroft, J., and Haldane, J. S. 1902. A method of estimating the oxygen and carbonic acid in small quantities of blood. *J. Physiol.* 28, 232–40.

Bowes, G. W. 1969. Carbonic anhydrase in marine algae. *Plant Physiol.* 44, 726–32.

Brodie, T. G. 1910. A new form of apparatus for blood gas analysis. *J. Physiol.* 39, 391–6.

Dawes, C. J. 1981. *Marine Botany.* Wiley, New York. 628 pp.

Dawes, C. J., LaClaire, J. W., and Moon, R. E. 1976a. Culture studies on *Eucheuma nudum* J. Agardh, a carrageenan producing red alga from Florida. *Aquaculture* 7, 1–9.

Dawes, C. J., Moon, R. E., and Davis M. A. 1978. The photosynthetic and respiratory rates and tolerances of benthic algae from a mangrove and salt marsh estuary: a comparative study. *Est. Coast. Mar. Sci.* 6, 175–185.

Dawes, C. J., Moon, R. E., and LaClaire, J. 1976b. Photosynthetic responses of the red alga *Hypnea musciformis* (Wulfen) Lamouroux (Gigartinales). *Bull. Mar. Sci.* 26, 467–73.

Dixon, M. 1952. *Manometric Methods.* Cambridge University Press, Cambridge. 165 pp.

Dromgoole, F. I. 1978. The effects of pH and inorganic carbon on photosynthesis and dark respiration of *Carpophyllum* (Fucales, Phaeophyceae). *Aquat. Bot.* 4, 11–22.

Everson, R. G. 1970. Carbonic anhydrase and CO_2 fixation in isolated chloroplasts. *Phytochemistry* 9, 25–32.

Graham, D., and Smillie, R. M. 1976. Carbonic dehydrogenase in marine organisms of the Great Barrier Reef. *Aust. J. Plant Physiol.* 3, 113–19.

Hoffman, W. E., and Dawes, C. J. 1980. Photosynthetic rates and primary production by two Florida benthic red algal species from a salt marsh and a mangrove community. *Bull. Mar. Sci.* 30, 358–64.

Littler, M. M. 1979. The effects of bottle volume, thallus weight, oxygen saturation levels, and water movement on apparent photosynthetic rates in marine algae. *Aquat. Bot.* 7, 21–34.

Lowry, O. H., Rosebrough, N. J., Farr, A. L., and Randall, R. J. 1951. Protein measurement with the Folin phenol reagent. *J. Biol. Chem.* 193, 265–75.

McLachlan, J. 1973. Growth media: marine. In Stein, J. R. (ed.), *Handbook of Phycological Techniques: Culture Methods and Growth Measurements*, pp. 25–51. Cambridge University Press, Cambridge.

Ogata, E. 1966. Photosynthesis in *Porphyra tenera* and some other marine algae as affected by tris(hydroxymethyl)aminomethane in artificial media. *Bot. Mag. Tokyo* 9, 271–82.

Pardee, A. B. 1949. Measurement of oxygen uptake under controlled pressure of carbon dioxide. *J. Biol. Chem.* 179, 1085–91.

Ramus, J. 1978. Seaweed anatomy and photosynthetic performance: the ecological significance of light guides, heterogenous absorptance and multiple scatter. *J. Phycol.* 14, 352–62.

Ramus, J., Beale, S. I., and Mauzeral, D. 1976. Correlation of changes in pigment content with photosynthetic capacity of seaweeds as a function of water depths. *Mar. Biol.* 37, 231–8.

Stephenson, T. A., and Stephenson, A. 1972. *Life between Tidemarks on Rocky Shores.* Freeman, San Francisco. 425 pp.

Umbreit, W. W., Burris, R. H., and Stauffer, J. F. 1972. *Manometric and Biochemical Techniques,* 5th ed. Burgess, Minneapolis. 387 pp.

Warburg, O. 1962. *New Methods in Cell Physiology.* Wiley-Interscience, New York. 644 pp.

17: Electrodes and chemicals

MARK M. LITTLER

Department of Botany, National Museum of Natural History, Smithsonian Institution, Washington, D.C. 20560

KEITH E. ARNOLD

Department of Biological Sciences, California State Polytechnic University, Pomona, California 91768

CONTENTS

I. Introduction

A. *Purpose of the methods*

Knowledge of the primary production rates of macroalgae is valuable not only from a descriptive point of view but also because it makes possible the rapid testing of hypotheses concerning the effects of environmental factors. Primary productivity is defined herein as the rate at which inorganic matter and free energy are converted to organic matter and bound energy per unit of algal material or unit of the earth's surface area. This is almost entirely due to photosynthesis, and sunlight is the energy source, as follows:

$$6H_2O + 6CO_2 \xrightarrow[\text{chlorophyll}]{\text{light}} C_6H_{12}O_6 + 6O_2 \qquad (1)$$

Equation 1 is essentially reversed for respiration. All parts of the equation potentially could be used to estimate primary production; however, this chapter focuses on CO_2 and O_2 fluxes. Some useful definitions are as follows:

Productivity = production per unit time
Gross primary production (GPP) = net photosynthesis (Ps) + respiration (R)
Net primary production (NPP) = GPP − R
NPP per day (24 h) = (NPP per daylight day) − (R per night)

Since all biological activity ultimately depends on net primary production, many ecologists require reliable methods for obtaining photosynthetic, respiratory, and other metabolic data relevant to their studies. Among the most widely used and credible (Czaplewski and Parker 1973) approaches to the assessment of productivity are those employing O_2 electrodes. The pH electrode technique is comparable but only half as sensitive (Marsh and Smith 1978) and not as frequently employed. The most time-tested dissolved-O_2 method is the modified Winkler chemical titration, which is quite labor intensive, comparatively inexpensive, and comparable in accuracy to the oxygen probe under most field conditions. An excellent

[350]

treatment of virtually all methods of O_2 measurement is available (Hitchman 1978) and will be of use to many readers.

B. Overview of incubation methods

Incubation methods involving any of the aforementioned techniques traditionally rely on the use of light and dark bottles (Gaarder and Gran 1927) deployed (1) in situ, (2) in sunlight incubators, or (3) in artificially illuminated chambers. Temperature control is achieved by water cooling, or more often by refrigeration/heating units in the last case. Metabolic studies in open ecosystems are dealt with by Kinsey (Chap. 21).

For in situ measurements, samples are incubated at or very near the same locations from which they were taken. This method has the advantage of most closely approximating natural conditions. However, site-specific and temporal variations of environmental parameters limit the degree to which the results can be generalized or compared with other habitats. This method requires much more effort and manpower than is needed when incubators are used.

Sunlight incubators are trays made of polycarbonate or other optically transparent materials that employ natural sunlight and colored filters (see Dring and Lüning 1977; Faust et al. 1982) to simulate the light at a given depth. This method approximates natural lighting somewhat, but local variations in conditions again limit the degree of generalization or comparison.

Artificially illuminated field incubators use combinations of fluorescent and incandescent lamps with color or neutral density filters in a portable environmentally controlled and mechanically stirred chamber (Doty and Oguri 1958; S. Blair, unpublished data). Environmental factors are kept consistent, so generalizations and comparisons among species or from one locality to another are more valid. However, these devices are by definition unnatural, and it is often difficult to extrapolate from data obtained in incubators to those that would pertain in the natural system. This method requires less manpower and effort, because incubations can be performed under a relatively controlled environment approximating that of the laboratory.

1. Laboratory incubation techniques. Laboratory experiments generally are conducted in large photoperiod incubators, typically maintained at or below $\sim 200 \ \mu E \cdot m^{-2} \cdot s^{-1}$ ($\sim 10,000$ lux) by using cool-white fluorescent bulbs, which is well above saturation but below light inhibition for most macrophytes (King and Schramm 1976). Higher light energies of different spectra (Lüning 1981) are often obtained with slide projectors having 500-W incandescent bulbs. Both the incubation water and associated equipment should be brought to the desired temperature before use to prevent the formation of small bubbles due to degassing on surfaces. All experimental material is collected submerged and immediately returned to the laboratory in insulated coolers.

a. Bottle experiments. Net photosynthesis is determined on selected whole plants by means of O_2 analyzers or pH meters and appropriate electrodes. Wide-mouthed, clamp-lid, rectangular canning jars (Snap Top Jars, Allied Trading Co.) serve well as incubation containers and can be readily and cheaply obtained nearly anywhere in the world. The seawater used during the incubations is taken at the time and place of algal collection, shaken vigorously to bring to O_2 saturation (Strickland and Parsons 1972), immediately filtered through a nannoplankton net (10-μm pore size) to remove most plankton organisms, and stored in the laboratory at the desired temperature in the dark. When individual bottles are being filled; they should be slowly submerged so as to exclude all air from the stocks of incubation water.

Comparative incubations are carried out between 0900 and 1500 h to reduce differences due to possible endogenous photosynthetic periodicities (e.g., see Hoffman and Dawes 1980). Incubation times for the larger macrophytes are usually restricted to about 2.0 h, since the representative thallus portions are much larger than those of smaller forms. During incubation, the water within the bottles (minimally four light and two dark replicates, since respiration is lower and less variable) is completely recirculated by means of magnetic stir bars and air- or water-driven magnetic stirrers (GFS Chemicals) at a minimum of 10-min intervals. Czaplewski and Parker (1973) described a useful technique for determining the required level of replication as a function of the confidence limits desired. Bottles are systematically rotated in position within the incubators to ensure uniform light conditions. A thin perforated acrylic sheet can be employed to increase the effectiveness of the stir bar and to keep it from contacting delicate algal material.

After the final O_2 or pH levels are recorded, individual thalli are carefully separated, spread, and photocopied or photographed; projected area determinations are made from each photocopy or print by overlaying a transparent gridwork of dots ($16 \cdot cm^{-2}$) and counting those intercepting the thalli. Care must be taken that the photocopier does not distort the specimens copied. If a uniform size change occurs, this is easily calculated and compensated. If a photocopier of sufficient quality is lacking, the plants can be spread and scored directly, although this slows the processing markedly and no permanent record is maintained. After photocopying, specimens are dried at 80°C until they reach constant weight. All O_2 of pH values are converted to grams carbon fixed per square meter of thallus per hour and to milligrams fixed per gram dry weight per hour as outlined in Strickland (1960). For calcareous algae, ash-free dry weight may be used after 24 h of combustion at 500°C (see Brinkhuis, Chap. 22, for additional discussion). Other workers report their data as milligrams carbon per milligram chlorophyll *a* per hour after determining the pigment contents on fresh material (see extraction methods suggested in Duncan and Harrison 1982). For conversion of O_2 data to carbon units, a photosynthetic quotient (PQ = ratio of moles O_2 liberated to CO_2 taken up) and respiratory quotient (RQ = moles CO_2/MO_2) of 1.00 are usually assumed to facilitate inter-conversion with other data when different values are used. This assumption is unnecessary when pH measurements of CO_2 flux are determined directly. However, because calcification also influences CO_2 flux, the pH method must be supplemented with total alkalinity titrations in the case of calcareous algae. The differences among various experimental treatments can be examined statistically by single-factor analysis of variance or the Newman–Keuls Multiple Range Test (Sokal and Rohlf 1969).

   b. Continuous-monitoring experiments. This design consists of simultaneously monitoring net photosynthesis via either O_2 or pH or both during controlled laboratory manipulations. Algal thalli are placed in specially constructed Plexiglas chambers, each fitted with pH and O_2 electrodes (Fig. 17–1; see Littler 1973a for details). Oxygen evolution can be read at the parts per 0.01 million level by means of a suitable O_2 analyzer and polarographic electrodes, whereas pH is monitored (to 0.001 pH unit) with a digital meter and combination electrodes. At the end of each experiment, thallus dry weights are determined as described earlier. The slopes of the curves for O_2 production and CO_2 uptake as a function of time are computed by regression statistics and compared for the different circumstances and measurement techniques. The incubation protocol is generally the same as that described for the bottle experiments, except that a cylindrical, magnet-driven stir bar under a perforated partition in each container is used to maintain a constant flow of medium past the alga and individual electrodes.

2. Specific field incubation techniques. Because our emphasis is on field O_2 studies, which we consider the method of choice in physiological–ecological work, we shall present the detailed step-by-step breakdown of the techniques developed by our group in the past decade. The emphasis will be on intertidal habitats, but the procedures are equally applicable subtidally (see Heine 1983).

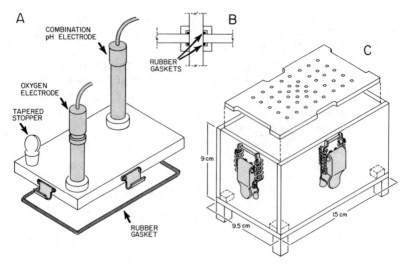

Fig. 17–1. Example of a gas-tight acrylic chamber used in photoperiod incubators for continuously monitoring CO_2 and O_2 flux. A, Design of the chamber lid with electrodes inserted; B, sectional view showing the design of the seals with an electrode inserted; C, view showing the component parts of the chamber.

a. Site selection. An area of the shoreline that is relatively flat and that receives direct sunlight during the complete incubation period (usually from 0900 to 1500) is selected as close as possible to the habitat from which collections are to be taken. The O_2 analyzer is set up in a shaded area out of direct sunlight. Bellows-type foot pumps (Zodiac boat inflators) and stirring racks are placed on hard level ground so as to prevent the contamination of the stirring turbines with particulate matter. If no hard substratum is present, flat pieces of plywood, cut to fit in the bottom of footlockers, are placed beneath the stirring apparatus.

b. Incubation water. Incubation jars are unpacked and placed upright in clear polycarbonate incubation trays (Curtin Matheson. Scientific) filled with ambient seawater. The lids are opened, and stir bars, plastic partitions, and clean ambient seawater are added. The seawater is collected during early morning on the day of the experiments. Water from tide pools or that appears otherwise contaminated is not used.

c. Algal collection. All algal collections are made by diving or wading to obtain submerged material. Approximately twice as much algal material is collected as will be needed for incubation. Separate whole individual thalli should be used when possible; however, productivity measurements of the larger species are conducted on representative blades or branches. Care is taken to select individuals

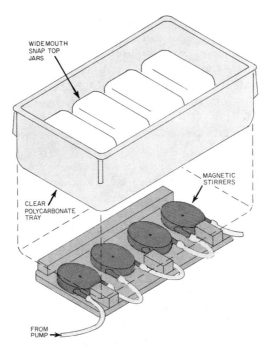

Fig. 17–2. Sunlight field incubation apparatus. It is important to connect all stir motors in series, and not in parallel, to provide uniformity and to reduce the volume of air or water required. The stir motor rack is placed inside the tray when water driven.

that are reasonably representative of populations within the system under study. If the specimens incubated are fertile, are wounded, or have unusual features, these observations are recorded. Macrophytes unduly exposed to air or to unnatural levels of sunlight or macrophytes otherwise injured are to be avoided. The collected plants are placed in separate, clear polycarbonate trays filled with ambient seawater and then hand-sorted and cleaned of epiphytes. Voucher samples can be taken from this material to provide taxonomic documentation (Tsuda and Abbott, Chap. 4). Identification notes, numbers, dates, substratum, depth, collector, and location should be included for all vouchers. Specimens are incubated at ambient water temperatures in approximately 0.5- to 4.0-ml widemouthed canning jars (volume is dependent on sizes and estimated productivity of the species), which are placed in clear polycarbonate trays (Fig. 17–2). Throughout the experiments, the average thallus dry weight per volume of water for a 2- to 4-h incubation period is maintained below the ratio of $0.04 \text{ g} \cdot \text{liter}^{-1}$ for high producers (e.g., sheetlike and filamentous forms). This is equal to about 0.2 to 0.4 g wet weight, which represents a very small quantity of material relative to that typically used in most previous studies. For lower-producing species (e.g., crustose algae), up to 10 times this amount can be used. Incubation of multiple small pieces should be avoided, since this increases chances of self-shading or overlapping diffusion gradients during incubation. Because O_2 flux is ~50–90% lower during respiration, dark bottles are normally incubated with about double the content of algal material as the light bottles.

d. Setup. To begin the incubation procedures, the cooling water is emptied from the jars, which are replaced upright in the polycarbonate trays. Each bottle should be immersed smoothly below the surface of the filtered incubation water in 25-liter buckets, allowing water to flow gently down the sides of the bottle until nearly full. Bubble formation on the walls of the bottle and partition is to be strictly avoided. Scrubbing bottles with distilled water reduces degassing problems. Once all bottles are partially filled, they are kept upright in the plastic trays, and external cooling is maintained with ambient seawater. The appropriate amount of material for each species is then selected, making sure that no bubbles adhere, and each thallus is slowly submerged into the proper incubation bottle. Dense algal mats tend to trap air bubbles and must be eliminated by gentle shaking or swirling. Bottles must be refilled sufficiently below the surface so that the lids can be securely clamped without trapping air. Bubbles are often squeezed from the rubber washer into the bottles when the lids are clamped and should be released before the lid is secured. If all components are not at the same temperature, degassing on surfaces will occur. Once the first small bubble forms, degassing will accelerate (Littler 1979) since oxygen is more soluble in air than in seawater.

For every six bottles incubated, one initial bottle (i.e., without algal material) should also be included. These bottles are treated in the same way as those containing algae and analyzed for O_2 content before reading of the associated light and dark bottles. Additional initial bottles are retained to serve as checks for plankton or bacterial metabolism as well as possible electrode drift. All bottles should be completely submerged on their sides in the polycarbonate incubation trays. Trays should not shade each other, and personnel must be kept away from the trays while samples are incubating to prevent shading. Dark bottles, produced by wrapping and taping two layers of heavy-duty aluminum foil, are always incubated separately to prevent the impingement of reflected light on the initial and light bottles. It is important to ensure that hardware around the tops of the bottles does not puncture the foil. It normally takes 3–5 min to set up completely the minimum four replicate light bottles and two replicate dark bottles per species. The "time in" recorded for each species is the time that the last bottle is completely set up.

e. Incubation. For studies designed to compare the performances of different species, sample incubation begins between 0900 and 1030 h, and final O_2 values are read between 1300 and 1500 h, with maximum incubation times lasting 3.5–4.5 h for light bottles and 5.0 h for dark bottles. Depending on the questions being asked, any midday depression in photosynthesis (Ramus and Rosenberg 1980) due to excessive light causing photoinhibition can be avoided by the reduction of the light with layers of neutral density screening (nylon window screening). Species that are high producers (e.g., filamentous or sheet forms) should be analyzed early to avoid bubble formation and nutrient depletion. At ~10-min intervals, each bottle is thoroughly mixed with stir bars and magnetic stirrers (Fig. 17–2) driven by an air pump or mixed continuously if an electric water pump (Pony Pump, Proven Pump Corporation) and gasoline generator are available. Our trays hold four 1-liter jars and fit onto racks with four appropriately placed motors for stirring. Cooling is accomplished by refilling the trays with ambient seawater at 10- to 15-min intervals or by the excess water from the electric pump.

f. Physical data. Physical measurements that are minimally required include light readings taken at 15-min intervals with three different sensors (Li-Cor model LI-185A, Lambda Instruments, Inc.): illumination (lux), photosynthetically active radiance (microeinsteins per square meter per second; $\mu E \cdot m^{-2} \cdot s^{-1}$), and power (watts per square centimeter; $W \cdot cm^{-2}$). The sensors are cosign-corrected and must be perfectly level during readings to ensure highest accuracy (see Ramus, Chap. 2). The sensors should not be shadowed by any object, particularly the measurer. When readings are taken during intermittent cloud cover, the approximate mean values occurring over 1-min intervals are recorded. A minimum of 30,000 lux is required for realistic natural field rates to ensure that measurements are conducted above light saturation (for typical light saturation values for marine macroalgae, see King and Schramm 1976; and Arnold and Murray 1980). Light approaching 120,000 lux is inhibitory for many seaweeds (see Lüning 1981), and the use of neutral density screening may be warranted in some cases. Air and water temperatures should be recorded to 0.2°C throughout the experimental run and out of direct sunlight, the time being noted. Periodically, the temperature of the incubation water should be checked and should not be allowed to rise more than 1.5°C above the ambient water temperature. Occasionally, dark bottles will be cooler than light bottles, but not by more than 1.0°C. If freshwater drainage is detected by salinity measurements (to 0.5 ppt, American Optical Corporation, model 10412 salinity refractometer), the salinity of the ambient water should be monitored periodically.

We record all data on waterproof plastic paper (PolyPaper, Nalge Company) that is stored in an all-plastic three-ring binder. Essentially, three types of records must be maintained in the field: (1) incubation data, including species, time in, date, bottle number, time out, and dissolved O_2 values; (2) physical data, incorporating light measurements, water and air temperature, and salinity; and (3) general notes concerning algal collection data, weather conditions, or any other pertinent information. Three persons operating at a high level of field efficiency can incubate about 60 bottles per day; with completely mechanized cooling and stirring, two persons can handle up to 80 bottles per day.

g. Harvesting. Bottles are assigned their numerical order at the time they are harvested, and the final readings are taken. The light bottles are grouped by species and processed first with the time recorded for each. Each dark bottle should be unwrapped only just before being read. Aluminum foil develops light leaks at the corners of folds and must be discarded after each experimental run. When the sequence of O_2 readings is begun, all other bottles being incubated must continue to be stirred and cooled at frequent intervals.

h. Reading. In preparation for the dissolved-O_2 readings, the bottles to be read are placed in an upright position in a polycarbonate tray on a separate stir motor rack near the analyzer. Sufficient ambient seawater is replenished in the tray to provide proper temperature control. The clamp on the jar is unfastened and the suction seal is broken by pulling the rubber tab at right angles. The plastic partition is quickly removed, as is the specimen. The O_2 electrode in a no. 14 stopper is then inserted rapidly and smoothly into the bottle so that all of the air is displaced. Each algal specimen is retained without excess water in a labeled (species, date, location, and sequence number) Whirl-pak bag. For filamentous algal material that is too fine to remove quantitatively by forceps, the O_2 level is recorded first and the contents of the jar are then poured through a fine-mesh tea strainer and the thalli are scraped into a bag. The bottle, positioned over one of the stir motors, should be mixed at the maximum possible spin rate. If electric self-stirring electrodes capable of mixing the entire contents of the bottles are available, then the turbine stirrers and foot pump are not required during the reading process.

For measurement of the dissolved O_2, the analyzer is set to the appropriate scale and turned on after about 30 to 45 s of mixing. When making O_2 readings, one must take care to prevent differential heating of the electrode or sample in direct sunlight. The O_2 readings in most light bottles will continue to change for ~ 2 min if stirring is at the maximal rate. It should be noted that the response becomes proportionately slower as the meter approaches the correct and final value. The final reading is recorded next to the sequence number of the bottle when the meter becomes stable.

i. Processing. Specimens should be stored in a dark, cool, and dry location. The samples are transported to the laboratory in an insulated cooler. Upon being returned to the laboratory, the samples are photocopied and weighed as soon as possible. We have found that materials can be left for no more than about 4 d, even under refrigeration, or they will begin to decompose partially. At humid, tropical field sites, we fix the specimens in 4% Formalin in buffered seawater, drain the excess liquid, and store them in the dark. The samples are sorted and processed in order of their incubation dates and times, the earlier materials being treated first.

Photocopying should be carried out by means of a rigorously standardized procedure. The contents of each Whirl-pak bag are shaken down, and the bag is cut with a razor blade near the bottom. The contents are placed onto the copier surface, and a forceps is used to spread the material to its natural configuration. The field label should be placed next to the specimen so that it appears legibly on the copy. The species numbers and identifications are recorded on the data sheet with the number of its tared aluminum foil container. After photocopying, the algae are dried to constant weight at 80°C in a drying oven. Final weighing to 0.001 g is generally done 2 or 3 d after the material has been dried and cooled in a desiccator.

Area determinations are made on the two-dimensional photocopies by placing a dot-matrix grid over the impression (see Brinkhuis, Chap. 22). The area is scored as cumulative hits (i.e., when the dots on the grid intercept the impression of the algal thallus), and the hits are directly converted to two-dimensional area. Lambda Instruments, Inc., also manufactures a leaf-area meter (model 3100-H.1) that is expensive but time saving in this application. Total surface area can be calculated for the appropriate geometric shapes involved (multiply two-dimensional area by 4 for spherical thallus portions, by $3.146 = \pi$ for cylinders, etc.). We reemphasize that care must be taken to ensure that distortion is negligible during the photocopying process. The point intercepts are calibrated to a specific area (e.g., $16 \cdot cm^{-2}$), and consequently hits are directly proportional to area.

j. Calculations. In calculating the photosynthetic data, we average the values for all initial bottles for a given species for the day of each experiment and contrast their mean with those of the light bottles, as well as with the dark bottle numbers, to obtain net photosynthesis

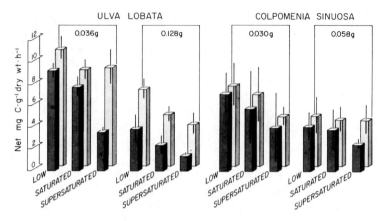

Fig. 17–3. Apparent photosynthesis per gram dry wt of *Ulva lobata* and *Colpomenia sinuosa* as a function of the interacting effects of bottle size, mean thallus weight, and initial dissolved O_2 tension (low, saturated, and supersaturated). The darker histograms are for 310-ml bottles, the lighter histograms for 1220-ml jars; ±95% confidence intervals are given by the straight lines at the top of each histogram. Modified from Littler (1979).

and respiration, respectively. After all calculations are completed, the average net productivities and respiration values, their standard deviations, 95% confidence intervals, and coefficients of variation are computed. These data are summarized on a separate data sheet and subsequently tabulated and plotted (Figs. 17–3 to 17–5). Calculations are as follows:

LB = O_2 content of the light bottle after incubation (milligrams O_2 per liter) × volume (liters)

DB = O_2 content of the dark bottle after incubation (milligrams O_2 per liter) × volume (liters)

IB = O_2 content of the water before incubation (milligrams O_2 per liter) × volume (liters)

t_x = incubation time

$$\text{NPP} \times t_x^{-1} = \text{LB} - \text{IB} \qquad (2)$$

$$\text{GPP} \times t_x^{-1} = (\text{LB} - \text{IB}) + (\text{IB} - \text{DB}) \qquad (3)$$

This provides a measure of net and gross photosynthesis and respiration for the period of incubation, expressed in units of dissolved O_2. To convert these O_2 units to units of assimilated carbon, that is, NPP or GPP, it is necessary to introduce not only PQs and RQs into the calculations, but also a factor to convert O_2 units to carbon units.

Dissolved O_2 is usually measured in units of milligrams O_2 per

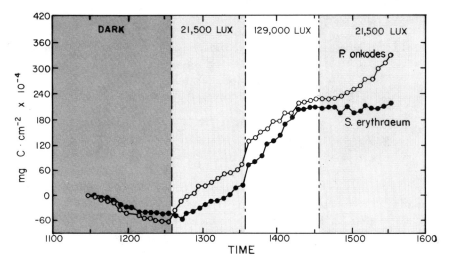

Fig. 17–4. Variations in the productivity of a light-adapted plant (*Porolithon onkodes*) and a shade-adapted macrophyte (*Sporolithon erythraeum*) to initial control, experimental, and final control light intensities.

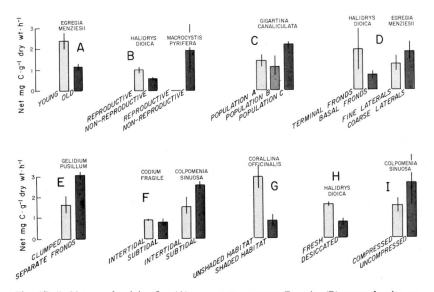

Fig. 17–5. Net productivity for (A) young vs. mature *Egregia*, (B) reproductive vs. nonreproductive *Halidrys* and *Macrocystis*, (C) wiry (populations A and B) vs. fleshy (population C) of *Gigartina*, (D) terminal vs. basal fronds of *Halidrys* and fine vs. coarse laterals of *Egregia*, (E) clumped vs. separate thalli of *Gelidium*, (F) intertidal vs. subtidal *Codium* and *Colpomenia*, (G) *Corallina* from unshaded vs. shaded habitats, (H) desiccated vs. fresh *Halidrys*, and (I) compressed vs. uncompressed *Colpomenia*. Modified from Littler and Arnold (1980).

liter (= ppm), milligram-atoms O_2 per liter, or milliliters O_2 per liter, depending on the equipment available. The first unit is the most common for polarographic dissolved-O_2 analyzers. Since most meters read in parts per million, one must multiply the readings by the bottle volume in liters to get actual milligrams O_2 per bottle. To convert milligrams O_2 to milligrams C, a factor of 0.375 is used:

$$\text{mg C per algal unit} \times t_x^{-1} = 0.375 \frac{\text{LB (mg } O_2) - \text{IB (mg } O_2)}{\text{PQ}}$$

$$+ \text{ IB (mg } O_2) - \text{DB (mg } O_2) \times \text{RQ} \tag{4}$$

In addition, the factor converting milligram-atoms O_2 to milligrams of assimilated carbon is 12, and the factor for milliliters O_2 is 0.536. The resultant milligrams carbon value is divided by the incubation time in hours and hundredths of hours and then again by thallus weight, chlorophyll content, or area. One can convert milligrams carbon per square centimeter per hour to grams carbon per square meter per hour by multiplying by 10. If standing stock determinations have been made concomitantly with photosynthesis, the average net production for each macrophyte at a given site can be estimated by use of the net productivity per unit of area or weight of thallus per unit of time in conjunction with the overall percentage cover value per unit of substratum area or population weight.

   k. Metabolic quotients. Values for PQ and RQ often vary significantly from unity; however, unless precise measurements are made, assumed values become necessary. The numbers most commonly recommended in the literature are 1.00–1.25 for PQ and 1.00 for RQ. Because these values are uncertain to at least ±10%, we prefer to use 1.00 in both cases to facilitate interconvertibility in comparisons with other studies. For more information on metabolic quotients and assumptions, see Kinsey (Chap. 21).

   l. Daily rates. When the measurements of primary production have been made in situ or by using a simulated in situ sunlight incubator, they are sometimes extrapolated to a 24-h day. There are considerable pitfalls in extending short-term rates to daily rates due to unaccounted variations caused, for example, by possible endogenous periodicities (Hoffman and Dawes 1980), environmental fluctuations, and, in some cases, photoinhibition near midday (Ramus and Rosenberg 1980). If extrapolation is deemed necessary, respiration must be normalized to a 24-h day. The calculations would be as follows:

NPP mg C per algal unit per 24-h day

$$= (\text{NPP} \times \text{h per daylight day} - (\text{R} \times \text{h per night}) \tag{5}$$

Short-term or even diurnal rates cannot be used to derive yearly rates adequately.

II. Equipment and materials

Analytical equipment for primary productivity measurements must be of the highest precision and accuracy available to detect the extremely small O_2 and CO_2 fluxes normally encountered. Precision is most critical because changes in parameters are of primary importance; however, if the instruments are not accurately calibrated initially, slope problems (Hitchman 1978) will occur. Consequently, instrumentation must be both accurate and capable of maximum precision (see Hitchman 1978; Polgreen and Coker 1981; Smith and Horner 1982).

A. Oxygen analyzers

1. Amplifiers. There are many analytical systems of sufficient quality, the best of which are polarographic instruments that can be precisely read at the 0.1- to 0.01-ppm level. We have had considerable experience with the Beckman Instruments Fieldlab oxygen analyzer, Yellow Springs Instruments model 57, and Orbisphere Laboratories model 2610. The last two are temperature-compensated systems and provide accuracy and precision at the required levels while being sufficiently reliable under rigorous field conditions. The O_2 electrode method exhibits close agreement with Winkler-determined O_2 values (Beckman Instruments, Inc. 1972; Hitchman 1978) and has a linear response throughout all dissolved-O_2 concentrations (including supersaturated levels). The Winkler method is much more labor intensive but is generally thought to provide increased sensitivity when used in the laboratory, although Orbisphere Laboratories manufactures an amplifier (model 2713) with a special flow-through electrode capable of indicating O_2 changes at the parts per hundred billion level. For both O_2 and pH, digital systems, such as the Orion digital pH analyzer (model 801) with the Orion digital printer (model 851) and the Orion automatic electrode switch (model 855), are preferred to analog meters because of the difficulty of reading a moving pointer due to boat motion and parallax. The best approach with both O_2 and pH analyzers is to record the millivolt output directly into a data logger. Also, an instrument interfacer (Analogue Devices) in combination with a microcomputer now makes it possible to program multiple electrodes for monitoring at selected intervals in the laboratory and to analyze, tabulate, and plot data in the absence of the investigator.

2. Membranes. Plastic (polyethylene) electrode membranes perform better than other materials because of their longer stability after calibration (Hitchman 1978). However, it should be noted that changing the membrane material from that specified by the electrode manufacturer will alter the thermal response characteristics, requiring appropriate compensation. Calibration of the O_2 electrode is required because the sensitivity of the analyzer depends on the tension in the sensor membrane. Typically, a plastic membrane relaxes during a period of about 1 h after mounting, with associated alterations in thickness and O_2 permeability. Therefore, it is essential to wait at least an hour before calibration. The sensitivities of different sensors, or the same sensor fitted with different membranes, lie within a range of $\pm 5.0\%$ around a mean. A newly refilled sensor should be immediately inspected for trapped air bubbles within the circumference of the sensor tip. Trapped air bubbles or folds in the membrane near the central area must be eliminated for satisfactory performance of the sensor at very low oxygen levels. Small folds in the membrane near the outer perimeter are unavoidable and have no influence. It is important to avoid touching the membrane of a calibrated sensor, since this or other disturbance can cause a change in membrane tension, which necessitates recalibration.

3. Cathodes. Electrodes with gold cathodes are preferable to platinum cathodes (Hitchman 1978) because of the resistance of gold to oxidation by free sulfide. Large cathodes and large electrolyte volumes are desirable because of their sensitivity; however, they tend to consume more O_2 than smaller units in closed systems. The latter have more rapid response times, which is important. The cathode is the most delicate part of the entire O_2-measuring system because it produces the signal that is eventually displayed and because it must maintain extreme sensitivity and selectivity for O_2 during the entire life of the system. It should never be touched by fingers or exposed to detergents or oily liquids.

4. Calibration. Calibration should preferably be conducted near the middle of the temperature range over which O_2 measurements are to be performed. Oxygen electrodes can be calibrated in water-saturated air or air-saturated distilled water, but it is slightly more precise to air-saturate the actual seawater to be incubated. The appropriate O_2 values for calibration are obtained from standard tables (Carpenter 1966) by interpolation from temperature, salinity, and barometric data. During calibration in air, a stable reading can be obtained only if the temperature of the air and that of the sensor are constant. Consequently, drafts should be avoided and the sensor dried so that evaporative cooling does not occur. Agitation of the sample is not necessary during gas-phase measurements.

The temperature measurement detected by the sensor must be constant and agree with that given by a good-quality thermometer (within 0.2°C). About 2 min are required for stabilization of the display upon immersion of the sensor in water. Much longer periods are required for stabilization upon removing the sensor from water into air at a different temperature, due to the lower thermal conductivity of air. At first reading, the sensor will produce an anomalously high signal due to the consumption of the O_2 dissolved in the interior filling solution. The signal will stabilize within about 3 min to a steady value when read in air. Air-saturated water for calibration can be produced either by passing fine air bubbles through the water or by shaking with air. It is important to note that this equilibration process normally requires as much as 15 min. Most O_2 electrodes will require relatively high mixing rates in water, equivalent to about 50 cm $\cdot$ s^{-1} linear velocities. One should always check the adequacy of stirring by noting any effect of movement of the sensor that will cause the signal to rise if stirring is inadequate. The YSI and Orbisphere electrodes are available with built-in stirring devices that are quite advantageous because they eliminate the need for stir motors and other external means of mixing during the time of reading. Once the output becomes constant, it may be adjusted by means of the calibration control.

B. pH meters

A thorough discussion of the use of pH in metabolic studies has appeared (Smith and Kinsey 1978) and should be consulted if further methodological detail is required. The finest-quality instruments in terms of stability and sensitivity are essential for determining CO_2 flux in seawater. For field use, the unit must be readable to 0.001 pH unit and reliable to ± 0.005 pH unit; it should be battery-operated, or a field generator should be available. The combination pH electrodes (e.g., Broadly-James no. 9061-18S) can be calibrated in standard buffer solutions, but buffers near the salinity of the seawater to be measured are preferred. See Kinsey (Chap. 21) for a detailed discussion of pH electrodes and buffers.

For calcareous macroalgae, the pH method must be used in conjunction with total alkalinity titrations, because calcification as well as organic carbon flux affects the pH. This represents an advantage over O_2 methods, which cannot discriminate calcification, for such organisms. If total alkalinity is not determined, changes in pH are converted to changes in CO_2 concentration by means of the standard procedures of Beyers (1970). This involves interpolating from a CO_2 versus pH function previously determined for a particular medium (i.e., by removing all CO_2 with bubbled N_2 and then titrating with CO_2-saturated distilled water to obtain a curve for pH as a function of millimolar CO_2) and then to grams carbon fixed per square meter of thallus, milligrams carbon per gram dry weight, or milligram carbon per milligram chlorophyll a. The CO_2/pH function is not linear and must be assessed for each different source of incubation water.

C. Winkler reagents and equipment

Winkler O_2 titration is a simple, accurate, nonautomated, low-cost, but labor intensive chemical means of dissolved-O_2 assessment. It requires only a limited amount of glassware including 300-ml biological oxygen demand (BOD) bottles, 150-ml Erlenmeyer flasks, reagent bottles, pipettes, a burette capable of reading to 0.01 ml, a stir bar, a stirring motor, and an incandescent light source. Special reagents required are manganous sulfate, sodium hydroxide, potassium iodide, soluble starch, sodium thiosulfate, hydrochloric acid, glacial acetic acid, sulfuric acid, sodium carbonate, carbon disulfide, and potassium iodate.

The reader is referred to Strickland and Parsons (1972) for basic detail, but for the sake of completeness, a working summary of the time-tested method is included here. Carpenter (1965a) described an improved version of the technique. A 1.0-ml manganous sulfate solution (480 g $MnSO_4^- \cdot H_2O$ per liter) is added to the BOD sample bottle with an automatic pipette, and the sample is restoppered and shaken. Then, 1.0 ml alkaline iodide solution (500 g NaOH per 0.5 liter + 300 g KI per 0.45 liter) (do not cross-contaminate pipettes) is added, and the sample contents are mixed thoroughly. After the precipitate settles slightly (2–3 min), it is mixed and resuspended. At this stage, the stoppered samples can be stored at constant temperature for up to 1 d if necessary.

After the precipitate again settles one-third of the way, 1.0 ml of concentrated sulfuric acid is added; the bottle is restoppered and then mixed until all of the precipitate dissolves. Within 1 h, 50.0 ml of this solution is pipetted into a flask and titrated at once over a mechanical stirrer with standard thiosulfate solution (145 g $Na_2S_2O_3$ $\cdot 5H_2O$ + 0.1 g Na_2CO_3 per liter + one drop CS_2) until a pale straw color remains. Five milliliters of starch indicator is added, and the titration is concluded to a colorless end point. A white background and good light source are needed to detect the end points accurately. The O_2 content is calculated from the following formula using the volume of the total titre (V) when a 50.0-ml aliquot is taken from a

300-ml BOD bottle:

$$\text{mg-atom } O_2 = 0.1006fV \qquad (6)$$

To determine the f value, a 300-ml BOD bottle is filled with seawater and 1.0 ml of concentrated sulfuric acid added. The contents are mixed, 1.0 ml alkaline iodide solution is added, and the contents are stirred again. Finally, 1.0 ml of manganous sulfate solution is added and mixed. Fifty-milliliter aliquots are then withdrawn into two titration flasks, and 5.00 ml of 0.0100 N iodate is added to each of these flasks with a calibrated, clean, 5-ml pipette. After a 2- to 5-min delay, during which the two solutions are kept out of the direct sunlight, the iodine is titrated (using 5.0 ml of starch indicator) with the standard thiosulfate solution. Considering the mean volume V of both titers in milliliters, f is obtained by

$$f = 5.00/V \qquad (7)$$

The milliliters of O_2 per liter in the water sample can be computed as follows:

$$\text{mg } O_2 \text{ per liter} = 16.00 \times \text{mg-atoms } O_2 \text{ per liter} \qquad (8)$$

Carpenter's (1965b) modifications improve the accuracy of the Winkler technique, and microtechniques involving dispensers can be advantageous (Fox and Wingfield 1938; Duedall et al. 1971).

D. Calcification methods

The pH/alkalinity method (see Kinsey, Chap. 21) and the [14]C (Borowitzka 1979) and [45]Ca (Böhm 1978) isotopic–kinetic methods are precise enough that one can determine changes in calcification, the last two being most sensitive. Calcium electrodes do not provide the required level of sensitivity or precision to detect the small changes that take place during biological calcification in seawater (Littler 1973a).

III. Critical evaluation

A. Oxygen electrode, pH, and Winkler methods

In the field, the O_2 electrode technique is preferable to the pH electrode method because it is more reliable and nearly twice as sensitive (Marsh and Smith 1978). Also, compared with Winkler equipment and supplies, O_2 analyzers are less bulky, faster, and easier to use under rugged field conditions. No matter which of the various techniques is chosen for primary productivity studies, uncertainties will exist concerning precisely what is being measured. Much of the problem lies in the confusion regarding (1) excretion of dissolved organic carbon and (2) whether respiration as measured by O_2 uptake in the dark is equivalent to that in the light, as well as the facts that (3) respiratory CO_2 or photosynthetic O_2 can be recycled during photosynthesis or respiration, respectively, without ever leaving the alga, (4) some plants can virtually "shut down" respiration in the dark or under nutrient deficiency, and (5) photorespiration, if present (Burris 1977; Kremer 1980), can involve the consumption of O_2 in the light during glycollate synthesis.

B. Units reported

We strongly suggest that productivity rates in ecological studies be reported in terms of area or weight of thallus (as opposed to pigment content) at saturating but not inhibiting light, because these are related to standing stocks and other more ecologically relevant factors. For example, space, light, and nutrients are known to be limiting resources in many benthic macrophyte communities (Dayton 1975), and algae compete for these by means of their surface area/cover. Biomass (organic dry weight) is also an ecologically significant parameter because it represents the standing stock or organically bound energy potentially available to higher trophic levels. Consequently, macrophyte cover and biomass are of primary ecological interest.

Furthermore, as Ramus et al. (1976, 1977) pointed out, the customary plot of photosynthetic performance versus chlorophyll content can be misleading. Only in optically thin plants can the chlorophyll concentration sometimes approximate linearity with O_2 production. A major problem with normalizing photosynthesis to the traditional parameter, chlorophyll a concentration, is that it brings into the calculations only one of several of the important light-harvesting pigments. Therefore, the expression of carbon flux per unit chlorophyll is not as appropriate in benthic ecology, as has been the convention in biological oceanography (mainly because phytoplankton biomass and area are relatively intractable parameters). Algae can also alter their pigment contents dramatically (Ramus et al. 1976; Ramus and van der Meer 1983) depending on the light environment, and this induces another source of variability. However, pigment data are certainly very appropriate for interpreting weight- and area-based photosynthetic differences or if questions concerning assimilation numbers and light-gathering capacities are posed.

C. Incubation conditions

Many of the studies on primary productivity of marine macrophytes have not adequately considered the effects of (1) incubation conditions, (2) antecedent environmental differences, and (3) intrinsic aspects of variation within the organisms themselves.

1. Thallus-weight/bottle-volume ratios. Generally speaking, ratios of thallus weight to bottle volume should be optimized while ensuring that the specimens used are representative of the organisms being investigated and that enough material is incubated to produce reliable, measurable metabolic changes (Fig. 17–3; Littler 1979). Even in relatively large containers, numerous small specimens can clump, shade each other, and result in overlapping diffusion gradients, a situation that will lead to markedly lower apparent production rates. For smaller algae, it is far better to incubate one individual in a small bottle than to use several or many thalli in a larger bottle. In the case of larger algae, the proportions of the container should be commensurate with the size and metabolic rate of an entire representative thallus, whenever practicable. Plastic bags and domes of various materials have occasionally been used (Towle and Pearse 1973; Hatcher 1977; Smith and Harrison 1977) as field incubation chambers. These are useful if provision is made for (1) adequate mixing, (2) prevention of gaseous exchange with the surrounding medium, and (3) adequate replication – all very difficult criteria to satisfy with large containers. In order that valid comparisons can be made with other studies, it is mandatory that ratios of thallus dry weight to bottle volume be reported.

2. Length of incubation. The length of the incubation period also should be optimized (Littler 1979), because this factor interacts with the thallus-weight/bottle-volume ratios. Too long an incubation period can be problematic in that autoinhibitory substances might accumulate (Curl and McLeod 1961), bacterial populations tend to increase on the surfaces within the bottles, or nutrients and inorganic carbon may be depleted. A short period can induce errors due to possible daily photosynthetic periodicities and transient CO_2 or O_2 carried over in response to previous holding conditions. Consequently, it is important to report the incubation interval so that the magnitude of changes in O_2, CO_2, and pH can be evaluated when comparisons are desired.

3. Continuous monitoring. Because the gas exchange rates are likely to change during the course of an experiment, the continuous-monitoring method, utilizing individual electrodes within chambers, has a definite advantage over bottle experiments in which only initial and final values are used to calculate rates. Continuous monitoring also permits the comparison of regression slopes (analysis of covariance; Sokal and Rohlf 1969) during initial control, experimental, and final recovery periods (Fig. 17–4) to assess the impact of environmental factors on the natural homeostatic capabilities of algae (Littler 1973b).

Table 17–1. *Slopes determined for regressions during continuous pH monitoring experiments of net photosynthesis and respiration*

Species and experiment number	Light		Dark	
	No Stirring	Stirring	No stirring	Stirring
Ulva lobata				
Exp. 1	3.72	16.10	1.65	3.72
Exp. 2	4.80	12.45	1.72	2.53
Exp. 3	4.07	16.59	4.80	4.42
Mean	4.20	15.04[a]	2.72	3.56
Colpomenia sinuosa				
Exp. 1	0.85	1.23	0.72	0.54
Exp. 2	1.40	1.36	1.11	0.89
Mean	1.13	1.29	0.92	0.72

Note: Slopes expressed as milligrams carbon per gram dry weight each hour.
[a] Mean for stirring is significantly different from mean for no stirring at $P < .05$

4. Wounding. The use of cut disks or fragments, as well as rough handling, generally is to be avoided, depending on the nature of the experiment, because such phenomena as wound respiration and oxidation of organic exudates often result in unnaturally low net photosynthetic rates (Hatcher 1977; Dromgoole 1978).

5. Mixing. It has been shown consistently that some means of stirring is required to obtain realistic production rates in closed containers. However, mixing is much more critical for the more productive thallus forms having relatively high surface/weight ratios (Table 17–1). To ensure adequate mixing, a growth or NPP versus mixing rate curve can be generated to establish the water-motion saturation level (Santelices 1978). Since shading is difficult to avoid during hand shaking of bottles, we have found magnetic turbines operated by a water pump or a bellows-type air pump to be most effective for mixing. If shaking by hand is the only method available for agitation, we suggest that rectangular bottles be used, since rotating them results in the generation of considerable movement and momentum within the contained water mass.

6. Bottle type. The environmentally significant ultraviolet portion of the solar spectrum should be considered (Ohle 1958; Findenegg 1966) when incubation bottles are chosen. The bottles recommended here are made of the Wheaton "800"-type glass, which transmits less of the photosynthetically harmful ultraviolet spectrum than does Pyrex glass (Worrest et al. 1980). The rectangular versions are preferable (Fig. 17–2) because of economy in packing, the flat bottle sides do not restrict stir bar action, and greater water motion can be produced by hand rotation than is the case with cylindrical jars. The insides of the jars should be cleaned in aqua regia, rinsed, and then aged in distilled water for 30 d before service. In our opinion, one should never expose incubation glassware to detergents, preservatives, chromic acid, or other toxic chemicals.

7. Metabolic quotients. We strongly recommend supplementing the O_2 electrode technique with some other method such as the pH electrode technique or ^{14}C labeling when experiments are to be conducted at high O_2 levels. Supplementation with the pH technique has the additional advantage of providing an estimate of the PQ and RQ. The PQ can be an extremely important parameter to measure because it provides a useful index of (1) the type of nutrition (e.g., NO_3^{2-} or NH_4^+ as nitrogen sources), (2) changes in physiological state (e.g., due to stress or changes in light energy; see Kindig and Littler 1980), and (3) the type of material stored (i.e., carbohydrates, proteins, or lipids). The RQ is influenced primarily by the compounds being metabolized and tends to increase at lower temperatures.

8. O_2 supersaturation. Procedures that might result in bubble formation must be avoided because this can be problematic depending on the bubble volume and the extent to which the gas and liquid phases are in equilibrium. Oxygen supersaturation poses problems (Burris 1977; Littler 1979; Kremer 1980) other than bubble formation in the measurement of productivity of marine algae in closed systems. Dromgoole (1978) has shown that macroalgae liberate O_2 more rapidly at low initial O_2 levels than at high levels. Estimates of productivity based on light and dark closed containers will vary as a function of the initial O_2 concentration and the relative sensitivities to O_2 tensions of photosynthesis and dark respiration in the organisms under study (Fig. 17–3; Littler 1979).

In remote field situations, it is not always possible to use N_2 or other gases to reduce the supersaturated levels of O_2 characteristically present near highly productive nearshore environments, and available CO_2 might also be decreased in the process. As a consequence, we recommend the vigorous pouring or shaking of a stock incubation batch for at least 15 min (Strickland 1960) until air saturation is obtained. Incubation water below O_2 saturation levels also can be obtained before sunrise when O_2 levels have been naturally reduced by night respiration. Depending on the objectives of the experiment, field comparisons can be made reasonably only under uniform conditions of initial dissolved O_2. The initial O_2 tensions of both the light and dark bottles must be given.

D. *Intrinsic variability*

Other important considerations relevant to many experimental designs are the intrinsic causes and ranges of photosynthetic variability (see Littler and Arnold 1980). Such variations can be quite pronounced owing to differences in season, age, reproductive condition, morphology, and thallus portion incubated, as well as to previous conditions of crowding, macrohabitat, microhabitat, desiccation, and physical stress (Fig. 17–5). These factors, unless controlled, can result in inordinate within-species discrepancies and certainly must be taken into consideration in the design of studies on marine macroalgal productivity. An understanding of sources of photosynthetic variation, such as those outlined earlier and documented in Figs. 17–3 and 17–5, is of paramount importance in order to (1) make accurate estimates of individual photosynthetic values, (2) determine seaweed contributions to marine productivity, and (3) analyze evolutionary strategies of carbon allocation. This knowledge is equally essential for both field and laboratory studies.

IV. Acknowledgments

Sincere appreciation is extended to D. S. Littler, who provided the artwork and contributed significantly to the production of this treatment. The research leading to this analysis was supported by the Office of Water Research and Technology, USDI, under the Allotment Program of Public Law 88-379, as amended, and by the University of California, Water Resources Center, as a part of Office of Water Research and Technology Project A-054-CAL and Water Resources Center Project W-491.

V. References

Arnold, K. E., and Murray, S. N. 1980. Relationships between irradiance and photosynthesis for marine benthic green algae (Chlorophyta) of differing morphologies. *J. Exp. Mar. Biol. Ecol.* 43, 183–92.

Beckman Instruments, Inc. 1972. *In Situ Dissolved Oxygen Measurements.* Industrial Product Data Sheet 7601, Beckman Scientific Instruments Division, Fullerton, Calif. 4 pp.

Beyers, R. J. 1970. A pH–carbon dioxide method for measuring aquatic primary productivity. *Bull. Georgia Acad. Sci.* 28, 55–68.

Böhm, E. L. 1978. Application of the ^{45}Ca tracer method for determination of calcification rates in calcareous algae: effect of calcium exchange and differential saturation of algal calcium pools. *Mar. Biol.* 47, 9–14.

Borowitzka, M. A. 1979. Calcium exchange and the measurement of calcification rates in the calcareous coralline red alga *Amphiroa foliacea. Mar. Biol.* 50, 339–47.

Burris, J. E. 1977. Photosynthesis, photorespiration, and dark respiration in eight species of algae. *Mar. Biol.* 39, 371–9.

Carpenter, J. H. 1965a. The Chesapeake Bay Institute technique for the Winkler dissolved oxygen method. *Limnol. Oceanogr.* 10, 141–3.

Carpenter, J. H. 1965b. The accuracy of the Winkler method for dissolved oxygen analysis. *Limnol. Oceanogr.* 10, 135–40.

Carpenter, J. H. 1966. New measurements of oxygen solubility in pure and natural water. *Limnol. Oceanogr.* 11, 264–77.

Curl, H., Jr., and McLeod, G. C. 1961. The physiological ecology of a marine diatom, *Skeletonema costatum* (Grev.) Cleve. *J. Mar. Res.* 19, 70–88.

Czaplewski, R. L., and Parker, M. 1973. Use of a BOD oxygen probe for estimating primary productivity. *Limnol. Oceanogr.* 18, 152–4.

Dayton, P. K. 1975. Experimental evaluation of ecological dominance in a rocky intertidal algal community. *Ecol. Monogr.* 45, 137–59.

Doty, M. S., and Oguri, M. 1958. Selected features of the isotopic carbon primary productivity technique. *Cons. Int. Expl. Mer* 144, 47–55.

Dring, M. J., and Lüning, K. 1977. Significance of enhancement for calculations of photosynthesis of red algae from action spectra. *J. Phycol.* 13(suppl.), 18.

Dromgoole, F. I. 1978. The effects of oxygen on dark respiration and apparent photosynthesis of marine macroalgae. *Aquat. Bot.* 4, 281–97.

Duedall, I. W., Coote, A. R., Knox, D. F., and Connolly, G. F. 1971. A dispenser for Winkler's dissolved oxygen reagents. *J. Fish. Res. Bd. Can.* 28, 1815–16.

Duncan, M. J., and Harrison, P. J. 1982. Comparison of solvents for extracting chlorophylls from marine macrophytes. *Bot. Mar.* 25, 445–7.

Faust, M. A., Sager, J. C., and Meeson, B. W. 1982. Response of *Prorocentrum mariae-lebouriae* (Dinophyceae) to light of different spectral qualities and irradiances: growth and pigmentation. *J. Phycol.* 18, 349–56.

Findenegg, I. 1966. Die Bedeutung kurzwelliger Strahlung für die planktische Primärproduktion in Seen. *Verb. Int. Verein. Limnol.* 16, 314–20.

Fox, H. M., and Wingfield, C. A. 1938. A portable apparatus for the determination of oxygen dissolved in a small volume of water. *J. Exp. Biol.* 15, 437–45.

Gaarder, T., and Gran, H. H. 1927. Investigations of the production of plankton in the Oslofjord. *Rapp. Proc. Verb. Cons. Int. Expl. Mer* 42, 3–48.

Hatcher, B. G. 1977. An apparatus for measuring photosynthesis and respiration of intact large marine algae and comparison of results with those from experiments with tissue segments. *Mar. Biol.* 43, 381–5.

Heine, J. N. 1983. Seasonal productivity of two red algae in a central California kelp forest. *J. Phycol.* 19, 146–52.

Hitchman, M. L. 1978. *Measurement of Dissolved Oxygen.* Wiley, New York. 255 pp.

Hoffman, W. E., and Dawes, C. J. 1980. Photosynthetic rates and primary production by two Florida benthic red algal species from a salt marsh and a mangrove community. *Bull. Mar. Sci.* 30, 358–64.

Kindig, A. C., and Littler, M. M. 1980. Growth and primary productivity of marine macrophytes exposed to domestic sewage effluents. *Mar. Environ. Res.* 3, 81–100.

King, R. J., and Schramm, W. 1976. Photosynthetic rates of benthic marine algae in relation to light intensity and seasonal variations. *Mar. Biol.* 37, 215–22.

Kremer, B. P. 1980. Photorespiration and β-carboxylation in brown macroalgae. *Planta* 150, 189–90.

Littler, M. M. 1973a. The productivity of Hawaiian fringing-reef crustose Corallinaceae and an experimental evaluation of production methodology. *Limnol. Oceanogr.* 18, 946–52.

Littler, M. M. 1973b. The population and community structure of Hawaiian fringing-reef crustose Corallinaceae (Rhodophyta, Cryptonemiales). *J. Exp. Mar. Biol. Ecol.* 11, 103–20.

Littler, M. M. 1979. The effects of bottle volume, thallus weight, oxygen saturation levels, and water movement on apparent photosynthetic rates in marine algae. *Aquat. Bot.* 7, 21–34.

Littler, M. M., and Arnold, K. E. 1980. Sources of variability in macroalgal primary productivity: sampling and interpretive problems. *Aquat. Bot.* 8, 141–56.

Lüning, K. 1981. Light. In Lobban, C. S., and Wynne, M. J. (eds.), *The Biology of Seaweeds,* pp. 326–55. University of California Press, Berkeley.

Marsh, J. A., and Smith, S. V. 1978. Productivity measurements of coral reefs in flowing water. In Stoddart, D. R., and Johannes, R. E. (eds.), *Coral Reefs: Research Methods,* pp. 361–77. UNESCO, Paris.

Ohle, W. 1958. Diurnal production and destruction rates of phytoplankton in lakes. *Rapp. Cons. Expl. Mer* 144, 129–31.

Polygreen, M. C., and Coker, P. D. 1981. A portable oxygen meter and thermometer for field measurements. *J. Appl. Ecol.* 18, 827–33.

Ramus, J., Beale, S. I., and Mauzerall, D. 1976. Correlation of changes in pigment content with photosynthetic capacity of seaweeds as a function of water depth. *Mar. Biol.* 37, 231–8.

Ramus, J., Lemons, F., and Zimmerman, C. 1977. Adaptation of light-harvesting pigments to downwelling light and the consequent photosynthetic performance of the eulittoral rockweeds *Ascophyllum nodosum* and *Fucus vesiculosus. Mar. Biol.* 42, 293–303.

Ramus, J., and Rosenberg, G. 1980. Diurnal photosynthetic performance of seaweeds measured under natural conditions. *Mar. Biol.* 56, 21–8.

Ramus, J., and van der Meer, J. P. 1983. A physiological test of the theory of complementary chromatic adaptation. 1. Color mutants of a red seaweed. *J. Phycol.* 19, 86–91.

Santelices, B. 1978. Multiple interaction of factors in the distribution of some Hawaiian Gelidiales (Rhodophyta). *Pac. Sci.* 32, 119–47.

Smith, D. F., and Horner, S. M. J. 1982. Laboratory and field measurements of aquatic productivity made by a minicomputer employing a dual oxygen electrode system. *Mar. Biol.* 72, 53–60.

Smith, S. V., and Harrison, J. T. 1977. Calcium carbonate production of the

Mare Incognitum, the upper windward reef slope, at Enewetak Atoll. *Science* 197, 556–9.

Smith, S. V., and Kinsey, D. W. 1978. Calcification and organic carbon metabolism as indicated by carbon dioxide. In Stoddart, D. R. and Johannes, R. E. (eds.), *Coral Reefs: Research Methods,* pp. 469–84. UNESCO, Paris.

Sokal, R. R., and Rohlf, F. J. 1969. *Biometry.* Freeman, San Francisco.

Strickland, J. D. H. 1960. *Measuring the Production of Marine Phytoplankton.* Bulletin 122, Fisheries Research Board of Canada, Ottawa. 172 pp.

Strickland, J. D. H., and Parsons, T. R. 1972. *A Practical Handbook of Seawater Analysis,* 2nd ed. Bulletin 167, Fisheries Research Board of Canada, Ottawa. 310 pp.

Towle, D. W., and Pearse, J. S. 1973. Production of the giant kelp, *Macrocystis,* estimated by in situ incorporation of ^{14}C in polyethylene bags. *Limnol. Oceanogr.* 18, 155–9.

Worrest, R. C., Brooker, D. L., and Van Dyke, H. 1980. Results of a primary productivity study as affected by the type of glass in the culture bottles. *Limnol. Oceanogr.* 25, 360–4.

18: The carbon-14 method for measuring primary productivity

KEITH E. ARNOLD

Department of Biological Sciences, California State Polytechnic University, Pomona, California 91768

MARK M. LITTLER

Department of Botany, National Museum of Natural History Smithsonian Institution, Washington, D.C. 20560

CONTENTS

[377]

I. Introduction

A critical evaluation of the ^{14}C method (Steemann-Nielsen 1952) as applied to the measurement of benthic macroalgal photosynthesis (primary productivity) is given in this chapter. At present, there are no standard, universally applicable isotopic techniques for measuring the photosynthesis of all marine macroalgae, because this approach has had only limited application to seaweeds in the past decade. Before attempting any experimentation, one must have a thorough understanding of the photosynthetic metabolism of macroalgae, as well as a familiarity with radiotracer techniques and safety measures.

Several reviews (Harris 1978; Peterson 1980; Dring and Jewson 1982) have discussed the technical pitfalls of this method in the measurement of planktonic primary productivity, as well as the problems associated with the interpretation of whether this technique actually measures net or gross photosynthesis. Many of these practical and theoretical problems are also relevant to the measurement of macroalgal photosynthesis (Kremer 1981a). On a theoretical basis, gross or "real" photosynthesis is measured by the ^{14}C method only if the following conditions are met (Thomas 1963). (1) ^{14}C is assimilated at the same rate as ^{12}C; (2) no ^{14}C is incorporated into cellular material by nonmetabolic processes; (3) no ^{14}C is lost by dark respiration and/or photorespiration (Burris 1977), which may accompany photosynthesis; and (4) no ^{14}C is lost by excretion. In reality, none of these conditions is completely satisfied; therefore, with the ^{14}C method, one measures (Peterson 1980) something between net and gross photosynthesis.

A discrimination factor of 1.05 (see Peterson 1980) is often used to correct for differences in the metabolism of ^{14}C compared with that of ^{12}C (^{12}C is incorporated 1.05 times faster than ^{14}C).

Estimates of ^{14}C incorporation by processes other than photosynthesis are often obtained from specimens incubated in the dark. Dark ^{14}C fixation (DCF) in macroalgae is significant, and it has been demonstrated (see Kremer 1979, 1981a,b) that active ^{14}C fixation in the dark occurs via carboxylating enzymes, primarily phosphoenol-

[378]

pyruvate carboxykinase (EC 4.1.1.32) and, to a lesser extent, phosphoenolpyruvate carboxylase (EC 4.11.31). The amount of nonmetabolic uptake via adsorption, absorption, and isotopic exchange is thought to be small in comparison with that via DCF, except for calcareous macroalgae (Littler 1973). In the past, values of DCF have been subtracted from the light carbon fixation (LCF) rates, recorded separately, or omitted entirely.

Direct loss of recent photosynthetically fixed ^{14}C can occur either through active excretion (as well as wounding) or from respiratory processes that occur in the light. Rates of organic ^{14}C excretion are normally low in marine macroalgae and represent less than 5% of the LCF rates (Brylinsky 1977). The rate of refixation of respired (dark mitochondrial respired and/or photorespired) $^{14}CO_2$ that occurs in the light cannot be accurately estimated. This point proves to be the major (Peterson 1980) theoretical question concerning this method.

Depending on the question(s) being asked and the level of accuracy of information needed, the choice of using the ^{14}C method rather than the conventional (Gaarder and Gran 1927) O_2 light and dark bottle techniques should be carefully evaluated (see Littler and Arnold, Chap. 17). Although the general equation for photosynthesis seems relatively simple, the light-dependent incorporation of ^{14}C into photosynthetic products is an extremely complex process. Researchers who attempt to use this method, under either field or laboratory conditions, must be aware of the theoretical and practical difficulties involved or enormous errors in the estimation of macroalgal photosynthesis will occur. Reviews on macroalgal photosynthesis (Kremer 1978, 1981a; Ramus 1981) and radiotracer techniques (Neame and Homewood 1974; Peng 1981) should be consulted for appropriate background information.

II. Equipment and materials

A general list of equipment and supplies is given in Table 18–1. The following sections provide an overview of the types of equipment and supplies that have been used in past studies and the limitations that various macroalgae and incubation conditions impose on the use of such items.

A. Incubation equipment

1. Incubation vessels. Thallus size, metabolic activity, and incubation time are important criteria for the selection of optimal incubation vessel size. Preliminary linearity runs for *each* algal species should be conducted before experimentation to avoid "bottle effects" (see Littler

Table 18 − 1. *General list of equipment and supplies for measuring the*
productivity of marine benthic macroalgae by the ^{14}C *method*

I. *Incubation*
  A. Incubation vessels: canning jars (variable sizes), BOD bottles, custom plexiglass chambers, polyethylene or polyamide–polyethylene bags
  B. Stirring: air-driven magnetic stirrers, foot pump, Teflon-coated magnetic stir bars, battery-operated electric stirrers or submersible electric pumps
  C. Incubation trays: clear polycarbonate animal cages
  D. Miscellaneous: hand-held 10-μm plankton net, large-volume buckets, thermometers, salinity refractometer, aluminum foil, water-resistant tape, water-resistant labels, plastic bags, ice chest, quick-freeze Freon, analytical balance
II. ^{14}C *and accessories*
  A. ^{14}C: working stock solution (550 μCi·ml^{-1}), syringes, micropipettes with disposable tips, rubber gloves, radioactive identification labels, metal waste disposal can
  B. Activity determination: scintillation vials, liquid scintillation counter, CO_2-trapping agent, saltwater-compatible LSC cocktail, quenched standards
  C. Alkalinity determination: 250-ml plastic jars, pH meter, standardized HCl, 50-ml pipettes, buffers
  D. Tissue extraction: 80% ethanol, 6 N HCl, 30% H_2O_2, NCS (Amersham), variable-temperature (up to 80°C) water bath, other solvents for fractionation

and Arnold, Chap. 17). Most commonly, seaweeds have been enclosed in biological oxygen demand (BOD) bottles of either 300- or 500-ml capacity (Wassman and Ramus 1973; Bach and Josselyn 1979; Arnold and Murray 1980) or wide-mouthed canning jars, the size of which ranges from 0.5 to 4.0 liters (Littler and Arnold 1980, 1982; Heine 1983). Specially constructed plexiglass or polyvinyl chloride chambers with accessible sampling ports have been used for both in situ (Brylinsky 1977; Hatcher 1977; Rogers and Salesky 1981) and laboratory (Littler 1973; Darley et al. 1976) experiments. For in situ experiments on large kelps (Towle and Pearse 1973; Lobban 1978) or rockweeds (Guterstam 1976), polyethylene or polyamide–polyethylene plastic bags of various sizes can be used. The O_2 and CO_2 permeability properties of these bags should be known before experimentation.

2. *Mixing.* Adequate mixing during incubation is necessary to ensure that photosynthesis is not diffusion-limited. The method of stirring is dependent on the type of incubation vessel chosen and the species incubated. For in situ experiments employing plexiglass chambers (Brylinsky 1977) or jars (Littler and Arnold 1980; Heine 1983), air-

driven magnetic stirrers (A. H. Thomas Company) have been used with propulsion coming from compressed air tanks or manually operated foot pumps. For subtidal incubations (Heine 1983) it is important to consider whether the emitted bubbles interfere with the incident light field. Battery-operated stirrers (Bittaker and Iverson 1976; Marker 1976) and water pumps (Hatcher 1977; Rogers and Salesky 1981) have also been used successfully. Stirring within plastic bags by the buffeting action of oceanic swells is thought to be adequate (R. Zimmerman, personal communication); however, this probably varies with specific incubation conditions.

B. Incubation materials

Radioactive ^{14}C is normally obtained in the bicarbonate form ($NaH^{14}CO_3$) as a dry solid or as a sterile aqueous solution in a sealed ampoule (New England Nuclear or Amersham). Specific activities are variable, but for productivity studies they should be $\sim 5-50$ mCi $\cdot$ mmol^{-1}. The powdered $NaH^{14}CO_3$ is recommended, and stock solutions are prepared by adding an appropriate volume of sterile filtered (0.2-μm pore size) seawater or a synthetic substitute (Strickland and Parsons 1972). The stock solution should be adjusted to pH 9.5 (Peterson 1980) to prevent loss of activity due to $^{14}CO_2$ blowoff that occurs at a lower pH. Working solutions of small volume (5 ml) can be prepared from the concentrated stock. The stock and working solutions should be stored in a freezer in separate sterile glass vials until used (see Neame and Homewood 1974).

III. Methods

A. Incubation of sample

An incubation vessel of appropriate size is chosen and filled with filtered ambient seawater (10-μm plankton net) to exclude air, and a defined amount (usually 1–10 μCi per liter of incubation water) of radioactivity from the ^{14}C working solution is introduced. This can be done by injection of the contents of a syringe through rubber serum caps installed in plexiglass chambers, polyethylene bags, or the glass lids of canning jars. For BOD bottles, the activity can be easily added directly through the aperture with a fixed-volume pipette. Subtidal incubations are most difficult, and Drew (1973) and Smith (1981) give an overview of the methods of collecting and incubating seaweeds without bringing them to the surface. If subtidal specimens are brought to the surface so that they can be more easily placed in incubation vessels and cleaned, they should not be exposed to air or unnatural levels of sunlight. After introduction of the ^{14}C, the incubation medium should be stirred for at least 2 min, and then three 0.5-ml subsamples (the volume depends on the specific activity and seawater capacity of the cocktail) are taken for initial activity. It is best to introduce the alga into the vessel at this point, rather than before the label is added as is necessary in most subtidal incubations (Drew 1973; Towle and Pearse 1973). Initial activity samples should always be taken, since the activity of working ^{14}C solutions can change during storage (Neame and Homewood 1974). Great care should be taken at this point to prevent pipette error, since any inaccurate estimates of initial activity translate into inaccuracies in the estimate of productivity. Initial activity samples should be immediately mixed with liquid scintillation counting (LSC) cocktail and stored in a cool, dark place until counted.

The LSC system should be compatible with seawater; it should not show strong color or chemical quench, nor cause acidic-induced changes in the chemical CO_2 equilibria of the sample. Iverson et al. (1976) found that Aquasol II (New England Nuclear) behaves as a Lewis acid and causes loss of activity as $^{14}CO_2$ gas, which escapes from solution into the head space of the vial, with the result that initial counts are lowered. This is a time-dependent process and can cause a significant reduction in the count rate of initial activity samples. With acidic cocktails such as Aquasol II, a CO_2-trapping agent such as phenethylamine (Iverson et al. 1976), ethanolamine, or hyamine is added to prevent lowered count rates. Cheaper alternatives to expensive commercial cocktails can be easily made in the laboratory (Waite et al. 1973). Whatever LSC cocktail is chosen, it should be tested for its capacity to mix with seawater and to retain inorganic ^{14}C (HCO_3^- and CO_2). K. E. Arnold and S. Manley (unpublished data) routinely use a cocktail of 10 ml of Aquasol II and 3.0 ml of NCS (Amersham) for each 0.5-ml initial water sample. The trapping agent is added first, followed by the seawater sample, and the vial closed and shaken vigorously. The cocktail is then added with further mixing.

Initial incubation water samples are also taken for determination of total "CO_2," the volume of which depends on the technique used (Strickland and Parsons 1972). Smith and Kinsey (1978) give a thorough guide to the practical problems of measuring pH and carbonate alkalinity in seawater.

B. Dark carbon fixation

Traditionally, incubation of experimental thalli in the dark provided a "blank" or a control for nonbiological mechanisms of ^{14}C uptake (Thomas 1963). More recent studies (for a review see Kremer 1981a) have shown that DCF is associated with the activity of carboxylating enzymes other than ribulose-1,5-bisphosphate carboxylase-oxygenase.

Dark carbon fixation rates can be determined using the same general incubation procedures as outlined earlier. Rates of DCF for Rhodophyta and Chlorophyta are almost always (Kremer 1979, 1981a,b) less than 3% of the corresponding LCF rates. In Phaeophyta, especially the kelps and rockweeks, DCF rates can range upward to 30 to 50% of LCF. Highest DCF rates are associated (Kremer 1981a) with the young meristematic growing regions (transition zone) of the laminae. Most DCF occurs primarily through the activities of the anapleurotic enzyme phosphoenolpyruvate carboxykinase, and the distribution of its activity has been found (Kremer 1981a) to exhibit longitudinal profiles similar to those of DCF along kelp blades.

Since DCF is much lower in the red and green seaweeds, greater specific activities of $H^{14}CO_3^-$ are needed in the incubation water to obtain measurable rates. Generally up to 50 μCi $\cdot$ liter^{-1} are needed, depending on the incubation time and metabolic activity. Patterns of ^{14}C fractionation into end products of fixation are likely to be much different from those found with LCF experiments. In young growing kelp tissues, for instance, there is a much higher dark incorporation of ^{14}C into the insoluble fraction in short-term experiments (1–3 h) (Willenbrink et al. 1979). It should be emphasized that all incubation manipulations must be performed in the strict absence of light; otherwise, inaccurate rates of DCF will be obtained. This obviously gives rise to very special problems for field incubations, and for this reason in situ DCF experiments have been largely ignored.

C. Extraction of experimental thalli

Upon termination of incubations, the algal thalli are carefully removed and rinsed for several minutes in unlabeled seawater. Material should *not* be rinsed in acidified seawater, as some have done in the past (Wassman and Ramus 1973), because this may begin the extraction process prematurely. Thalli are blotted and dried, and their fresh weights are determined (fresh-weight/dry-weight ratios are determined on similar thalli). This can be done in the field with a battery-operated balance (see Brinkhuis, Chap. 22). Thalli are then placed immediately into an extraction solvent [volume is variable, but normally 10–20 times (milliliters) the fresh weight (grams) is sufficient] and stored in the dark until processed. The solvent routinely used (Brinkhuis and Jones 1974; Brinkhuis 1977a,b) is 80% ethanol acidified to pH ~2.0 with concentrated HCl; this removes any residual cell-wall-bound $H^{14}CO_3^-$ (Willenbrink et al. 1979). Depending on the experimental design, the use of other solvents may be more appropriate. For larger thalli such as kelps, productivity can be estimated from tissue subsamples taken down the longitudinal axis (Towle and Pearse 1973; Johnson et al. 1977; Küppers and Kremer 1978; Arnold 1980) of the lamina and from "representative" subsamples of stipe and holdfast. If extraction cannot be accomplished in the field, whole thalli or tissue subsamples can be quick-frozen on dry ice (Penhale and Smith 1977) or with liquid Freon (Brinkhuis and Jones 1974) and stored until processed for LSC. Long time intervals between termination of incubation and fixation of the algal tissue should be avoided.

D. Excretion of ^{14}C-labeled dissolved organic matter

Final water samples for estimating the quantity of ^{14}C excreted as dissolved organic matter (DOM) may also be taken upon termination of incubation. Sample size can range from 1 to 100 ml depending on the initial activity, the metabolic rate, and the percentage of LCF excreted (normally less than 5%). These samples should be prefiltered (0.2 μm) and kept on ice (to avoid bacterial degradation) until processed. It may also be worthwhile to dissolve and count the prefilters after an acid wash to assess possible particulate ^{14}C loss (Fankboner and de Burgh 1977).

E. Precautions and waste disposal

In general, all containers in contact with ^{14}C should be clearly labeled with radioactivity stickers. Most academic institutions have specific health and safety regulations concerning the handling and disposal of radioactivity, and these should be consulted before experimentation. Disposal of radioactivity is one of the most costly aspects of this type of research, principally because of the large amount of activity remaining in the incubation water. The remaining $H^{14}CO_3^-$ can be precipitated with barium hydroxide, and this effectively reduces the total volume of activity to be disposed.

F. Sample work-up and calculations

General aspects of LSC procedures can be found in the monographs by Neame and Homewood (1974) and Peng (1981). A familiarity with these techniques is mandatory for accurate measurements of ^{14}C. Only specific aspects of these procedures as they apply to ^{14}C productivity measurements are discussed here.

1. Initial activity. Initial activity samples should be dark-adapted for at least 24 h (as should all LSC samples) to avoid problems of chemical luminescence and photoluminescence (Peng 1981). These samples should then be counted as soon as possible thereafter, since most

LSC cocktails will evaporate with time, thus changing the ratio of water sample to cocktail and enhancing the formation of precipitates. A separate quench curve obtained by one of the three major (Peng 1981) methods – internal standard, channels ratio, and external standard channels ratio (ESCR) – should be constructed for water samples. The simplest and least costly is the ESCR method, which necessitates having a liquid scintillation counter with an external standard (most commercial counters have this feature). A consistent procedure should be adopted on the basis of preliminary experiments designed to provide information on the optimal sample/cocktail volume ratio, total cocktail–sample volume, and need for CO_2-trapping agents. All of these variables influence the amount of quench that occurs in each sample, and once established, they should not be altered because this will change the quench characteristics.

2. *Release of* ^{14}C-*labeled DOM.* The technique employed by Smith (1975) for measuring excreted ^{14}C-labeled DOM by phytoplankton is easily modified for examining macroalgal excretion rates. After incubation, a known volume (10–100 ml) of the filtered incubation medium is placed in a special apparatus (see Smith 1975) and acidified to pH 2.5 to 3.0 by the addition of concentrated HCl or H_3PO_4. The volume added (usually less than 1 ml) will depend on the sample size and acid strength. Upon acidification, the solution is bubbled with air for at least 10 min. The $^{14}CO_2$ that is driven off is collected in an NaOH trap. A functional apparatus can be easily designed from a pair of side-arm filtration flasks, Tygon tubing, stoppers, and Pasteur pipettes (Fig. 18–1). Other possible designs can be found in the papers by Gachter and Mares (1979) and Wessels and Birnbaum (1979). The activity remaining in solution represents the amount of excreted (acid-stable) organic carbon. For low excretion rates, higher sample/cocktail volume ratios can be used to give significant counts above background. For accurate measurements, a separate quench curve should be prepared. Blanks prepared by the addition of a known amount of working stock $NaH^{14}CO_3$ solution to sterile filtered seawater and acidified to pH 2.5 to 3.0 should also be run to test the effectiveness of the method and to control for any possible contamination (Williams et al. 1972) of the original stock activity with acid-stable ^{14}C.

3. *Sample extraction and solubilization.* Information regarding the fractionation of photosynthetically fixed ^{14}C into different products can be conveniently obtained from laboratory or field ^{14}C fixation experiments. Studies involving the measurement of flow of carbon into major end products suggest that the resultant patterns of carbon

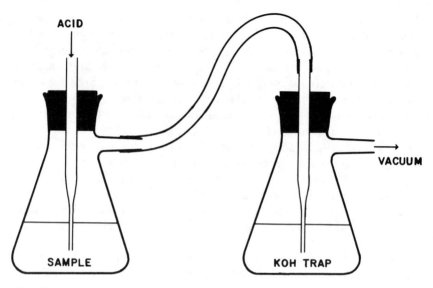

Fig. 18–1. Apparatus for the production of acid-stable ^{14}C-labeled dissolved organic matter.

allocation are altered dramatically by such factors as light intensity (Kirst 1981), light quality (Bird et al. 1981), O_2 levels (Burris 1977; Kirst 1981), and nitrogen supply (Bird et al. 1981). Fractionation studies require significantly more bench work, and a good general discussion of these procedures can be found in Howard et al. (1975) and Kremer (1978) as well as in a paper on phytoplankton fractionation by Hitchcock (1983). Generally, when incorporating a fractionation design into productivity studies, one must employ higher activities of $H^{14}CO_3^-$ and in the initial sample fixation step, use a solvent that is compatible with the chosen fractionation scheme.

For routine productivity measurements, a convenient fractionation scheme involving ethanol-soluble and ethanol-insoluble fractions (Brinkhuis and Jones 1974) or acetone-soluble and acetone-insoluble fractionation (Wassman and Ramus 1973) is sufficient. In highly productive and fast-growing seaweeds such as *Ulva* and *Enteromorpha* (Brinkhuis 1977b; K. E. Arnold, unpublished data), ^{14}C uptake into ethanol-insoluble fractions may approach 50% or more of the total uptake, whereas in rockweeds and kelps (Brinkhuis 1977a,b; Buggeln 1979; K. E. Arnold and S. Manley, unpublished data), the proportion of the ethanol-insoluble activity is normally less than 15%. The ratio of extraction volume to sample weight should be optimized such that extraction is efficient and yet the count rate per milliliter of extract remains at least 1000 disintegrations per minute (dpm) above background.

Techniques routinely used for solubilization of the insoluble material are given by Lobban (1974) and Gagne et al. (1979). These techniques are relatively simple and inexpensive and can be used to solubilize unextracted tissues as well. Loss of some activity by oxidation to $^{14}CO_2$ remains a problem, and the extent appears to vary (Gagne et al. 1979) with different species. Routinely (K. E. Arnold and S. Manley unpublished data), 7–9% of the label is lost when kelp tissues are solubilized directly by either technique. The method of Gagne et al. (1979) is preferred (K. E. Arnold and S. Manley, unpublished data) because it produces less quench, produces lower chemiluminescence, and offers more complete digestion. Samples with a dry weight of up to 25 mg can be solubilized in a scintillation vial by the addition of 0.25 ml of 30% H_2O. The vial is capped and kept at 50°C for 18 h in the dark, followed by the addition of 4 ml NCS (Amersham) at 50°C for an additional 16 h in the dark. A solubilization technique used for aquatic angiosperms (Beer et al. 1982) makes it possible to measure both ^{14}C-labeled photosynthate and chlorophyll *a* on the same tissue sample. This technique might prove useful for macroalgal tissues as well. An alternative to solubilization involves the combustive oxidation of the total tissue to CO_2 (Brylinsky 1977; Penhale and Smith 1977; Kremer 1978; Adkin and Ho 1981), followed by trapping in an organic base. Wassman and Ramus (1973) suggested drying the tissue and grinding it into a fine powder followed by suspension into a gelled fluor. In this case, quench correction must be accomplished (Neame and Homewood 1974) by means of the channels ratio method.

4. Calculations. Rates of primary production are normalized to some parameter of biomass (fresh weight, dry weight, ash-free dry weight, or chlorophyll *a* content) or photosynthetic area (unit area of thallus surface or square meters of intertidal surface). The normalization parameter will depend on the particular ecological or physiological question(s) being asked. The most easily obtained biomass parameter is dry weight. For comparison purposes, we suggest that, even if other parameters are used, a conversion factor to productivity per unit dry weight be provided. Productivity per unit thallus area is also easily obtained for most sheetlike forms (e.g., *Ulva, Porphyra,* and *Laminaria*), and area may be expressed as square meters of one (Littler and Arnold 1980) or both (Wheeler 1980) sides of a thallus. Ramus (1978) suggested that seaweed photosynthesis, normalized to chlorophyll *a*, will often be incorrect, leading to gross over- or underestimation of net primary productivity (for a more complete discussion of this, see Littler and Arnold, Chap. 17).

Equation 1 can be used for the calculation of net particulate primary productivity on an hourly basis:

$$P_n = \frac{S_{dpm} \times CO_2 \times V \times 1.05}{W_{dpm} \times B \times H} \tag{1}$$

where P_n is the net particulate primary productivity, expressed as milligrams carbon per gram dry weight each hour or grams carbon per square meter each hour. S_{dpm} is the total corrected (including quench and background) activity (disintegrations per minute) in the sample, including both the activities from the ethanol-soluble and -insoluble fractions; CO_2 is the total carbon available for photosynthesis in terms of milligrams carbon per liter, calculated from pH and total alkalinity data; V is the volume of the incubation vessel in liters; 1.05 is the isotopic discrimination quotient; W_{dpm} is the total corrected activity (disintegrations per minute) in the incubation water; B is biomass (grams dry weight) or photosynthetic area (square meters of thallus or of intertidal surface); and H is the length of incubation in hours.

A sample calculation for the net particulate primary productivity of a juvenile sporophyte of *Eisenia arborea* (from Arnold 1980) is as follows:

$$P_n = 2.20 \text{ mg C} \cdot g^{-1} \cdot h^{-1}$$

$$= \frac{1.50 \times 10^5 \text{ dpm} \times 26.4 \text{ mg C} \cdot liter^{-1} \times 1.22 \text{ liters} \times 1.05}{1.89 \times 10^7 \text{ dpm} \times 0.035 \text{ g} \times 3.48 \text{ h}}$$

$$\tag{2}$$

If estimates of excretion of ^{14}C-labeled DOM are obtained, they should be reported separately and not added to the total sample activity, since by definition (Peterson 1980) the ^{14}C technique approximates net particulate primary productivity. These rates are normally presented as a percentage of the corresponding LCF rates.

Similarly, rates of DCF are also reported separately as milligrams carbon per gram each hour or grams carbon per square meter each hour or as a percentage of the LCF values. Kremer (1981b) suggests that DCF rates be included in gross productivity calculations, yet the theoretical reasons for this remain unclear.

In the initial design of ^{14}C experiments, it is wise to use calculations such as those just presented to estimate the initial activity necessary. This calculation is based on an approximate value for the expected productivity (this might be estimated from literature values) and the minimum value of activity needed in the tissue samples. These generated estimates are determined for a specific volume of incubation water, weight of tissue, and incubation time interval.

IV. Critical evaluation

A. Application considerations

With proper application, the ^{14}C technique offers very important advantages over the traditional O_2 light and dark bottle technique, particularly (1) when within-plant photosynthetic differences attributed to age, development, and reproduction are to be examined without the problems associated with excision of isolated tissue plugs from the parental thallus (Hatcher 1977; Arnold 1980; K. E. Arnold and S. Manley, unpublished data); (2) when greater sensitivity is needed for the measurement of photosynthesis on algal thalli that are either small and difficult to manipulate (such as laminarian gametophytes; Kremer and Markham 1979) or that form complex turf associations (Borowitzka et al. 1978) or when low metabolic rates are expected such as observed for many algal crusts (Littler and Arnold 1982); (3) when greater insight is needed with respect to the photosynthetic strategies of carbon allocation and how carbon flow into various end products varies with ontogenetic and environmental factors; and (4) when rates of photosynthesis in air (Kremer 1978; Darley et al. 1976) are to be compared with rates of photosynthesis under water. Alone or along with uptake of ^{45}Ca, the incorporation of $H^{14}CO_3^-$ can be used to estimate the rates of calcification in marine macroalgae. A review of methods of calcification rate measurement can be found in Borowitzka (1977).

The employment of the ^{14}C technique should be carefully weighed according to the questions being asked, since some applications, such as primary productivity measurements of large algal thalli and community productivity studies, may be unnecessarily laborious. Furthermore, it should be emphasized that specific techniques that work for one algal species may have to be modified for another, and the key to success is to conduct several preliminary experiments to ensure that label losses are kept to a minimum. The ^{14}C technique does, however, have some serious disadvantages as compared with other physiologically based (O_2 evolution) techniques. As pointed out in the introduction, this method measures something between net and gross photosynthesis. Furthermore, because dark respiration cannot be determined or accurately estimated (Steemann-Nielsen and Hansen 1959) by this technique, 24-h net primary productivity rates (see Littler and Arnold, Chap. 17) cannot be calculated. Economically, the method is much more costly, in terms of both initial equipment outlay and the inordinately large quantities of supplies needed for counting ^{14}C activity. In addition, it is much more labor intensive than either the O_2 or pH electrode technique, particularly when whole-plant rates are to be obtained on moderately large thalli.

B. Dark carbon fixation

The measurement of $^{14}CO_2$ uptake in the dark, as outlined earlier, will probably yield values that seriously underestimate the true rate of DCF, since simultaneously occurring respiratory processes are giving off CO_2. Some of the CO_2 is refixed by phosphoenolpyruvate carboxykinase while the rest leaves the thallus surface and enters solution. Kremer (1981b) found an overall 25% increase in DCF rates in kelp frond samples in which the action of the Krebs cycle (respiration) had been inhibited with monofluoroacetate. Since DCF partially compensates for dark respiratory release of CO_2, Kremer (1981b) believes that the inclusion of DCF rates in calculations of gross productivity is justified. Because DCF represents a more special case involving mainly members of the Phaeophyta (particularly kelps), we believe that it is inappropriate at this time to include DCF rates in productivity calculations for marine macroalgae in general. Furthermore, we suggest that, when such measurements are taken, they be reported separately. The real applicability of these analyses, however, lies in the ability (Arnold 1980) to fingerprint the actively growing sink (Lobban 1978) tissues within fronds of kelps.

C. Release of ^{14}C-labeled dissolved organic matter

Release of ^{14}C-labeled DOM can occur through active excretion such as that associated with photorespiratory release of glycolate (Fogg 1981) or passive leakage due to wounding or release of mucilage. Estimates of the release rates of ^{14}C-labeled DOM by marine macroalgae are important to the understanding of both the dynamics of carbon flow within plants and the flow of carbon between algae and other biotic components of the community. We therefore recommend that ^{14}C-labeled DOM release data be routinely taken during ^{14}C experiments.

Numerous interpretive errors, however, are associated with the estimation of DOM release by the methods outlined in this chapter. The analysis of the rate of release of ^{14}C acid-stable label into solution probably underestimates the true release of DOM, because in short-term incubations, released products have a low specific activity although the results of time course experiments (Brylinsky 1977) of release of acid-stable label and pulse–chase experiments suggest that, in the algae examined (*Acanthophora*, *Chondria*, *Dictyota*, and *Sargassum*), release rates were not seriously underestimated. This is likely, however, to vary with the species incubated. In seaweeds with low metabolic turnover of the photosynthetic products such as in kelps, release of mucilages (which are metabolically of greater distance from the site of fixation than the primary products) of very low specific activity occurs (K. E. Arnold, personal observation). Thus, in this case, the release of DOM would be dramatically underestimated.

Wounding of the thallus material before or during incubation causes significant release of DOM. In fact, the frequently cited work of Khailov and Burlakova (1969) on DOM release by macrophytes gives very high rates because the experimental thalli were in various stages of death and decomposition. In the incubation of thalli, great care should be taken to prevent wounding. Furthermore, experiments conducted on representative tissue plugs or thallus parts (Drew et al. 1982) should include an assessment of the effects of wounding and loss of ^{14}C-labeled DOM: otherwise, the results must be viewed with caution. Another complication of estimating DOM release arises from possible bacterial uptake (Brylinsky 1977) and metabolism of the released compounds back into $^{14}CO_2$.

D. Carbon budgets of kelps

Estimates of the primary productivity of large seaweeds such as kelps are difficult to obtain with ^{14}C techniques. In the past (Drew 1973; Towle and Pearse 1973; Johnston et al. 1977; Drew et al. 1982), isolated thallus parts or representative segments were incubated and the total plant productivity calculated from these rates after correction for the distribution of the various plant parts within the thallus. In the case of complex kelps such as *Macrocystis,* accurate estimates of whole frond productivity are thus difficult to obtain without elaborate subsampling regimes. Time and cost are key factors that detract from the use of ^{14}C methods on large, morphologically complex algae.

E. Comparisons with other methods

Remarkably few simultaneous measurements of O_2 evolution and ^{14}C uptake have been conducted on marine macroalgae. Calculated values of net particulate carbon fixation are almost always (Arnold 1980) lower than the corresponding rates of net photosynthesis calculated from O_2 experiments in which a PQ (photosynthetic quotient; ratio of O_2 evolved to CO_2 fixed) of unity is assumed. Good agreement between the two methods was found by Arnold (1980 and unpublished data) for photosynthetic measurements on *Enteromorpha intestinalis* and young sporophytes of *Eisenia arborea*. An average PQ of 1.14 (range, 0.96–1.32) was found in short-term experiments (2.5 h) for *Eisenia. Enteromorpha* collected from low-nitrogen habitats had higher PQ values (average of 1.91) than thalli collected from higher nitrogen habitats (average of 1.32). Hoffman and Dawes (1980) found PQs that ranged from 1.2 to 1.3 for *Gracilaria verrucosa* and 1.2 to 1.4 for *Bostrychia binderi*. Despite the uncertainties intrinsic in the ^{14}C and O_2 techniques, it is interesting to note how close these values are to the value of 1.25, suggested by Ryther (1956) for the conversion of O_2 values to carbon equivalents. Other values for PQs of benthic macroalgae can be found in a paper by Buesa (1980). Given the great sources of potential error in photosynthetic measurements, PQ values should be interpreted with great caution.

V. Acknowledgments

We would like to acknowledge G. C. Stephens, in whose laboratory much of the experimental work was conducted. Dr. Stephens also provided intellectual and financial support (to KEA) during this period. We also thank S. Manley, who contributed in a very large way to the refinement of the techniques presented in this chapter. Marilyn Steinle kindly typed the manuscript, and Larry Finkle prepared Fig. 18–1. Helpful discussions with J. N. Walker, D. L. Fong-Walker, M. Hill, and C. Beam improved the quality of the manuscript. We would also like to thank M. Arnold for assisting with the manuscript preparation.

VI. References

Adkin, M. P., and Ho, L. C. 1981. Simple conversion of a carbon–hydrogen analyser for ^{14}C determination. *Ann. Bot.* 48, 247–9.

Arnold, K. E. 1980. "Aspects of the Production Ecology of Marine Macrophytes." Ph.D. dissertation, University of California, Irvine. 149 pp.

Arnold, K. E., and Murray, S. N. 1980. Relationships between irradiance and photosynthesis for marine benthic green algae (Chlorophyta) of differing morphologies. *J. Exp. Mar. Biol. Ecol.* 43, 183–192.

Bach, S. D., and Josselyn, M. N. 1979. Production and biomass of *Cladophora prolifera* (Chlorophyta, Cladophorales) in Bermuda. *Bot. Mar.* 22, 163–8.

Beer, S., Stewart, A. J., and Wetzel, R. G. 1982. Measuring chlorophyll *a* and ^{14}C-labeled photosynthate in aquatic angiosperms by the use of a tissue solubilizer. *Plant Physiol.* 69, 54–7.

Bittaker, H. F., and Iverson, R. L. 1976. *Thalassia testudinum* productivity: a field comparison of measurement methods. *Mar. Biol.* 37, 39–46.

Bird, K. T., Dawes, C. J., and Romeo, J. T. 1981. Light quality effects on carbon metabolism and allocation in *Gracilaria verrucosa*. *Mar. Biol.* 64, 219–23.

Borowitzka, M. A. 1977. Algal calcification. *Oceanogr. Mar. Biol. Annu. Rev.* 15, 189–223.

Borowitzka, M. A., Larkum, A. W. D., and Borowitzka, L. J. 1978. A preliminary study of algal turf communities of a shallow coral reef lagoon using an artificial substratum. *Aquat. Bot.* 5, 365–81.

Brinkhuis, B. H. 1977a. Seasonal variations in salt-marsh macroalgae photosynthesis. 1. *Ascophyllum nodosum* ecad *scorpioides. Mar. Biol.* 44, 165–75.

Brinkhuis, B. H. 1977b. Seasonal variations in salt-marsh macroalgae photosynthesis. 2. *Fucus vesiculosus* and *Ulva lactuca. Mar. Biol.* 44, 177–86.

Brinkhuis, B. H., and Jones, R. F. 1974. Photosynthesis in whole plants of *Chondrus crispus. Mar. Biol.* 27, 137–41.

Brylinsky, M. 1977. Release of dissolved organic matter by some marine macrophytes. *Mar. Biol.* 39, 213–220.

Buesa, R. J. 1980. Photosynthetic quotient of marine plants. *Photosynthetica* 14, 337–42.

Buggeln, R. G. 1979. Photosynthesis and translocation in relation to growth in Laminariales. In Marcelle, R., Clijsters, H., and Van Poucke, M. (eds), *Photosynthesis and Plant Development*, pp. 251–61. Junk, The Hague.

Burris, J. E. 1977. Photosynthesis, photorespiration, and dark respiration of eight species of algae. *Mar. Biol.* 39, 371–9.

Darley, W. M., Dunn, E. L., Holmes, K. S., and Larew, H. G. 1976. A ^{14}C method for measuring epibenthic microalgal productivity in air. *J. Exp. Mar. Biol. Ecol.* 25, 207–17.

Drew, E. A. 1973. Primary production of large marine algae measured *in situ* using uptake of ^{14}C. In *A Guide to the Measurement of Primary Production under Some Special Conditions*, pp. 22–6. UNESCO, Paris.

Drew, E. A., Ireland, J. F., Muir, C., Robertson, W. A. A., and Robinson, J. D. 1982. Photosynthesis, respiration and other factors influencing the growth of *Laminaria ochroleuca* Pyl. below 50 metres in the Straits of Messina. *Mar. Ecol.* 3, 335–55.

Dring, M. J., and Jewson, D. H. 1982. What does ^{14}C uptake by phytoplankton really measure? A theoretical modelling approach. *Proc. R. Soc. Lond. Ser. B* 214, 351–68.

Fankboner, P. V., and de Burgh, M. E. 1977. Diurnal exudation of ^{14}C-labelled compounds by the large kelp *Macrocystis integrifolia* Bory. *J. Exp. Mar. Biol. Ecol.* 28, 151–62.

Fogg, G. E. 1981. The ecology of an extracellular metabolite of seaweeds. In Fogg, G. E., and Jones, W. E. (eds). *Proceedings of the Eighth International Seaweed Symposium*, pp. 46–53. University College of North Wales, The Marine Science Laboratories, Menai Bridge.

Gaarder, T., and Gran, H. H. 1927. Investigations of the production of plankton in the Oslofjord. *Rapp. Proc. Verb. Cons. Int. Expl. Mer* 42, 3–48.

Gachter, R., and Mares, A. 1979. Comments to the acidification and bubbling method for determining phytoplankton production. *Oikos* 33, 69–73.

Gagne, J. A., Larochelle, J., and Cardinal, A. 1979. A solubilization technique to prepare algal tissue for liquid scintillation counting, with reference to *Fucus vesiculosus* L. *Phycologia* 18, 168–70.

Guterstam, B. 1976. An *in situ* study of the primary production and the metabolism of a Baltic *Fucus vesiculosus* L. community. In Keegan, B. F., Ceidigh, P. O., and Boaden, P. J. S. (eds).; *Biology of Benthic Organisms*, pp. 311–19. Pergamon Press, New York.

Harris, G. P. 1978. Photosynthesis, productivity and growth: the physiological ecology of phytoplankton. *Arch. Hydrobiol. Beih. Ergebn. Limnol.* 10, 1–171.

Hatcher, B. G. 1977. An apparatus for measuring photosynthesis and respiration of intact large marine algae and comparison of results with those from experiments with tissue segments. *Mar. Biol.* 43, 381–5.

Heine, J. N. 1983. Seasonal productivity of two red algae in a central California kelp forest. *J. Phycol.* 19, 146–52.

Hitchcock, G. L. 1983. Photosynthate partitioning in cultured marine phytoplankton. 1. Dinoflagellates. *J. Exp. Mar. Biol. Ecol.* 69, 21–36.

Hoffman, W. E., and Dawes, C. J. 1980. Photosynthetic rates and primary production by two Florida benthic red algal species from a salt marsh and a mangrove community. *Bull. Mar. Sci.* 30, 358–64.

Howard, R. J., Gayler, K. R., and Grant, B. R. 1975. Products of photosynthesis in *Caulerpa simpliciuscula* (Chlorophyceae). *J. Phycol.* 11, 463–71.

Iverson, R. L., Bittaker, H. F., and Myers, V. B. 1976. Loss of radiocarbon in direct use of Aquasol for liquid scintillation counting of solutions containing ^{14}C-NaHCO$_3$. *Limnol. Oceanogr.* 21, 756–8.

Johnston, C. S., Jones, R. G., and Hunt, R. D. 1977. A seasonal carbon budget for a laminarian population in a Scottish sea-loch. *Helgo. Meeres.* 30, 527–45.

Khailov, K. M., and Burlakova, Z. P. 1969. Release of dissolved organic matter by marine seaweeds and distribution of their total organic production to inshore communities. *Limnol. Oceanogr.* 14, 521–7.

Kirst, G. O. 1981. Photosynthesis and respiration of *Griffithsia monilis* (Rhodophyceae): effects of light, salinity, and oxygen. *Planta* 151, 281–8.

Kremer, B. P. 1978. Determination of photosynthetic rates and ^{14}C photoassimilatory products of brown seaweeds. In Hellebust, J. A., and Craigie, J. S. (eds.), *Handbook of Phycological Methods: Physiological and Biochemical Methods*, pp. 269–83. Cambridge University Press, Cambridge.

Kremer, B. P. 1979. Light independent carbon fixation by marine macroalgae. *J. Phycol.* 15, 244–7.

Kremer, B. P. 1981a. Aspects of carbon metabolism in marine macroalgae. *Oceanogr. Mar. Biol. Annu. Rev.* 9, 41–94.

Kremer, B. P. 1981b. Metabolic implications of nonphotosynthetic carbon fixation in brown algae. *Phycologia* 20, 242–50.

Kremer, B. P., and Markham, J. W. 1979. Carbon assimilation by different developmental stages of *Laminaria saccharina*. *Planta* 144, 497–501.

Kuppers, U., and Kremer, B. P. 1978. Longitudinal profiles of carbon dioxide fixation capacities in marine macroalgae. *Plant Physiol.* 62, 49–53.

Littler, M. M. 1973. The productivity of Hawaiian fringing-reef crustose Corallinaceae and an experimental evaluation of production methodology. *Limnol. Oceanogr.* 18, 946–52.

Littler, M. M., and Arnold, K. E. 1980. Sources of variability in macroalgal primary productivity: sampling and interpretative problems. *Aquat. Bot.* 8, 141–56.

Littler, M. M., and Arnold, K. E. 1982. Primary productivity of marine macroalgal functional-form groups from southwestern North America. *J. Phycol.* 18, 307–11.

Lobban, C. S. 1974. A simple, rapid method of solubilizing algal tissue for scintillation counting. *Limnol. Oceanogr.* 19, 356–9.

Lobban, C. S. 1978. Translocation of ^{14}C in *Macrocystis pyrifera* (giant kelp). *Plant Physiol.* 61, 585–9.

Marker, A. F. H. 1976. The benthic algae of some streams in southern England. 11. The primary production of the epilithon in a small chalk-stream. *J. Ecol.* 64, 359–73.

Neame, K. D., and Homewood, C. A. 1974. *Liquid Scintillation Counting.* Wiley, New York. 180 pp.

Peng, C. T. 1981. *Sample Preparation in Liquid Scintillation Counting.* Amersham Corporation, Arlington Ill. 112 pp.

Penhale, P. A., and Smith, W. O. 1977. Excretion of dissolved organic carbon by eelgrass (*Zostera marina*) and its epiphytes. *Limnol. Oceanogr.* 22, 400–7.

Peterson, B. J. 1980. Aquatic primary productivity and the ^{14}C-CO_2 method: a history of the productivity problem. *Annu. Rev. Ecol. Syst.* 11, 359–85.

Ramus, J. 1978. Seaweed anatomy and photosynthetic performance: the ecological significance of light guides, heterogenous absorption and multiple scatter. *J. Phycol.* 14, 352–62.

Ramus, J. 1981. The capture and transduction of light energy. In Lobban, C. S., and Wynne, M. J. (eds.), *The Biology of Seaweeds,* pp. 458–92. University of California Press, Berkeley.

Rogers, C. S., and Salesky, N. H. 1981. Productivity of *Acropora palmata* (Lamarck), macroscopic algae, and algal turf from Tague Bay Reef, St. Croix, U.S. Virgin Islands. *J. Exp. Mar. Biol. Ecol.* 49, 179–87.

Ryther, J. H. 1956. The measurement of primary production. *Limnol. Oceanogr.* 1, 72–84.

Smith, S. V., and Kinsey, D. W. 1978. Calcification and organic carbon metabolism as indicated by carbon dioxide. In Stoddart, D. R., and Johannes, R. E. (eds.), *Coral Reefs: Research Methods,* pp. 469–84. UNESCO, Paris.

Smith, W. O. 1975. The optimal procedures for the measurement of phytoplankton excretion. *Mar. Sci. Commun.* 1, 395–405.

Smith, W. O. 1981. Photosynthesis and productivity of benthic macroalgae on the North Carolina continental shelf. *Bot. Mar.* 24, 279–84.

Steemann-Nielsen, E. 1952. The use of radioactive carbon (^{14}C) for measuring organic production in the sea. *J. Cons. Perm. Int. Expl. Mer* 18, 117–40.

Steemann-Nielsen, E., and Hansen, V. K. 1959. Measurements with the carbon-14 technique of the respiration rates in natural populations of phytoplankton. *Deep-Sea Res.* 5, 222–33.

Strickland, J. D. H., and Parsons, T. R. 1972. *A Practical Handbook of Seawater Analysis.* Bulletin 167, Fisheries Research Board of Canada, Ottawa. 310 pp.

Thomas, W. H. 1963. Physiological factors affecting the interpretation of phytoplankton production measurements. In Doty, M. S. (ed.), *Proceedings of the Conference on Primary Productivity Measurements, Marine and Freshwater,* pp. 147–62. U.S. Atomic Energy Commission, Washington, D.C.

Towle, D. W., and Pearse, J. S. 1973. Production of the giant kelp *Macrocystis,* estimated by *in situ* incorporation of ^{14}C in polyethylene bags. *Limnol. Oceanogr.* 18, 155–9.

Waite, D. T., Duthie, H. C., and Matthews, J. R. 1973. A note on two liquid scintillation fluors useful for primary production work. *Hydrobiologia* 43, 231–4.

Wassman, E. R., and Ramus, J. 1973. Primary production measurements for the green seaweed *Codium fragile* in Long Island Sound. *Mar. Biol.* 21, 289–97.

Wessels, C., and Birnbaum, E. 1979. An improved apparatus for use with the ^{14}C acid-bubbling method of measuring primary production. *Limnol. Oceanogr.* 24, 187–8.

Wheeler, W. N. 1980. Pigment content and photosynthetic rate of the fronds of *Macrocystis pyrifera. Mar. Biol.* 59, 97–102.

Willenbrink, J., Kremer, B. P., Schmitz, K., and Srivastava, L. M. 1979. Photosynthetic and light carbon fixation in *Macrocystis, Nereocystis,* and some selected Pacific Laminariales. *Can. J. Bot.* 57, 890–7.

Williams, P. J. L., Berman, T., and Holm-Hansen, O. 1972. Potential sources of error in the measurement of low rates of planktonic photosynthesis and excretion. *Nature* 236, 91–2.

19: Measurement of photosynthesis by infrared gas analysis

JOHN A. BROWSE

Plant Physiology Division, Department of Scientific and Industrial Research, Palmerston North, New Zealand

CONTENTS

I. Introduction

Infrared gas analyzers (IRGAs) are a convenient means of continuously measuring CO_2 concentration in the gas phase. Because of their sensitivity, reliability, and widespread commercial production they are by far the most widely used instruments for measuring net photosynthesis in terrestrial plants. The sensitivity of the IRGA is an advantage in studies of submerged aquatic plants also, because the relatively large changes in plant environment required for an estimate of photosynthesis using an oxygen electrode are no longer necessary, and open-circuit experiments (in which the plant is kept in steady state) can be carried out conveniently.

In this chapter, I first review field techniques used on terrestrial plants and the adaptation of these to the measurement of gas exchange of intertidal macroalgae during aerial exposure. Then I consider the more complex problem of measuring photosynthesis by submerged plants. This requires the adoption of techniques that ensure that the gas and water phases are in equilibrium (i.e., $[CO_2(g)]$ is a measure of $[CO_2(aq)]$) and that overcome the problems associated with the pH-dependent equilibrium between $[CO_2(aq)]$ and other dissolved inorganic carbon species:

$$CO_2 + H_2O \rightleftharpoons H_2CO_3 \qquad (1)$$

$$H_2CO_3 \rightleftharpoons HCO_3^- + H^+ \qquad (2)$$

$$CO_2 + OH^- \rightleftharpoons HCO_3^- \qquad (3)$$

$$HCO_3^- \rightleftharpoons CO_3^{2-} + H^+ \qquad (4)$$

Because of the sophisticated equipment required and the type of data obtained, the use of IRGA techniques in phycological research has been confined largely to laboratory studies. I describe some of the circuit arrangements and systems that have been used and discuss the suitability of these systems for adaptation to use in the field.

[398]

II. Equipment

A. Infrared gas analyzer

Detailed information on the principle and operation of IRGAs can be found in Sestak et al. (1971). A large variety of models are currently manufactured with the specifications of individual instruments adjusted to suit the purchaser's requirements. Users are advised to consider carefully the features they desire before ordering an analyzer.

1. General features. Analyzers vary mainly in the type and arrangement of the absorption tubes on the optical bench. The simplest analyzers have a sealed reference cell containing CO_2-free air. They can measure only the absolute concentration of CO_2 in the sample gas stream.

With a flow-through reference cell, absolute CO_2 concentrations can be measured when the reference cell is purged with CO_2-free air, or differential measurements are possible relative to a CO_2-containing gas mixture (normally atmospheric air) passed through the reference cell.

The measuring range obtained at a given amplifier gain depends on the length of the absorption path, and longer cells are required for differential measurements than for absolute measurements. For this reason the most versatile format for an analyzer is an optical bench incorporating a number of flow-through cells of different lengths. This allows absolute or differential measurements to be made by using appropriate cells and purging the rest with CO_2-free air.

Many analyzers incorporate features that simplify calibration in the differential mode with either an optical shutter (Mahon and Domey 1979) or a divided absorption tube (Parkinson and Legg 1971). Regular calibration in these instruments is considerably easier than it is with gas mixing pumps and more accurate than calibration achieved by the use of two standard gas cylinders.

For studies on macroalgae, it is particularly important to have an analyzer with a low sensitivity to water vapor, because it is generally not convenient to remove water from the gas streams. Sensitivity to water vapor is reduced in some analyzers by the use of optical interference filters to remove radiation of wavelengths at which the absorption spectra of CO_2 and water overlap or by the use of a filter cell filled with water vapor. In any case, unless calibration and measurements are made with similar water vapor concentrations, some correction will be necessary (Parkinson 1971).

2. Additional features of instruments used in the field. The majority of IRGAs in production are designed as laboratory instruments and are very poorly suited for field use. Compact dimensions, light weight, and the capability to be run from batteries are obvious advantages in a field instrument. Far more importantly, the analyzer chosen should be designed and constructed to reduce the sensitivity to vibration and to environmental temperature fluctuations typically found in IRGAs intended for laboratory use.

3. Mobile laboratory or portable system. Enclosing the plant material in an assimilation chamber during aerial studies results in temperature, humidity, and air flow that are different from ambient conditions. The air conditioning required to minimize these changes and the need for a stable environment and power supply for the IRGA led to the development in the 1960s and 1970s of mobile laboratories for the measurement of gas exchange in the field (e.g., Mooney et al. 1971; Sinclair et al. 1979). Such systems have been used with minimum modification to measure photosynthesis by marine ma-croalgae under emergent conditions (Johnson et al. 1974), and similar measurements have been made in base laboratories (Brinkhuis et al. 1976; Quadir et al. 1979; Hodgson 1981).

The trend in terrestrial plant physiology has been toward fully portable gas exchange systems (Mahon and Domey 1979; Griffiths and Jarvis 1981; McPherson et al. 1983). In these, the plant material is enclosed only briefly in a clamp-on chamber so that air conditioning is no longer required. Considerably more measurements can be taken (albeit with lower precision), and the results may therefore be more representative of the population being studied. In the future, portable systems will be used for studies on algae because of their ease of operation and because they can be used at inaccessible intertidal sites.

4. Suitable models. Many manufacturers produce analyzers that can be used in mobile or base laboratories. The most popular include the ADC series 225 (Analytical Development Co.) and the Beckman 315 (Beckman Instruments). However, if a fully portable system is to be built, either the AR 401 (Anarad Inc.) or the BINOS 1 (Leybold-Heraeus GMBH) should be considered, since both have been incor-porated into systems for measuring CO_2 exchange by terrestrial plants in the field (Mahon and Domey 1979; McPherson et al. 1983).

B. Assimilation chamber

A variety of chambers suitable for measurements on terrestrial plants and exposed algae are described in Sestak et al. (1971; see also references in Sec. II.A.3). Chamber conditions should be defined and as close as possible to those in the natural environment. For an exposed algal thallus, this will normally mean that illumination and contact with the air are restricted to the upper side. Thallus temperature should be monitored by a thermocouple and controlled by rapid air flow within the chamber (to reduce the boundary layer) and air conditioning if necessary. Measurement and control of water vapor pressure and possibly oxygen concentration may also be necessary.

For studies on submerged plants, quite different features are required (Browse 1979). To ensure rapid equilibration of the gas and water phases, one should divide the gas stream entering the chamber into fine bubbles by passing it through a sintered glass or perforated stainless steel disk forming the chamber floor. Under these conditions water flow in the chamber will be rapid, and (because of diffraction and reflection at bubble surfaces) the light regime fairly diffuse. Light intensity is most easily measured at the inside face of the chamber, but measurements taken at both frond surfaces provide a more precise description of conditions. Light should be measured as a photon flux density (in micromoles photons per square meter per second) for wavelengths from 400 to 700 nm with a quantum sensor (e.g., LI 190S Lambda Instruments; see Ramus, Chap. 2).

In studies on submerged algae, photosynthesis or respiration measurements should be made at constant temperature not only for physiological reasons but because the dissociation constants of carbonic acid are temperature dependent. Especially at high bicarbonate concentrations, relatively small variations in temperature ($\sim 1.0°C$) may alter the chemical equilibrium and cause uptake or release of CO_2 by the medium. The chamber should therefore have some means of temperature control such as a water jacket fed by seawater or from a thermostatically controlled bath.

As outlined later (Sec. III.B), pH can be used in calculating the rate of photosynthesis or respiration. If it is, the pH electrode and meter should be of high accuracy and repeatability (both ± 0.01 pH unit or better; see Littler and Arnold, Chap. 17) since the calculation of total inorganic carbon is extremely sensitive to errors in pH.

An injection port for removing water samples and a means of holding the plant material in a fixed orientation may also be desirable.

C. Components of the gas circuit

Tubing in the gas circuit should be of a material with a low permeability to CO_2. Rubber tubing should never be used, but medium-wall plastic tubing is often satisfactory. Nylon, glass, or stainless steel are preferable when diffusion and adsorption are

potential problems. Diaphragm pumps are normally used because
they are gas tight. These are available from IRGA manufacturers.
Leyhold-Heraeus, manufacturers of the BINOS IRGA, supply a
small battery-operated vane pump that is convenient for field applications.

Open-circuit systems also require wash bottles to bring the incoming
gas stream to water saturation and at least one accurate flow meter.
An electronic (mass) flow meter may be situated before the assimilation chamber; this minimizes error caused by any leaks in the
system. More convenient rotameter-type meters should be situated
at the end of the gas line, because they are affected by backpressure.

D. Calibration gases

Analyzers used in the absolute mode require one calibration gas of
accurately known CO_2 concentration. Use of a cascading series of
gas mixing pumps (Sestak et al. 1971) to produce the calibration gas
is the method of choice, but accurately calibrated gas cylinders are
satisfactory. Unless the analyzer incorporates features to simplify
differential calibration (Sec. II.A), the same methods must be used
to provide the two CO_2 concentrations required for differential
calibration. In this case (because the analyzer scale will represent a
small difference between two high CO_2 concentrations), the absolute
CO_2 concentrations in the calibration mixtures must be known to
$\pm 0.1\%$ to avoid large errors (Sestak et al. 1971).

III. Methods

A. Studies on algae in air

IRGAs have been used to measure terrestrial plant photosynthesis
in closed-, semiclosed-, or open-circuit systems. The principle and
advantages of each of these configurations are discussed in Sestak et
al. (1971). Techniques used with terrestrial plants are easily adapted
for measurements on exposed macroalgae (Johnson et al. 1974;
Brinkhuis et al. 1976; Hodgson 1981). Because semiclosed systems
are seldom used, only the closed and open types are discussed here.

Closed-circuit experiments simply require the air to be circulated
between the assimilation chamber and IRGA. The analyzer is calibrated in the absolute mode according to the manufacturer's instructions; with no air entering or leaving the circuit, the change in CO_2
concentration is a measure of plant metabolism. The photosynthesis
or respiration rate is calculated from

$$F = V \, \Delta C / tA \qquad (5)$$

where F is the rate of CO_2 uptake or release (cubic centimeters per square centimeters per second, $cm^3 \cdot cm^{-2} \cdot s^{-1}$), $\Delta C/t$ is the rate of change in CO_2 concentration (cubic centimeters per cubic centimeter per second, $cm^3 \cdot cm^{-3} \cdot s^{-1}$), V is the volume of the circuit (cubic centimeters), and A is the leaf area (one or both sides, square centimeters). Disadvantages of the closed circuit include the fact that the plant is not in steady state and that the circuit must be essentially leak free. Any minor leakage can be calculated after the circuit is filled with CO_2-free air, and the measured gas exchange rates should be corrected accordingly.

In open systems, air of known CO_2 and water concentration is continuously passed through the assimilation chamber. The gas may be supplied from compressed cylinders, gas mixing pumps, or atmospheric air. If atmospheric air is used, it should always be from a remote source (human breath contains 4% CO_2). The air will show large short-term fluctuations in CO_2 concentrations, and these should be damped by passing the air through a container with a capacity of at least 20 liters and preferably 200 liters. It is an advantage to mix the air in this container with an electric fan.

The IRGA is used in the differential mode and measures the difference in CO_2 concentration between a sample of the incoming airstream in the reference cell and air leaving the chamber through the analysis cell. In this case,

$$F = J \, \Delta C/A \qquad (6)$$

where J is the flow rate through the assimilation chamber (cubic centimeters per second) and ΔC is the difference in CO_2 concentration across the assimilation chamber (cubic centimeters per cubic centimeter).

B. Studies with submerged algae

Complications introduced by phase and chemical equilibria require the use of modified procedures in the measurement of photosynthesis and respiration by submerged algae. Both closed- and open-circuit systems have been used for submerged plants, and in both, CO_2 equilibrium between the water and gas phases may be more easily maintained by recirculating the gas rapidly through the assimilation chamber via a closed loop in the gas circuit (Fig. 19–1). Apart from the modified assimilation chamber, the gas circuits are similar to those used for aerial studies. A filter should be included to prevent water droplets from entering the IRGA or flow meter. In open-circuit experiments the incoming air for both reference and analysis streams is saturated with water vapor by bubbling of the air through distilled water adjusted to pH 4 to 4.5 with HCl.

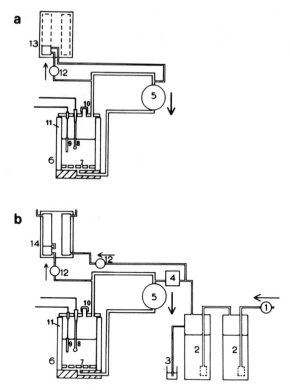

Fig. 19–1. IRGA circuit configurations for (a) closed- and (b) open-circuit experiments. (1) Cylinder or pump supplying incoming gas mixture; (2) wash bottles of distilled water, pH 4.5; (3) vent to atmosphere; (4) mass flow meter; (5) large-capacity pump for the closed loop; (6) assimilation chamber with (7) scintered glass floor; (8) pH electrode; (9) thermistor probe; (10) injection port; (11) water jacket; (12) sampling pump; (13) IRGA in absolute mode; (14) IRGA in differential mode. Arrows show the direction of gas flow through the pumps.

Calibration of the IRGA is most simply carried out using gases saturated with water vapor at the temperature of the assimilation chamber. The temperature of the analyzer must be higher than this to prevent water from condensing in the absorption tubes. For maximum accuracy, the pH meter is calibrated by means of a two-buffer technique.

The gas and water phases are equilibrated (constant $[CO_2(g)]$ in a closed-circuit; zero CO_2 differential in an open circuit) before the plant material is placed in the chamber, and the plant material should be exposed to the experimental conditions either in the assimilation chamber or in a separate container for at least 10 min before measurements of photosynthesis are started.

1. Closed-circuit techniques. The closed system described here incorporates a water-jacketed assimilation chamber (Browse 1979) containing 170 ml of water and about 0.8 g fresh weight (0.9 mg chlorophyll *a*; 95 cm^2 leaf surface area) of plant material. The gas stream is cycled through the chamber at 4 liters min^{-1} and through the IRGA at 0.8 liter min^{-1}. The volume of the gas phase is 450 ml.

Once the plant material is added, the circuit closed and the pumps started, inorganic carbon abstraction by the plants will cause the aqueous and (because they are in equilibrium) the gaseous CO_2 concentrations to fall. However, the IRGA will underestimate photosynthesis whenever bicarbonate is present, because CO_2 will be generated from bicarbonate as the dissolved inorganic carbon pool shifts toward a new equilibrium. The problem is to estimate the change in the total inorganic carbon content of the system:

$$C_T = [CO_2(g)]V_g + [TIC]V_1 \tag{7}$$

where V_g and V_1 are the volumes of the gas and water phases, respectively, $[CO_2(g)]$ is the gas-phase CO_2 concentration, and $[TIC]$ is the total inorganic carbon concentration in the water.

Calculation of $[TIC]$ requires the solution of the equilibrium equations (1–4). If $CO_2(aq)$ is taken as the sum of dissolved CO_2 and undissociated H_2CO_3, then the dissolved inorganic carbon pool can be described in terms of two equilibrium constants,

$$K_1 = \frac{[H^+][HCO_3^-]}{[CO_2(aq)]} \tag{8}$$

$$K_2 = \frac{[K^+][CO_3^{2-}]}{[HCO_3^-]} \tag{9}$$

and a concentration condition,

$$[TIC] = [CO_2(aq)] + [HCO_3^-] + [CO_3^{2-}] \tag{10}$$

Substituting Equations 8 and 9 in Equation 10 gives

$$[TIC] = [CO_2(aq)] \left(1 + \frac{K_1}{[H^+]} + \frac{K_1K_2}{[H^+]^2}\right)$$

$$= K_H[CO_2(g)] \left(1 + \frac{K_1}{[H^+]} + \frac{K_1K_2}{[H^+]^2}\right) \tag{11}$$

where K_H is Henry's law constant for CO_2. A more complete discussion and values of K_1, K_2, and K_H for fresh and salt water can be found in Riley and Skirrow (1965) and Stumm and Morgan (1970).

Experiments can be performed in one of three ways:

1. Experiments can be carried out at 1.5 to 2 pH units below pK_1 (Brown and Tregunna 1967). Bicarbonate and carbonate are then essentially absent ([TIC] = [CO_2(aq)] = K_H[CO_2(g)]). Under these conditions closed-circuit IRGA techniques are simple to use and show maximum sensitivity. However, the absence of bicarbonate and the low pH will normally bear no relation to the conditions of inorganic carbon supply experienced by the plants in the field.

2. The pH can be held constant by the use of additional buffers. Under these conditions the expression

$$1 + \frac{K_1}{[H^+]} + \frac{K_1 K_2}{[H^+]^2}$$

becomes a constant, and calculation of [TIC] is simplified. Provided that only small changes in [CO_2(g)] are measured, the plant experiences a stable inorganic carbon supply during the experiment, since the relative proportions of CO_2, HCO_3^-, and CO_3^{2-} remain constant. Since the added buffer must be in excess of the bicarbonate present, the method is normally limited to pH values rather less than pK_1. In addition, the added buffer may produce artifacts by affecting the plants or by enhancing HCO_3^- conversion to CO_2 within the boundary layer surrounding the plants. Use of a pH stat instead of buffers would eliminate these problems, but such an application has not yet been reported. Within the vicinity of (and above) pK_1, the maintenance of pH within very narrow limits is critical since small changes will result in considerable uptake or release of CO_2 by the medium. The difficulty of meeting this criterion limits the usefulness of the technique.

3. If the pH is allowed to rise during the experiment, it can be monitored and [TIC] calculated using Equation 11. In this type of experiment a small initial change in [TIC] is accompanied by a large change in the ratio of [CO_2(aq)] to [HCO_3^-]. Estimation of [TIC] from pH and pCO_2 becomes prone to error at high pH. As a check, [TIC] is often measured in water samples removed from the assimilation chamber during the course of the experiment. As was pointed out in the first description of this technique (Tregunna and Thomas 1968), measurement of [TIC] by acid release will be in error if suspended but undissolved $CaCO_3$ is present. Therefore, the use of the two methods together is beneficial. If calcification occurs during the experiment, this process will also cause a reduction in [TIC]. When calcification (or solution) is suspected, the bathing medium should be checked for any change in total aklalinity and a correction

Table 19–1. *Halftime ($t_\frac{1}{2}$) with which inorganic carbon equilibrium is reestablished in an open-circuit IRGA system at different CO_2 and HCO_3^- concentrations*

Background $[CO_2(g)]$ ($cm^3 \cdot cm^{-3} \times 10^6$)	$[HCO_3^-]$ (M)	$t_\frac{1}{2}$ (min)
2320	10^{-2}	1.2
870	10^{-2}	2.8
	10^{-3}	0.83
280	10^{-2}	15
	10^{-3}	1.7
	10^{-4}	0.61
193	10^{-2}	25
	10^{-3}	4
	10^{-4}	0.81

Note: Flow rates were 4 liters $\cdot$ min^{-1} in the recirculating loop and 0.5 liter $\cdot$ min^{-1} net flow. Water volume was 130 ml and total circuit volume 470 ml.
Source: Browse (1979).

made to the calculated photosynthesis (or respiration) rate if necessary (Smith and Kinsey 1978).

When closed-circuit IRGA techniques are used, the sensitivity decreases linearly as the ratio of CO_2 to [TIC] increases. Whereas measurements at low pH are considerably more sensitive than measurements by oxygen techniques, as the pH rises a larger and larger change in [TIC] is required to produce a given change in $[CO_2(g)]$. Consequently, by pH 8.5 the closed system has no more sensitivity than is typical of an oxygen electrode.

2. *Open-circuit technique.* The preceding discussion shows how much more complicated closed-circuit experiments are with submerged than with exposed algae. In contrast, open-circuit experiments with submerged algae are relatively simple. Because the dissolved inorganic carbon system reaches a stable equilibrium, photosynthesis or respiration by the algae can be calculated from Equation 6. The result will be accurate, provided that the total alkalinity of the medium does not change during measurements. Either net acid or net base excretion by the plants (Thomas and Tregunna 1968) or calcification/ solution reactions will alter total alkalinity and cause an error in the calculated photosynthesis rate. If any of these processes is occurring, the pH of the bathing medium will change, and for this reason it is sensible to ensure that the pH remains constant during the period

of constant CO_2 differential. As with closed-circuit methods, corrections can be made if the shift in total alkalinity is measured (Smith and Kinsey 1978).

In contrast to that of the closed-circuit techniques, the sensitivity of the open-circuit method remains high at high pH and [TIC]. However, as pH and [TIC] rise, the open circuit becomes sensitive to minor perturbations of temperature (which alter the pK values) and pH. In addition, the time taken to reach a stable differential inorganic carbon equilibrium increases. The approach to equilibrium is approximately exponential, and the halftime ($t_{1/2}$) depends on the experimental conditions. A set of typical $t_{1/2}$ values is shown in Table 19–1. The response time can be reduced by decreasing the amount of water in the assimilation chamber and, to a lesser extent, by reducing the volume of the whole circuit. Increasing the flow rate in the recirculating loop will reduce the response time only when the phase equilibrium, rather than the chemical equilibrium, is limiting.

IV. Sample data and calculations

Figure 19–2 shows the changes in pH and gas-phase CO_2 concentration recorded in the closed-circuit IRGA system during light-saturated photosynthesis by the aquatic angiosperm *Egeria densa*. Water samples were removed at intervals and [TIC] estimated either from these or from pH and [CO_2(g)] using Equation 11. Equation 7 was used to produce the curve of C_T shown, and tangents drawn to this curve give the photosynthesis rate. The inorganic carbon environment of the plant changes during the course of the experiment, with free CO_2 decreasing proportionally much more than bicarbonate. In the example shown, photosynthesis rate is not a simple function of [TIC] or [CO_2] or [HCO_3^-] because both CO_2 and HCO_3^- are substrates for photosynthesis, but the plant responds more to an increase in [CO_2] than to an increase in [HCO_3^-].

Figure 19–3 shows three sections from a recorder trace of the IRGA output during open-circuit measurements. The system was in phase and chemical equilibrium at the start of each measurement. Entry of air into the circuit when the plant material was placed in the assimilation chamber caused a temporary positive differential. At the same reference [CO_2(g)], the time for the CO_2 differential to reach a constant value after introduction of the plant material increased with [TIC] of the medium. The flow rates and water and circuit volumes are recorded in Table 19–1. The chamber contained only 0.5 g fresh weight of plant material (0.54 mg chlorophyll *a*; 58 cm^2 leaf area), which shows the sensitivity of the technique. Because

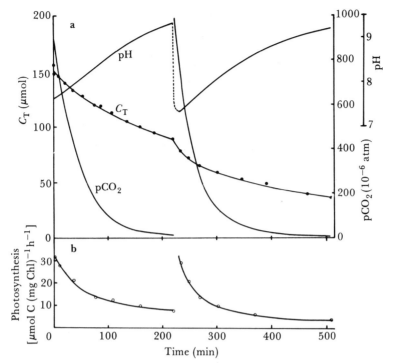

Fig. 19–2. Changes in a closed-circuit IRGA system during photosynthesis by *Egeria densa* (a). The rate of photosynthesis during the experiment shown in (a) was calculated from tangents to the C_T curve (b). At 220 min, ~0.4 ml of 0.1 M HCl was injected into the assimilation chamber (Chl, chlorophyll). From Browse et al. (1979).

CO_2 is held at nearly the same concentration, the experiment shown in Fig. 19–3 demonstrates the contribution of bicarbonate to photosynthesis by *Egeria*. In the absence of HCO_3^- (pH 4.5), the rate of photosynthesis is

$$F = \frac{8.33 \times 2.3 \times 10^{-6}}{58}$$

$$= 0.33 \times 10^{-6} \text{ cm}^3 \cdot \text{cm}^{-2} \cdot \text{s}^{-1}$$

and this increases to 1.3×10^{-6} and 3.2×10^{-6} cm$^3 \cdot$ cm$^{-2} \cdot$ s^{-1} at 0.8 and 8.2 mM HCO_3^-, respectively.

V. Critical evaluation

A. Studies on algae in air

Taking measurements with IRGAs has been the standard method of investigating terrestrial plant photosynthesis and respiration for more than 20 years. The continuous, nondestructive, and accurate mea-

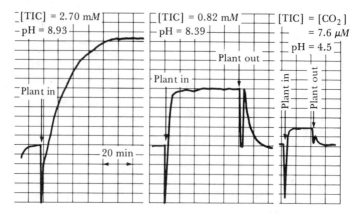

Fig. 19–3. Chart recorder tracings of IRGA output during measurements of photosynthesis by *Egeria densa* at the same background CO_2 concentration (inlet $[CO_2(g)] = 193 \times 10^{-6}$ cm^3·cm^{-3}) but different [TIC]. From Browse (1979).

surement of CO_2 concentration that they provide, coupled with their reliability and ease of operation, makes them ideal laboratory instruments. Advances in instrument technology and design of gas circuits have greatly increased the usefulness of the technique in the field. In studies on exposed algae, detailed measurements of water vapor exchange are not required, since stomatal resistance is not a consideration. This means that less complex systems can be used, with simple measurements of frond water content (by weight) and ambient humidity being sufficient when gas exchange is measured during desiccation. If a measure of net photosynthesis or respiration is the only datum required, there are now several systems that allow one or two operators to make measurements at the rate of 20 or more per hour, each measurement on a separate piece of plant material.

B. Studies with submerged algae

The main advantage of IRGA methods in studies of submerged algae is that one of the carbon species used as a substrate for photosynthesis is measured directly, and bicarbonate and carbonate concentrations can also be estimated quite simply. If these three species are considered potential substrates for photosynthesis, then at least four types of response have been recorded for different submerged species (e.g., Thomas and Tregunna 1968; Jolliffe and Tregunna 1970):

1. Only free CO_2 is used for photosynthesis; HCO_3^- and CO_3^{2-} have no effect.

2. Free CO_2 is used considerably more efficiently than HCO_3^-; CO_3^{2-} is inhibitory or has no effect.

3. Free CO_2, HCO_3^-, and possibly CO_3^{2-} are used with equal facility for photosynthesis.

4. Bicarbonate is used slightly more efficiently than free CO_2.

For this reason, it is obviously useful to know the concentrations of the individual carbon species rather than simply [TIC]. The inorganic carbon pool (and particularly the CO_2/HCO_3^- ratio) may change significantly during measurements of photosynthesis in closed containers, so the continuous record provided by the IRGA and pH meter offers a useful check of the conditions. IRGA methods give a direct estimate of carbon fixation and hence productivity. Oxygen techniques, in contrast, must asume a photosynthetic quotient, and this may give rise to errors (Burris 1981; Johnson et al. 1981; Kinsey, Chap. 21).

Open-circuit IRGA methods are highly sensitive. An additional advantage is that the water in the assimilation chamber is held constant in $[CO_2]$, $[HCO_3^-]$, and pH so that the effect of changes in other factors such as light can be monitored continuously, although rapid changes (times of the order of $t_{1/2}$ in Table 19–1) will be damped by the slow equilibration of the dissolved inorganic carbon system. As already noted, one can reduce $t_{1/2}$ by decreasing the water and gas volumes of the system.

The vigorous aeration of the medium required by IRGA methods induces water flows around the plants that may be much higher than those found in nature. Under these high flow rates, the stagnant boundary layer around the algae will be minimized. Since the boundary layer is often the major resistance to CO_2 uptake (Wheeler 1980), photosynthesis may be higher than under natural flow conditions.

Problems caused by $CaCO_3$ precipitation or solution and acid or base excretion by the plants require special precautions. It is often convenient to avoid situations in which these processes occur, but corrections can be made if necessary.

Adaptation of an IRGA for field measurements on submerged algae requires only slightly more equipment than is required for field measurements on exposed algae. However, because of the time required to reach chemical and phase equilibrium, only one to four measurements per hour are possible. For this reason, the use of IRGAs to measure productivity of submerged algae in the field is confined to situations of particular physiological interest rather than simple bioassays.

VI. Alternative techniques

Manometric and ^{14}C methods can be used to measure gas exchange by exposed algae (see Dawes, Chap. 16; Arnold and Littler, Chap. 18; Sestak et al. 1971). For submerged algae, oxygen, ^{14}C, pH, and manometric methods described in other chapters of this volume all provide alternatives to closed-circuit IRGA techniques. Steady-state open-circuit measurements can be performed by the use of duplicate oxygen determinations (e.g., Bottom 1981), but the methods are considerably less sensitive than those involving differential IRGA as described in this chapter.

VII. References

Bottom, D. L. 1981. A flow-through system for field measurements of production by marine macroalgae. *Mar. Biol.* 64, 251–7.

Brinkhuis, B. H., Tempel, N. R., and Jones, R. F. 1976. Photosynthesis and respiration in exposed salt-marsh fucoids. *Mar. Biol.* 34, 349–59.

Brown, D. L., and Tregunna, E. B. 1967. Inhibition of respiration during photosynthesis by some algae. *Can. J. Bot.* 45, 1135–43.

Browse, J. A. 1979. An open circuit infrared gas analysis system for measuring aquatic plant photosynthesis at physiological pH. *Aust. J. Plant Physiol.* 6, 493–8.

Browse, J. A., Dromgoole, F. I., and Brown, J. M. A. 1979. Photosynthesis in the aquatic macrophyte *Egeria densa*. 3. Gas exchange studies. *Aust. J. Plant Physiol.* 6, 499–512.

Burris, J. E. 1981. Effects of oxygen and inorganic carbon concentrations on the photosynthetic quotient of marine algae. *Mar. Biol.* 65, 215–9.

Griffiths, J. H., and Jarvis, P. G. 1981. A null balance carbon dioxide and water vapour porometer. *J. Exp. Bot.* 32, 1157–68.

Hodgson, L. M. 1981. Photosynthesis of the red algae *Gastroclonium coulteri* (Rhodophyta) in response to changes in temperature, light intensity, and desiccation. *J. Phycol.* 17, 37–42.

Johnson, K. M., Burney, C. M., and Sieburth, J. McN. 1981. Enigmatic marine ecosystem metabolism measured by direct diel CO_2 and O_2 flux in conjunction with DOC release and uptake. *Mar. Biol.* 65, 49–60.

Johnson, W. S., Gigon, A., Gulmon, S. L., and Mooney, H. A. 1974. Comparative photosynthetic capacities of intertidal algae under exposed and submerged conditions. *Ecology* 55, 450–3.

Jolliffe, E. A., and Tregunna, E. B. 1970. Studies on HCO_3^- ion uptake during photosynthesis in benthic marine algae. *Phycologia* 9, 293–303.

McPherson, H. G., Green, A. E., and Rollinson, P. L. 1983. The measurement within seconds of apparent photosynthetic rates using a portable instrument. *Photosynthetica* 17, 395–406.

Mahon, J. D., and Domey, J. 1979. A light-weight battery operated infrared gas analyzer for field measurements of photosynthetic CO_2 exchange. *Photosynthetica* 13, 459–66.

Mooney, H. A., Dunn, E. L., Harrison, A. T., Morrow, P. A., Bartholomew, B., and Hays, R. L. 1971. A mobile laboratory for gas exchange measurements. *Photosynthetica* 5, 128–32.

Parkinson, K. J. 1971. Carbon dioxide infra-red gas analysis. Effects of water vapour. *J. Exp. Bot.* 22, 169–76.

Parkinson, K. J., and Legg, B. J. 1971. A new method for calibrating infra red gas analysers. *J. Phys. E: Sci. Instr.* 4, 598–600.

Quadir, A., Harrison, P. J., and De Wreede, R. E. 1979. The effects of emergence and submergence on the photosynthesis and respiration of marine macrophytes. *Phycologia* 18, 83–8.

Riley, J. P., and Skirrow, G. (eds.). 1965. *Chemical Oceanography*, vol. 1. Academic Press, London. 712 pp.

Sestak, Z., Catsky, J., and Jarvis, P. G. (eds.). 1971. *Plant Photosynthetic Production: Manual of Methods.* Junk, The Hague. 818 pp.

Sinclair, T. R., Johnson, M. N., Drake, G. M., and Van Houtte, R. C. 1979. Mobile laboratory for continuous long-term gas exchange measurements of 39 leaves. *Photosynthetica* 13, 446–53.

Smith, S. V., and Kinsey, D. W. 1978. Calcification and organic carbon metabolism as indicated by carbon dioxide. In Stoddart, D. R., and Johannes, R. E. (eds.), *Coral Reef Research Methods*, pp. 469–84. UNESCO, Paris.

Stumm, W., and Morgan, J. J. 1970. *Aquatic Chemistry.* Wiley-Interscience, New York. 583 pp.

Thomas, E. A., and Tregunna, E. B. 1968. Bicarbonate ion assimilation in photosynthesis by *Sargassum muticum. Can. J. Bot.* 46, 411–15.

Tregunna, E. B., and Thomas, E. A. 1968. Measurement of inorganic carbon and photosynthesis in seawater by pCO_2 and pH analysis. *Can. J. Bot.* 46, 481–5.

Wheeler, W. N. 1980. Effect of boundary layer transport on the fixation of carbon by the giant kelp *Macrocystis pyrifera. Mar. Biol.* 56, 103–10.

20: Carbon allocation

RICHARD G. BUGGELN

*Marine Sciences Research Laboratory, Memorial University of Newfoundland,
St. John's, Newfoundland, Canada A1C 5S7*

CONTENTS

[415]

I. Introduction

In some algae, notably members of the Laminariales and Fucales, photosynthesis in the meristoderm does not support the metabolic carbon demand in the subtending cells of the cortex and medulla. Such regions of carbon deficit, so-called sinks, include meristematic zones of blades, stipes, sporophylls, and holdfasts. Portions of the thallus where CO_2 is photosynthesized in excess relative to the local carbon requirement are termed "sources." Products of photoassimilation, namely, mannitol, amino acids, and organic acids, are transported from source to sink regions through intercellular pathways in the cortex and medulla by mechanisms that are poorly understood (Moorby 1981). Qualitative evidence suggests that kelp fronds have multiple sources and multiple sinks, with interactions between sources supplying one particular sink as well as "competition" between sinks (Buggeln 1983). Quantitative aspects of photoassimilate transport involve the determination of the rate (mass) and the velocity (speed) of translocation. Described here are methods of exploring source–sink relationships and other translocation phenomena. For the most part, these techniques have been developed for genera within the Laminariales. Members of this order are large and convenient to manipulate. In addition, morphological divisions of labor within a frond present numerous "physiological handles" that tempt the investigator to formulate hypotheses related to the mechanism of photoassimilate transport within the thallus. Ecologists interested in carbon budgets of single thalli or entire populations of a species or in seasonal growth "strategies" may apply some of the techniques described here. Many of these techniques have been tailored to a specific alga in order to exploit particular morphological and/or anatomical features. Specific genera are therefore mentioned for each protocol with the knowledge that application of the method to other species may not be feasible.

General experimental format

When fed to a portion of a frond as $NaH^{14}CO_3$, $^{14}CO_2$ is incorporated into photoassimilates. Some of these ^{14}C-labeled solutes are transported from the pulsed site to various sinks within the frond during the subsequent "chase" period, which may be hours or days in duration. The progress of tracer movement between source and sink regions in the frond can often be monitored with a Geiger–Müller (GM) detector probe. At the conclusion of an experiment, the frond is dried, and radioautography can be used to locate regions accumulating the ^{14}C-labeled photoassimilate. Chemical analyses may characterize and quantify the distribution of ^{14}C in various organic constituents in source, sink, or loci along the transport pathway.

II. Methods

A. Incubation chambers

The design of the following chambers reflects the morphology of the alga under study and/or specific modifications made to the material for the experiments. Whole lamina of *Macrocystis* spp. have been enclosed in polyethylene bags (Lobban 1978; Schmitz and Srivastava 1979a). A plexiglass chamber was used to encase one end of a 200-mm-long dumbbell-shaped segment of an *Alaria esculenta* blade (Buggeln and Lucken 1979). Plexiglass chambers have been glued to the flat surface of *Laminaria* spp., *A. esculenta*, and *Saccorrhiza dermatodea* blades (Lüning et al. 1973; Buggeln 1981; Emerson et al. 1982; Buggeln and Varangu 1983) using the following tissue adhesives: Sicomet 800 (Lüning et al. 1973) or Histoacryl-N-blau, a cyanobutyl acrylate that polymerizes on contact with water (Trihawk International; see Buggeln 1976). For corrugated blades such as *Macrocystis* spp., the area around the site to be pulsed with ^{14}C is first coated with Histoacryl-N-blau, and a layer of petroleum jelly is applied over the glued area to fill in the irregular depressions. Then a cylindrical plexiglass chamber with a wide square base (having overall dimensions at least 1 cm per side greater than the diameter of the cylindrical chamber) can be seated firmly on the bed of petroleum jelly (R. G. Buggeln, unpublished data).

B. Tracer preparation and application

The [^{14}C]bicarbonate stock solution is made up in HA (0.45-μm-diameter) Millipore-filtered seawater. The incubation seawater medium should first be filtered as above and then purged with N_2 (15 min) to reduce the O_2 tension in the water before addition of the required amount of ^{14}C from the stock solution.

To maximize the incorporation of ^{14}C, vigorous stirring of the incubation medium is required; otherwise, boundary layers, depleted of tracer, form at the blade surface and tracer uptake is considerably reduced. Stirring can be achieved by connecting the chamber to a recirculating pump (Lüning et al. 1973; Buggeln and Lucken 1979). Unfortunately, the use of a pump requires a large volume of incubation medium and a correspondingly large amount of tracer. The volume of the incubation medium can be kept small, and a vigorous movement of the tracer over the blade surface (surging action) can be achieved by the use of a syringe to draw up the medium and then vigorously expel it in rapid succession throughout the incubation period.

At the end of the ^{14}C pulse, the tracer is withdrawn and the chamber and tissue are thoroughly rinsed with seawater. It should be noted that unfixed [^{14}C]bicarbonate, presumably in the intercellular spaces of the meristoderm layers (the apparent free space), continues to exchange with [^{12}C]bicarbonate in the seawater mileu for 6 to 12 h after the pulse (R. G. Buggeln, E. G. Young, and L. K. Varangu, unpublished data). Thus, the frond should be positioned so that the water flow over the ^{14}C-pulsed locus is directed away from any sink regions. If the pulsed area remains enclosed by a chamber, a flow of nonlabeled seawater (open system) is continuously fed over the area.

C. Monitoring ^{14}C-photoassimilate translocation in vivo

In kelp blades, the ability to detect ^{14}C-photoassimilate movement between the ^{14}C-pulsed source and a distant sink is feasible in laminae with a thin cortex, for example, *Alaria* (Buggeln and Varangu 1983); however, success was also obtained in blades of *Saccrorhiza dermatodea*, which have a thick cortex (Emerson et al. 1982). In other thick-bladed lamina (e.g., *Laminaria digitata*), it is difficult to detect ^{14}C en route, although the arrival of ^{14}C-photoassimilates in the meristem at the blade base can be monitored (Buggeln 1981).

To follow the movement of labeled materials, a ^{14}C-pulsed blade is quickly and gently blotted and covered with a piece of aluminum foil in which one or more slits (about 1 × 4 cm) or holes (~2–4 cm in diameter) have been cut. The foil is positioned with the hole(s) over the putative route of ^{14}C-solute transport and repositioned in the identical location each time a measurement is to be made. A GM detector probe (e.g., probe model P-11, Technical Associates) is placed over a window in the foil, and the radioactivity in the blade is recorded on a scaler (e.g., model FS-8W, Technical Associates) in several short counting periods (e.g., three 30-s counts) and averaged. Measurements should be made as rapidly as possible to avoid excessive drying of the blade.

1. Determination of the GM probe counting efficiency. It is assumed – but has not yet been experimentally verified – that in the course of a 5-

to 7-d chase period, the bulk of the ^{14}C-labeled photoassimilate remains in the medulla of the kelp blade sink. We may therefore assume that the counting efficiency of the GM probe does not change during this time. A few days after the pulsing of a kelp blade with [^{14}C]bicarbonate, disks of tissue (~1.5 cm in diameter) are punched from the blade sink. Radioactivity in each disk is determined with a GM probe, and then the disks are individually digested in scintillation vials by the method of Lobban (1974). A secondary ^{14}CO$_2$ absorbing trap (a scintillation vial containing a 1-cm-diameter phenylethyl-amine-soaked paper disk) is attached with glass tubing to the scintillation vial containing the kelp tissue and the HClO$_4$/H$_2$O$_2$ digestion mixture (Lobban 1974). This ensures complete recovery of ^{14}CO$_2$ released during the digestion, which ordinarily escapes when the scintillation vial is opened for the addition of scintillation cocktail (R. G. Buggeln and L. K. Varangu, unpublished data). The radioactivity in disintegrations per minute (dpm) is determined in both vials by liquid scintillation spectroscopy and summed. The counts per minute (cpm) in the disk, as measured by the GM probe, are calculated as a percentage of the disintegrations per minute in the digested tissue. The resulting counting efficiency of the GM probe (typically <1%) will depend on cortex thickness.

2. Calculation of carbon equivalent from ^{14}C detected along transport pathway. The counts per minute detected in a window in the aluminum foil located over part of the source–sink transport pathway or over part of the kelp blade sink can be converted to a carbon equivalent by use of the specific activity of the [^{14}C]bicarbonate–seawater incubation medium. In the following example, the specific activity of the ^{14}C stock solution is 0.23×10^{-1} μCi per millimolar CO$_2$ with a titer of 100 μCi ^{14}C and 4.3×10^{-6} M CO$_2$ per milliliter of stock seawater solution. One milliliter of stock solution was used in the ^{14}C pulse. The GM probe, with a counting efficiency of 0.1% (1 μCi = 37,000 dps), detects 2500 cpm at some point along the transport pathway.

$$\frac{2.5 \times 10^3 \text{ cpm}}{0.001} = 2.5 \times 10^6 \text{ cpm}$$

$$\text{or} \quad 1.13 \text{ μCi} \quad (1)$$

$$\frac{1.3 \text{ μCi} \times 4.3 \times 10^{-6} \ M \ \text{CO}_2}{100 \text{ μCi}} = 4.8 \times 10^{-8} \ M \text{ carbon}$$

$$\text{or} \quad 4.8 \times 10^{-2} \ \mu M \text{ carbon} \quad (2)$$

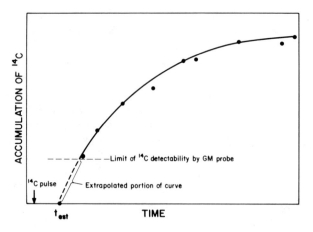

Fig. 20–1. Hypothetical accumulation of ¹⁴C at a locus in a kelp blade. Extrapolated portion of curve (– – –) intersects the abscissa at t_{est}, the estimated time of arrival of the front of the ¹⁴C-labeled pool of photoassimilates.

D. Translocation velocity

The velocity, or speed, of photoassimilate translocation can be determined by a method applicable to all kelp blades (Buggeln 1981) or by a special technique that is uniquely suited to *Macrocystis* stipes (Schmitz and Srivastava 1979b). In the more general method, [¹⁴C]bicarbonate pulse is fed to the distal, nongrowing portion of a kelp blade, and a GM probe monitors the arrival of ¹⁴C-solutes in the blade meristem (as described in Sec. II.C). The arrival data are plotted as a function of time (hours) from the beginning of the ¹⁴C pulse and extrapolated to intersect with the abscissa (time axis) of the graph (Fig. 20–1). The intersection of the extrapolated graph and abscissa gives the estimated time t_{est} of arrival of the first ¹⁴C-labeled solutes in the area under the GM probe. The distance (centimeters) between the pulsing chamber and the area where ¹⁴C accumulation is being measured is divided by the time (t_{est}) to give the estimated speed of the advancing tracer front.

In some cases, the velocity of ¹⁴C-solute movement may be determined at intermediate positions along the transport pathway by means of an aluminum foil shield with a linear sequence of holes or slits, as described in Sec. II.C (Emerson et al. 1982; Buggeln and Varangu 1983).

E. Rate of translocation

The width of the medullary pathway (sieve cells) translocating ¹⁴C-photoassimilate between the pulsed site on the blade and the blade sink is equal to the width (or diameter) of the pulsing chamber itself (Buggeln 1983; Buggeln and Varangu 1983). The rate or mass of carbon delivered to the sink should be computed on the cross-sectional area of the *actively conducting sieve cells* (i.e., milligrams carbon per hour and square millimeter of sieve cells). Such area estimates have been reported for *Alaria* and *Macrocystis*, but they are highly uncertain. At present, it is more realistic (and conservative) to use the cross-sectional area of the section of the medulla through which the ^{14}C-labeled solutes travel. For example, the cross-blade width of the pulsing chamber may be 20 mm, and the thickness of medulla determined microscopically from blade cross sections may be 1 mm, which gives a medullary transport area of 20 mm^2. Using a 2-d accumulation of carbon in a small area of a sink, say 0.10 μM carbon, we determine, the rate of translocation to be

$$0.10 \ \mu M \ \text{carbon} \cdot 2 \ \text{d}^{-1} \cdot 20 \ \text{mm}^{-2} \ \text{medulla}$$

or

$$2.5 \times 10^{-3} \ \mu M \ \text{carbon} \cdot \text{d}^{-1} \cdot \text{mm}^{-2} \ \text{medulla}$$

If we could compute the transport rate based on the actual cross section of the sieve cells with the ^{14}C-labeled solute, it would be much higher because the mass of carbon, 0.10 μM, would be divided by a smaller number, that is, the pooled diameters of the individual sieve cells. The rate of translocation calculated in this manner is called the *specific mass transport* (SMT; Canny 1973).

There are several more accurate ways to calculate the SMT, but they require quantitative information on several physiological variables that are difficult to measure in most kelp. In the first method (Canny 1973) the concentration of organic solutes in sieve cell exudate and the speed (velocity) of translocation are needed:

$$\text{SMT} = \text{concentration of solutes} \times \text{speed of translocation}$$

$$g \cdot h^{-1} \cdot cm^{-2} = g \cdot cm^{-3} \times cm \cdot h^{-1} \tag{3}$$

The second method involves the prolonged photoassimilation of ^{14}C into organic matter in order for solutes in the translocate pool to reach a constant specific activity. The specific activity of the ^{14}C in the incubation medium must be controlled and kept constant during the continuous incubation period (possibly lasting several hours). Labeling of the translocated solutes is monitored in the medullary exudate, and steady-state labeling of the solute pool is achieved when radioactivity in the exudate becomes constant. The plateau value (microcuries) can be multiplied by the constantly maintained specific activity of ^{14}C in the source incubation medium to obtain a direct measurement of the translocation rate of carbon (Geiger and Swanson 1965; Buggeln 1983). This technique may be feasible only with genera such as *Macrocystis* that exude at high rates (Schmitz and Srivastava 1979b), but a carefully controlled ^{14}C-labeling protocol has not yet been attempted with any alga.

F. Carbon exported from the source

How much photosynthetically fixed carbon is retained by the source and how much is exported to sinks in the thallus? For kinetic experiments of this kind, "destructive" techniques (i.e., chemical analyses) require the use of many plants to determine the fate(s) of photoassimilate during the chase period. The two methods presented here are based on different concepts of what constitutes a kelp blade source. The first method tacitly assumes that the meristoderm receiving the ^{14}C pulse, together with the underlying cortex and medulla, is part of the source. The second method is predicated on the proposal that the source is restricted to the photosynthetic meristoderm (see Buggeln 1983) with the underlying cortex as the primary, or short-distance, transport pathway and the medullary sieve cells as the secondary pathway for long-distance transport.

Method 1. A weighed sample of the ^{14}C-pulsed source is extracted three times in 80% methanol containing 1% HCl, and total alcohol-soluble ^{14}C in the pooled extracts is determined by scintillation spectroscopy. The alcohol-extracted residue is dried (60°C, 48 h), weighed, and then digested (as described in Sec. II.C) for total ^{14}C in insoluble matter or further treated to determine the ^{14}C content in protein, laminarin, and other substances (see Hellebust and Craigie 1978). The methanol extract is passed through a cation-exchange resin (H^+ form), which retains amino acids; mannitol and organic acids are washed through the column and collected. The amino acids are eluted with NH_4OH in ethanol. Mannitol and organic acids are separated on an anion-exchange resin (OH^- or formate form), in which the former compound passes through whereas the latter are retained and eluted with 1 N formic acid. The ^{14}C content of the amino acid, organic acid, and mannitol fractions can then be determined by liquid scintillation spectroscopy.

Method 2. In this method (R. G. Buggeln, E. G. Young, and L. K. Varangu, unpublished data), the area of the thallus that received the ^{14}C pulse is blotted dry and flattened between two pieces of Parafilm; a flat piece of dry ice is then firmly appressed to the blade–Parafilm sandwich. The frozen blade and the underlying Parafilm are transferred to a precooled sheet of galvanized metal (approximately 10

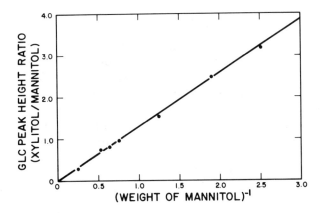

Fig. 20–2. Standard curve for determining mannitol concentrations in algal extracts using the xylitol/mannitol peak height ratio obtained by gas–liquid chromatography.

× 20 × 0.5 mm thick) resting on a slab of dry ice. The ^{14}C-pulsed meristoderm is very gently scraped off the thallus with a scalpel into a preweighed vessel. The scrapings are dried (60°C), weighed, and extracted three times with 80% methanol, and the pooled extracts are analyzed for distribution of ^{14}C in photoassimilates, as described earlier.

G. Mannitol determination: gas–liquid chromatography (peak height ratio method)

The mannitol fraction from Sec. II.F.1. plus 0.5 ml of xylitol (internal standard) in anhydrous, redistilled pyridine (20 mg xylitol/100 ml pyridine) is dried under an N_2 stream at 40°C. One milliliter of a silylating reagent (e.g., Tri-sil-Z, Pierce Chemical Co.) is added to the dried residue and heated for a few minutes on a water bath (70°C) or until the residue is dissolved. Gas–liquid chromatographic analysis of 3 μl of silylated mannitol/xylitol is carried out under the following conditions: 2 m × 5 mm (i.d.) column packed with 3% Silar-5-CP on Gaschrom Q (80–100 mesh); 154°C isothermal; retention times: xylitol, ~11–13 min; mannitol, ~34–37 min. The quantity of mannitol in an extract is obtained from the standard curve of xylitol/mannitol peak height ratios (ordinate, Fig. 20–2) plotted against the reciprocal of mannitol concentration (abscissa, Fig. 20–2). To derive a standard curve, a series of mannitol/xylitol mixtures is prepared in which mannitol concentration ranges from 1.6×10^2 to 1.6×10^3 mg · liter^{-1}, whereas the concentration of xylitol, the internal standard, is held constant (e.g., 8.0×10^2 mg · liter^{-1}).

H. Radioautography: whole thallus

At the end of a ^{14}C pulse–chase experiment, the experimental material is spread out and dried in air on a piece of herbarium paper. In the case of wide kelp blades, shrinking during drying can be reduced by placement of weights along the blade edges. The dry frond is mounted under a piece of No-Screen Medical x-ray film (NA-45T, Eastman Kodak Co.). The duration of exposure of the film to the frond (in darkness) must be determined empirically, but may typically require anywhere from 18 h to 10 d, depending on the level of radioactivity in the material. The x-ray film is developed to reveal the location(s) of radioactive areas on the kelp frond.

III. Acknowledgments

I am indebted to Linda K. Varangu, L. H. K.-ret., for her valuable assistance in developing and modifying some of the techniques described in this chapter. Maureen Janes typed the manuscript, which was prepared through funding by National Science and Engineering Research Council of Canada Operating Grant A-6866. Marine Sciences Research Laboratory Contribution 519.

IV. References

Buggeln, R. G. 1976. Auxin, an endogenous regulator of growth in algae? *J. Phycol.* 12, 355–8.

Buggeln, R. G. 1981. A general method for determining translocation velocity in Laminariales. *Can. J. Bot.* 59, 132–6.

Buggeln, R. G. 1983. Photoassimilate translocation in brown algae. In Round, F. E., and Chapman, D. J. (eds.). *Advances in Phycological Research*, vol. 2, pp. 283–332. Amsterdam, Elsevier.

Buggeln, R. G., and Lucken, S. 1979. Kinetic characteristics of photoassimilate translocation in *Alaria esculenta* (Laminariales, Phaeophyceae). *Planta* 147, 214–45.

Buggeln, R. G., and Varangu, L. K. 1983. The cross-wing translocation pathway in the blade of *Alaria esculenta* (Laminariales, Phaeophyta). *Phycologia* 22, 205–9.

Canny, M. J. 1973. *Phloem Translocation.* Cambridge University Press, Cambridge. 301 pp.

Emerson, C. J., Buggeln, R. G., and Bal, A. K. 1982. Translocation in *Saccorhiza dermatodea* (Laminariales, Phaeophyceae): anatomy and physiology. *Can. J. Bot.* 60, 2164–84.

Geiger, D. R., and Swanson, C. A. 1965. Evaluation of selected parameters in a sugar beet translocation system. *Plant Physiol.* 40, 942–7.

Hellebust, J. A., and Craigie, J. S. 1978. *Handbook of Phycological Methods: Physiological and Biochemical Methods.* Cambridge University Press, Cambridge. 512 pp.

Lobban, C. S. 1974. A simple, rapid method of solubilizing algal tissue for scintillation counting. *Limnol. Oceanogr.* 19, 356–9.

Lobban, C. S. 1978. Translocation of ^{14}C in *Macrocystis pyrifera* (giant kelp). *Plant Physiol.* 61, 585–9.

Lüning, K., Schmitz, K., and Willenbrink, J. 1973. CO_2-Fixation and translocation in benthic marine algae. 3. Rates and ecological significance of translocation in *Laminaria hyperborea* and *L. saccharina. Mar. Biol.* 23, 275–81.

Moorby, J. 1981. *Transport in Plants.* Longman, London. 169 pp.

Schmitz, K., and Srivastava, L. M. 1979a. Long distance transport in *Macrocystis integrifolia.* 1. Translocation of ^{14}C-labelled assimilates. *Plant Physiol.* 63, 995–1002.

Schmitz, K., and Srivastava, L. M. 1979b. Long distance transport in *Macrocystis integrifolia.* 2. Tracer experiments with ^{14}C and ^{32}P. *Plant Physiol.* 63, 1003–9.

21: Open-flow systems

DONALD W. KINSEY

Australian Institute of Marine Science, Townsville, Queensland 4810, Australia

CONTENTS

[427]

I. Purpose of the method

A. The concept of open-flow, total-system monitoring

Biologists and many ecologists are often more concerned with the details of how particular organisms or groups of organisms function and what niche they fill in the ecosystem than in considering the in situ rates at which they function or the overall performance of the ecosystem at a holistic level. Thus, when estimates of the flux of energy, carbon, or nutrients through the system are required, it is common to attempt this by summing what is known of the components. Frequently, the components themselves have been assessed in isolation from the system. Clearly, risks are inherent in this approach. The most obvious of these are (1) that not all natural components necessarily will be included in the integration and (2) that estimates of an organism's activity made in isolation from the natural environment are liable to depart from the normal.

The study of unconstrained or "open" natural systems is clearly subject to neither of these problems. The concept requires the study of the net flux of any parameter of interest into or out of the biological system by considering the rate of exchange with the surrounding medium. There are some problems in carrying out this procedure in a medium such as air that may have no absolute boundary except with the system itself. On the other hand, the procedure is readily applicable to subaquatic systems in which there are two boundaries to the vertical extent of the surrounding medium. Of course, even here there can be problems if there is a rapid exchange with the ground water. However, in the majority of situations that will be considered, the influences of such exchange are very small compared with the rates of metabolism by the epibiotic community. Total-system, open-flow studies ensure a very reliable inference of gains and losses by the system in terms of either organic inputs and outputs or changes in standing stocks.

Similarly, it is clear that the in situ "open" approach sometimes can be used to obtain more reliable estimates of rates for individual organisms or components of the greater system provided that either

[428]

(1) they occur occasionally in sufficient isolation from other active components that the medium flowing past them can be considered to be influenced by them alone or (2) it is feasible to channelize the flow over them in order to ensure isolation from adjacent components. In the latter case, the "openness" may be questionable, but at least the organism or group of organisms is in most respects in its natural environment.

In summary, the principal advantages of the open approach are the certainty that all system components are included and that they are behaving in their "normal" manner under the environmental conditions prevailing. The disadvantages are a relative lack of biotic and spatial specificity, a generally lower level of analytical precision than can be achieved in laboratory or manipulative experimentation, and an inability to control the environment. The extent to which these are real disadvantages is very much a function of the complexity of the system under consideration. Suffice it to say that care must be exercised in equating, say, the study of an algal-covered tidal flat with the performance of the algae themselves. There is no such thing as a field of algae growing in sterile water on a biologically inert substratum! However, with a high enough standing stock one might approach that situation.

The approaches used in open-system studies were established largely by the efforts of H. T. Odum (Odum and Odum 1955; Odum 1956, 1957a,b; Odum and Hoskin 1958). However, there are numerous examples of earlier studies using various versions of upstream–downstream monitoring (e.g., Sargent and Austin 1949). Since these early studies, there have been many publications dealing with a more rigorous handling of environmental data to infer total community metabolic rates (e.g., Hornberger and Kelly 1975; Schurr and Ruchti 1977; Johnson et al. 1979; Kemp and Boynton 1980). The increased complexity described in these more recent publications has been required principally to handle deficiencies of the basic Odum approach with respect to physical mixing phenomena in deeper water columns. In the situation prevailing in most short-term, high-productivity, shallow-water macroalgal environments, the more basic approach with a great deal of upgrading of technology is very satisfactory.

This chapter is heavily influenced by my own preoccupation with coral reef environments (Kinsey and Davies 1979a), and the methodology strongly reflects that covered by the UNESCO handbook *Coral Reefs: Research Methods* (Stoddart and Johannes 1978), in which the relevant chapters are those by Kinsey, Marsh and Smith, Smith, and Smith and Kinsey. Although most techniques to be discussed have been developed for saltwater situations, they generally are equally relevant to freshwater environments. It is not feasible to consider metabolic monitoring only in the context of primary production. All aspects of CO_2 flux will be considered. Thus, photosynthesis, respiration, and calcification are the important system parameters to be elucidated. All techniques involve the monitoring of changes in the chemistry of the water overlying (and generally moving across) the benthic community but containing the planktonic community. The changes observed reflect the metabolic activities of both.

B. The system

There is no simple criterion for determining the suitability of a system for open monitoring. The ultimate determinant is whether we can measure a change in water chemistry over an acceptable time interval and whether the water flow and mixing patterns are sufficiently straightforward to allow the calculation of meaningful metabolic rates. Thus, we are considering the interactions among the following:

1. System extent (the area over which an operational uniformity may be assumed, suitable to our purpose)
2. System activity (higher metabolic activity obviously simplifies and shortens the time required for adequate monitoring of metabolic rates)
3. System physiography (a simple two-dimensional surface parallel to the sea surface is obviously easier to handle than a slope or a complex three-dimensional surface)
4. Water depth (determines how long we *need* to see adequate changes in water chemistry)
5. Water residence time (determines how long we *have* to see adequate changes in water chemistry)
6. Water flow characteristics (turbulent, well-mixed flow is much easier to work with than stratified or horizontally heterogeneous water masses)
7. Degree of "openness" (an extension of water flow; it is always much easier to work with a naturally channelized system such that flow is more easily described)

C. The techniques

Virtually all the classical studies have characterized total-system metabolism in terms of O_2 flux converted to its implied carbon equivalent or directly in terms of CO_2 flux. The latter can be determined from actual CO_2 concentrations or from the more common pH–alkalinity relationships of seawater. The pH–alkalinity approach is difficult to apply in freshwater systems. However, in marine systems it has a marked advantage over the use of O_2 or directly determined total CO_2 in that it provides data that allow the CO_2 flux to be partitioned into the calcification component (frequently quantitatively important in marine systems) and the component due to photosynthesis and respiration.

Unfortunately, neither CO_2 nor O_2 concentration is a conservative parameter within the benthic/pelagic/water systems under consideration. Departures in the dissolved concentration of either from that which is in atmospheric equilibrium will result in gas evasion or invasion. In the short term (minutes or hours) this exchange will be relatively small for O_2 and very small for CO_2, and quite accurate corrections can be made in analytical data. Hence, both parameters are excellent, sensitive, and accurate indicators of diel metabolism (which in turn can be mathematically integrated to predict long-term net trophic balance). However, neither parameter is suitable for the direct, long-term monitoring (days, weeks, or months between sampling intervals) of net trophic balance. The concept of using phosphate concentration as a "conservative" system parameter to monitor metabolic rates has been proposed and used very successfully (Smith and Jokiel 1975; Atkinson 1981; Smith and Atkinson 1983). This approach is unsuitable for short-term diel consideration of community metabolism because of the analytical limitations in detecting changes. However, the approach has very great value in considering long-term (weeks or months) net trophic balance in extensive enclosed or semienclosed systems (large embayments, some estuaries, and many coral atoll lagoons).

Although the relationship between community metabolism and light is of the utmost importance, the measurement of incident and photosynthetically available light is not covered in this chapter (however, the subject is discussed by Ramus, Chap. 2). It is recommended that all open-system metabolic studies be done in conjunction with the acquisition of meaningful light data and that the frequency of such acquisition be greater (preferably continuous) than that of the acquisition of data for metabolic parameters.

This chapter addresses primarily the consideration of diel metabolism in relatively shallow systems using the O_2/CO_2 conventional approaches.

II. Equipment

In general, equipment for monitoring CO_2 and O_2 changes for the derivation of metabolic information must be of the highest precision, since very small changes are normally encountered. If equipment is to function from small boats, it is very desirable to avoid simple analog meter readouts because these are excessively responsive to movement. Either digital readouts or positively driven outputs such as potentiometric recorders are preferable. All equipment for high-precision use in an open marine environment must be either water-proof or at least very securely protected from spray, and the enclosure of silica gel or other desiccant inside instrument cabinets is normally essential.

It should be stressed that, in the discussion of instrumentation that follows, changes in parameters, not the absolute values, are of prime importance. Therefore, precision is much more important than accuracy.

A. *Oxygen*

1. Oxygen analyzer: A good-quality instrument and electrode system is probably the most essential piece of equipment for straight-forward, accurate community metabolism studies. The choice of equipment, electrode handling, and electrode calibration are discussed by Littler and Arnold in Chap. 17.

2. Wind velocity: Hand-held anemometer-type instruments will do a very adequate job. Readout should be to $0.5 \text{ m} \cdot \text{s}^{-1}$ (1 knot).

3. Temperature measurement: By far the most practical and stable system for in situ temperature monitoring is the calibrated submersible thermistor. This instrument exhibits very good long-term stability and is very trouble free. Readout should be to $0.1°C$.

4. Recorders and data loggers: Very satisfactory, compact, potentiometric, battery-operated chart recorders (e.g., Goerz Minigor) are available and will do a very reliable job for many years of field service. The more recent data-logging technology has developed past the earlier need for tape-based systems. Solid-state storage devices that record many thousands of logs with total reliability are now available. In most respects these are much preferable to chart recorders. They also allow direct dumping of data into a computer.

5. Salinity: In most systems, continuous salinity information is not required, and adequate information can be obtained if samples are placed into well-sealed polyethylene bottles for later estimation. If a highly variable salinity pattern is to be encountered, then any good conductivity field system should be adequate. Induction systems frequently prove to be unsatisfactory for field use. The precision required is first decimal place.

B. *Carbon dioxide*

1. pH meter: A very stable, high-sensitivity battery-operated pH meter is essential. It *must* be reliable to the third decimal place.

Digital systems are preferable to analog systems, because inertial problems from boat movement and so on can be avoided. The best approach is probably to go through a very high resistance input circuit ($>10^{13}$ Ω) directly to a data logger and to allow the computer to handle the raw data.

2. pH electrodes: Only the systems of highest quality should be used. I generally prefer separate glass and reference electrodes to avoid streaming-potential problems associated with reference junctions being too close to the measuring electrode in combined systems. It is not necessary or desirable to use high-sodium-tolerance glass electrodes. Combined systems have been used very satisfactorily by many workers. The new Ross system (Orion Instruments) has proved to be very satisfactory in this marine application. The equilibration to changes in temperature and changes in ionic strength is extremely rapid. Nevertheless, there is no system on the market that allows a stable reading to be made less than 10–20 min after transfer from dilute buffers into seawater.

3. pH buffers: Standard commercial dilute buffers are satisfactory for checking the slope and general responses of the pH meter. As a field reference system, I have found a seawater-based buffer to be much more satisfactory. With such a buffer it is possible to go from buffer to seawater and take a reading within a few seconds. The most reliable system is 0.05 M Tris in seawater with HCl added to give a final pH of ~8.35. The seawater should be boiled and passed through a 0.45-μm filter before the other reagents are added. This buffer is very stable, if preserved with thymol, but has a very high temperature dependence (~0.027 pH unit decrease per 1°C). Because of this dependence, it is necessary to prepare a pH–temperature regression and to use such temperature-corrected values to the nearest 0.1°C.

4. pH sampling and measurement: Continuous in situ pH measurement is quite difficult at the required precision. When used, it is critical that the electrodes be subjected to sufficient positive head that some outflow is maintained at the reference junction. Thus, significant submergence requires careful pressurization. When short-term monitoring is being done, it is usually simpler to sample with a device capable of taking nonturbulent (to avoid gas exchange) samples. These samples are then read against a seawater buffer using a water bath to ensure that both are at the same temperature.

5. Salinity and temperature: Both are required for CO_2-based studies (see comments in Sec. II.A). Temperature must be read to 0.1°C precision. Salinity should preferably be determined to a precision of 0.01, although the first decimal place is satisfactory in systems where no change is anticipated.

6. Alkalinity reagents and equipment: In its simplest, most accurate, and nonautomated form, the alkalinity estimation requires only a pH meter as accurate as that required for general pH work; 0.015 N HCl, glass-stoppered bottles (polyethylene satisfactory but not preferred) of ~125-ml capacity, and A-grade pipettes of 10 ml (for acid) and 50 ml (for seawater) are required.

C. Flow

1. Dyes: Fluorescein is best in very clear water and is easiest to handle as a 20% solution. Rhodamine B is usually better in turbid water.

2. Drogues: Curtain drogues work very well in shallow lagoon-type environments of 3- to 20-m depth. Small current crosses made of plexiglass are generally best in shallow situations (0.5–2-m), although plastic bottles full of water with a small weight to achieve a very small negative buoyancy work well. With all drogues it is critical to minimize the size of the surface float so that small wind waves will not "push" the drogue downwind.

3. Depth measurement: In shallow waters, simple calibrated rods or lead lines are all that is required.

4. Enclosures: In situations in which some enclosure is justified, the form of the enclosure may be one of a very large range: plastic bags (minimum 2-liter, up to 1000-liter or more), plexiglass hemispheres, semirigid plastic sheeting attached to fence posts, and so on. A number of these are discussed in Sec. III K and by Foster et al. in Chap. 10. It is important that the transmission spectrum and percentage of transmission be such that minimal interference with photosynthesis occurs.

All the equipment referred to in this section is discussed in somewhat more detail by Kinsey (1978a) and Smith and Kinsey (1978).

III. Methods: details, evaluations, and alternatives

A. Site selection and suitability

Although it is true that any community will influence the chemistry of the water in contact with it, it is clear that the most credible and quantitatively significant information will be obtained for horizontal, homogeneous, high-activity communities in shallow water. The ideal case would consist of something like macroalgal beds, seagrasses, or coral reef flat organisms distributed uniformly (or with active patches alternating in a consistent pattern with low-activity exposed substratum) over at least 100 m or more in each direction and with a uniform water depth of 0.5 to 2 m (Kinsey and Davies 1979a).

Furthermore, the water would be well mixed by virtue of its movement across the zone and/or wind action and remain within the zone for at least 0.5 to 1 h. It would also be much simpler if the water entering the zone were quite uniform in its composition (e.g., from a low-activity environment such as the open sea or other deep water).

Exceptions to each of these ideal conditions can certainly be handled, but the departure from any one of them will place a greater demand on at least one of the others. Reasonable data can be obtained in water depths of up to 10 m or more, but a residence time of at least 6 to 8 h is then necessary for the water mass. Very narrow zones can be monitored well, but only if the water flow is very linear with minimal lateral mixing. Very small zones can be monitored, but only if the water is virtually standing. A sloping environment can be handled quite well if the flow is along the sloping face rather than across it. However, the methods detailed here will fall short of the rigorous mathematics needed for flow across slopes. Extremely uneven bottom topography (say 1–2 m relief) poses no problems in a system where the average water depth is not progressively changing, but the complex mixing patterns caused will require that the residence time within such a zone be greater than 1 h. Of course, nonuniform biotic composition is not in itself a problem. It simply requires that the purpose of the study be more generalized – for example, the study of a reef flat rather than the study of an *Acropora palmata* zone or the study of a nearshore coastal environment rather than the study of a *Zostera* bed. Low-activity zones place greater demands on analytical precision or require longer residence times. Lack of effective turbulent mixing requires the acquisition of vertical profiles for all data and sometimes the introduction of an extensive sampling regime (say 10- to 20-m radius from the station locations) rather than the normal spot determinations.

B. Monitoring rationale and sampling regime

1. Upstream–downstream: two fixed stations. This is the classical approach of most of the early coral reef studies. It assumes that a unidirectional water flow occurs between two points. Flow rate is accurately determined, and the time taken for the water to move between the two stations is estimated. Because of the fixed locations of the stations, long-term monitoring with tower-mounted or floating anchored instrumentation is feasible. The obvious liability of the approach is that very few situations have such uniformity of flow with respect to either direction or rate. Although rate can be measured with current meters, this does nothing to reposition the downstream station to accommodate changes in direction. However, the approach can be a very satisfactory one for shallow stretches of channelized flow such as that in rivers. A sophisticated version of this approach was developed for rivers by Hornberger and Kelly (1974).

2. Fixed station, flowing water. This approach can be a specialized case of the preceding one. Here it is assumed that the upstream station is the open ocean (or equivalent) with a virtually constant water composition that does not require constant or frequent monitoring. In cases in which this assumption is valid, the method is very simple and again can have the advantage of a long-term tower-mounted monitoring station. It still, however, requires an accurate projection of the time that the water took to flow across the transect to the station (i.e., the time during which the water was exposed to the community under consideration). The need for this information usually requires that current and direction information be obtained.

In a very different variant of the single-station, flowing-water approach, Schurr and Ruchti (1977) pointed out that many rivers exhibit sufficiently adequate biotic homogeneity over long stretches to ensure that upstream and downstream water composition is the *same* at any one instant throughout the diel cycle. Thus, it is possible to monitor such a river from a single fixed station. It is clearly not necessary to monitor flow rate in this approach, but it is necessary to establish the validity of the uniformity assumption.

3. Fixed station, standing water. In many ways this is the ideal simple situation. It requires that the community under examination be exposed to the same water mass over extended time (at least during the extent of the monitoring period). Since the biological influences on the water will reflect only the "local" community, careful site selection is required if data are to be typical of the environment at large. The approach can be very satisfactory in some coral reef locations (Kinsey 1978a; Kinsey and Domm 1974; Kinsey and Davies 1979b) and in most pond or small landlocked lagoon situations (Welch 1968). Clearly, the ideal case is a tide pool in the high intertidal zone where isolation occurs over long periods in the tidal cycle. As with all these aproaches, this standing-water situation allows the use of fixed, untended monitoring stations. Even when the standing water is a short-term situation related to slack tide periods, the fixed monitor is still a realistic option, with only the standing-water data being used.

4. Moving station, flowing water. This is certainly the most versatile of all approaches and the one requiring the least setup time. The water entering the chosen system is marked with dyes or drogues, and the sampling and/or instrument monitoring follow the water along whatever route it may take across the system. Dyes are much preferred in shallow water because they *must* indicate the actual status of the water flow and mixing patterns, whereas drogues are very prone to moving downwind because of wave impulse effects on their floating component. The instruments may be boat-mounted (Kinsey 1978a). However, in a very sophisticated variation, Barnes (1983) used a "buoy"-mounted instrument package that floated across the transect on a long floating tether such that it could be kept in the marked water mass and finally recovered. The obvious advantage of this system is that it can be "fed" into places to which a boat may have no access.

C. Temporal considerations

In the planning of an open-system study, the prime consideration is its purpose. If broad patterns of diel community metabolism are required, sampling periods representing full transect transit times are perfectly adequate. It may be possible, after initial consideration of full diel cycles, to reduce the study to specific periods, such as sunrise, sunset, midday, and midnight. This kind of abbreviated study has worked well in pond situations (Welch 1968). It is usually necessary to consider both day and night periods, the former to determine *net* photosynthesis and the latter total community respiration. Day-to-day variability can be quite large and will certainly reflect protracted periods of cloudy weather or abnormal temperatures. These effects extend well beyond the time of their direct influence (Kinsey and Domm 1974).

Seasonality is quite marked in even the most tropical systems (Kinsey 1977, 1979, 1983) and must be considered if the annual performance of the system is to be assessed.

Year-to-year variability has generally been found to be small for coral reefs (Kinsey 1977), but there will be many systems in which marked variability would be expected (particularly those in studies of response to, or recovery from, stress).

The advent of data-logging technology has introduced a new possibility, particularly when a mobile instrument station is used to follow water mass through the system. Sampling frequencies of as little as 1 s are now perfectly feasible, particularly when very high capacity solid-state data storage is used. This technique has been utilized by D. J. Barnes (unpublished data) to discriminate at the level of identifying the activity of each major coral head and low-activity sand area, and so on, across a shallow coral reef flat transect. High-frequency sampling, together with continuous light monitoring also allowed Barnes to correlate community response accurately with the light reduction caused by individual clouds. This new technology will certainly revolutionize and revitalize total-system metabolic studies in shallow water.

D. Community considerations

The significance of two components in the community structure must be identified to facilitate the analytical approaches and overall sampling strategy to be used.

First, if any significant level of calcification *or* decalcification (boring into limestone substratum, etc.) is associated with the community, any CO_2 monitoring must be designed so as to permit partitioning of the total CO_2 flux (Sec. III.G). It should be noted that typical coral reef communities may transfer up to 20% of their total CO_2 fixation into calcification, with only the remaining 80% representing photosynthesis (Kinsey 1979). If O_2 concentration is to be the only parameter monitored and no attempt will be made to estimate the community respiratory or photosynthetic quotient (see Sec. III.I), calcification need not be considered.

Second, the significance of the planktonic component of the total community metabolic activity and the need to be able to separate this component must be determined (Kinsey 1983). The only way this can be done is to isolate samples of the overlying water. This has been accomplished reasonably effectively with polyester bags (for low gas transfer and high clarity; Kinsey 1978a, 1979) of not less than 2-liter capacity or large plexiglass hemispheres (250-liter) with the flat side uppermost. These devices are submerged on site to about average depth or to a series of depths and monitored for the same variables as used for the complete system. Some form of mixing to simulate that occurring external to the enclosure is normally desirable. One advantage of plastic bags is that they flex substantially in turbulent environments, thus giving a reasonable form of internal turbulence. Unfortunately, all enclosure techniques impose a degree of modification on the system, causing a departure from the certainty that one is monitoring the natural performance of the environment. In very enriched systems such as sewer outfall areas, the contribution of plankton may reach 30% of the total photosynthesis and 15% of the total respiration (Kinsey 1979). This can occur even when the water is overlying a flat of 1-m depth with reasonable light penetration to a rubble substratum.

E. Metabolic variables

The following variables and conventions will be used:

y	Hourly rate of net photosynthesis
r	Hourly rate of respiration, where r is a specific case of y in

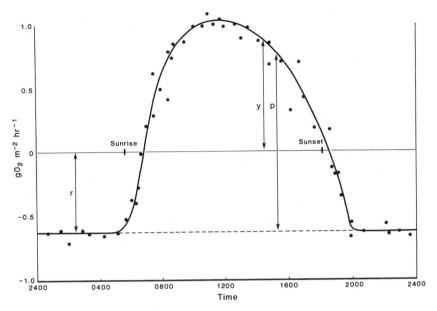

Fig. 21–1. Diel oxygen curve for a 1-m-deep coral reef flat. Correction for atmospheric exchange has been applied. The curve is a composite of 50 low-tide standing-water periods in October but from several different years. In all periods there was complete insolation. A single value of y or r has been plotted for each period. There is a small deviation of values from the curve, principally because of the very discrete environment sampled by a single station in standing water.

	which gross photosynthesis is equal to zero (see Fig. 21–1)
y_{max}	Peak hourly rate of net photosynthesis
p	$= y + r =$ hourly rate of gross photosynthesis
p_{max}	Peak hourly rate of gross photosynthesis
c	Hourly rate of net calcification
c_d	Hourly rate of net calcification in daytime
c_n	Hourly rate of net calcification at night
R	Diel respiration $= 24r$ (assumes no photorespiration)
P	Diel gross photosynthesis
E	Diel net trophic gain ($= P - R$)
G	Diel net gain from calcification

The most practical units for most biologists are those expressed in the carbon flux forms: grams carbon per square meter per day or gram-atoms carbon per square meter per day or millimoles CO_2 per square meter per day. This allows direct comparison of data whether they are derived from CO_2, O_2, or even calcium flux and whether they represent organic metabolism or calcification. Obviously, the time scale could be anything from seconds to years depending on the purpose of the study. The consideration of community metabolism on the projected plane area basis (per square meter) is by far the most usual.

F. The oxygen approach

1. Introduction to the concept. In general, the procedures outlined here are a condensed version of those covered by Kinsey (1978a). Oxygen flux can be used to derive rates for primary production or respiration provided that (1) sufficient CO_2 data are available to establish the respiratory quotient (RQ) and photosynthetic quotient (PQ) or that credible values for the quotients are already available, (2) corrections for atmospheric exchange can be made using either standard coefficients from the literature or coefficients determined during the period of study, and (3) loss of photosynthetically produced oxygen as bubbles is negligible.

In the case of communities in which gas loss is appreciable, there is probably no alternative but to measure directly the volume of escaping O_2 per unit area of community. However, most studies of this phenomenon have revealed that the proportion of total O_2 lost in this manner is very small. Even in shallow, warm, standing water with extreme dissolved-O_2 concentrations approaching 200% of air saturation, gas bubble loss has been found to be of little quantitative significance (Kinsey 1978a). However, there may be systems, particularly areas dominated by algal mats, in which bubble loss is significant (see Sec. III.I). The retention of O_2 gas within tissues and its reabsorption at night is another phenomenon that might be very significant in some vascular plants, fleshy algae, and corals (Bellamy and Risk 1982; W. M. Kemp, personal communication). However, it should be noted that the potential for underestimating both photosynthesis and respiration from this cause is not a problem unique to the O_2 approach. The same retention of metabolic dissolved gases is at least as great a problem when CO_2 concentration is the prime monitored variable.

Oxygen concentration can be estimated directly with electrodes to a high precision and, as such, is a very suitable parameter for continuous long-term recording or data logging. Oxygen data tell us nothing about calcification. Parameters to be measured are oxygen concentration, temperature, salinity, and wind velocity (or other requirements for the calculation of diffusion constants). All parameters must be checked for vertical stratification or patchy horizontal distribution, and if either is significant, monitoring of more than one point in the relevant plane(s) at the station location may be necessary. In addition, records must be kept of depth, light (assuming correlations are required), and whatever parameters of community structure and zonation may facilitate the meaningful expression of the final derived metabolic information (e.g., percentage of bottom cover and biomass estimates).

Oxygen can be measured in terms of concentration (milligrams per liter, microgram-atoms per liter, etc.) or saturation relative to a standard atmosphere (percentage). This will depend on the equipment available. In either case, accurate temperature (to 0.1°C) and salinity (to 0.1‰ or better) will be necessary for the interconversion because both concentration and percentage of saturation will be used in the final calculations. Both temperature and salinity can be measured continuously with recording devices although in most systems occasional salinity estimates on retained samples will be adequate. All O_2 electrodes require some movement past the membrane, and this must be provided in any long-term monitoring devices. Hand-induced movement is adequate in short-term, spot-check approaches.

Oxygen electrodes are always best calibrated in the medium under examination and at a temperature close to the ambient temperature in the system regardless of assurances by manufacturers of adequate temperature compensation and of the expediency of calibration in damp air. Electrode calibration procedures are discussed at length by Littler and Arnold in Chap. 17.

2. Atmospheric diffusion coefficient. I still prefer the simple H. T. Odum (1956) expression of diffusion coefficients in terms of grams O_2 evading or invading per square meter of sea surface per unit of saturation deficit. Although this is simplistic and somewhat imprecise in terms of rigorous physical considerations, it gives very adequate precision for the short-term metabolic studies being considered here.

The following conventions and variables are used:

K Areal diffusion coefficient for O_2
S Saturation of O_2 with respect to air at atmospheric pressure (expressed as percentage of equilibrium value)
C Concentration in any suitable units (say grams per cubic meter)
t Time
Z Water depth
D Mean value of the saturation excess (positive value) or deficit (negative value) at the water surface during the period used for experimental determination of dC/dt, that is,

$$D = \frac{[(S' + S'')/2] - 100}{100} \tag{1}$$

Odum's (1956) technique for estimating diffusion coefficients is based on the assumption that respiration rates of the community are the same just after dusk as they are just before dawn. Thus, any decrease in apparent O_2 consumption rate reflects a greater rate of invasion into the predawn, O_2-depleted waters. Solving for a maintenance of constant rate gives a value for diffusion coefficient. The assumptions in this technique are suspect, and I do not recommend its use.

An approach that I have used in coral reef areas where standing water occurs on each low tide is as follows. Advantage is taken of the very substantial differences in O_2 concentrations in the water between the low-tide standing water situation and those during or immediately after the "flushing" of the system by high tides. Unlike earlier methods, this approach allows the calculation of K equally well from daytime or nighttime data. It also is not dependent on the assumption of constant respiration rates throughout the night. However, like all approaches to this problem, it is dependent on certain suspect assumptions, namely, production and respiration are assumed to be unaffected by O_2 tension within the range 70 to 150% of air saturation, and values for K are calculated from data within this range. The reasoning applied to the determination of K is as follows:

1. High-tide O_2 range is much smaller than the low-tide slack water range.

2. Low-tide O_2 concentration for any particular time is therefore dependent on the number of hours since the preceding high tide.

3. Given constant temperature, insolation, and wind velocity, the low-tide O_2 saturation level will be different at any particular time during a series of four to five consecutive days or nights, due to the decreasing time elapsed since the preceding high tide.

4. Therefore, values obtained for the rate of change of O_2 concentration (dC/dt) at that time on each of the days will differ, but because of differences in diffusion alone (i.e., real metabolic rates are assumed to be constant).

5. Thus, a value of K relevant to the wind velocity chosen can be calculated by the procedure of Odum and Hoskin (1958) on the premise that net production rate y or respiration rate r must have remained constant.

6. To summarize, strictly within the experimental framework outlined under points 1–5,

$$(dC/dt)'Z + D'K = (dC/dt)''Z + D''K = (dC/dt)'''Z + D'''K \text{ etc.} \quad (2)$$

for any particular time over a period of several days at the one site. Here, K is a constant for any wind velocity. Thus, by solving any pair where a constant wind velocity applies, a value for K is obtained relevant to the site under consideration at its low-tide depth and to the particular wind velocity. The units for K will be grams O_2 per square meter per hour at 100% air saturation deficit or similar, depending on the units of O_2 concentration and time chosen.

Constant wind conditions usually occur sufficiently often to enable one to calculate such values from data accumulated during one field trip of a few weeks. The value of K is a function of the type of environment involved (e.g., general depth, topography, and wave fetch) but is independent of time of day and temperature; K should exhibit temperature dependence when expressed in the units just cited because of the temperature dependence of the amount of O_2 present at 100% saturation (i.e., the solubility of O_2). However, in practice, the values obtained from coral reef environments over the temperature range 18–30°C seem to have exhibited little correlation with temperature.

A regression analysis of data from various coral reef environments has led to the following formulas for the derivation of K.

Shallow (to 1-m) reef flat and back reef environments with very limited wave action:

$$K = 3.7 \times 10^{-3}V^2 + 0.32 \qquad (3)$$

Shallow lagoonal situations or open water with wave fetch restricted to 500 m:

$$K = 4.45 \times 10^{-3}V^{2.095} + 0.50 \qquad (4)$$

where V is wind speed in knots.

Values obtained from Equations 3 and 4 have been found to have good general application (Kinsey 1979). Equation 3 should be used only in areas of shallow flats in which there is virtually no wave fetch. Equation 4 seems to work well in all open water where the depth does not exceed ~3 m or in deeper water where the wave fetch is restricted to no more than ~500 m. Considerable caution should be used in applying these coefficients outside the temperature range for which they were determined.

Marsh and Smith (1978) and Barnes (1983) used coefficients that combined the earlier data of Kinsey and Domm (1974) and Kinsey (1978a) with those of Kanwisher (1963). However, the end result differed appreciably from values obtained from Equations 3 and 4 only in the range of wind velocities above ~18 knots.

Obviously, the values discussed here are only a guide, although in warm marine environments it is probably better to use these values than to use casually determined field values. The only alternative is to carry out a careful and detailed study such that specific values can be determined for a particular environment.

Table 21–1. *Constants for determining oxygen concentration (C)*

Constant	For C in cm^3 (STP) liter^{-1}	For C in μmole kg^{-1}
A_1	-173.4292	-173.9894
A_2	249.6339	255.5907
A_3	143.3483	146.4813
A_4	-21.8492	-22.2040
B_1	-0.033096	-0.037362
B_2	0.014259	0.016504
B_3	-0.0017000	-0.0020564

Note: Values of C can readily be converted to milligrams per liter.
Source: Kester (1975).

A totally different, nonbiological approach to the determination of diffusion coefficients has been proposed by P. F. Roques and S. W. Nixon (unpublished data). In this procedure, a floating gas-filled dome with a reduced O_2 level is floated over the surface of the water in the environment under consideration. The rate of equilibration of the partial pressure of O_2 in the gas space with that of the underlying water mass allows the mathematical determination of the exchange coefficient. The authors claim that the reduction of surface waves or ripples inside the chamber does not modify the result.

The application of diffusion correction to data from freshwater environments is not considered specifically here. However, Hart (1967) proposed an interesting approach to the situation in streams.

3. Determination of metabolic rate. All O_2 saturation data should be converted to concentrations if not directly measured as such. This requires measured temperature, salinity, and saturation values as well as standard solubility values for O_2. There have been many updates of O_2 solubility tables, but for marine waters the values of Carpenter (1966) are still very acceptable and closer than 1% to any other published values. Carpenter's values, with only minor modification, have been fitted (Weiss 1970) to the thermodynamically consistent equation

$$\ln C = A_1 + A_2(100/T) + A_3 \ln(T/100) + A_4(T/100)$$
$$+ S\text{‰}[B_1 + B_2(T/100) + B_3(T/100)^2] \quad \cdot \qquad (5)$$

where C is the solubility of O_2 with respect to an atmosphere of air at 100% relative humidity and total atmospheric pressure of 760 mm

Hg, and T is the absolute temperature. The constants used are given in Table 21-1. If vertical stratification has been found, then a weighted vertical mean concentration for the entire water column must be determined.

Time is now taken into account to determine dC/dt. The time used is the interval between readings taken by the monitoring system while it is in the area of the benthic environment under consideration. It could be the time taken for the flow of the water mass right across a transect or between two fixed stations, the time between two data sets from a fixed monitor, or the time intervals in a rapid series of automatically logged data. In a simple sampling regime with appreciable time intervals between estimations of O_2 concentration, it is assumed that

$$dC/dt = \Delta C/\Delta t \qquad (6)$$

The dC/dt is converted to a rate of change in the quantity of oxygen (M) overlying a unit of bottom area by introducing the mean depth. Thus,

$$dM/dt = dC/dt \, Z \qquad (7)$$

This rate of change can now be converted to real metabolic rate by introducing the correction for atmospheric exchange. Thus,

$$y \text{ (or } r) = dM/dt + DK \qquad (8)$$

It should be stressed that the value D must be based on values for S at the surface only, in cases where dC/dt is based on vertical means.

The values for y (or r) obtained are still in O_2 units. The conversion of these to carbon units can, in the most simple form, be made on the assumption that $RQ = PQ = 1$. Thus, the rate would be identical on a molar basis. However, the metabolic quotients cannot, with any confidence, be assumed to equal unity. This subject is discussed further in Sec. I.

The whole simplified approach to data handling outlined here is summarized in Table 21–2, which is based on Kinsey (1978a). It should be stressed, however, that the development of a more rigorous mathematical approach is preferable. The data in Table 21–2 are based on a standing-water environment and a single fixed station. The changes in variables are over longer times and are therefore greater than would normally be encountered in flowing-water situations.

G. The carbon dioxide approach

1. Introduction to the concept. In general, the procedures outlined here are a condensed version of those covered by Smith and Kinsey (1978)

Table 21–2. *Processing short-term low-tide oxygen data from a rich coral area of the reef flat at One Tree Island: standing-water, fixed-station conditions*

1	2	3	4	5	6	7	8	9	10	11	12	13	14	15	16
Time	Hour	Wind (knots)	Total depth (cm)	Depth of reading Z (cm)	Temp (°C)	O_2 satu-ration S (%)	O_2 concen-tration C (g·m^{-3})	O_2 concen-tration C (vertical mean)	dC/dt (g·m^{-3}·h^{-1})	dM/dt (g·m^{-2}·h^{-1})	K	Mean surface satu-ration (%)	Satu-ration excess D	y (g·m^{-2}·h^{-1})	Time
0900	0	10	81	2	19.6	94	7.00	6.68							
	0	10	81	10	19.6	91	6.78								
	0	10	81	70	19.4	87	6.51								0930
1000	1	10	81	2	20.5	108	7.92	7.85	1.17	0.95	0.70	101	0.01	0.95	
	1	10	81	10	20.5	108	7.92								
	1	10	81	70	20.5	106	7.77								1100
1200	3	15	80	2	21.4	140	10.11	10.11	1.13	0.91	0.84	124	0.24	1.11	
	3	15	80	10	21.4	140	10.11								
	3	15	80	70	21.4	140	10.11								

Note: Insolation: 100%; *S‰*: 35.6; Tides: 0305, 186 cm (ponding below 155 cm; thus, standing water from soon after 0700 to 1200); 0907, 104 cm; 1555, 247 cm. Columns 1–7 are field data; 8–16 are derived data. Special points, relevant to column numbers, are as follows: 4, Slight change in depth over 3 h causes negligible volume transport to interfere with standing-water concept. 5, When no stratification is detected, only one monitoring depth need be used. 7, Depressed levels initially reflect standing-water conditions near dawn when respiration dominant. Increasing wind responsible for better mixing later in the series. 8, Derived from values in column 7 using O_2 solubilities after Carpenter (1966). 9, Vertical means calculated after Kinsey (1972); i.e., assume 80 to 10 cm band is at the mean value of the O_2 concentrations (8) for 70 and 10 cm and assume the 0 to 10 cm band is at the mean value of the O_2 concentrations for 2 and 10 cm. 10, Difference between two consecutive vertical means over the time interval between them. 11, $dM/dt = dC/dt \cdot Z$ (mean relevant values from column 4). This value is the rate of change in the quantity of O_2 overlying unit plane bottom area. 12, Values using mean wind velocity, i.e., 12.5 knots for 1000 to 1200. May differ slightly from values given by Equations 3 and 4 in Sec. III.F.2. 13, Mean value at 2 cm from column 7 for the time period concerned. 14, [(value in column 13) − 100]/100. 15, $y = dM/dt + DK$. This is the metabolic rate for net photosynthesis; it is still in O_2 units. 16, Time at which the value of y in column 15 is considered to apply.

in the *Coral Reef: Research Methods* handbook. The outline considers only the CO_2 system in seawater.

The measurement of CO_2 flux has three marked advantages over the measurement of O_2 flux. First, it is a direct indication of carbon flux, and hence knowledge of the metabolic quotients is not required. Second, it gives an estimate of calcification as well as of respiration and photosynthesis. Third, the properties of the CO_2 system, in seawater at least, are such that exchange with the atmosphere can be ignored over the time periods used to determine diel metabolism. Such diffusion exchange cannot be ignored in studies of long-term net effects (Sec. I.C).

Parameters to be measured for the derivation and partitioning of CO_2 data are pH, temperature, salinity, and total alkalinity (TA) together with all the general site data outlined in Sec. III.F. Wind velocity is not necessary, because no diffusion correction will be applied. All parameters except total alkalinity can be monitored readily from on-site instrument packages. However, alkalinity, at least up to the present time, has required the taking of samples and a laboratory-based estimation. Hillbom et al. (1983) reported a method for the automated in situ determination of alkalinity in natural waters. Such a procedure could represent a major break-through for CO_2 monitoring; however, I have not had an opportunity to apply this technique to shallow-water short-term metabolic moni-toring. Overall, the disadvantages of the CO_2 approach are (1) that it is analytically indirect and less precise than the O_2 approach, (2) that it has not allowed the use of continuous recording or logging of data because of the need for the laboratory-based alkalinity estimation (Sec. III.H), and (3) that it is subject to interferences in some systems (Sec. III.G.4).

All parameters measured are required for the determination of a value for the change in total carbon dioxide ($\Delta\sum CO_2$). The total alkalinity measurements alone allow the determination of the change in CO_2 due to calcification (ΔCO_{2_c}). By difference, a value is obtained for the change in CO_2 due to photosynthesis and/or respiration (ΔCO_{2_y}). Thus,

$$\Delta\sum CO_2 = \Delta CO_{2_c} + \Delta CO_{2_y} \tag{9}$$

2. Total CO$_2$ ($\sum CO_2$). Total CO_2 concentration in seawater can be estimated from the following relationships (Smith and Kinsey 1978; Skirrow 1975):

$$CA = TA - BA - HA \tag{10}$$

where CA is the carbonate alkalinity; TA is estimated; BA, the borate alkalinity, can be determined from salinity ($S‰$), temperature, and pH; and HA, the hydroxyl alkalinity, can be determined from temperature and pH.

$$\Sigma\, CO_2 = \frac{K_{1c}a_H + K_{1c}K_{2c} + a_H^2}{K_{1c}a_H + 2K_{1c}K_{2c}} \tag{11}$$

where all factors are functions of pH, S‰, and temperature [a_H is the hydrogen ion activity; K_{1c}, K_{2c} and K_{1b}, K_{2b} (Equation 12) are the first and second dissociation constants of carbonic and boric acids, respectively].

The following additional relationships are necessary to derive these factors:

$$BA = \Sigma\, B\, \frac{a_H K_{1b} + 2K_{1b}K_{2b}}{a_H^2 + a_H K_{1b} + K_{1b}K_{2b}} \tag{12}$$

where $\Sigma\, B$ is total borate concentration.

$$HA = \frac{1}{a_H}(10^{-14} - a_H^2) \tag{13}$$

$$\Sigma\, B = 2.06 \times 10^{-5} Cl‰ \tag{14}$$

$$Cl‰ = (S‰ - 0.03)/1.805 \tag{15}$$

where $Cl‰$ is chlorinity.

$$a_H = 10^{-pH} \tag{16}$$

The dissociation constants K_{1b}, K_{2b} and K_{1c}, K_{2c} are derived as reported in Skirrow (1965, 1975). Readily computerized equations are given for these. All are temperature and salinity dependent.

As an alternative to this relatively rigorous and programmable approach to $\Sigma\, CO_2$ calculation, one can use the empirical approach reported by Strickland and Parsons (1972). The same variables are monitored.

The change in CO_2 caused by all biological (and chemical) functions ($\Delta\Sigma\, CO_2$) is simply the change with time in the value of $\Sigma\, CO_2$ as just estimated.

3. Change in CO_2 due to calcification (ΔCO_{2c}). As already indicated, system calcification can be determined from total alkalinity alone. The theory of this has been adequately covered by Smith and Key (1975) and Smith and Kinsey (1978). The basis of the derivation of calcification is that photosynthesis and respiration (or any other process involving just CO_2) can have no effect on the total alkalinity of seawater because that value is maintained as long as the cation composition remains unaltered. However, calcification (the precipitation of calcium carbonate) or carbonate dissolution has a simple, stoichiometric influence on total alkalinity:

$$\Delta CO_{2_c} = \Delta TA/2 \qquad (17)$$

Interferences with this simple relationship are discussed in Sec. III.G.4.

4. Monitored variables. For satisfactory results from the approach outlined so far, each of the required variables must be determined with a high degree of precision. The pH must be measured to 0.001 unit for all but the highest-activity systems in very shallow water (say 0.5 m). The equipment and buffers have already been listed. It should be noted that a seawater reference buffer has proved to be a much more convenient field system than conventional dilute buffers because of the very short time needed to recheck electrode calibration. It is essential that both the in situ temperature and the temperature at which the pH is measured be recorded if pH is not actually measured in situ. In this case, the pH estimation should be made within a few minutes of sampling. The correction of seawater pH to its in situ value can be made according to the following relationship:

$$pH \text{ in situ} = pH \text{ measured} + (T \text{ measured} - T \text{ in situ})0.0103 \qquad (18)$$

Temperature should be measured to 0.1°C and $S\text{‰}$ to at least 0.1‰.

The alkalinity method used is a modified version of the Culberson et al. (1970) technique, which in turn is based on the Anderson and Robinson (1946) fixed acid addition approach. The method involves the measurement of the pH of a seawater/acid mixture, with the highest precision possible, preferably to within 0.001 pH unit. The most common combination is 50 ml of seawater (V_s), prefiltered soon after sampling through a 0.45-μm filter, added to 10 ml of 0.015 N HCl (V_a) in a glass-stoppered bottle that has been equilibrated to a similar mixture for several months. The final pH must lie in the range 3.2–3.9, or a different acid volume or strength must be used. The mixture is shaken very vigorously with at least two changes of head space air or bubbled with water-saturated air. Many small but important details must be observed if high precision is to be achieved in this method [Smith and Kinsey (1978) consider the majority of these]:

$$TA = \frac{1000}{V_s} V_a N - \frac{1000}{V_s}(V_s + V_a)10^{-pH} f \qquad (19)$$

where N is the normality of the acid, and f is an empirical constant for which the most satisfactory value in seawater has been found by experimental verification (D. W. Kinsey, unpublished data) to be that given by Anderson and Robinson (1946): 0.758. Although TA was indicated earlier to consist of CA + BA + HA, there are actually a number of other anions and cations that may influence it appreciably in shallow or enriched systems. Notably, these are nitrate, ammonia, sulfide, and organic anions. Kinsey (1978b) has established that, in coral reef environments, the effects of changes in these components are insignificant compared with the change in TA effected by calcification. In contrast, in systems of low or negligible calcification, particularly when the system is eutrophic and exhibits a significant level of anaerobic processes in the bottom sediments, it may be very important to estimate any or all of these interfering ions separately. The change in TA will then have to be corrected accordingly before any realistic estimate of calcification or ΔCO_2 can be made. In such systems, it is probably of very doubtful value to attempt CO_2 flux studies using the pH–alkalinity approach.

5. Determination of metabolic rate. The procedures are essentially the same as those already outlined for O_2 data (Sec. III.F.3. and Table 21–2), except that no correction for atmospheric diffusion is applied and no interconversion to saturation levels is required. Thus, all concentrations for $\sum CO_2$ and TA are first converted to vertical weighted means. These in turn can be converted to rates of change in concentration, dC/dt, by simply dividing the total changes by the time intervals involved.

Remembering that $\sum CO_2 = CO_{2_y} + CO_{2_c}$ (Equation 9), we can readily establish dC/dt for the two metabolic variables:

$$dCO_{2_y}/dt = [\Delta \sum CO_2 - (\Delta TA/2)]/\Delta t \tag{20}$$

$$dCO_{2_c}/dt = (\Delta TA/2)\Delta t \tag{21}$$

When these rates are multiplied by the depth, we have a direct expression (in carbon units) of the metabolic rates on an areal basis:

$$y \text{ or } r \text{ (the net photosynthetic rate)} = (dCO_{2_y}/dt)Z \tag{22}$$

$$c \text{ (the net calcification rate)} = (dCO_{2_c}/dt)Z \tag{23}$$

H. The combined oxygen/carbon dioxide approach

Most studies are likely to involve the acquisition of at least some metabolic rates by both the O_2 and the CO_2 approaches. This is done so that correct PQ and RQ can be determined for the particular system (Sec. I). However, Barnes (1983) proposed an approach involving both but avoiding the repetitive need for precise determinations of TA. Thus, because routine sampling for TA is not required, the approach allows continuous recording or logging of data together with the estimation of calcification.

This new method takes advantage of the fact that (1) ΔCO_{2_y} can be determined from O_2 data, provided that credible metabolic quotient data are available, and (2) $\Delta\sum CO_2$ can itself be determined with only relatively imprecise values for TA, which can be determined from occasional samples. Thus,

$$\Delta CO_{2_e} = \Delta\sum CO_2 - \Delta O_2 \times RQ \quad \text{(or 1/PQ)}$$

$$= \Delta\sum CO_2 - \Delta CO_{2_y} \quad \text{(O_2-derived)} \quad (24)$$

This allows the determination of calcification changes (ΔCO_{2_c}) without requiring the highly precise TA data series that would have been used in the more conventional approach. Even the values for PQ and RQ, which become so critical to this approach, can be cross-checked by periodic checks on the calcification rates using the more direct TA approach. For this new method to have any real advantage over previous procedures, it is necessary that highly satisfactory and very precise continuous in situ pH monitors be available.

I. Metabolic quotients

Oxygen concentration may be a very precise variable for metabolic studies, notwithstanding the substantial atmospheric exchange corrections required. However, its value as an indicator of carbon flux is no better than the precision of the values available for the metabolic quotients, RQ and PQ. By convention,

$$RQ = \Delta CO_2/\Delta O_2 \quad (25)$$

$$PQ = \Delta O_2/\Delta CO_2 \quad (26)$$

It is important to stress that PQ is the quotient relevant to gross photosynthesis p, not net photosynthesis y. This distinction is much more important when one is dealing with complete systems in which r may be very large than it is when one is dealing with the plants themselves. Of course, there is no reason that an empirical *net* photosynthetic quotient could not be applied in some cases; however, it would then be very important that a predictable relationship be maintained between system respiration and system photosynthesis. Table 21–3 stresses the differences in the quotient determined in these two ways.

Although the application of metabolic quotients to O_2 data to obtain their true carbon equivalent will frequently change the rates only marginally, these changes may make a very large difference to the value obtained for E, the net gain or loss in the system over a 24-h period.

When O_2 is to be the routine monitored variable, one of two approaches must be taken: (1) Reliable values for the metabolic

Table 21–3. *Metabolic quotients for a hypothetical system typical of
shallow marine environments*

Metabolic parameter	ΔO_2	ΔCO_2	Metabolic quotient
r (measured respiration rate)	0.8	0.8	1.0 (RQ)
y (measured net photosynthesis rate)	0.4	0.2	2.0 (net PQ)
p (true gross photosynthesis rate)	1.2	1.0	1.2 (PQ)

quotients must be available from the literature or previous studies, or (2) sufficient CO_2-based metabolic rate determinations must be made to estimate confidently the quotients for the system under consideration. If the estimates are to be made experimentally, the derivation will be

$$RQ = r(\text{CO}_2\text{-based})/r(\text{O}_2\text{-based}) \qquad (27)$$

$$PQ = p(\text{O}_2\text{-based})/p(\text{CO}_2\text{-based}) \qquad (28)$$

where $p = r + y$, and *y* is preferably near y_{max}.

Many factors influence the values obtained for RQ and PQ. In general, a fully aerobic, carbohydrate-based system tends toward an RQ of unity. In systems with a high trophic input of protein- and fat-based materials (systems feeding on high-grade plankton), RQ tends to be less than unity. In systems with a substantial anaerobic component in the sediments, RQ exceeds unity. The PQ for most systems tends to lie between 1.1 and 1.5 (Ryther 1956; Strickland 1960), although it has been stressed (Williams et al. 1979) that total dependence on nitrate may give a system an apparent PQ as high as 1.5–1.8. Most early studies reported values obtained with phytoplankton. Nixon et al. (1976) reported RQ values of 0.75 to 0.9 and PQ values of 1.1 to 1.2 for a heterotrophic community in which the dominant plants were macrophytes. These values appear to be based on net photosynthesis.

Table 21-4 lists a range of values (Kinsey 1978a, 1979) obtained by the present methodology for tropical coral-reef-related environments. These are useful indications of results obtained. It would be unwise to use them indiscriminantly, because they reflect not only the communities under examination but the minor idiosyncrasies of the application of the methodology. The anomalous low values found for PQ in the algal flats and for the algal mats among filter-feeding communities almost certainly indicate a substantial loss of O_2 directly to the atmosphere in bubble form. The occasional high RQ values indicate environments with some major level of associated anaerobic decomposition.

Table 21–4. *Values for metabolic quotients in coral reef environments*

Community	RQ	PQ
Coral heads and associated community	0.85–0.90	1.1–1.15
Seaward reef slopes	0.8	1.1
Reef flats dominated by corals and coralline algae	0.9–1.05	1.05–1.1
Reef flats with high foliose and filamentous algal cover	1.1–1.2	0.95–1.05
Algal flats	1.1–3.0	0.6–1.1
Areas dominated by filter-feeding, heterotrophic communities	1.1–1.6	0.5–0.9

It is clear that considerable care must be exercised in applying the O_2 approach in many systems, notwithstanding the simple technological advantages.

J. Diel patterns

1. Diel rate curve. Instantaneous or short-term metabolic rates (y, r, or c) give useful information concerning the rates at which systems function. However, they tell us little about the overall turnover or net gains and losses within the system. It is usually the latter kind of information that we are seeking in studying an open complete system.

The most basic unit of system function is the diel performance. All other extended considerations are simply integrations of diel units with superimposed seasonality, perturbations, and so on. If the rate estimates y and r are plotted over a full diel cycle, the form of the curve will be something like that given in Fig. 21–1. It is important to note that such a curve will be given whether the plot is for a series of rates estimated within one diel cycle or for a series of discontinuous time periods within the same general time of the year (say 1 mo). The latter approach is generally superior because it will smooth any day-to-day variability in the system.

Carbon dioxide-based curves, of course, have a form that is the approximate inverse of O_2-based curves. The departure from precise equivalence is that imposed by the RQ or PQ when these parameters differ from unity.

In the case of the photosynthesis/respiration curves, nighttime sections *always* indicate consumption of O_2 or production of CO_2. The daytime sections always tend toward positive O_2 production or CO_2 consumption, although in highly heterotrophic systems they may never actually cross the zero rate line.

In the case of calcification curves, it is possible for both day and night rates to be negative (indicating net dissolution) or for both day and night rates to be positive (indicating continuous net calcification). However, the most negative (or least positive) rates are almost always found at night, since there is usually at least some component of photo-locked calcification in carbonate-based communities. The most common situation in fully developed coral reef communities is that the net nighttime rate c_n is slightly negative and the net daytime rate c_d is positive.

Most diel curves are likely to be asymmetric (as indicated in Fig. 21–1), although the major period of photosynthesis will vary from afternoon in coral reefs to morning for most phytoplankton-based communities. The peak rates y_{max} are usually found close to true solar midday, although it is worth noting that many areas are working to clock time (even without daylight savings), which is a half-hour or more away from true solar time.

The daily net gain or loss, E for trophic balance, G for carbonate balance, will be obtained if the diel curve is integrated over the full 24-h cycle. The values obtained are not influenced by the validity of the assumption concerning the maintenance of nighttime respiration (or carbonate dissolution) rate throughout the day (with or without photorespiration) since E and G are simply the integration of the curve with respect to the zero rate line. This net gain information is therefore the most reliable information that can be derived from an accurate diel metabolic rate curve. It should be stressed that positive values for E or G do not necessarily indicate that the system is gaining. Such values can be interpreted in two ways: (1) There is an in-place gain in standing stocks (from E) or a positive substratum growth rate (from G), or (2) there is an export of organic matter, dissolved or particulate, which exceeds any import of organic matter (from E), or that there is an export of carbonate gravels, silt, and so on, which exceeds any import of such materials (from G). Most often, the actual status of the system is a combination of these two interpretations. For instance, a coral reef slope may exhibit an overall positive value for E over a year, indicating that macroalgal material is physically broken away and exported downstream onto the reef flat. In contrast, the value for E in the summer may be much higher than the value during winter indicating a real seasonal cycle in standing stocks of macroalgae (Kinsey 1977).

If we assume that photorespiration is not a factor to be considered and that daytime respiration will be a continuation of the nighttime baseline (dashed line in Fig. 21–1), then the integration of the area below the zero rate line may be considered the 24-h respiration for the community and is given by

$$R \approx 24r \tag{29}$$

where r is the average value.

By the same reasoning, if we integrate the curve with respect to p, where $p = y + r$, then the diel photosynthesis for the community is given by

$$P \approx \sum p \tag{30}$$

The gross parameters P and R are those by which most ecosystems are assessed. If photorespiration is a factor to be considered, then P and R will both increase by an identical amount. The methods here do nothing to elucidate this parameter. Of course,

$$P - R = E \tag{31}$$

and this holds regardless of photorespiration or the validity of the assumption concerning the maintenance of the nighttime respiration rate throughout the day.

One of the most common parameters quoted in the description of an ecosystem is the dimensionless P/R ratio. This is a measure of the proportional autotrophic self-sufficiency of the system. In general, it is assumed that $P/R > 1$ implies a system with no requirement for outside trophic support. Similarly, $P/R < 1$ implies a heterotrophic balance with a definite need for an organic input. As discussed earlier for E, these simplistic assumptions may be suspect, but they are a good first approximation.

2. *Systematic shortcuts.* It has been found (Kinsey and Domm 1974; Kinsey 1979) that a relatively fixed relationship exists, for any one system, between the peak rate of photosynthesis p_{max} and the diel gross photosynthesis P. This relationship holds only for a particular time of the year but is maintained even with significant modification or perturbation to the nutrient forcing functions of the system and so on. This allows for a very expedient monitoring procedure, for any one ecosystem in which only typical nighttime respiration r and the noon-time net photosynthesis y_{max} must be measured routinely. The need for the development of a full diel curve is thus avoided. Similarly, the shape of a diel curve for any one system is typically sufficiently reproducible that y_{max} can be predicted from values for y obtained between about 1000 and 1400 h. It is important that the values for r be reliable, because they influence not only R ($R = 24r$) but P (since $p_{max} = y_{max} + r$) and then E. Because of this, it is unwise to obtain respiration always during the early postdusk period. In many systems, r tends to be higher than average at that time. Around midnight is a more reliable time, with the safest time range being 2100 to 0300 h in low-latitude areas.

Some typical relationships found in practice to approximate very closely the results obtained by integrating an experimentally determined full diel curve are the following.

One Tree Island reef flat (latitude 23°S):

$$\text{Mid-May, } P = 9p_{max} \qquad R = 24r$$
$$\text{Late June, } P = 8p_{max} \qquad R = 24r$$
$$\text{Late September, } P = 10p_{max} \qquad R = 24r \qquad (32)$$
$$\text{Mid-December, } P = 11p_{max} \qquad R = 24r$$

Lizard Island reef flat (latitude 15°S)

$$\text{Mid-May, } P = 9.5p_{max} \qquad R = 24r$$
$$\text{Late June, } P = 9p_{max} \qquad R = 24r$$
$$\text{Late September, } P = 10.5p_{max} \qquad R = 24r \qquad (33)$$
$$\text{Mid-December, } P = 10.7p_{max} \qquad R = 24r$$

Equivalent relationships can be developed between hourly calcification rates and the integrated diel calcification gain G.

In summary, the approach is based on the assumption that there is a "standard" form for the diel curve developed for *any one* system and that this varies primarily as a function of day length and insolation intensity. Thus, the determination of an amplitude change (r or y) at any reasonable time on the curve can be simplistically assumed to have influenced the whole curve (nighttime for r or daytime for y) in a proportionally equivalent manner. I have found this approach to be very reliable in manipulative experimentation on open systems but recommend considerable caution in its application.

K. Inevitable reduction of "openness"

Although the methods and conceptual approaches described in this chapter have stressed the value of studying open systems to obtain holistic values for the integrated ecosystem, there frequently comes a time when some advantage can be gained by constraining the system in some way.

Because the planktonic community is transitory and varies frequently with time and tide (except in permanent standing-water environments), it cannot be considered an integral part of the total community. Thus, in systems where the benthic activity is relatively low and the planktonic activity is relatively high, it may be desirable to separate the system into these two basic components. The activity of the benthic community can be assessed by difference if (1) the total combined system is studied by the general open-system approach, and (2) the water is isolated into large bags or rigid stirred enclosures, as discussed in Sec. III.D, to allow estimates of planktonic activity. Consider that increased water depth alone will have the effect of increasing the plankton component of the total open-system data.

A second situation in which even less interference to the system is involved is that in which a narrow band of a particular community type occurs and the objective is to monitor this separately from its adjoining systems. This can be achieved readily in all but the highest-energy environments, provided that there are periods in which the water depth is no more than ~ 1 m. At such times, transparent plastic "fences," emergent at the top, can be used to channelize the natural flow so as to ensure that only the environment required is exposed to the water mass between two sampling stations. Conversely, a floating single station monitor can be allowed to follow the water through the flume. In cases in which there is naturally a considerable period of slack water at low tide, this fence approach may be extended to the form of a full non-flow-through fence enclosure (Kinsey 1972, 1978a) of several meters in diameter. Such an enclosure does little to modify any conditions normally applying at such times but ensures the discrete monitoring of a particular subsystem and so on.

The third situation in which enclosure is desirable (in fact essential) is that in which the water depth is sufficiently great that no appreciable changes in water chemistry will be effected by the benthic system under normal open conditions. Such cases do not really belong in a discussion of open systems. However, it has proved possible to monitor such a benthic community by the use of open-ended tunnels of plastic over metal frames (Rogers 1979) such that a relatively natural water flow-through occurs. In this case, the only major interferences with the natural conditions are some reduction in turbulence and the forced increase in the magnitude of chemical anomalies within the water column over those normally occurring. The easier, but even more interfering variation on the deep-water enclosure approach is that of using rigid domes (Wells 1974; Smith and Harrison 1977). These are usually sufficiently small (250 liters for a 1-m-diameter hemisphere) that they substantially restrict the types of community that can be contained. They also considerably modify both the physical and chemical environment from that occurring naturally. Nevertheless, they have proved to be invaluable for in situ community metabolism studies of many deeper high-energy environments.

IV. References

Anderson, D. H., and Robinson, R. J. 1946. Rapid electrometric determination of the alkalinity of sea water using a glass electrode. *Ind. Eng. Chem. (anal. ed.)*. 18, 767–9.

Atkinson, M. J. 1981. "Phosphorous Metabolism of Coral Reef Flats." Ph.D. dissertation, University of Hawaii, Honolulu. 90 pp.

Barnes, D. J. 1983. Profiling coral reef productivity and calcification using pH and oxygen electrodes. *J. Exp. Mar. Biol. Ecol.* 66, 1–13.

Bellamy, N., and Risk, M. J. 1982. Coral gas: oxygen production in *Millepora* on the Great Barrier Reef. *Science* 215, 1618–19.

Carpenter, J. H. 1966. New measurements of oxygen solubility in pure and natural water. *Limnol. Oceanogr.* 11, 264–77.

Culberson, C., Pytkowicz, R. M., and Hawley, J. E. 1970. Seawater alkalinity determination by the pH method. *J. Mar. Res.* 28, 15–21.

Hart, I. C. 1967. Nomograms to calculate dissolved-oxygen contents and exchange (mass-transfer) coefficients. *Water Res.* 1, 391–5.

Hillbom, E., Linden, J., and Pettersson, S. 1983. Probe for in situ measurement of alkalinity and pH in natural waters. *Anal. Chem.* 55, 1180–2.

Hornberger, G. M., and Kelly, M. G. 1974. A new method for estimating productivity in standing waters using free oxygen measurements. *Water Res. Bull.* 10, 265–71.

Hornberger, G. M., and Kelly, M. G. 1975. Atmospheric reaeration in a river using productivity analysis. *J. Environ. Eng. Div. ASCE* 101, 729–39.

Johnson, K. S., Pykowicz, E. M., and Wong, C. S. 1979. Biological production and the exchange of oxygen and carbon dioxide across the sea surface in Stuart Channel, British Columbia. *Limnol. Oceanogr.* 24, 474–82.

Kanwisher, J. 1963. On the exchange of gases between the atmosphere and the sea. *Deep-Sea Res.* 10, 195–207.

Kemp, W. M., and Boynton, W. R. 1980. Influence of biological and physical processes on dissolved oxygen dynamics in an estuarine system: implications for measurement of community metabolism. *Est. Coast. Mar. Sci.* 11, 407–31.

Kester, D. R. 1975. Dissolved gases other than CO_2. In Riley, J. P., and Skirrow, G. (eds.), *Chemical Oceanography*, vol. 1, pp. 498–553. Academic Press, London.

Kinsey, D. W. 1972. Preliminary observations on community metabolism and primary productivity of the pseudo-atoll reef at One Tree Island, Great Barrier Reef. In Mukundan, C., and Gopinadha Pillai, C. S. (eds.), *Proceedings of the First International Symposium on Corals and Coral Reefs*, pp. 13–22. Marine Biological Association, Cochin, India.

Kinsey, D. W. 1977. Seasonality and zonation in coral reef productivity and calcification. In *Proceedings of the Third International Coral Reef Symposium*, pp. 383–8. Rosenstiel School of Marine and Atmospheric Science, University of Miami, Miami.

Kinsey, D. W. 1978a. Productivity and calcification estimates using slack-water periods and field enclosures. In Stoddart, D. R., and Johannes, R. E. (eds.), *Coral Reefs: Research Methods*, pp, 439–68. UNESCO, Paris.

Kinsey, D. W. 1978b. Alkalinity changes and coral reef calcification. *Limnol. Oceanogr.* 23, 989–91.

Kinsey, D. W. 1979. "Carbon Turnover and Accumulation by Coral Reefs." Ph.D. dissertation, University of Hawaii, Honolulu. 248 pp.

Kinsey, D. W. 1983. Standards of performance in coral reef primary production and carbon turnover. In Barnes, D. (ed.), *Perspectives of Coral Reefs*, pp. 209–20. Australian Institute of Marine Science; Brian Clouston, Publisher, Manuka, ACT (Australia).

Kinsey, D. W., and Davies, P. J. 1979a. Inorganic carbon turnover, calcification and growth in coral reefs. In Trudinger, P. A., and Swaine, D. J. (eds.), *Biogeochemical Cycling of Mineral Forming Elements*, pp. 131–62. Elsevier, Amsterdam.

Kinsey, D. W., and Davies, P. J. 1979b. Effects of elevated nitrogen and phosphorus on coral reef growth. *Limnol. Oceanogr.* 24, 935–40.

Kinsey, D. W., and Domm, A. 1974. Effects of fertilization on a coral reef environment: primary production studies. In *Proceedings of the Second International Symposium on Coral Reefs*, vol. 1, pp. 49–66. Great Barrier Reef Committee, Brisbane, Australia.

Marsh, J. A., and Smith, S. V. 1978. Productivity measurements of coral reefs in flowing water. In Stoddart, D. R., and Johannes, R. E. (eds.), *Coral Reefs: Research Methods*, pp. 361–78. UNESCO, Paris.

Nixon, S. W., Oviatt, C. A., Garber, J., and Lee, V. 1976. Diel metabolism and nutrient dynamics in a salt marsh embayment. *Ecology* 57, 740–50.

Odum, H. T. 1956. Primary production in flowing waters. *Limnol. Oceanogr.* 1, 102–17.

Odum, H. T. 1957a. Trophic structure and productivity of Silver Springs, Florida. *Ecol. Monogr.* 27, 55–112.

Odum, H. T. 1957b. Primary production measurements in eleven Florida springs and a marine turtle grass community. *Limnol. Oceanogr.* 2, 85–97.

Odum, H. T., and Hoskin, C. M. 1958. Comparative studies on the metabolism of marine waters. *Public Inst. Mar. Sci. Texas* 5, 16–46.

Odum, H. T., and Odum, E. P. 1955. Trophic structure and productivity of a windward coral reef community on Enewetak Atoll. *Ecol. Monogr.* 25, 291–320.

Rogers, C. S. 1979. The productivity of San Christobal Reef, Puerto Rico. *Limnol. Oceanogr.* 24, 342–9.

Ryther, J. H. 1956. The measurement of primary production. *Limnol. Oceanogr.* 1, 72–84.

Sargent, M. C., and Austin, T. S. 1949. Organic productivity of an atoll. *Trans. Amer. Geophys. Union* 30, 243–9.

Schurr, J. M., and Ruchti, J. 1977. Dynamics of O_2 and CO_2 exchange, photosynthesis, and respiration in rivers from time-delayed correlations with ideal sunlight. *Limnol. Oceanogr.* 22, 208–25.

Skirrow, G. 1965. The dissolved gases: carbon dioxide. In Riley, J. P., and Skirrow, G. (eds.), *Chemical Oceanography*, vol. 1, pp. 312–16. Academic Press, London.

Skirrow, G. 1975. The dissolved gases: carbon dioxide. In Riley, J. P., and Skirrow, G. (eds.), *Chemical Oceanography*, vol. 2, pp. 1–192. Academic Press, London.

Smith, S. V., and Atkinson, M. J. 1983. Mass balance of carbon and phosphorus in Shark Bay, western Australia. *Limnol. Oceanogr.* 28, 625–39.

Smith, S. V., and Harrison, J. T. 1977. Calcium carbonate production of the *Mare Incognitum*, the upper windward slope, at Enewetak Atoll. *Science* 197, 556–9.

Smith, S. V., and Jokiel, P. L. 1975. Water composition and biogeochemical gradients in the Canton Atoll lagoon. 2. Budgets of phosphorus, nitrogen, carbon dioxide, and particulate materials. *Mar. Sci. Commun.* 1, 165–207.

Smith, S. V., and Key, G. S. 1975. Carbon dioxide and metabolism in marine environments. *Limnol. Oceanogr.* 20, 493–5.

Smith, S. V., and Kinsey, D. W. 1978. Calcification and organic carbon metabolism as indicated by carbon dioxide. In Stoddart, D. R., and Johannes, R. E. (eds.), *Coral Reefs: Research Methods,* pp. 469–84. UNESCO, Paris.

Stoddart, D. R., and Johannes, R. E. (eds.). 1978. *Coral Reefs: Research Methods.* UNESCO, Paris. 581 pp.

Strickland, J. D. H. 1960. *Measuring the Production of Marine Phytoplankton.* Bulletin 122, Fisheries Research Board of Canada, Ottawa. 172 pp.

Strickland, J. D. H., and Parsons, T. R. 1972. *A Practical Handbook of Seawater Analysis.* Bulletin 167, Fisheries Research Board of Canada, Ottawa. 310 pp.

Weiss, R. F. 1970. The solubility of nitrogen, oxygen and argon in water and seawater. *Deep Sea Res.* 17, 721–35.

Welch, H. E. 1968. Modified diurnal curves. *Limnol. Oceanogr.* 13, 679–87.

Wells, J. M. 1974. The metabolism of tropical benthic communities: in situ determination and their implications. *Mar. Technol. Soc. J.* 8, 9–11.

Williams, P. J. L., Raine, R. C. T., and Bryan, J. R. 1979. Agreement between the ^{14}C and oxygen methods of measuring phytoplankton production: reassessment of the photosynthetic quotient. *Oceanol. Acta* 2, 411–16.

22: Growth patterns and rates

BOUDEWIJN H. BRINKHUIS

*Marine Sciences Research Center, State University of New York,
Stony Brook, New York 11794*

CONTENTS

[461]

I. Objectives

A. Introduction

This chapter outlines procedures for determining growth patterns and rates of macroalgae in field and laboratory experiments. Growth measurements undertaken in the field can be quite different from those in the laboratory, not only because of physical and equipment limitations in the field, but also in the matter of definition. One should first ask, What is growth?

The term "growth" applies to the physiological and biochemical processes as well as the product of these processes. Monod (1949) designated G as *total growth*; however, his definition is really a measure of *yield*, since it is the balance between anabolism and catabolism. The total outcome, or *potential growth*, is not considered in G. There may be additional losses due to cell death, grazing, exudation, and so on, and they must be considered for the time interval over which growth is measured. These losses are bound to be different under field and laboratory conditions, as well as a function of species-specific conditions.

Yield as an indicator of growth is what most investigators measure during routine sampling. Yield can be expressed as some net increase in wet or dry weight, area, length, volume, or cellular component (e.g., carbon). The limitations of using yield must be clearly understood. The time course over which the net increase is measured is usually arbitrarily chosen or is determined by the cessation of growth. The latter is a valid method for organisms with limited growth, that is, determining yield in annual or perennial algae at the end of their vegetative or reproductive phases. However, the measurement of growth during some arbitrary interval may contradict the basic objectives of the research. During some periods in the assessment interval, growth rate may be lower, or net growth may be absent due to some cyclical biochemical processes or physiological responses to environmental changes. Therefore, it is imperative to understand the growth phases of the subjects fully and to evaluate the kinetics of growth as a process within the period of assessing yield.

[462]

The kinetics of growth can be described by examining growth rates calculated from increases in biomass or dimension, as delineated in this chapter, or by measuring net or gross carbon fixation (see Dawes, Chap. 16; Littler and Arnold, Chap. 17; Arnold and Littler, Chap. 18; Browse, Chap. 19; and Kinsey, Chap. 21). To obtain a better understanding of the process of growth and to interpret growth rates calculated as described later, it is advisable to use some estimator of gross productivity to predict potential growth. Such comparisons often provide valuable insight into factors (biotic and abiotic) affecting net growth, or yield.

B. Definitions

Growth rate is a measure of the velocity of growth, or an index relating the multiplication of organic mass to time. Traditional measures of growth rates of microbial cells in batch culture (i.e., constant volume of unchanged medium) led to descriptions of universal sigmoidal growth curves. A description of the various phases can be found in Sorokin (1973). Sigmoidal growth curves during growth of macroalgae may be expected under certain conditions in both the field and the laboratory (Fig. 22–1). It is important to note that most estimates of growth rate should be based on the period called the *exponential* (Monod 1949) or *logarithmic growth phase* (Buchanan 1918), during which the mass of cells doubles over each successive, equally spaced time interval. The doubling time and growth rate are constant throughout this period. The reader is directed to Sorokin (1973) for further discussion of *lag, declining,* and *stationary* growth phases.

The *exponential growth rate* is determined by the following mathematical expression:

$$k = \frac{\log_2 X_2 - \log_2 X_1}{t_2 - t_1} \tag{1}$$

where k is growth rate in units of increase per unit of time (e.g., divisions per day); X_1 and X_2 are the values of wet or dry weight, length, or some other growth index at the beginning and the end, respectively, of the period during which the rate is constant; and t_1 and t_2 are corresponding times at which values were determined, in minutes, hours, days, weeks, months, or years.

The *doubling (generation) time* can be calculated by dividing the time period for which the rate was determined by k. In Equation 1, base 2 logarithm was used because this form yields k as divisions $\times t^{-1}$ (t is the time interval of measurement). If exponential growth rates have been calculated using common (K_{10}) or natural logarithms (K_e), appropriate conversion factors ($K_e \div 0.6931$ or $K_{10} \times 3.322 = k$) can be used to express growth rate in terms of base 2.

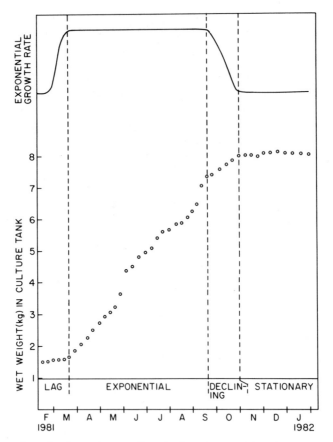

Fig. 22–1. Growth phases in a *Codium fragile* culture. This culture received daily additions of ammonium nitrate to maintain [N] at 30 μ*M*. Tank surface area and volume are 0.25 m² and 150 liters, respectively. Unfertilized cultures (not shown) achieved a biomass of only 6 kg and reached this peak earlier than fertilized cultures (B. H. Brinkhuis, unpublished data).

Confusion in terminology is created by investigators reporting exponential growth rates as *specific growth rates,* using symbols such as *k* and *u*. Although there is nothing wrong with using the term "specific growth rate," it should not be referred to by these symbols, because they are reserved for use with reaction rates. Specific growth rate is customarily reported as percentage per day (Evans 1972) and is based on calculations using natural logarithms in Equation 1. To avoid further confusion, investigators should specify the formula used in their calculations when reporting any growth rate data. The reader is also referred to Kain (1982) for the use of growth rate measurements and terminology.

Growth can sometimes by determined by counting the actual increase in cell number (e.g., in uniseriate filaments or monostratose thallus disks). The term "growth constant" is commonly used to describe growth of unicellular microorganisms. This constant (K_e) must be calculated with natural logarithms in Equation 1. Counts are usually made at two times to ensure that the population of cells is in fact growing exponentially (Guillard 1973). The *division rate* is calculated as

$$\text{divisions per day} = \frac{K_e}{\ln 2} \tag{2}$$

The *generation time* can be calculated as

$$T_\text{d} \text{ (days per division)} = 1 \cdot \text{(divisions per day)}^{-1} \tag{3}$$

This can be multiplied by 24 to obtain hours per division and so on. The reader is referred to Guillard (1973) for further details regarding the computation of K_e and the use of graphic estimates of division rate from linear and semilog plots.

II. Equipment

Determination of growth rates in macroalgae does not require specialized equipment. Such materials as plastic rulers, compound and dissectng microscopes, specimen trays and buckets, and collecting bags are among the obvious items. Other equipment depends on the method chosen to record dimensional or weight increases of plants. Knowledge of the growth regions of the macroalgae under study (apical, basal, or intercalary meristems) may largely determine the appropriate methods and equipment.

A. Field measurements

Many investigators have used cameras to record area coverage (see Littler and Littler, Chap. 8) and size increases when removal of specimens for measurement is not desirable. An underwater camera or one protected with a housing is normally used in both the intertidal and subtidal. A tape measure to record the distance from the focal plane to the subject or an object of known size placed in the photographic view should be used to determine scale factors. The camera also can be attached to a permanent frame of constant size. Maintenance of a perpendicular view toward the image being viewed is essential to avoid size distortion. The film used can be of any type, but 35-mm color slides are useful if the investigator has access to color/digital imaging equipment to determine area coverage (size) of the plants.

Other techniques for measuring macroalgal growth in situ center on the use of markers. Zieman (1974) used a modified stapler to mark individual seagrass blades, where meristematic activity is confined to the base of leaves. Repeated measurements of staple locations yielded growth and dehiscence rates. Cork borers can be used to punch holes in flattened thalli, and the distance these holes move over time can be used to calculate growth rate. Individual plants that are to be repeatedly measured can be tagged with plastic flagging tied to the base of the plant and marked with an identification code by means of indelible ink. Nicholson (1970) made tattoo marks by dipping a needle in carbon ink and puncturing the epidermal layers of *Nereocystis lutkeana*. Some species can be marked with dyes to identify old tissue. More details on these methods are given in Sec. III.A.3.

Weighing macroalgae in the field has always presented problems for investigators. Repeated weighings of individual plants require that the plants be removable; this can be accomplished by attachment of the plants to pieces of pipe, brick and stone, or rope (see Lüning 1969, 1971; Duncan 1973; Waaland 1973; Brinkhuis and Jones 1976). Also, until recently, no reliable balance that was insensitive to wind and other problems could be easily transported to the field. Spring and triple-beam balances work only if there is no wind or vertical motion. Liquid crystal display, battery-powered balances have become available in several weighing and taring ranges (e.g., Ohaus Scale Corp.). These balances are not as sensitive to vertical motion, and they can be easily protected from wind by a small shelter. The use of battery-powered balances with red or blue LED (light-emitting diode) light displays should be avoided, since the numbers can hardly be seen in daylight outdoors.

B. Laboratory measurements

Laboratory studies require suitable culturing systems. Fortes and Lüning (1980) and Markham et al. (1980) describe typical culturing setups for a variety of macroalgal species. Laboratory measurement of size increase (number of cells, length and diameter, or area) for smaller macroalgae generally centers on the use of microscopes equipped with stage or ocular micrometers for measurement calibration. Camera attachments are most useful in maintaining permanent records. For larger specimens, a camera is mounted on a tripod. Photocopy procedures also can be used. The lenses in these machines are not of high quality, and considerable distortion (5–7 mm) may occur at the edges of the field (a photocopy is almost never super-imposable on the original). This is not a significant problem for larger algae, but researchers should calibrate the copies with a scale bar or ruler placed with the material being photocopied if accurate measurements are desired.

III. Methods, sample data, and evaluations

A. Length

1. Flattened thalli. The length of flattened thalli can be expressed in different ways. The overall length of the longest blade is commonly used. In repeated measurements to calculate growth rate, overall length determination can lead to an underestimate of growth rate due to losses from grazing and blade erosion. Such calculations certainly do not represent growth rate in meristematic regions. In some single-bladed forms (e.g., Laminariales) hole punching just above the meristematic region (Parke 1948) has been used extensively to determine growth rate. In a modification of this much practiced technique, small holes (<5 mm) are punched at fixed distances along the blade (every 10 or 20 cm). Large holes should not be used since these weaken the blade, as well as interfere with normal photosynthate transport that occurs along the blade.

At each measurement time, the number of holes remaining, the overall blade length, the stipe length, and the distance from the blade–stipe junction to the basipetal hole are recorded. This information enables one to calculate growth rate from the acropetal movement of the punched holes. Crude estimates of tissue loss due to erosion and grazing can be determined by recording the number of holes lost or by differences between meristematic growth projections and actual measurements of overall blade length increases (better estimate). By measuring distances between holes along the blade, one can also determine whether growth has occurred in the older parts of the blade. In *Laminaria saccharina,* most of the growth occurs in the basal 20 cm, and very little occurs above that (B. H. Brinkhuis, unpublished data). Of course, the growth occurring in nonblade portions should not be ignored. Holdfasts of *Macrocystis* may represent 30–40% of the total plant biomass (Neushul et al. 1982).

Figure 22–2 compares the growth of *Laminaria saccharina* as measured by overall length and by growth in the meristematic region. Note that episodic events can be observed in the overall length determination of growth. The apparent growth rates decrease constantly over time when calculated from overall length (Fig. 22–2a), but not when determined from meristematic activity (Fig. 22–2b).

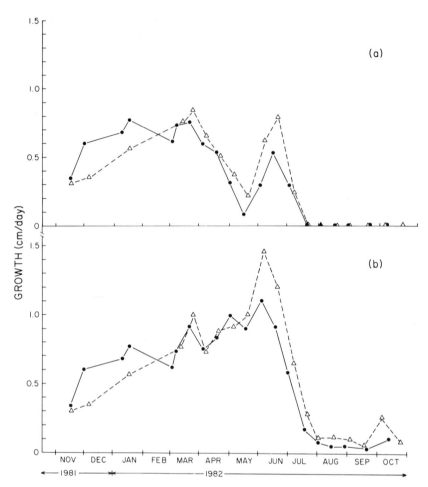

Fig. 22–2. Growth of *Laminaria saccharina* L. on rafts submerged 1 m (●——●) and 2.5 m (△‒‒‒△) below mean low water, as measured by (a) overall blade length changes and (b) hole punch method for determining length changes near the meristem. Overall length measurements suggest a decline in growth rate starting in late March, but meristematic growth does not decline until June. The difference is attributable to apical blade loss caused by storm activity (B. H. Brinkhuis, unpublished data).

The thickness of flattened thalli can be determined from cross-section measurement (microscope or calipers) of disks removed for hole punching (e.g., Kain 1982). Growth models relating these dimensions can then be constructed to calculate volume increases. Many species become thicker at maturity, and this growth is not accounted for by simple length measurement. If these data are further correlated to some biomass (weight) factor, a productivity

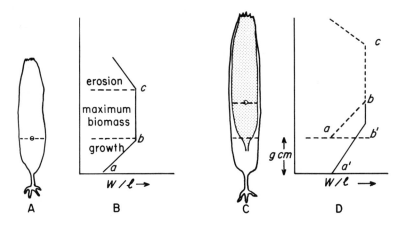

Fig. 22–3. Diagram showing (A) size and shape of *Laminaria saccharina* L. blade with hole punched ~10 cm above the blade base at $t = 0$; (B) distribution of wet weight along the blade ($W \times l^{-1}$); (C) increased size of blade due to growth in meristem and the location of the hole at $t = 1$; and (D) new distribution of wet weight. Note that the curve has shifted to the right due to some width increase as well. After Mann and Mann (1981).

model can be constructed to determine growth rate; however, some macroalgae exhibit large variations in thickness along the thallus. Mann et al. (1979) and Mann and Kirkham (1981) have used the expression

$$(W_{\max}) \times l^{-1} \times G \tag{4}$$

to determine production in morphologically complex Laminariales, *Ecklonia radiata,* where $W_{\max} \times l^{-1}$ is the fresh weight in a section of maximum biomass where no further growth is occurring, and G is the distance the punched hole moved over the measurement interval. The derivation of this geometric expression for *Laminaria saccharina* is shown in Fig. 22–3. Mann (1972) found that a total year's growth of *Laminaria* amounted to one to five times the initial length and that the biomass increase was roughly proportional to the square of the length increase. Power curve fits ($y = ax^b$) are also useful in relating length to weight increases (see Mann 1972; Khailov and Firsov 1976).

2. Filamentous and branched forms. Determining the number of branches and length of each branch to calculate the growth rate of the entire plant is laborious, but not impossible. Khailov and Firsov (1976) provided a good example of relating growth in branched forms (*Cystoseira barbata*) to changes in branch number and size in both annual and perennial forms, but there are disappointingly few such complete observations for macroalgae. There are, however, procedures for modeling tree architecture that could be adapted for use with macroalgae. Borchert and Slade (1981) described the determination of bifurcation ratios and the problems inherent in determining adaptive geometries of tree canopies. They concluded that models based solely on branching ratios in different branch orders do not adequately predict canopy structure and biomass at maturity. Some investigators have supplemented ratio data with physical dimensions of branches in each branch order (e.g., Honda 1971; Barker et al. 1973; Fisher and Honda 1977). Models based on deterministic branching rules alone cease to be predictable when trees reach a certain relatively small size (Borchert and Slade 1981). Additional information on changes in branch dimensions at various developmental stages should yield fairly accurate predictions of canopy structure in macroalgae. One of the problems associated with gathering the necessary data for trees is the long maturation period and life span; this should not be an impediment with macroalgae.

Most estimates of growth rate in filamentous forms are for small macroalgae in which the number of cells or dimensional increase have been determined microscopically. The counting of cells is adequately covered by Guillard (1973) and Green (1973).

3. Calcareous forms. Photographic records of branch elongation can be used to determine growth rate in branched calcareous seaweeds (e.g., Johansen and Austin 1970) or crustose forms (Paine et al. 1979). The vital stain alizarin red, applied at a concentration of 0.25 g · liter^{-1}, has also proved to be a valuable marker for subsequent growth of apical tissue (Andrake and Johansen 1980). The dye labels actively growing tissue depositing carbonate at the time of exposure. Even in the field, the dye was distinguishable after 8 mo.

B. Weight

1. Wet and dry weight. Determination of growth rate from weight increases is the most common method used. Wet weight is supposedly simple to measure, but it is not very meaningful as a biomass (organic tissue) indicator unless some attempt is made to relate wet weight to size or total water content. Morphological and tissue structural differences as a function of age and size or season often affect the total water content and, therefore, apparent wet weight (Black 1950; Baardseth 1970). A more reliable indicator is dry weight, but one often finds that the relationship between wet and dry weight is not linear at higher biomass levels of individual plants (usually exhibited as a decrease in slope).

Many investigators have used different methods to assess wet weight, including standardized blotting, slinging in mesh bags, or other routines to remove "excess" water. Of course, these are not precise techniques, but such standardization can significantly enhance reproducibility of weight measurements. Care should be taken to wash material in clean seawater to remove sediments.

Determination of "dry weight" has also not received the careful attention that it should. Some authors dry samples at temperatures as low as 20°C, whereas others use temperatures in excess of 110°C. Clendenning (1962) discussed the determination of water content in a variety of macroalgae and the consequences of using different temperatures. The low temperature cited is usually referred to as "air dry weight," but frequently the drying temperature and other conditions are not stated. Air dry weight probably should be avoided altogether, because reproducing humidity, wind, or temperature conditions between repeated measurement times is very difficult. All these conditions affect the apparent weight recorded. Furthermore, different species have different chemical compositions; algae containing large amounts of sulfated polysaccharides (e.g., *Chondrus*) are hydrophilic (Clendenning 1962). The use of air dry weight makes interspecies comparisons difficult, if not impossible.

By definition, "dry weight" means the weight of the sample with all water removed. To accomplish this water removal, temperatures in excess of the boiling point must be used. However, temperatures that are too high may volatilize organic compounds in the tissues, especially in algae containing sulfated polysaccharides (Clendenning 1962). It is best to use a temperature between 100 and 110°C to determine dry weight, but one should be aware that the samples may have undergone some chemical alteration, rendering samples unsuitable for other intended analyses. Furthermore, even these higher temperatures may not remove all water since some is organically bound in extremely hydrophilic molecules (e.g., agar and carrageenin). If preservation for such analyses is desired as well, temperatures should not exceed 60°C. Authors should always specify the drying temperatures. One way to obtain dry weight without the use of high temperatures is to lyophilize, or freeze-dry, samples. This reportedly does not alter chemical structure or result in significant losses by volatilization (Clendenning 1962; B. H. Brinkhuis, unpublished data).

In all dry-weight determinations, measurements should be repeated over time until no further weight changes occur. This may be as short as 12 h for small amounts of material in a good convection oven or as long as several days for bulkier samples. Upon removal from the oven, samples should be placed in a desiccator for cooling (minutes) and weighed as quickly as possible thereafter. Many seaweed tissues are very hygroscopic and can remove moisture from a desiccator more effectively than the desiccant used. Substantial weight changes can be observed while the sample sits on the balance. If a high-precision balance is used, a petri dish containing desiccant should be placed in the weighing chamber. It may seem superfluous to mention these precautions concerning such a simple task as weighing material, but it is surprising how often they are ignored by students and professionals.

Dry weight in calcareous forms can be determined after dissolution of carbonate by dilute (5–10%, v/v) HCl. This process may take several days, and the acid solutions should be changed frequently. A subsequent wash with clean seawater should be used. Exposure of organic material to acid results in some losses of organic matter (~10%; M. M. Littler, personal communication). A saturated solution of EDTA can also be used to remove $CaCO_3$ by chelation. The solution is changed daily until all the carbonate has disappeared.

One last point to be made regarding the determination of dry weight is that samples of dried seaweed contain appreciable amounts of salt. It may not always be desirable to wash out the salts before drying, but one should be aware that the recorded dry weight is not all organic weight.

2. Ash-free dry weight. Ash-free dry weight is the dry weight of organic matter, not including the weight of inorganic materials, or "ash." The use of this index results in a better estimate of growth, since it does not include the weight of sediment, carbonate, salts, or other inorganic compounds. Samples should be dried to constant dry weight, as described earlier, followed by combustion in a muffle furnace at 500 to 550°C for at least 3 to 4 h, depending on the amount of tissue. Samples should be combusted in ceramic crucibles that have been acid-cleaned and dried. After combustion, the residue is weighed and its weight is subtracted from the initial dry weight to yield ash-free dry weight. Growth rate can then be calculated from changes in organic weight. Since it is not always practical to determine ash-free dry weight of entire samples, subsampling is frequently performed. However, one should attempt to use similar amounts and types of dry tissue for each subsample.

One should exercise some caution when combusting specimens containing carbonates. The maximum recommended combustion temperature for calcareous forms is 550°C, because higher temperatures can lead to carbonate and salt combustion (see Clendenning 1962). Lower temperatures may lead to incomplete combustion of organic components in heavily calcified samples. It is advisable to perform ash-free analysis on tissues first treated with acid or EDTA to dissolve carbonates. One should assay ash-free yield versus combustion time and temperature. Samples to be combusted should not be placed directly in a hot oven; rapid ignition may follow and create "hot spots" in the crucible, resulting in losses of inorganic material. Oven temperatures should be increased slowly, the samples being placed in the oven when temperatures are less than 200°C, and the combustion timing should start when the furnace has reached the desired combustion temperature.

C. Carbon

In the previous section, dry and ash-free dry weight were indicated to be the most usual base from which to calculate growth rate. However, the carbon content of a given mass of plant matter can also be used as a base. In fact, it may be desirable to determine carbon-specific growth rate, or the carbon-specific carbon uptake measured in primary production assays.

Carbon content can be determined by a variety of methods, most of which were adequately covered by Menzel and Dunstan (1973). Since that time, more modern, automatic carbon/hydrogen/nitrogen analyzers have been developed that permit 100–150 determinations in 24 h. One in particular has a high sensitivity with samples weighing less than 1 mg (Carlo Erba Model 1106 Elemental Analyzer, Erba Instruments).

D. Area

Growth rate also can be derived from an increase in the surface area of a blade. Relating physiological activity (photosynthesis, nutrient uptake, etc.) to surface area rather than weight is preferred by many investigators (e.g., Hatcher et al. 1977; Lüning 1979).

The surface area of cylindrical or elliptical cross-sectioned thalli can most easily be determined from appropriate length and width (diameter) measurements made under a dissecting or compound microscope. However, areas of flattened thalli are not always easily determined from such simple measurements because the blade does not have a convenient geometric shape. This is particularly true for algal crusts that cannot be easily removed. In these cases, fresh material can be placed on a suitable tracing medium and the outlines traced (or photocopied) or photographed in situ (see Paine et al. 1979). The area of the outline is determined after the fact with a planimeter. Marsh (1970) molded aluminum foil over surfaces harboring algal crusts to fit into depressions and over projections. Some investigators have reported cutting out the outline with scissors and developing a weight versus area relationship for the tracing medium. The recorded weight of the cutout can then be equated to area.

A more sophisticated method relies on the use of an area meter (e.g., Lambda Instruments Model LI-3100 Area Meter). An area meter can provide high precision with up to 0.1-mm^2 areas. Several microcomputer companies also offer software–hardware systems that use optical pens to trace outlines on a digitizing tablet (e.g., Apple tablet). The area of the trace, or parts of the thallus, can then be calculated. A more expensive method for determining area came with the advent of color-enhanced video computer systems (e.g., Carl Zeiss). Any image (videotape, 35-mm slide or negative, print, etc.) projected on a monitor can be quantified in the context of area – even areas of slightly different contrasts can be determined.

In all determinations of area of flattened thalli, it is important to remember that the area must be multiplied by 2 to include the area of the other side. Partially desiccated material should not be used.

E. Volume

The volume of a seaweed can be determined and used to calculate growth rate. Volume can be assessed by calculations from physical measurements (e.g., Kain 1982). It can also be determined by displacement in a graduated cylinder (least accurate) or by the collection of water displaced via an overflow and weighing of the water. One should ensure that overflow tubes or channels are kept clean so that water flows easily. Treatment of glassware with surfactants helps. The temperature should also be recorded. Most volumetric measuring devices are made of glass, which has a small temperature coefficient (0.003% $\cdot$ °C^{-1}). Heat-resistant glass has a coefficient of ~0.001%. Thus, the volume changes of glassware due to temperature differences need be considered only for the most exacting measurements. However, the volume of the liquid in the container is highly dependent on temperature and must be corrected to standard conditions (usually 20°C) by standard inorganic chemistry procedures.

IV. Acknowledgments

Some of the data in this chapter were the result of work sponsored by the Gas Research Institute, the New York State Energy Research and Development Authority, New York Gas Group, and the New York Sea Grant Institute. This is Contribution 358 of the Marine Sciences Research Center, State University of New York at Stony Brook.

V. References

Andrake, W., and Johansen, H. W. 1980. Alizarin red dye as a marker for measuring growth in *Corallina officinalis* L. (Corallinaceae, Rhodophyta). *J. Phycol.* 16, 620–2.

Baardseth, E. 1970. Seasonal variation in *Ascophyllum nodosum* (L.) Le Jol. in the Trondheimsfjord with respect to the absolute live and dry weight and the relative contents of dry matter, ash, and fruiting bodies. *Bot. Mar.* 13, 13–22.

Barker, S. B., Cumming, G., and Horsfield, K. 1973. Quantitative morphometry of the branching structure of trees. *J. Theoret. Biol.* 40, 33–43.

Black, W. A. P. 1950. The seasonal variation in weight and chemical composition of the common British Laminariaceae. *J. Mar. Biol. Assoc. U.K.* 29, 45–72.

Borchert, R., and Slade, N. A. 1981. Bifurcation ratios and the adaptive geometry of trees. *Bot. Gaz.* 142, 394–401.

Brinkhuis, B. H., and Jones, R. F. 1976. The ecology of temperate salt-marsh fucoids. 2. *In situ* growth of transplanted *Ascophyllum nodosum* ecads. *Mar. Biol.* 34, 339–48.

Buchanan, R. L. 1918. Life phases in a bacterial culture. *J. Infect. Dis.* 23, 109–25.

Clendenning, K. 1962. Determination of fresh weight, solids, ash, and equilibrium moisture in *Macrocystis pyrifera*. *Bot. Mar.* 4, 204–18.

Duncan, M. J. 1973. *In situ* studies of growth and pigmentation of the phaeophycean *Nereocystis luetkeana*. *Helgo. Meeres.* 24, 510–25.

Evans, G. C. 1972. *The Quantitative Analysis of Plant Growth: Studies in Ecology*, Vol. 1. Blackwell, Oxford. 734 pp.

Fisher, J. B., and Honda, H. 1977. Computer simulation of branching pattern and geometry in *Terminalia* (Compretaceae), a tropical tree. *Bot. Gaz.* 138, 377–84.

Fortes, M. D., and Lüning, K. 1980. Growth rates of North Sea macroalgae in relation to temperature, irradiance and photoperiod. *Helgo. Meeres.* 34, 15–29.

Green, P. B. 1973. Intracellular growth rates. In Stein, J. R. (ed.), *Handbook of Phycological Methods: Culture Methods and Growth Measurements*, pp. 369–74. Cambridge University Press, Cambridge.

Guillard, R. R. L. 1973. Division rates. In Stein, J. R. (ed.), *Handbook of Phycological Methods: Culture Methods and Growth measurements*, pp. 289–311. Cambridge University Press, Cambridge.

Hatcher, B. G., Chapman, A. R. O., and Mann, K. H. 1977. An annual carbon budget for the kelp *Laminaria longicruris*. *Mar. Biol.* 44, 85–96.

Honda, H. 1971. Description of the form of trees by the parameters of the tree-like body: effects of the branching angle and the branch length on the shape of the tree-like body. *J. Theoret. Biol.* 31, 331–8.

Johansen, H. W., and Austin, L. F. 1970. Growth rates in the articulated coralline *Calliarthron* (Rhodophyta). *Can. J. Bot.* 48, 125–31.

Kain, J. M. 1982. Morphology and growth of the giant kelp *Macrocystis pyrifera* in New Zealand and California. *Mar. Biol.* 67, 143–57.

Khailov, K. M., and Firsov, Yu. K. 1976. The relationship between weight, length, age and intensity of photosynthesis and organotrophy in the thallus of *Cystoseira barbata*. *Bot. Mar.* 19, 329–34.

Lüning, K. 1969. Growth of amputated and dark-exposed individuals of the brown alga *Laminaria hyperborea*. *Mar. Biol.* 2, 218–23.

Lüning, K. 1971. Seasonal growth of *Laminaria hyperborea* under recorded underwater light conditions near Helgoland. In Crisp, D. J. (ed.), *Proceedings of the Fourth European Marine Biology Symposium*, pp. 347–61. Cambridge University Press, Cambridge.

Lüning, K. 1979. Growth strategies of three *Laminaria* species (Phaeophyceae) inhabiting different depth zones in the sublittoral region of Helgoland (North Sea). *Mar. Ecol. Progr. Ser.* 1, 195–207.

Mann, K. H. 1972. Ecological energetics of the sea-weed zone in a marine bay on the Atlantic coast of Canada. 2. Productivity of the seaweeds. *Mar. Biol.* 14, 199–209.

Mann, K. H., Jarman, N. and Diekmann, G. 1979. Development of a method for measuring productivity of kelp *Ecklonia maxima* (Osbeck) Papenf. *Trans. R. Soc. S. Afr.* 44, 27–41.

Mann, K. H., and Kirkham, H. 1981. Biomass method for measuring productivity of *Ecklonia radiata*, with the potential for adaptation to other large brown algae. *Aust. J. Mar. Freshw. Res.* 32, 297–304.

Mann, K. H., and Mann, C. 1981. Problems of converting linear growth increments of kelps to estimates of biomass production. In Levring, T. (ed.), *Proceedings of the Tenth International Seaweed Symposium*, pp. 699–704. De Gruyter, Berlin.

Markham, J. W., Kremer, B. P., and Sperling, K. R. 1980. Effects of cadmium on *Laminaria saccharina* in culture. *Mar. Ecol. Progr. Ser.* 3, 31–9.

Marsh, J. A., Jr. 1970. Primary productivity of reef building calcareous red algae. *Ecology* 51, 255–63.

Menzel, D. W., and Dunstan, W. M. 1973. Growth measurements by analysis of carbon. In Stein, J. R. (ed.), *Handbook of Phycological Methods: Culture Methods and Growth Measurements*, pp. 313–20. Cambridge University Press, Cambridge.

Monod, J. 1949. The growth of bacterial cultures. *Annu. Rev. Microbiol.* 3, 371–94.

Neushul, M., Harger, B. W., and Woessner, J. W. 1982. Laboratory and nearshore field studies of the giant California kelp as an energy crop plant. In *1981 International Gas Research Conference*, pp. 699–708. Government Institutes, Inc., Rockville, Md.

Nicholson, N. L. 1970. Field studies on the giant kelp *Nereocystis*. *J. Phycol.* 6, 177–82.

Paine, R. T., Slocum, C. J., and Duggins, D. O. 1979. Growth and longevity in the crustose red alga *Petrocelis middendorffii*. *Mar. Biol.* 51, 185–92.

Parke, M. 1948. Studies on British Laminariaceae. 1. Growth in *Laminaria saccharina* L. Lamour. *J. Mar. Biol. Assoc. U.K.* 27, 651–709.

Sorokin, C. 1973. Dry weight, packed cell volume and optical density. In Stein, J. R. (ed.), *Handbook of Phycological Methods: Culture Methods and Growth Measurements*, pp. 321–43. Cambridge University Press, Cambridge.

Waaland, J. R. 1973. Experimental studies on the marine algae *Iridaea* and *Gigartina. J. Exp. Mar. Biol. Ecol.* 11, 71–80.

Zieman, J. C. 1974. Methods for the study of the growth and production of turtle grass, *Thalassia testudinum* Konig. *Aquaculture* 4, 139–43.

23: Calorimetry

THOMAS H. CAREFOOT

Department of Zoology, University of British Columbia, Vancouver, British Columbia, Canada V6T 2A9

CONTENTS

I. Application of calorimetry to research on macroalgae

The development of fast, accurate, and relatively inexpensive techniques for measuring heats of combustion of biological materials has contributed to the proliferation of research papers on ecological energetics over the past two decades. Terrestrial and freshwater ecosystems have received the most attention, whereas less emphasis has been placed on the marine ecosystem, particularly on marine macroalgae. It is not the purpose of this chapter to review the field of energetics of algae, but only to emphasize that calorimetry can be profitably used, as in other areas of terrestrial and aquatic trophodynamics, in studies of production, conversion efficiency, and partitioning of energy in seaweeds and freshwater algae and in the energetics of herbivore–algal interactions.

The calorie also has been used frequently by ecologists in "time, energy, and risk" theorizing of foraging strategy and has been proposed as a major factor in governing feeding preferences. In this regard, its ease of measurement and status as a fundamental unit in thermodynamics may have exaggerated its importance in such nutritional considerations. The calorific value of a food, although unquestionably an important nutritional component, is unlikely to be any more important in governing feeding preferences of an animal than such characteristics as availability, palatability, or the presence of necessary quantities of essential amino acid, vitamin, and mineral nutrients. These principles apply to fish and invertebrate macroalgivores just as to other terrestrial hervibores and predators.

The *calorie*, a unit of heat energy, has been superseded by the *joule*, a unit of mechanical energy, in countries that have adopted the International System (SI) of Units. Joule determined the mechanical equivalent of heat and showed that 1 calorie equaled 4.187 $\times$ 10^7 ergs or 4.187 joules. The joule has not gained common acceptance in biology, even in countries subscribing to the International System of Units, and it may be a long time before the calorie is displaced as a unit of energy. *Calorific value, caloric value, calorific equivalent, caloric equivalent,* and *caloric content* are used interchangeably to denote calories per unit weight, although the last term may sometimes signify the total number of calories (i.e., calorific value times weight).

II. Equipment

The following equipment is used for nonadiabatic semimicro- or microbomb calorimetry. It is assumed that the researcher has a normal complement of laboratory equipment, including microbalances, chemicals, glassware, distilled water, and so on. Adiabatic calorimetry is briefly treated in Sec. VI.

1. Semimicro- or microbomb calorimeter for small samples (e.g., Parr semimicrobomb calorimeter: 50-mg sample; Phillipson oxygen microbomb calorimeter: 5-mg sample)
2. Potentiometer chart recorder (1 mV full deflection for Phillipson model)
3. Oxygen cylinder with regulating valves
4. Quick-release valves and connectors to join bomb to oxygen regulator (these are normally provided with the calorimeter)
5. Platinum sample holder pans and fuse wire
6. Pellet press
7. Benzoic acid standard
8. Drying oven
9. Freeze-dryer
10. Muffle furnace

III. Method

The simplest part of calorimetry of macroalgae is the actual measurement of the heat of combustion of a sample in a calorimeter, but before this can be done, the algae must be collected, cleaned, dried, pulverized, and formed into pellets. The following suggestions are related primarily to marine macroalgae but also apply to freshwater forms.

A. Collection of specimens

Since most calorimetric analyses are done on algae collected from the field, the researcher will have had to decide where and when to collect and how to standardize samples among different habitats. I could find only two articles (Paine and Vadas 1969; Littler and Murray 1978) in which calorific values were compared for algae from different habitats. Each of these described possible pollution-related differences in calorific value for marine algae collected from two different areas. Despite the lack of published information, however, known differences in morphology, physiology, and reproductive periods between subtidal and intertidal forms of the same species of marine algae or among the same species at different intertidal heights would suggest correlative differences in calorific values based on such habitat differences. Even if the height of collections were standardized for intertidal marine algae, there might be differences relating to aspect, orientation, and other microhabitat conditions that would affect growth form and physiology and thus also calorific value. The brown alga *Fucus distichus*, for example, adopts a markedly different growth form in areas of muddy, as opposed to rocky, substratum. Although time of day is not known to have an important effect on calorific value, time of year must be taken into account for marine and possibly freshwater algae (Himmelman and Carefoot 1975).

The portion of the alga collected should also be noted and standardized in each set of samples collected. Stipe, frond, and reproductive structures differ in their chemical constituents and thus presumably also differ in their calorific values [see Littler and Murray (1978), although such differences were not shown in Paine and Vadas's (1969) study on several species of kelp].

B. Cleaning and drying specimens

Carefully clean the algae of epiphytes and of attached and associated animals. The most desirable specimens to collect are completely free of visible contamination, although even these will harbor some microscopic surface growth such as encrusting diatoms, protozoans, and bacteria. Most of this surface film can be cleaned by gentle rubbing of the algal tissues between thumb and forefinger, then shaking of the alga in water obtained from the field sampling site. (This is especially important for marine algae if collections are from areas of different salinities; kelps, in particular, lose mucilage if washed in low-salinity water or freshwater.) Most of us at one time or another have had to discard a sample, usually in the grinding process, on the unexpected appearance of a tiny dried amphipod or isopod. In order to obtain the most accurate and consistent results possible, the importance of cleaning algae completely, even of the diatom epiflora, cannot be overemphasized.

Dry the plants as soon as possible after collecting and cleaning. Any prolonged delay (i.e., several hours) will increase the risk of chemical changes due to bacterial and internal enzymatic action. For large, bulky algae, at some intermediate point in drying when the tissues are still soft but not wet, the algal tissue can be sliced finely to increase the surface area (this will help in the later grinding process). There are several methods of drying the samples, each varying in convenience, time required, and extent to which chemical constituents of the tissues are destroyed or driven off. The gentlest treatments in this regard are freeze-drying (lyophilization) or vacuum-drying at moderate temperature (e.g., 50°C). Do not oven-dry at temperatures above 90°C owing to the risk of volatilizing such materials as fatty acids and other organic compounds. The use of temperatures lower than 90°C, however, prolongs the drying process and may lead to enzymatic and other processes of degradation. When freeze-drying is employed, the samples may require further drying to remove the last vestiges of water (~5%). This can be of short duration (2 h or to constant weight) at moderate temperature (e.g., 50°C). Alternatively, since the purpose of freeze-drying is to avoid heating the tissues, a correction factor can be applied to obtain the true weight. (This is achieved by heating other samples of the same plant material to constant weight.) Under field conditions, preliminary drying over calcium chloride or another desiccant, or even by the sun, can be a stopgap remedy until proper drying facilities can be employed. In the same manner, one can improvise drying ovens in the field from propane burners, Coleman lamps, and light bulbs run from portable generators. None of these give as consistent values as would be obtained from the use of a proper oven or freeze-dryer. Inconsistencies in drying lead to inconsistencies in calorific values. One should avoid paper toweling when drying and never use press-dried algae. A final caution is not to use preserved samples, because leaching and chemical effects can cause measurable changes in calorific values. The final dried and weighed samples can be stored in air-tight vials or desiccator in a freezer until they can be prepared for calorimetry. Samples should not be stored in a drying oven. Even at moderate temperatures, prolonged storage (e.g., 3–6 mo) in a drying oven can significantly change the calorific value of biological material (Sisula and Virtanen 1977).

If live/dry-weight conversion values are required, replicate portions of the samples can be gently damp-dried and weighed prior to drying the samples to completeness. Absorbent cloth is usually sufficient for this drying, with attention being paid to water trapped in interstices of the alga. A freshwater rinse for marine algae should be avoided because of undesirable osmotic effects. Plants having excessive mucus, such as certain large kelps, can usually be cleaned and damp-dried satisfactorily over all but the cut edges.

C. Preparation of pellets

The final step before calorimetry is to grind the sample to a homogeneous powder and form it into pellets. All but the most woody algae can be easily ground in a mortar and pestle. The stubborn ones must be treated individually and may require mechanical grinding. Pellet presses can be purchased but, once the principle is understood, can easily be made in a workshop. Motor-driven presses are also available. These have the advantage of consistent compression during formation of the pellets. With hand-operated presses, the amount of pressure used to make the pellet has to be learned. A packing that is too tight may lead to incomplete combustion, and the pressure required to form the pellet may cause lipids to be squeezed out; one that is too loose may cause the pellet to crumble during weighing or during attachment of the fuse wire. The packing quality of the algal powder will vary with its physical and chemical makeup, and each species will be slightly different. Sometimes it is easier to form pellets from an algal powder that is slightly moist and then to dry the completed pellet immediately afterward. Pellet size is determined by the constraints of the calorimeter and recording equipment but ranges from 5 to 20 dry milligrams, with a diameter of 3 to 5 mm, for a microbomb calorimeter of the Phillipson type. The pellets should be dried briefly at low temperature (40°C for 30 min), cooled, and weighed before being combusted in the calorimeter.

D. Ashing

If ash-free conversion values are required, replicate portions of the dry powdered sample or pellets can be ashed at a temperature of 500°C to constant weight in a muffle furnace. Temperatures lower than this may lead to incomplete incineration, whereas higher temperatures may cause volatilization of sodium, potassium, and other salts (Grove et al. 1961) and cause calcium carbonate to decompose into calcium oxide and carbon dioxide. An alternative, but not recommended method of obtaining ash weight is to collect and weigh the ash residues remaining in the bomb after combustion. This practice is not uncommon but requires that the bomb be dried and that all the ash be carefully collected from the bomb and separated from any fragments of partially burned fuse wire. It may be difficult to collect all the ash if there has been explosive movement of material during combustion. This method may also lead to inconsistencies in values for ash equivalents for replicate samples, because the actual combustion temperatures may vary from one sample to another. In this regard, Reiners and Reiners (1972) showed small but consistent discrepancies between ash values obtained from bomb combustions and those obtained from muffle furnace combustions for a variety of plant materials, leading to 1 to 2% underestimates of ash-free calorific values. In order to restrict this potential error to less than 1%, these authors recommend that muffle furnace combustions be used to obtain ash values when total ash contents are greater than 5% dry weight.

E. Calorimetry

Follow the methods outlined in the instruction manual for the calorimeter being employed. There is a certain knack to manipulating the pellet on its pan, attaching the fuse wire, assembling the bomb, charging and flushing the bomb with oxygen, and cooling it without dislodging the pellet, but once this is learned an operator can normally perform two or three determinations per hour. To minimize the error associated with acid formation (discussed in the following section), flush the bomb at least once with oxygen to expel residual nitrogen gas before combusting the sample.

The calorimeter and potentiometer–recorder system is calibrated by combusting reference standards, usually benzoic acid (6.318 kcal·g^{-1}). Calibration can be done as a standard reference curve [potentiometer output is linearly related to the weight of benzoic acid combusted over a range of 4 to 85 mg in a Phillipson microbomb calorimeter (Phillipson 1964)] or periodically interspersed with the algal samples. For the latter procedure, one suggestion is to alternate four or five unknowns with one reference standard and to calculate a running average for the standards. In our laboratory, we use an arbitrary "two-behind, one-ahead" pattern, in which two previous standard values, each separated by five unknowns, are averaged with the standard value to come, to give the calibration value to be used for a given unknown. Variation of these standard reference values or of replicate determinations of unknowns (usually three) should be no more than 3–5%.

IV. Sample data

A. Example of a potentiometer recording

Figure 23–1 is a potentiometer record showing potentiometer output over time for the combustion of 4.84 mg of dry *Ulva lactuca* in a Phillipson microbomb calorimeter. Note that both axes are in arbitrary units corresponding, on an original potentiometer recording, to the grid units on the tracing paper. Since the bomb is not fully insulated, heat flux with the surrounding air requires that both pre- and postfiring corrections be made. Some researchers wait for the prefiring temperature of the bomb to equilibrate exactly with air temperature before initiating combustion, but this is not necessary. In the example given, then, some of the apparent temperature rise of 15.7 units (estimating tenths on the tracing paper record) repre-

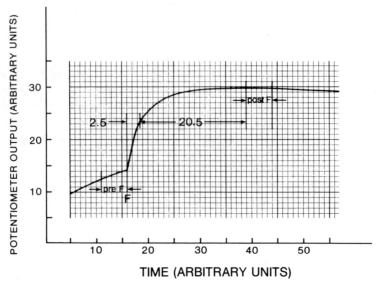

Fig. 23–1. Sample potentiometer record from the combustion of 4.84 mg dry *Ulva lactuca* in a Phillipson microbomb calorimeter. Symbols: F, firing of bomb; pre F, prefiring estimate of temperature change; post F, postfiring estimate of temperature change. The numbers indicate time units required to reach 60% of the peak temperature (2.5 time units) and time units required for the remaining 60–100% of the peak temperature (20.5 units).

sents a continuation of the prefiring gain of 1.6 temperature units over 5 time units ("pre F" in Fig. 23–1; 5 time units are used simply to permit a more reliable estimate of the prefiring gain to be made). Later in the combustion process, a postfiring loss of 0.1 temperature unit over 5 time units has reduced the overall temperature rise. Since we do not know precisely when one effect stops and the other begins, we arbitrarily apply a prefiring correction to 60% of the total apparent rise and a postfiring correction to the remaining 40% of the rise. In this sample it took 2.5 time units to reach 60% of the rise, necessitating a correction of -0.8 temperature unit [(2.5/5.0 $\times$ 1.6], and 20.5 time units to reach the peak rise, necessitating a correction of $+0.4$ temperature unit [(20.5/5.0) $\times$ -0.1].

Upon ignition, a small increment of heat is produced by the fuse wire. Although it is sometimes ignored, a further correction can be made for this additional heat by calculating it from the amount of wire between the electrodes (the calorific equivalent varies with the type of wire used) or measuring directly by firing a bomb with the fuse wire connected but without a sample under conditions in which the bomb is perfectly equilibrated with the air temperature outside of the bomb. Our fuse wire value for a Phillipson microbomb calorimeter is 0.2 temperature unit, which is subtracted from the total apparent temperature rise.

With these corrections, the temperature rise becomes 15.1 units. This rise is converted to calories or joules using the data from combustion of known amounts of benzoic acid reference standards.

B. Sample calculations

1. Reference standards
   a. 10.96 mg benzoic acid gave a corrected rise of 50.6 units (corrected also for acid; see Sec. V) at

$$6.318 \ \text{cal·mg}^{-1} = \frac{6.318 \times 10.96}{50.6} = 1.368 \ \text{cal·unit}^{-1}$$

   b. Similarly, 12.22 mg benzoic acid = 57.9 units = 1.333 cal·unit^{-1}
   c. 12.10 mg benzoic acid = 55.1 units = 1.387 cal·unit^{-1}; mean = 1.363 cal·unit^{-1}
2. Unknown sample (4.84 mg dry *Ulva lactuca*)
   a. Total caloric content = 15.1 units × 1.363 cal·unit^{-1} = 20.58 cal; corrected for acid (see Sec. V) = 20.58 cal − 0.13 cal = 20.45 cal
   b. Energy values
      i. Calories per dry gram = (1000/4.84) × 20.45 cal = 4.225 kcal·dry g^{-1}; Joules per dry gram = 4.225 × 4.187 = 17.690 kJ·dry g^{-1}
      ii. Calories per ash-free dry gram (20% dry weight ash) = (100/80) × 4.225 = 5.281 kcal·g^{-1} ash free or (100/80) × 4.225 × 4.187 = 22.113 kJ·g^{-1} ash free
      iii. Calories per live gram (18% dry weight) = 0.18 × 4.225 = 0.761 kcal·live g^{-1} or 0.18 × 4.225 × 4.187 = 3.184 kJ·live g^{-1}

V. Critical evaluation

During the combustion process, small amounts of nitric and sulfuric acids are produced, nitric acid from the residual nitrogen gas remaining in the bomb after flushing with oxygen and from nitrogenous components in the sample, and sulfuric acid from various sulfur compounds in the sample. (A small quantity of hydrochloric acid may also be produced.) These reactions are exothermic and thus increase the apparent heat output from the combustion of a sample. An acid correction is easy to make by assuming that all acid is HNO_3 and titrating the bomb washings with 0.0725 N sodium carbonate, yielding a -1.0 cal correction per milliliter of titrate on the caloric content of the sample combusted, but it is not always done in microbomb calorimetry. Its effect is considered to be negligible (1.0% or less for most biological materials, including algae) and, in comparative studies, is in any case considered to be consistent among samples of similar size and chemical makeup. It is interesting that Schroeder (1977) showed that, whereas the overall effect of acid correction on the calorific value of plant material was small (e.g., 0.27–0.31% for leaves of several species of deciduous trees), it nevertheless accounted for a large proportion of the overall variability in replicate determinations for both the Parr semimicrobomb and Phillipson microbomb calorimeters (20–30 and 10–20% of the variations, respectively). Lieth (1975) reported correction values of less than 0.1% for a variety of plant materials, and Paine (1971) reported correction values of 0.2 to 0.9% for three species of marine algae, in each case well within the overall variability of the technique. Our own values for acid correction for marine algae are mainly less than 1.0% (using a Phillipson microbomb calorimeter) and are thus consistent with the aforementioned values. Clearly, if acid corrections are not applied, the amount of potential error can be minimized by careful and consistent multiple flushings of the bomb with oxygen before combustion.

High concentrations of inorganic substances in the sample have two main effects, both of which lead to error in determining calorific value. The first is that combustion may be poor, evidenced by black carbon ash on the sample pan and in the bomb. One solution to this problem is to augment the sample with a measured amount of combustible "filler" material of known caloric content, usually benzoic acid or mineral oil, and to combust the sample and additive together. Although this can be effective, it can also lead to error, depending on the type of filler material used and possible uncertainty as to the exact proportion of filler substance to biological material in the pellet (Phillipson 1964). Alternatively, the wet-oxidation method of calorimetry can be used (see following section).

The second effect of inorganic content, particularly in samples rich in carbonates or other decomposable salt, is endothermy. This problem may occur when samples of coralline algae or other algae rich in calcium carbonate are combusted. Paine (1966) discussed this problem in detail and provided graphs showing the extent of endothermy at different concentrations of calcium carbonate (endothermy equivalent to 0.14 cal per milligram of $CaCO_3$). Generally speaking, no correction is necessary for algae with calcium carbonate levels of less than ~40% dry weight. Unfortunately, most coralline

algae have 75–90% dry weight calcium carbonate, for which corrections are therefore necessary. For this purpose, a reasonable estimate
of the amount of calcium carbonate in a sample can be obtained by
measuring weight loss after acid treatment.

VI. Alternative techniques

A. Iodate/sulfuric acid wet-oxidation method

If samples with high ash value combust poorly (even with addition
of known proportions of benzoic acid) or if unrealistically low calorific
values are obtained (e.g., less than 4.18 kcal per gram of ash-free
material, equivalent to the calorific value of cellulose), even when
·corrections for endothermy (for $CaCO_3$) are applied, then the iodate/
sulfuric acid wet-oxidation procedure of Karzinkin and Tarkovskaya
(1964) and Hughes (1969) can be an alternative to bomb calorimetry.
The method is simple, requires no special equipment, and can be
performed on small samples (8–15 mg dry weight). However, since
both Hughes (1969) and Lilly (1975) found that protein is not
completely oxidized by this technique, a correction factor (ranging
from 1.13 to 1.69 kcal per gram of protein in the foregoing studies)
must be applied to obtain the correct calorific value. Since protein
must be determined separately (usually as an estimate from total
nitrogen in the sample), the entire procedure takes a long time.

B. Proximate analysis

An indirect method for obtaining the calorific value of a sample
involves converting the component weights of protein, ether-soluble
substances ("fats"), and carbohydrates in the sample to equivalent
calorific values. Because the conversion values represent averages
from the literature for the major foodstuff components of the sample
(see Brody 1945) and because the determinations of the components
themselves are at best approximations, the resulting calorific values
are, at most, estimates. I could find no published accounts on
macroalgae in which techniques of bomb calorimetry, wet oxidation,
and proximate analysis are compared. However, Lilly (1975) provided
such comparative data in an analysis of sea urchin gonads. Table
23–1 shows these data for sea urchins and indicates good agreement
of calorific values obtained by bomb calorimetry and those obtained
by the wet-oxidation method and, not unexpectedly, poorer agreement with those obtained by proximate analysis technique.

C. Adiabatic calorimetry

Another method of direct combustion of samples involves the adiabatic bomb calorimeter. This requires the presence of an insulating

Table 23–1. *Comparison of three methods of calorimetry*

	Calorific value (kcal · g^{-1} dry wt)		
Sample	Bomb	Wet oxidation	Proximate analysis
1	5.84	5.87	5.88
2	5.22	5.29	5.37
3	5.54	5.53	5.63
4	5.54	5.61	5.75
5	5.71	5.76	5.93
6	4.81	4.82	5.04

Note: Data are for sea urchin gonads (*Tripneustes ventricosus*).
Source: Lilly (1975).

temperature-controlled jacket around the bomb or around the liquid container in which the bomb rests. This jacket is maintained as precisely as possible at the temperature of the bomb throughout the combustion sequence ("zero thermal head"). Its use, therefore, eliminates the heat-leakage corrections necessary in nonadiabatic calorimetry. For this reason, the technique is more accurate than the nonadiabatic method, but because of the insulating jacket the equipment is larger, requires plumbing and/or auxiliary heating devices (thus, is less portable), and is slower to operate than the nonadiabatic kind.

VII. References

Brody, S. 1945. *Bioenergetics and Growth.* Hafner, New York. 1023 pp.

Grove, E. L., Jones, R. A., and Mathews, W. 1961. The loss of sodium and potassium during the dry ashing of animal tissue. *Anal. Biochem.* 2, 221–30.

Himmelman, J. H., and Carefoot, T. H. 1975. Seasonal changes in calorific value of three Pacific coast seaweeds, and their significance to some marine invertebrate herbivores. *J. Exp. Mar. Biol. Ecol.* 18, 139–51.

Hughes, R. N. 1969. Appraisal of the iodate–sulphuric acid wet-oxidation procedure for the estimation of the caloric content of marine sediments. *J. Fish. Res. Bd. Can.* 26, 1959–64.

Karzinkin, G. S., and Tarkovskaya, O. I. 1964. Determination of caloric value of small samples. In Pavlovskii, E. N. (ed.), *Techniques for the Investigation of Fish Physiology*, pp. 122–4. Israel Program for Scientific Translations, Jerusalem.

Lieth, H. 1975. Measurement of caloric values. In Lieth, H., and Whittaker, R. H. (eds.), *Primary Productivity of the Biosphere*, pp. 119–29. Springer-Verlag, New York.

Lilly, G. R. 1975. "The Influence of Diet on the Growth and Bioenergetics of the Tropical Sea Urchin, *Tripneustes ventricosus* (Lamarck)." Ph.D. dissertation, University of British Columbia, Vancouver. 216 pp.

Littler, M. M., and Murray, S. N. 1978. Influence of domestic wastes on energetic pathways in rocky intertidal communities. *J. Appl. Ecol.* 15, 583–95.

Paine, R. T. 1966. Endothermy in bomb calorimetry. *Limnol. Oceanogr.* 11, 126–9.

Paine, R. T. 1971. The measurement and application of the calorie to ecological problems. *Annu. Rev. Ecol. Syst.* 2, 145–64.

Paine, R. T., and Vadas, R. L. 1969. Calorific values of benthic marine algae and their postulated relation to invertebrate food preference. *Mar. Biol.* 4, 79–86.

Phillipson, J. 1964. A miniature bomb calorimeter for small biological samples. *Oikos* 15, 130–9.

Reiners, W. A., and Reiners, N. M. 1972. Comparison of oxygen-bomb combustion with standard ignition techniques for determining total ash. *Ecology* 53, 132–6.

Schroeder, L. A. 1977. Caloric equivalents of some plant and animal material: the importance of acid corrections and comparison of precision between the Gentry–Wiegert micro and the Parr semi-micro bomb calorimeters. *Oecologia* 28, 261–7.

Sisula, H., and Virtanen, E. 1977. Effects of storage on the energy and ash content of biological material. *Ann. Zool. Fennici* 14, 119–23.

24: Nutrient uptake

MARILYN M. HARLIN

*Department of Botany, University of Rhode Island,
Kingston, Rhode Island 02881*

PATRICIA A. WHEELER

*School of Oceanography, Oregon State University,
Corvallis, Oregon 97331*

CONTENTS

I. Objectives

Several objectives underlie the assessment of nutrient uptake and uptake kinetics by both laboratory and field measurements. Major considerations include (1) the mechanism of nutrient uptake, (2) the primary environmental and physiological factors affecting nutrient uptake, (3) the identification of nutrient limitation of growth, (4) the prediction of the relationship between nutrient supply, uptake rates, and growth rates, (5) the calculation of nutrient turnover rates and the evaluation of the contribution of each species of a nutrient element to total plant requirements, and (6) the use of kinetic parameters of uptake to predict the competitive advantage of one species over another when nutrient supplies are limiting.

Although interest in nutrient uptake by macrophytic algae has increased dramatically in the past decade, this research field is still relatively undeveloped both theoretically and experimentally in comparison to similar studies with cultured and natural phytoplankton populations. Furthermore, the great morphological diversity of macroalgae may preclude the development of "standard" procedures for the determination of nutrient uptake. Here, we attempt to provide a general starting basis for initiating uptake studies. In addition to describing basic procedures that should be adapted to each alga or experimental question of concern, we have tried to point out the major issues that must be considered for sound interpretation of the data in physiological and ecological terms.

II. Equipment

A. Enclosed chambers

To quantify the amount of nutrient supplied to a known volume of water containing a specific algal mass requires a chamber that encloses either an entire plant or a part of a plant. In the laboratory, flasks or aquaria are used within a growth chamber or greenhouse with controlled light and temperature. Water movement is essential for proper nutrient uptake, and continuous water circulation or movement of the algal tissue should be maintained. Water circulation can be achieved with a shaker, pump, magnetic stir plate, or bubbler. Glass bottles or other inflexible containers used in the field (e.g., to test nutrient uptake at different depths) can be shaken from a frame. To assess nutrient uptake in situ, small-scale flexible but rugged plastic photosynthesis chambers are useful because water turbulence inside can be made to equal that outside (Gust 1977). Large-scale microcosms such as land-based tanks and submerged plastic spheres in which nutrient dynamics in ecosystems have been examined (reviewed in Pilson and Nixon 1980) could also serve as containers for assessing nutrient uptake in large masses of algae. The extent to which any chamber reduces light must be considered in the analysis of the data.

B. Enrichment dispensers

Nutrients can be dispensed either as a pulse or in a continuous flow. For either system, the nutrient concentration and volume of water are needed to quantify the supply. Pulses are supplied through a syringe into a closed plastic container or from a graduated cylinder. In a continuous-flow scheme, nutrients drip into the medium regulated by a valve or flow from a peristaltic pump.

C. Nutrient analyses

For a discussion of nutrient analyses see Wheeler (Chap. 3).

D. Biomass measurements

The equipment required for making biomass measurements consists of an oven for attaining constant dry weight (70–90°C) and a top-loading balance with sensitivity to distinguish changes in the biomass of individual alga or subset of a community (0.01–1000 g).

III. Methods

A. Sample preparation

In the laboratory, segments of thalli are generally used. These or small whole plants are cleaned of major epiphytes and allowed to equilibrate to the experimental conditions overnight. In the field, small whole plants (e.g., *Fucus*) are enclosed with the least possible disturbance, or individual blades of large macroalgae (e.g., *Macrocystis*) are isolated from the rest of the plant.

B. Containment and sampling procedure

Trial runs are required to establish the size of the chamber and the amount of tissue suitable for an individual experiment. The larger the plant-to-volume ratio, the more likely it is that uptake will be observed, but excessive plant matter can remove the nutrient too rapidly for uptake to be measured. Incubation times must also be tested with preliminary time course experiments. Too short a period may result in an underestimation of uptake rates if an initial lag phase was present. Too long a period may also result in an under-estimation of uptake rates if either nutrients become depleted or the algal tissue becomes nutrient-saturated. Incubation conditions (in either the field or the laboratory) should mimic in situ conditions as closely as possible. This includes substrate concentrations.

C. Disappearance of nutrients

The disappearance of nutrients can be assayed in both field and laboratory experiments. Whereas the former approach the "real world" more closely, the latter can be controlled more easily. Both types of measurements are needed. Replicate experiments are re-quired to determine physiological variability among plants. A known quantity of nutrient is supplied to algae in a container (e.g., Harlin and Craigie 1978), and after complete mixing, the nutrient concen-tration of the medium is measured. Thereafter, concentrations of nutrient in the medium are again determined at progressive intervals (e.g., 15 or 30 min). The absolute amount of nutrient removed is calculated from the average concentration and the volume of water for the particular time interval used. Because the rate of uptake is an average for the interval tested, the *average* nutrient concentration must be used in calculations. Correction for any significant changes in water volume during the course of an experiment is essential, particularly when the volume is small compared with plant tissue mass. An alternative procedure relies on the use of multiple flasks with different initial concentrations of nutrients and equivalent algal biomasses (Hanisak and Harlin 1978; Topinka 1978). The algae are removed from all flasks at the end of one interval (e.g., 2 h). The primary advantage of this method is that the possibility of an initial high nutrient concentration altering the subsequent uptake rate is avoided. In addition, there is no need to calculate changes in water volume. The disavantage of this approach is that it requires measuring different plants. In the field, any measurement of nutrient uptake by macroalgae also requires a measurement of uptake by phytoplank-ton and bacteria. This control can be achieved by setting up a separate container without macroalgae. Procedures for measuring nutrient concentration in water are described by Wheeler in Chap. 3.

D. Biomass determination

At the end of an experimental run, the plant or plant portion that was exposed to nutrients is rinsed with distilled water and dried to

constant weight. This step provides the last laboratory data needed
to calculate uptake rate. (A plausible alternative to standarization
might be to base uptake data on area, chlorphyll *a* content, or the
specific amount of tissue nutrient, e.g., grams NO_3 per gram nitrogen
per hour.)

E. Calculation of uptake rates and kinetic parameters

Uptake is defined as the net accumulation in plant tissue or net
disappearance from the medium of the nutrient element of interest.
As such, uptake rates should be expressed as the change in nutrient
concentration over time (ds/dt) normalized to some measure of
biomass. When possible, uptake rates should be calculated from the
slope of the linear regression of external or internal nutrient concentration versus time, using at least three or four points for each
rate determination.

In some cases, an initial rapid phase of uptake may occur and has
been attributed to the movement of solutes into the cell wall space.
Subsequent uptake is characterized by a slower phase, which continues at a constant rate for several hours. The initial rapid uptake has
been termed movement into the apparent free space (AFS). Methods
for the kinetic determination and physiological characterization of
AFS are given in Luettge and Higinbotham (1979). A portion of
solute uptake in the cell wall space is due to binding of charged
particles to fixed-charge sites and is most readily seen with divalent
cations (Haug and Smidsrød 1967, Myklestad and Haug 1981,
Skipnes et al. 1975). Similar biphasic kinetics have been observed for
ammonium and nitrate uptake by marine phytoplankton and can be
explained in part by an uncoupling of membrane transport and
subsequent metabolic assimilation (see Wheeler 1983). If nonlinear
time courses are obtained, interpretation of nutrient uptake by
macrophytes will be facilitated if the investigator can identify the
process(es) responsible for various phases.

The uptake of nutrients for a given set of environmental conditions
tends to show a hyperbolic relation to external nutrient concentra-
tion(s) commonly referred to as the Michaelis–Menten equation:

$$v = \frac{V_{max}S}{K_s + S} \qquad (1)$$

where V_{max} is the maximum uptake rate, S is the external nutrient
concentration, and K_s is the concentration at which $v = V_{max}/2$.
Some deviations have been found, the most common of which is a
linear increase in uptake rates at high external concentrations (D'Elia
and DeBoer 1978; Haines and Wheeler 1978).

Several procedures can be followed to estimate the parameters V_{max} and K_s that define the hyperbolic curve for saturable uptake kinetics. The best methods provide a direct fit to a hyperbola and also provide confidence intervals for both V_{max} and K_s (e.g., see Wilkinson 1961; Neame and Richards 1972). Alternative methods make use of linear transformations of the Michaelis–Menten equation. The two most common are the $1/v$ versus $1/S$ (Lineweaver–Burk plot), where V_{max} is the slope and $-K_s$ the intercept, and the v versus v/S transformation (Eadie–Hofstee plot), where V_{max} is the intercept and K_s the slope. The linear transformations tend to distort the low or high end of the curves, thus biasing the estimates of either V_{max} or K_s. For a comparison of the merits of these transformations, see Dowd and Riggs (1965).

Regardless of the procedure followed to estimate the parameters for saturation uptake kinetics, several assumptions are made and must be adequately fulfilled for valid determinations of the rates used for the kinetic analysis:

1. *Initial* uptake rates at each substrate concentration must be determined; that is, no significant changes in external or internal concentrations should occur during the incubation period.

2. External nutrient concentration should be the only factor limiting the rate of uptake, or all other factors should be held constant (Sec. IV.A).

3. There should be sufficient mixing to prevent depletion of nutrient at the tissue surface (Sec. IV.B).

The first assumption listed is the least likely to be valid; for example, in disappearance experiments, the external (and internal) nutrient concentrations are clearly not constant. The approach to be taken to minimize the effects of violating this assumption depends on the nutrient under examination and the specific question being asked. For the uptake of divalent metal cations, the slow rate of accumulation across the plasmalemma may be of much more ecological significance than the rapid exchange or filling of the cell wall space. For the uptake of major nutrients such as ammonium and nitrate, the investigator must determine whether membrane transport and internal accumulation or subsequent metabolism is the most appropriate process to examine for the question being asked. These processes are not always tightly coupled in marine phytoplankton (Wheeler 1983) or macrophytes (e.g., Chapman and Craigie 1977), yet the effects of uncoupling on kinetic parameters of uptake have not been adequately examined in either case.

IV. Environmental and physiological factors affecting uptake

A. General conditions

The uptake of nutrients is a dynamic process and can be affected by myriad environmental and physiological factors (for a brief review, see DeBoer 1981). The effects of particular factors (e.g., light intensity, temperature, pH, oxygen concentration, salinity, desiccation, presence of competing nutrients, as well as nutrient preconditioning) may be the specific focus and main variable for a particular study; alternatively, they should be held constant at an environmentally realistic level. Several additional factors that affect experimental determinations of macrophytic nutrient uptake are described in this section.

B. Flow rate

The uptake of solutes by a cell or plant tissue can be restricted by the rate of diffusion of the solute through the boundary layer between the bulk medium and the actual site of the membrane transport mechanism. For cells and tissues suspended in a fluid medium, the boundary layer is proportional to cell or tissue size and inversely proportional to flow or mixing rates. Diffusion of solutes through this boundary layer can have a significant effect on measured uptake rates, particularly at low substrate concentrations or whenever the mediated uptake mechanism is faster than diffusion of the solute. The width of the boundary layer can be decreased by increasing fluid motion around the plant tissue. Experimental effects of mixing and flow rates on nutrient uptake by macrophytes have been demonstrated in the laboratory (Wheeler 1980) but may not limit nutrient uptake in the field (Gerard 1982a).

C. Nutrient uptake by contaminating organisms

When uptake is measured by the disappearance of the nutrient from the medium, the possible uptake by "contaminating" microorganisms (epiphytes, phytoplankton, and bacteria) should be eliminated. When it is not feasible to remove these organisms, their uptake rate should be measured separately and subtracted from the total to determine the amount of uptake that can be attributed to the macrophyte. This precaution is particularly important when one is determining kinetic parameters for uptake, since affinity for a substrate seems in general to be inversely related to cell size.

D. Age and wounding of plant tissue

Macrophytic plant tissue can be quite heterogeneous in both composition and function. Both age and type of tissue sampled can significantly affect uptake rates determined for a given species. Growing tips and reproductive portions of *Fucus* exhibit rates of nitrogen uptake that differ from those in the remaining vegetative tissue (Topinka 1978). Similarly, mature and immature blades of *Macrocystis pyrifera* exhibit different uptake kinetics for nitrate (Gerard 1982b). Such differences should be considered in the extrapolation of results to estimate total uptake by a plant.

When the size of macrophytes precludes the determination of nutrient uptake by the entire plant, a portion of the plant is used. It can be, for example, a detached blade or a series of disks of plant tissue. The use of segments has been criticized because of the potential for wounding, considered to be the excretion of copious amounts of mucus by the plant tissue. Whenever segments of tissues are used in an experiment, it is advisable to postpone measurements until mucus excretion has terminated and the suspending medium can be replaced. (For *Macrocystis*, this process takes 30–60 min.) Simultaneous measurements with intact plants would permit one to estimate the magnitude of wounding effects on the measured uptake rates.

V. Sample data

A. *Disappearance of nutrient from the medium*

Uptake rates can be determined by following the disappearance of nutrient from the medium (Fig. 24–1). The data are obtained by periodic medium sampling and nutrient analysis. Enough data points should be collected for a regression analysis, and the slope of the line gives ds/dt. A lag period or nonlinear uptake suggests departure from ideal conditions. Significant departures from linearity over time could indicate a change in the rate-limiting step for uptake (see Sec. III), feedback inhibition, or nutrient depletion. In some cases these effects can be circumvented by adjusting experimental parameters (e.g., shortening the incubation period). Separating and identifying particular phases of uptake (both kinetic and physiological) are a great asset to subsequent interpretation.

B. *Uptake kinetic curves*

An idealized Michaelis–Menten hyperbolic curve and a linear increase at high [S] are illustrated in Fig. 24–2. A simultaneous plot of a hyperbolic curve and a linear transformation is illustrated in Fig. 24–3. To a certain extent, these curves provide reasonable approximations of observed uptake kinetics. However, it should be cautioned that uptake kinetics may not necessarily exhibit these kinetic patterns,

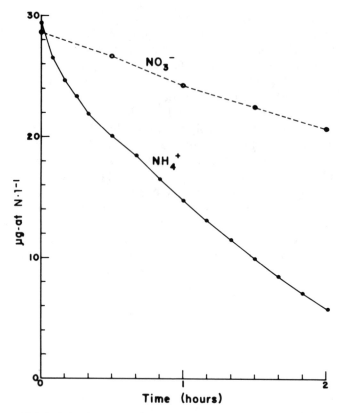

Fig. 24–1. Sample curve for measuring uptake rates by following the disappearance of nutrient from the medium. Only the initial, linear portion of the curve should be used to calculate uptake rates, that is, all points for NO_3^-, but only three points for NH_4^+. From Haines and Wheeler (1978).

and the occurrence of saturable uptake kinetics does not necessarily imply a certain uptake mechanism.

VI. Critical evaluation

A. *Limitations of enclosures*

Although enclosure of plant tissue provides the most sensitive method of measuring uptake rates, it also entails some experimental limitations. The two most serious of these are restrictions of flow rate and the possibility of nutrient depletion. Flow rate is known to affect uptake rates (Wheeler 1980; Gerard 1982a) but does not seem to have a particularly large effect on carbon fixation (Hatcher 1977) or nitrate uptake (Gerard 1982a) in the field. One can avoid nutrient depletion by adjusting the duration of the incubation period.

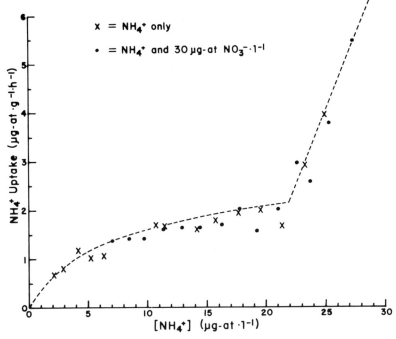

Fig. 24–2. Sample curve for the determination of uptake kinetics showing a hyperbolic relation at low [S] and a linear increase at high [S]. From Haines and Wheeler (1978).

There has been considerable debate concerning the demonstration and interpretation of multiphasic kinetics (Nissen et al. 1980; Borstlap 1981). The usual interpretation of the type of data shown in Fig. 24–2 is a combination of mediated uptake at lower concentrations (2–22 μg-atoms nitrogen per liter) and diffusion at higher concentrations (>23 μg-atoms nitrogen per liter). However, kinetic analysis is insufficient to explain the basis of multiphasic uptake and independent physiological data; for example, temperature effects or the use of inhibitors or competing substrates (see Luettge and Higinbotham 1979) would help distinguish between mediated and non-mediated transport processes. Similarly, for major nutrients such as nitrogen and phosphorus sources, a distinction between membrane transport per se and subsequent metabolic processes (see Wheeler 1983) would clarify the interpretation of multiphasic uptake kinetics.

B. *Correct incubation period and nutrient concentration*

The most critical aspect of valid determination of uptake rates and uptake kinetics is demonstrating that an initial or constant rate has

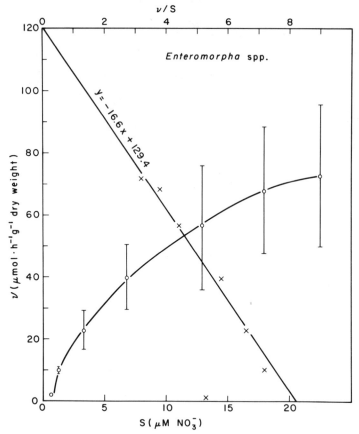

Fig. 24–3. Sample curve for the determination of kinetic parameters for uptake from the linear transformation v vs. v/S. The x's indicate the rate of uptake (v vs. S) shown on the same graph; the ○'s indicate that saturation was not reached. Data from Harlin (1978).

been determined, that is, that no significant substrate depletion and no significant internal accumulation resulting in feedback limitation have occurred. A safe limit to use as a guideline is $\leq 10\%$ change in external and internal considerations (Eppley et al. 1969). However, a ≤ 20–30% limit is often used.

C. Interpretation of uptake kinetics

If nutrient uptake is limited by the external concentration, then determination of the kinetic parameters $V_{\max}$ and K_s for a given set of conditions allows one to predict the uptake rate for a given external nutrient concentration. Likewise, for a constant nutrient supply, one can determine a $V_{\max}$ and K_s for nutrient-limited growth

(Chapman et al. 1978). It should be noted, however, that a simple analysis may not be forthcoming. Since kinetic parameters are highly dependent on a variety of environmental factors, determination of ecologically useful parameters and relationships depends on the ability of the macrophytes to adapt to particular aspects of the environment as well as on our ability to characterize the natural environment.

Although a known relationship between uptake rates and growth rates greatly facilitates determination of the physiological significance of a given nutrient supply, this relationship is relatively easy to establish only for steady-state or constant nutrient supply conditions. When the rate of nutrient supply is variable, uptake rates and growth rates may become uncoupled, resulting in either an internal accumulation of nutrient without a corresponding increase in growth rates (Gerard 1981) or a utilization of internal storage pools for growth when external supplies are depleted (Chapman and Craigie 1977; Gerard 1981).

In ecological terms, the determination of nutrient uptake kinetics has two major applications. The measurement of external nutrient concentrations and uptake rates allows the calculation of turnover rates and evaluation of the relative contribution of a given form of nutrient to growth requirements for the relevant element. This also allows one to estimate the impact of particular macrophytic species on nutrient availability. Similarly, the kinetic parameters for uptake can be used to predict the competitive advantage of one species over another when nutrient supplies are limiting.

VII. Alternative techniques

A. *Isotopes*

Isotopes provide a greatly increased sensitivity that makes it possible to measure uptake rates at low external nutrient concentrations and over shorter incubation periods. Both radioisotopes and stable isotopes are used, the former more frequently. Compounds labeled with ^{14}C ^{32}P, ^{33}P, or ^{3}H are readily available, and radioisotopes of various metals have served as tracers in macrophytes (Penot and Videau 1975; Penot et al. 1976; Manley 1981). Stable isotopes (e.g., ^{15}N) require analysis by mass or emission spectrometry and have only recently been applied to macrophytes (Cowper 1981; Haxen and Lewis 1981).

Although the increased sensitivity of isotopes makes tracer studies a preferred method for uptake measurements, the equipment and expense involved may preclude routine use. Several additional specific precautions must be taken in tracer uptake studies, namely, isotope dilution and equilibration, as well as metabolism and recycling of label. Standard texts on tracer kinetics (e.g., Sheppard 1962) should be consulted for proper use and interpretation of tracer uptake rates.

B. Analogs

Another aspect of tracer uptake is the use of nonmetabolized analogs of nutrients. The advantage of using such analogs lies in the possibility of measuring transport and accumulation capacities separately from metabolism. This separate analysis of various components of the uptake process allows one to examine kinetic parameters and regulation of transport systems. Examples of ^{14}C isotopes of nutrient analogs include aminoisobutyric acid and cycloleucine for amino acids, thiourea for urea, and methylamine for ammonia (Wheeler 1979; Smith and Walker 1978; Walker et al. 1979a,b; MacFarlane and Smith 1982). The major disadvantage of using analogs is that uptake of the nonmetabolized compounds gives a measure of transport and accumulation in contrast to total net uptake, which also includes metabolic assimilation.

C. Uptake measured by accumulation of the relevant element

When enclosure of plant tissue is not feasible, uptake measurements may be possible only by the determination of accumulation of the nutrient element in the plant tissue. Although this method is less sensitive than short-term measurements in enclosed chambers, it can still provide valuable information about uptake rates, especially in fertilization studies (Gerard and North 1981).

VIII. References

Borstlap, A. C. 1981. Invalidity of the multiphasic concept of ion absorption in plants. *Plant Cell Environ.* 4, 189–95.

Chapman, A. R. O., and Craigie, J. S. 1977. Seasonal growth in *Laminaria longicruris*: relations with dissolved inorganic nutrients and internal reserves of nitrogen. *Mar. Biol.* 40, 197–205.

Chapman, A. R. O., Markham, J. W., and Lüning, K. 1978. Effect of nitrate concentration on the growth and physiology of *Laminaria saccharina* (Phaeophyta) in culture. *J. Phycol.* 14, 195–8.

Cowper, S. W. 1981. Uptake of NH_3 by rhizoids of *Caulerpa cupressoides* and translocation: use of sediment nutrient sources. *J. Phycol.* 17(suppl.), 4.

DeBoer, J. A. 1981. Nutrients. In Lobban, C. S., and Wynne, M. J. (eds), *The Biology of Seaweeds*, University of California Press, Berkeley. pp 386–92.

D'Elia, C. F., and DeBoer, J. A. 1978. Nutritional studies of two red algae. 2. Kinetics of ammonium and nitrate uptake. *J. Phycol.* 14, 266–72.

Dowd, J. E., and Riggs, D. S. 1965. A comparison of estimates of Michaelis–Menten kinetic constants from various linear transformations. *J. Biol. Chem.* 240, 863–69.

Eppley, R. W., Rogers, J. N., and McCarthy, J. J. 1969. Half saturation constants for uptake of nitrate and ammonium by marine phytoplankton. *Limnol. Oceanogr.* 14, 912–20.

Gerard, V. A. 1981. Growth and utilization of internal nitrogen reserves by the giant kelp *Macrocystis pyrifera* in a low-nitrogen environment. *Mar. Biol.* 62, 27–35.

Gerard, V. A. 1982a. *In situ* water motion and nutrient uptake by the giant kelp *Macrocystis pyrifera*. *Mar. Biol.* 69, 51–4.

Gerard, V. A. 1982b. *In situ* rates of nitrate uptake by giant kelp (*Macrocystis pyrifera* (L.) C. Agardh): tissue differences, environmental effects, and predictions of nitrogen-limited growth. *J. Exp. Mar. Biol. Ecol.* 62, 211–24.

Gerard, V. A., and North, W. N. 1981. Kelp growth on an ocean farm in relation to fertilizing. In Levring, T. (ed.), *Proceedings of the Tenth International Seaweed Symposium*, pp. 581–6. De Gruyter, Berlin.

Gust, G. 1977. Turbulence and waves inside flexible-wall systems designed for biological studies. *Mar. Biol.* 42, 47–53.

Haines, K. C., and Wheeler, P. A. 1978. Ammonium and nitrate uptake by the marine macrophytes *Hypnea musciformis* (Rhodophyta) and *Macrocystis pyrifera* (Phaeophyta). *J. Phycol.* 14, 319–24.

Hanisak, M. D., and Harlin, M. M. 1978. Uptake of inorganic nitrogen by *Codium fragile* subsp. *tomentosoides* (Chlorophyta). *J. Phycol.* 14, 450–4.

Harlin, M. M. 1978. Nitrate uptake by *Enteromorpha* spp. (Chlorophyceae): applications to aquaculture systems. *Aquaculture* 15, 373–6.

Harlin, M. M., and Craigie, J. S. 1978. Nitrate uptake by *Laminaria longicruris* (Phaeophyceae). *J. Phycol.* 14, 464–7.

Hatcher, B. G. 1977. An apparatus for measuring photosynthesis and respiration of intact large marine algae and comparison of results with those from experiments with tissue segments. *Mar. Biol.* 43, 381–5.

Haug, A., and Smidsrød, O. 1967. Sr, Ca, and Mg in brown algae. *Nature* 215, 1167–8.

Haxen, P. G., and Lewis, O. A. M. 1981. Nitrate assimilation in the marine kelp, *Macrocystis angustifolia* (Phaeophyceae). *Bot. Mar.* 24, 631–5.

Luettge, U., and Higinbotham, N. 1979. *Transports in Plants.* Springer-Verlag, New York, 468 pp.

MacFarlane, J. J., and Smith, F. A. 1982. Uptake of methylamine by *Ulva rigida*: transport of cations and diffusion of free base. *J. Exp. Bot.* 33, 195–207.

Manley, S. L. 1981. Iron uptake and translocation by *Macrocystis pyrifera*. *Plant Physiol.* 68, 914–18.

Myklestad, S., and Haug, A. 1981. The content of polyamionic groups and cation binding in some brown algae. In Fogg, G. E., and Jones W. E. (eds.), *Proceedings of the Eighth International Seaweed Symposium,* The Marine Science Laboratories, Menai Bridge. pp. 589–95.

Neame, K. D., and Richards, T. G. 1972. *Elementary Kinetics of Membrane Carrier Transport.* Wiley, New York. 120 pp.

Nissen, P., Fageria, N. K., Hassan, A. J., and Tang Van Hai. 1980. Multiphasic accumulation of nutrients by plants. *Physiol. Plant.* 49, 222–40.

Penot, M., Floc'h, J. Y., and Penot, M. 1976. Étude comparée de l'absorption et de la rédistribution de ^{45}Ca chez divers groupes de végétaux. *Planta* 129, 7–14.

Penot, M., and Videau, C. 1975. Absorption de ^{86}Rb et du ^{99}Mo par deux alques marines: le *Laminaria digitata* et le *Fucus serratus*. *Z. Pflanzenphysiol. Bd.* 765, 285–93.

Pilson, M. E. Q., and Nixon, S. W. 1980. Marine microcosms in ecological research. In Giesey, J. P., Fr. (ed.), *Marine Microcosms in Ecological Research*, pp. 724–41. Symposium Series 52, U.S. Technical Information Center, U.S. Department of Energy, Washington, D.C.

Sheppard, C. W. 1962. *Basic Principles of the Tracer Method*. Wiley, New York. 282 pp.

Skipnes, O., Roald, T., and Haug. A. 1975. Uptake of zinc and strontium by brown algae. *Physiol. Plant.* 34, 314–20.

Smith, F. A., and Walker, N. A. 1978. Entry of methylammonium and ammonium ions into *Chara* internodal cells. *J. Exp. Bot.* 29, 107–20.

Topinka, J. A. 1978. Nitrogen uptake by *Fucus spiralis* (Phaeophyceae). *J. Phycol.* 14, 241–7.

Walker, N. A., Beilby, M. J., and Smith, F. A. 1979a. Amine uniport at the plasmalemma of charophyte cells. 1. Current–voltage curves, saturation kinetics, and effects of unstirred layers. *J. Membr. Biol.* 49, 21–5.

Walker, N. A., Smith, F. A., and Beilby, M. J. 1979b. Amine uniport at the plasmalemma of charophyte cells: 2. Ratio of matter to charge transported and permeability of free base. *J. Membr. Biol.* 49, 283–96.

Wheeler, P. A. 1979. Uptake of methylamine (an ammonium analogue) by *Macrocystis pyrifera* (Phaeophyta). *J. Phycol.* 15, 12–17.

Wheeler, P. A. 1983. Phytoplankton nitrogen metabolism. In Carpenter, E. J., and Capone, D. C. (eds.), *Nitrogen in the Marine Environment*, pp. 309–46. Academic Press, New York.

Wheeler, W. N. 1980. Effect of boundary layer transport on the fixation of carbon by the giant kelp *Macrocystis pyrifera*. *Mar. Biol.* 56, 103–10.

Wilkinson, G. W. 1961. Statistical estimations in enzyme kinetics. *Biochem. J.* 80, 324–32.

Section IV

Biological interactions

25: Competition among macroalgae

E. J. DENLEY* AND P. K. DAYTON

Scripps Institution of Oceanography, La Jolla, California 92093

CONTENTS

* Present address: Hopkins Marine Station, Pacific Grove, California 93950

[511]

One of the most important factors which will be at once met with in this field is competition, the competition of plants of the same species and of different species growing in the same community. . . . In order to determine the powers of the different species, we must resort to experiment. . . . The simplest way in which this can be done is by clearing a patch of ground of some or all of the species present and seeing what happens. By suitable modification of this procedure, we should be able to disentangle the factors which have led to the actual distribution of species in – that is, to the structure of – the closed association. We shall find that it depends on several factors. In the first place the general habitat conditions . . . determine what species *can* exist in a given spot. Secondly, we have to determine what species can actually gain access to the spot by seed or other propagative organs, and having gained access whether they can germinate and establish themselves.

Tansley (1914)

I. Objectives

The objectives of studies of competition include, first, a demonstration that competition occurs, second, identification of the mechanism by which it occurs, and, third, determination of the importance of competition to the ecology of species or communities. Although this chapter is about macroalgae, a useful background is provided by the literature on competition among other organisms, notably terrestrial plants (e.g., Tansley 1914; Salisbury 1929; Harper 1977; Jackson 1981) and sessile invertebrates (e.g., Hatton 1938; Crisp and Southward 1958; Connell 1972, 1974, 1975; Dayton and Oliver 1980; Underwood and Denley 1984). Many of the concepts discussed here have been applied with various degrees of success and elegance in these fields. An understanding of both the successes and failures in previous studies will aid in any new study of competition. We subscribe to Birch's (1957) second definition of "competition," which stresses a common resource potentially in short supply. We identify three important resources for macroalgae: light, nutrients, and space. There is a need for competitive interactions among species of macroalgae to be defined and quantified in the field. The main aim of this chapter is therefore to advocate and discuss methods for doing this. Higher levels of sophistication in studying competition also will be mentioned here. At present, however, methods for adequately establishing the existence and importance of competition among macroalgae in different environments have precedence. Environments in which algae occur are not all equally amenable to ecological study. Some are physically inhospitable, in others visibility is poor, and in others the organisms may be otherwise difficult to work with. Consequently, techniques possible in one habitat may be difficult or impossible to use in others. Here, we provide a general protocol for studying competition that is conceptually possible. The diversity of citations associated with many of the ideas indicates that such a protocol is practically, as well as conceptually, possible in many habitats.

There are two different forms of competition, although the distinction between them is not always clear. Exploitative or scramble competition refers to the joint usage of the limiting resource. The interaction is directly mediated by the resource in question, and as it becomes increasingly scarce, those individuals that exploit more efficiently dominate the competition. In exclusively exploitative competition, there are no direct, observable antagonistic interactions among individuals. In contrast, interference competition results from interactions among individuals and focuses indirectly, if at all, on the resource. When interference competition occurs, exploitative competition must, however, be potentially possible. Exploitative competition would involve, for example, preemption of a resource, whereas interference competition would result from whiplash or allelopathy, in which individuals were harmed but the resource was not necessarily in short supply.

There are some situations in which both types of competition may occur simultaneously. For example, when one alga overgrows another, any resulting detrimental effects on the survival or growth of the overgrown individual may be due to both physical interference and preemption of light, space, or nutrients. When possible, however, separation of exploitative and interference mechanisms is advised (e.g., Menge 1979), because the ecological and evolutionary consequences of these two types of competition may be very different. These differences are manifested in competition theory (Pianka 1976); overlap in the utilization of habitats is neither necessary nor sufficient for interference competition, but it is necessary although not sufficient for exploitative competition. These definitions point to one of the difficult problems of formulating hypotheses about competition among organisms. Rarely do observations reveal an adequate understanding of competitive processes, because the ability or inability of organisms to coexist does not provide direct evidence of competition. Consequently, we stress the need for field experimentation (see also the epigraph of this chapter).

Both exploitative and interference competition may affect reproductive output, settlement, germination success, survival, or growth of organisms. Reproductive output of macroalgae is usually very difficult to measure in the field but is likely to be proportional to the size of plants of a given species. Data on sessile invertebrates (Barnes and Barnes 1968; Wu et al. 1977) demonstrate, however, that the relationship between size and reproductive output of individuals varies with local and geographical variations in environment. Also, competition with species of red algae affects reproduction in *Durvillea* in New Zealand (Hay and South 1979). Thus, one must be careful when inferring a relationship between these life history characteristics. Because density, sizes, and reproductive output of plants determine the number of propagules released into a habitat, there is a need for techniques to estimate reproductive output adequately so that the population dynamics of species of algae will be understood. Once such techniques are defined, the methods discussed here should apply to determining competitive effects on reproductive output as well as on settlement, survival, or growth.

This discussion is further focused on interspecific competition, despite the fact that intraspecific competition occurs. The methods discussed here are applicable to both, but, again, it is important to remember that interspecific competition and intraspecific competition have different ecological and evolutionary consequences. For example, interspecific competition can lead to restrictions on limits of distribution of species (Hatton 1938; Burrows 1947; Burrows and Lodge 1952; Vadas 1968; Black 1974; Schonbeck and Norton 1980; Foster 1982a) or the domination by one or a few species in successional stages (Foster 1975a; Sousa 1979a,b; Duggins 1980). Intraspecific competition is more likely to result in changes in the density and size distribution of a species within its limits of distribution (Black 1974; Hay and South 1979). The relative importance of inter- and intraspecific competition in a community depends on the dynamics of densities and size distributions of species and also on the differences in the resource requirements of the species (e.g., Tilman 1981).

Macroalgae potentially face different forms of competition, partially because of their different morphologies. For simplicity, we identify three categories of adult algal forms:

1. Flat, encrusting algae, which are often considered to be two-dimensional, for example, encrusting corallines, species of *Hildenbrandia, Peyssonnelia, Pseudolithoderma,* and many blue-green algae.

2. Mat-forming algae, for which the amount of secondary space occupied is often similar to the amount of primary space occupied by the holdfast, for example, species of *Corallina*, *Gelidium*, *Endocladia*, and *Rivularia*.

3. Frondose algae, for which there can be a much greater secondary cover of fronds (and stipe) than the area of holdfast occupying primary space, for example, fucoid and laminarian algae.

These are broad categories, and within each there are substantial differences in morphology. For example, the surface textures of encrusting algae vary from hard and calcareous (e.g., *Lithophyllum*) to soft and slimy (e.g., *Codium lucassi* and blue-green algae). Similarly, the texture, height, and density of mat-forming algae vary considerably. The most apparent variations occur within the category of frondose algae, particularly in size. Included here are the surface canopy kelps, such as *Macrocystis pyrifera* and *Nereocystis luetkeana*, and smaller species, such as *Hormosira banksii*. Frondose species may form canopies, subcanopies. or understories, and the differentiation of these roles is most apparent in subtidal communities (e.g., the large kelp beds near the west coast of North America). The problems of dealing with such vertical differentiation in a community have been discussed by Neushul (1971), Dayton (1973), and Foster (1975a,b) and are akin to those found for terrestrial forests (e.g., Connell et al. 1984).

Some species are characterized by only one of these morphological types as adults. Other species have more than one morphology during their life history. The relative importance of resources and competitive processes among species would be expected to differ among these morphological types. For example, encrusting algae can preempt space, inhibiting the settlement of spores of another species, but could not cause a reduction in recruitment of the other species by whiplash effects. In contrast, a frondose alga exploits space for settlement of another species only at its holdfast, but may inhibit recruitment of the other species over a much wider area by whiplash effects.

There are also other attributes of algae that potentially affect their competitive abilities. Spores of some species may settle only on hard substrata, such as bare rock or encrusting algae, whereas spores of others may settle only epiphytically on upright algae. Yet others may attach to both primary and secondary substrata. Even among species that settle on hard substrata, some may subsequently grow epiphytically on other species. The distinction among these requirements for establishment and growth is integral to an understanding of competitive interactions among species. Also, the dispersive properties of spores of different species and the timing of their release affect their ability to colonize areas (Hutchins 1952; Foster 1975a). In fact, information on all aspects of the biology of species will enhance any study of competition.

Where competition among algae occurs, it is unlikely to be an isolated ecological process. An example is the interaction between competition and grazing (e.g., Foster 1975b; Lubchenco 1978; Underwood 1980). Other environmental factors also affect the importance and outcome of competition among algae (e.g., Foster 1975a, 1982a; Schonbeck and Norton 1980). These include competition for space with animals (Dayton 1971; Foster 1975b); more complex interactions among algae, sessile invertebrates, grazers, and predators in a community (e.g., Underwood et al. 1983); and the effects of physical factors (e.g., Schonbeck and Norton 1980). Some of these factors can be manipulated experimentally (e.g., the effects of grazers, predators, and sessile invertebrates); others cannot (e.g., the effects of many physical factors). When they cannot, there is a need for experiments to be performed under as many different conditions as possible to realize the full importance of competition for species.

Obviously, it is not necessary to have such a thorough understanding of the biology of a species, or of the ecological interactions in which it is important, to demonstrate that it is involved in competition. Ecologists, however, are interested not only in whether a process such as competition occurs, but also in its importance to the structure and dynamics of communities. Studies synthesizing the relative importance of ecological processes, including competition, and the interactions among them are most likely to provide this information.

II. Methods

A. Does important competition exist?

In some situations, the existence of competition is observable, for example, when one plant overgrows or shades another. Even in such situations, it is possible that factors other than competition, or a combination of these and competition, result in the observation. For example, we can imagine situations of apparent exploitative competition in which experiments reveal that resources are never limiting, and the decline of one species results from preferential grazing, sedimentation, or parasitization. Many other cases exist, however, in which competition is not directly observable. For example, exploitative competition can occur when one species preempts the space available for the settlement of the spores of another; interference competition can occur when one species prevents the establishment of another allelopathically. Even when one species occupies all suitable settling space for another, exploitative competition for space will occur only when propagules of the potentially excluded species are present and viable. This is commonly difficult to determine for algae but can be demonstrated if recruits occur in artificially cleared areas while remaining absent or reduced in number in adjacent natural controls (e.g., Chapman 1973; Black 1974; Dayton 1975b; Lubchenco and Menge 1978; Sousa 1979a; Schonbeck and Norton 1980; Underwood 1980; Kastendiek 1982; Reed and Foster 1984). In some cases, the results from experiments may seem obvious in hindsight, but given the complex and interactive nature of many communities, the need to demonstrate causality is often great (Underwood and Denley 1984).

It is well to remember also that the three identifiable resources for which algae compete (light, nutrients, and space) can also be modified by other environmental factors. Even when such factors are controlled or measured, the interpretation of experimental results is not always straightforward. For example, the clearing of an algal mat may lead to the absence, or a reduction in the number, of grazers on a substratum (Dayton 1975b). Alternatively, microalgal grazers from surrounding areas may find clearings more suitable for survival or for grazing (Underwood and Jernakoff 1981). When grazers are easily manipulated, these problems are solved by experiments containing treatments with and without grazers, as well as with and without potential competitors, in an orthogonal design (see Underwood 1981). For small grazers, such as amphipods and isopods, traditional methods of exclusion such as caging, fencing, and removal by hand are impractical, and new techniques must be developed.

B. Does one species affect the settlement and recruitment of another?

The recruitment of a macroalga includes settlment, either direct germination or fertilization and subsequent germination, and growth to a size at which it can be observed in the field. The most commonly observed exploitative competition affecting settlement is preemption of suitable space [e.g., the effects of *Enteromorpha* mats on the settlement of spores in the laboratory (Hruby and Norton 1979)]. Exploitative competition may also occur when species preempt light or nutrients, inhibiting the settlement, subsequent survival of newly settled individuals, or success of germination and survival until visible. Unfortunately, it is difficult to distinguish exploitative and interference competition for light and nutrients in the field. How does one tell whether an alga uses light and nutrients (exploitative) or simply prevents them from reaching another alga (interference)? A comparison for species B of treatments with artificial and real algae (species A) would distinguish the effects of exploitation and interference. It has been demonstrated that the presence of a canopy or subcanopy can reduce or prevent the establishment of algal recruits in intertidal and subtidal habitats (e.g., Dayton 1975a; Pearse and Hines 1979; Reed and Foster 1984). It is likely that at least part of this effect is due to the exploitation of light or nutrients by the adult plants present. Correlative data on the levels of light and nutrients and the recruitment of sporophytes of *Macrocystis* from known, manipulated densities of settling spores also suggest that the levels of these resources, particularly light, are important determinants of recruitment in a kelp bed in southern California (Dean and Deysher 1983). Laboratory studies on *Enteromorpha* (Christie and Evans 1962) demonstrate that the settlement of zoospores is affected by light.

Interference competition can be a result of a number of mechanisms. Even when suitable substrata are available for the settlement of spores, species A may provide a physical barrier to settlement (e.g., Hruby and Norton 1979). This mechanism is ecologically similar to preemption, although space is not limiting and competition results from interference. Interference by species A may also result from allelopathic agents such as poisons or slime, inhibitng either settlement or subsequent early survival. When species B settles on species A, interference competition may result from the sloughing of the outer epidermal cells of A following settlement. In addition, various types of abrasion and whiplash by A can kill recruits of B. In each case, it is assumed that A is established before B, and competition is related to the prevention of invasion. Removal experiments or treatments in which the density of A is manipulated to different proportions of the natural density (e.g., natural density, three-fourths, one-half, and one-quarter of the natural density, and zero), with natural controls for comparison, are the most straightforward way of demonstrating the presence or absence of, and the effect of, competition. Again, preferably the experiments should be done in the presence and absence of grazers and in as many different physical environments as possible.

Other forms of competition may also affect the settlement and survival of algae before the time they become visible. For example, when the settlement of spores of A occurs concurrently with that of spores of B, both exploitative competition and interference competition between them are potentially possible. Given that algal spores of some species do not appear to disperse widely, it is likely that intraspecific competition among them may occur. Present field work investigating the relationship between settlement and recruitment of kelps uses initial, artificial densities of spores that are great (e.g., Dean and Deysher 1983). Such work demonstrates the possibility of doing field experimentation on the smallest life history stages of macroalgae. The even more difficult task of determining natural initial densities of settling spores is necessary before results of these studies can be interpreted for natural communities. If, for example, natural densities of settlers are substantially smaller than the artificial densities used, and the proportion of settlers surviving to be visible recruits is affected by competition among them, then results from "seeding" experiments may not present a realistic picture of the importance of competition. We suggest that any seeding experiments consist of treatments with different initial concentrations of spores (e.g., Schiel and Choat 1980). Also, studies on experimental settling plates, which can be removed, observed in the laboratory under magnification, and returned to field sites, provide a way of determining the relationship between natural densities of settling spores, germination success, and survivial to a stage of visible recruitment (Neushul et al. 1976). Again, the effects of grazing on juvenile stages of algae may reduce their recruitment and should be controlled experimentally if possible.

C. Does one species affect the subsequent survival and growth of another?

Assuming that we can adequately quantify density, survivorship, sizes, and rates of growth of algae, it is possible to test for correlations between the density and sizes of species A and the rate of survival and growth of B. Again, experimentation is necessary to demonstrate causality in these relationships. For any hypothesis about the effect of A on the survival of B to be worth testing, the species must coexist in some places at some times, even if their distributions are only partially overlapping. In many cases, it is useful to monitor the effects of each species on the other, rather than simply the effect of A on B (e.g., Schonbeck and Norton 1980). For simplicity, the latter case is discussed here; the extrapolation of methods to reciprocal experiments on competition is straightforward. Natural densities, sizes, and morphologies of species A and B largely determine the experimental design used to test hypotheses about competition between the two species. Examples of different scenarios are given in Table 25–1 with appropriate experimental designs. We urge that the distributions, abundances, and size distributions of both species be sampled before experimentation to determine the most useful procedure. Table 25–1 is not an exhaustive list of situations in which competitive interactions affecting survival and growth of algae occur, but rather a set of examples elucidating the need to design experiments suitable to the species being studied. Some considerations are common to all examples; the overall densities of plants in treatments must be known and preferably manipulated. For example, in situation

Table 25–1. *Treatments for examining the potential competitive effects of species A on species B*

Morphologies and densities of A and B	Treatments
1. Morphologies and densities similar	a. Natural, $\frac{1}{2}A + \frac{1}{2}B$ b. Overall density halved, $\frac{1}{4}A + \frac{1}{4}B$ c. A removed, B present
2. Morphologies similar, densities different, e.g., $\frac{2}{3}A + \frac{1}{3}B$ naturally	a. Natural $\frac{2}{3}A + \frac{1}{3}B$ b. Overall density reduced to $\frac{1}{2}$ natural, $\frac{1}{3}A + \frac{1}{6}B$ c. A removed, B present
3. Morphologies different, densities similar, e.g., A – mat species; B – encrusting	a, b, and c, as for 1
4. Morphologies different, densities different, e.g., A – mat species; B – encrusting; $\frac{1}{4}A + \frac{3}{4}B$ naturally	a. Natural, $\frac{1}{4}A + \frac{3}{4}B$ b. Overall density reduced to $\frac{2}{3}$ natural, $\frac{1}{6}A + \frac{3}{6}B$ c. A removed, B present
5. Morphologies different, densities different, A much larger, but less abundant than B, e.g., A – canopy species; B – subcanopy or mat; $\frac{1}{5}A + \frac{4}{5}B$ naturally	a. Natural, $\frac{1}{5}A + \frac{4}{5}B$ b. A removed, B present
6. Morphologies and densities similar, A – epiphyte on B	a. Natural, $\frac{1}{2}A + \frac{1}{2}B$ b. A removed, B present
7. Morphologies similar, A and B occupy equal primary substratum, but A also grows on B, i.e., $\frac{2}{3}A + \frac{1}{3}B$ naturally	a. Natural, $\frac{2}{3}A + \frac{1}{3}B$ b. Epiphytic A removed, $\frac{1}{2}A + \frac{1}{2}B$ c. A on primary substratum removed, $\frac{1}{2}A + \frac{1}{2}B$ d. All A removed, B present e. Epiphytic A removed, overall density of A and B on primary substratum reduced to half natural, $\frac{1}{3}A + \frac{1}{3}B$

Note: In many cases the reciprocal experiments (effects of B on A) would also be done.

(1), in which A and B are similar in morphology and coexist in similar densities, the simplest experiment testing the effects of A on the survival and growth of B would contain two different treatments (each replicated):

$$\text{(i) A + B} \quad \text{and} \quad \text{(ii) B}$$

Such treatments can be set up in two ways. In the first, overall densities in both treatments are similar, and in the second, A is simply removed from treatment (ii), such that overall densities in (ii) are half that in (i). The effects of A on B in the latter experiment cannot be interpreted purely in terms of interspecific competition, because they may be due to an overall difference in density rather than to the presence of A. However, the overall difference in density is due to A, so the results of this experiment do tell us about the effects of A on B at their naturally occurring densities. If we do want to know if A specifically has a competitive effect on B, the former experimental design is more appropriate. When species A and B are very different in size, the problem of determining densities for estimating reciprocal competitive effects is more difficult. Usually, experiments containing treatments with different proportions of the natural densities of A and B are most appropriate.

D. Field methods of identifying resources in short supply

The presence of an algal species may affect other species negatively by reducing the amount of light, space, or nutrients available to them. Because experiments evaluating competition for nutrients are difficult to control, the relative importance of these three resources to competitive interactions is rarely known. The problem is compounded because the amounts of available space, light, and nutrients are also affected by other factors. Amounts of sedimentation, suspended particles, and phytoplankton in the water column, depth, incident light levels, strength of water currents, stratification of water masses, and abundance of sessile invertebrates occupying space are examples of factors that may affect the amounts of one or more of these resources. Yet again, the need for suitably controlled experiments, in conjunction with measurements of the amounts of space, light, and nutrients in different treatments, is required. If it is possible to control light with filters or shades and develop artificial algae, a general scheme such as that outlined in Table 25–2 will allow important distinctions to be made about the relative importance of the three resources to competitive interactions.

The protocol described in Table 25–2 can be used to test for the competitive effects of A on settlement, recruitment, survival, growth, or reproductive output of B. Such experiments are conceptually

Table 25–2. *Field experiments designed to distinguish the relative*
importance of the resources, space, light, and nutrients, on the competitive
effects of species A on species B

Treatment	Light	Space	Nutrients
1. Live algae (A) present; natural situation	Reduced	Reduced	Reduced
2. Artificial algae (A); interfere with light and space, not with nutrients	Reduced	Reduced	Not reduced
3. Light filter, algae (A) removed	Reduced	Not reduced	Not reduced
4. Artificial algae (A) made of clear material that does not reduce light	Not reduced	Reduced	Not reduced
5. Removal of all A	Not reduced	Not reduced	Not reduced

Note: These treatments should be set up in initially similar areas, in which both A and B potentially occur. Previous comments on densities of plants, the possibility of doing reciprocal experiments, and the need to control for other factors also apply here.

possible, but their feasibility depends on the algae being manipulated. In some studies, artificial algae have been made to mimic real plants successfully in their physical properties (e.g., Harlin 1973). Artificial shades have also been used in experiments, although not specifically for examining competition among species of macroalgae, for example, domes used by Connell (1974), roofs and cages by Underwood (1980), and shades by Backman and Barilotti (1976) subtidally. Both Connell and Underwood discussed the practical problems of shading areas, and their papers should be consulted before experiments are undertaken. Even when all five treatments listed in Table 25–2 are not possible, subsets of them (e.g., 1, 3, and 4) provide useful information. Care is required in executing the experimental design described in Table 25–2. For example, shades must cut out similar amounts of light as do plants of species A; artificial algae must be similar to actual plants in physical properties, such as flotation and movement in water currents and wave action. When it is not possible to do manipulations to distinguish among the resources for which algae are competing, acquiring correlative evidence is the only recourse.

As noted, the scheme of Table 25–2 is general and in many ways preliminary in approach. For example, some algae can store nutrients and compromise effects of competition on survival with flexible growth rates (Chapman and Craigie 1977). It is therefore advisable to measure rates of survival and growth concurrently in experiments. Scott and Hayward (1955) found that the amount of light affected the uptake and retention of sodium and potassium ions in *Ulva lactuca* and *Valonia macrophysa*. In this case, light and nutrients are likely to have an interactive effect on algae, rather than being totally independent resources. Also, species that commonly occur at different depths presumably have different physiological requirements for light. Work by Kanwisher (1966) and McLeod and Rhee (1970) indicates that different algae utilize different parts of the light spectrum. Lüning and Neushul (1978) determined that, for Californian species of Laminariales, low light levels were required for survival of gametophytes but that two to three times the light intensity was necessary to induce fertility in the same species. Schonbeck and Norton (1980) found differences in the ability of species of fucoid algae to tolerate periods with no light and suggested that this may affect the observed competitive interactions among them on the shore. Requirements for suitable settling space may also be qualitative as well as quantitative; for example, Foster (1975b) found that the establishment of *Macrocystis* sporophytes on artificial blocks in the field was greater near the edges than in the centers of the blocks; Luther (1976) found that *Enteromorpha* and *Porphyra* preferred different granule sizes in rock on which they settled; Harlin and Lindbergh (1977) found that *Corallina* attached to substrata of different granule size but that development was optimal on granules of certain sizes.

In fact, algae of different species and life history stages would be expected to respond to qualitative as well as quantitative changes in all three resources. Whenever possible, qualitative distinctions in resources should be made in field experiments. For example, it is possible to determine the effect of surface texture as well as amount of available substrata for settlement in the field by performing experiments on naturally different types of rock, on rock manipulated to contain pits or bumps, or on texturally different settling plates. Such differences in substrata may affect the outcome of experiments on competition among species from sporeling stages to adult plants. In many instances, however, the cruder experiments simply detecting and quantifying the range of competitive interactions among species are first necessary. Despite the inherent difficulties in determining the relative importance of the resources for which algae compete, this does not mean that the task is impossible. Ingenuity and creativity are important skills in field ecology, and inventive efforts to provide new types of manipulations and avenues of experimentation are needed.

III. Habitats

In this section, we briefly consider the two types of habitat in which ecological research on macroalgae is commonly done and the types of experimental analysis of competition that have been useful in each.

A. Rocky intertidal habitats

It is generally assumed that the main resource for which species compete on rocky shores is space, although the importance of light and nutrients as limiting resources has not been adequately explored. Furthermore, an overabundance of light may be detrimental to the establishment of some species in higher intertidal levels, so the physiological tolerances of individual species must be considered carefully. Certainly, competition among species of algae occurs in intertidal habitats (e.g., Hatton 1938; Chapman 1973; Black 1974; Lubchenco 1978; Lubchenco and Menge 1978; Sousa 1979a,b; Schonbeck and Norton 1980; Underwood 1980; Foster 1982b), but its relative importance in structuring communities on the shore is not well understood. Preemption of suitable settling space for spores is likely to be a very important form of competition (as it is for sessile invertebrates) and one of the factors causing the competition among algae described by Hatton (1938), Black (1974), Hay and South (1979), Foster (1982b) and Sousa (1979a,b) (see also Foster and Sousa, Chap. 13).

Interference competition for space is probably more common in intertidal habitats than we think, and more work is necessary to evaluate its role in intertidal communities. The importance of interference competition in which the fronds of one alga prevent the settlement of spores of another and in which fronds whiplash spores and newly settled plants can be tested by "seeding" the understory with small bags of fertile algal material (e.g., Paine 1979). As discussed earlier, however, the importance of natural levels of competition cannot be ascertained from such seedings. Allelopathic interference competition for space may also be important in some intertidal habitats. Simple removal experiments in the field test the null hypothesis that the alga potentially containing inhibitors has no effect on the settlement or recruitment of other species. To determine the mechanisms of competition, however, more refined experiments are required. For example, a comparison of the settlement of spores on settling plates that are cleared, covered with the potentially allelopathic alga, and soaked in an extract from the alga will often determine whether competition results from chemical inhibition or physical inhibition (e.g., preemption or whiplash). In order to identify chemical inhibitors and the mechanisms by which they are transferred to the surface of the alga, laboratory experiments are required.

Distinctions among different effects of canopies and subcanopies on understory species have not been adequately examined for most intertidal algae. There are myriad mechanisms for the changes that can be observed after the removal of a canopy or subcanopy (as discussed in Sec. II). Thus, there is a need for more extensive manipulative experimentation in this habitat.

B. Subtidal habitats

Much of the material discussed so far is relevant to subtidal habitats. Again, there is already a literature demonstrating that competition among species of algae is important in some subtidal habitats (see Sec. III. A). In these habitats, some recent work has focused on artificial dispersal of spores and inoculation of plates with spores as methods to examine interactions between species (e.g., Dean and Deysher 1983). The relevance as well as the problems of such approaches are discussed in Sec. II. Probably the most difficult problem of understanding algal competition in subtidal habitats is unraveling the interplay between competition for light and nutrients. This may also be a significant problem in intertidal habitats but has not been considered as important there. One approach has been that of adding nutrients to habitats (e.g., Cheng 1969). Nutrient enrichment presents two major problems. One is dilution due to water movements from currents and surge, internal tides, and the interplay between surface flow and turbulence over the fronds of algae. Even if one assumes some success with localizing added nutrients, the second major problem is related to the ability of kelps to store nutrients for utilization during periods of adequate light but low nutrient level (Chapman and Craigie 1977). The experimental scheme described in Table 25–2 provides a better way of evaluating the relative importance of resources in subtidal habitats. When that scheme is not applicable, competitive effects should be investigated under as many different nutrient regimes as possible and for suitable periods to detect any long-term effects on plants.

Although there are often difficult logistic problems in subtidal work (e.g., poor visibility and limited field time), studies have shown that it is possible to do sophisticated experimental manipulations in this habitat. Hence, all the designs for experiments discussed in Sec. II potentially apply to intertidal and subtidal research.

IV. Critical evaluation

The fact that references are associated with many of the ideas in this chapter indicates that methods are already available for investigating competition among species of macroalgae. Some of these methods have been more successful than others, and many problems still exist in the study of competition in the field. The experimental designs discussed in this chapter are simple, involving only two species. Many of the natural situations are complex, and more complex designs will be appropriate. The methodology of more complex experiments is conceptually no different from these simple situations. One of the greatest problems is designing experiments that adequately distinguish competition from the effects of other environmental factors and that also look at competition under different conditions. As mentioned previously, many natural competitive situations are likely to be influenced by grazing. In some cases, a dominant plant species is locally eliminated by preferential grazing (Foster 1975b; Lubchenco 1978), and in other cases, the observed recruitment and rates of survival of algae are at least influenced by grazing (e.g., Paine and Vadas 1969; Dayton 1973; Mann 1973; Lubchenco 1978; Norton 1978; Duggins 1980, 1981; Nicotri 1980; Underwood 1980; Underwood and Jernakoff 1981). Almost any manipulation of algae also influences grazers (e.g., Underwood and Jernakoff 1981), and we cannot overemphasize the need to control or examine such variables. Any other manipulable factors that may affect competitive interactions should be similarly controlled. Environmental factors that cannot be manipulated in experiments may also affect competitive interactions. for example, siltation (Foster 1975b; Norton 1978), temperature (Norton 1977), water movements, and storms (Dayton 1971; Paine 1979). Consequently, experiments should be repeated in as many sets of environmental conditions as possible (i.e., should be repeated in different sites or at different times or both) before general statements are made as to the importance of competition in communities. These other factors not only may far outweigh the importance of direct competition, but may interact with competitive effects so that competition is important under some conditions and not under others. We therefore urge that field studies on competition be synthetic, combining manipulative experiments of potentially important factors in communities with a suitable knowledge of the biology of the algae under investigation.

V. Acknowledgments

Financial support during the writing of this chapter was provided to EJD by a Harkness Fellowship from the Commonwealth Fund of New York. We are grateful to Dr. M. S. Foster and Dr. D. R. Schiel for discussion of the ideas and comments on the chapter and to the editors and reviewers for comments on an earlier draft.

VI. References

Backman, T. W., and Barilotti, D. C. 1976. Irradiance reduction: effects on standing crops of the eelgrass *Zostera marina* in a coastal lagoon. *Mar. Biol.* 34, 33–40.

Barnes, H., and Barnes, M. 1968. Egg numbers, metabolic efficiency of egg production and fecundity: local and regional variations in a number of common cirripedes. *J. Exp. Mar. Biol. Ecol.* 2, 135–53.

Birch, L. C. 1957. The meanings of competition. *Amer. Natur.* 91, 5–18.

Black, R. 1974. Some biological interactions affecting the intertidal populations of the kelp *Egregia laevigata. Mar. Biol.* 28, 189–98.

Burrows, E. M. 1947. "A Biological Study of *Ascophyllum nodosum.*" Ph.D. dissertation, London University, London.

Burrows, E. M., and Lodge, S. 1952. Autecology and the species problem in *Fucus. J. Mar. Biol. Assoc. U.K.* 30, 161–75.

Chapman, A. R. O. 1973. A critique of prevailing attitudes towards the control of seaweed zonation on the seashore. *Bot. Mar.* 16, 80–2.

Chapman, A. R. O., and Craigie, J. S. 1977. Seasonal growth in *Laminaria longicruris:* relations with dissolved inorganic nutrients and internal reserves of nitrogen. *Mar. Biol.* 40, 197–205.

Cheng, T.-H. 1969. Production of kelp: a major aspect of China's exploitation of the sea. *Econ. Bot.* 23, 215–36.

Christie, A. O., and Evans, L. V. 1962. Periodicity in the liberation of gametes and zoospores of *Enteromorpha intestinalis* Link. *Nature* 193, 193–4.

Connell, J. H. 1972. Community interactions on marine rocky intertidal shores. *Annu. Rev. Ecol. Syst.* 3, 169–92.

Connell, J. H. 1974. Field experiments in marine ecology. In Mariscal, R. (eds), *Experimental Marine Biology*, pp. 21–54. Academic Press, New York.

Connell, J. H. 1975. Some mechanisms producing structure in natural communities: a model and evidence from field experiments. In Cody, M. L., and Diamond, J. (eds.), *Ecology and Evolution of Communities*, pp. 460–90. Belknap, Cambridge, Mass.

Connell, J. H., Tracey, J. G., and Webb, L. J. 1984. Compensatory recruitment, growth, and mortality as factors maintaining rain forest tree diversity. *Ecol. Monogr.* 54, 141–64.

Crisp, D. J., and Southward, A. J. 1958. The distribution of intertidal organisms along the coasts of the English Channel. *J. Mar. Biol. Assoc. U.K.* 37, 157–203.

Dayton, P. K. 1971. Competition, disturbance and community organization: the provision and subsequent utilization of space in a rocky intertidal community. *Ecol. Monogr.* 41, 351–89.

Dayton, P. K. 1973. Dispersion, dispersal and persistence of the annual intertidal alga, *Postelsia palmaeformis* Reprecht. *Ecology* 54, 433–8.

Dayton, P. K. 1975a. Experimental studies of algal canopy interactions in a sea-otter dominated kelp community at Amchitka Island, Alaska. *U. S. Nat. Mar. Fish. Serv. Fish. Bull.* 73, 220–37.

Dayton, P. K. 1975b. Experimental evaluation of ecological dominance in a rocky intertidal community. *Ecol. Monogr.* 45, 137–59.

Dayton, P. K., and Oliver, J. S. 1980. An evaluation of experimental analysis of population and community patterns in benthic marine environments. In Tenore, K. R., and Coull, B. C. (eds.), *Marine Benthic Dynamics*, pp. 93–120. University of South Carolina Press, Columbia.

Dean, T. A., and Deysher, L. E. 1983. The effects of suspended solids and thermal discharges on kelp. In Bascom, W. (ed.), *The Effects of Waste Disposal on Kelp Communities*, pp. 114–35. Southern California Coastal Water Research Project, Long Beach.

Duggins, D. O. 1980. Kelp beds and sea otters: an experimental approach. *Ecology* 61, 447–53.

Duggins, D. O. 1981. Interspecific facilitation in a guild of benthic marine herbivores. *Oecologia* 48, 157–63.

Foster, M. S. 1975a. Algal succession in a *Macrocystis pyrifera* forest. *Mar. Biol.* 32, 313–29.

Foster, M. S. 1975b. Regulation of algal community development in a *Macrocystis pyrifera* forest. *Mar. Biol.* 32, 331–42.

Foster, M. S. 1982a. The regulation of macroalgal associations in kelp forests. In Srivastava, L. (ed.), *Synthetic and Degradative Processes in Marine Macrophytes*, pp. 185–205. De Gruyter, Berlin.

Foster, M. S. 1982b. Factors controlling the intertidal zonation of *Iridaea flaccida* (Rhodophyta). *J. Phycol.* 18, 285–94.

Harlin, M. M. 1973. Transfer of products between epiphytic marine algae and host plants. *J. Phycol.* 9, 243–8.

Harlin, M. M., and Lindbergh, J. M. 1977. Selection of substrata by seaweeds: optimal surface relief. *Mar. Biol.* 40, 33–40.

Harper, J. L. 1977. *The Population Biology of Plants.* Academic Press, London, 892 pp.

Hatton, H. 1938. Essais de bionomie explicative sur quelques especes intercotidales d'algues et d'animaux. *Ann. Inst. Oceanogr. Monaco* 17, 241–348.

Hay, C. H., and South, G. R. 1979. Experimental ecology with particular reference to proposed commercial harvesting of *Durvillea* (Phaeophyta, Durvilleales) in New Zealand. *Bot. Mar.* 22, 431–6.

Hruby, T., and Norton, T. A. 1979. Algal colonization on rocky shores in the Firth of Clyde. *J. Ecol.* 67, 65–77.

Hutchins, L. W. 1952. Relations to local environment. In *Marine Fouling and Its Prevention*, pp. 102–17, U.S. Naval Institute, Annapolis, Md.

Jackson, J. B. C. 1981. Interspecific competition and species distributions: the ghost of theories and data past. *Amer. Zool.* 21, 889–901.

Kanwisher, J. W. 1966. Photosynthesis and respiration in some seaweeds. In Barnes, H. (ed.), *Some Contemporary Studies in Marine Science*, pp. 407–22. Allen & Unwin, London.

Kastendiek, J. 1982. Competitor-mediated coexistence: interactions among three species of benthic macroalgae. *J. Exp. Mar. Biol. Ecol.* 62, 201–10.

Lubchenco, J. 1978. Plant species diversity in a marine intertidal community: importance of herbivore food preference and algal competitive abilities. *Amer. Natur.* 112, 23–39.

Lubchenco, J., and Menge, B. A. 1978. Community development and persistence in a low rocky intertidal zone. *Ecol. Monogr.* 48, 67–94.

Lüning, K., and Neushul, M. 1978. Light and temperature demands for growth and reproduction of laminarian gametophytes in southern and central California. *Mar. Biol.* 45, 297–309.

Luther, G. 1976. Bewuchsuntersuchungen auf Natursteinsubstraten im Gezeitenbereich des nordsylter Wattenmeeres: Algen. *Helgo. Meeres.* 28, 318–51.

McLeod, G. C., and Rhee, C. 1970. Physical measurements of our natural resources. In *Symposium on the Microbiological Aspects of Seawater Pollution*, pp. 135–40. Kingston, Rhode Island.

Mann, K. H. 1973. Seaweeds: their productivity and strategy for growth. *Science* 182, 975–81.

Menge, B. A. 1979. Coexistence between the seastars *Asterias vulgaris* and *Asterias forbesi* in a heterogeneous environment: a non-equilibrium explanation. *Oecologia* 41, 245–72.

Neushul, M. 1971. The kelp community of seaweeds. *Nova Hedwigia* 32, 265–7.

Neushul, M., Foster, M. S., Coon, D. A., Woessner, J. W., and Harger, B. W. W. 1976. An *in situ* study of recruitment, growth and survival of subtidal marine algae: technique and preliminary results. *J. Phycol.* 12, 397–408.

Nicotri, M. E. 1980. Factors included in herbivore food preference. *J. Exp. Mar. Biol. Ecol.* 42, 13–26.

Norton, T. A. 1977. Experiments on the factors influencing the geographical distributions of *Saccorhiza polyschides* and *Saccorhiza dermatodea*. *New Phytol.* 78, 625–35.

Norton, T. A. 1978. The factors influencing the distribution of *Saccorhiza polyschides* in the region of Lough Ine. *J. Mar. Biol. Assoc. U.K.* 58, 527–36.

Paine, R. T. 1979. Disaster, catastrophe, and local persistence of the sea palm, *Postelsia palmaeformis*. *Science* 205, 685–7.

Paine, R. T., and Vadas, R. L. 1969. The effects of grazing by sea urchins, *Strongylocentrotus* spp., on benthic algal populations. *Limnol. Oceanogr.* 14, 710–9.

Pearse, J. S., and Hines, A. H. 1979. Expansion of a central Californian kelp forest following the mass mortality of sea urchins. *Mar. Biol.* 31, 83–91.

Pianka, E. R. 1976. Competition and niche theory. In May, R. M. (ed.), *Theoretical Ecology: Principles and Applications*, pp. 114–41. Saunders, Philadelphia.

Reed, D. C., and Foster, M. S. 1984. The effects of canopy shading on algal recruitment and growth in a giant kelp forest. *Ecology* 65, 937–48.

Salisbury, E. J. 1929. The biological equipment of species in relation to competition. *J. Ecol.* 17, 197–222.

Schiel, D. R., and Choat, J. H. 1980. Effects of density on monospecific stands of marine algae. *Nature* 285, 324–6.

Schonbeck, M. W., and Norton, T. A. 1980. Factors controlling the lower limits of fucoid algae on the shore. *J. Exp. Mar. Biol. Ecol.* 43, 131–50.

Scott, C. T., and Hayward, H. R. 1955. Sodium and potassium regulation in *Ulva lactuca* and *Valonia macrophysa*. In Schones, A. M. (ed.), *Electrolytes in Biological Systems*, pp. 35–64. American Physiological Society, Washington D. C.

Sousa, W. P. 1979a. Experimental investigations of disturbance and ecological succession in a rocky intertidal algal community. *Ecol. Monogr.* 227–54.

Sousa, W. P. 1979b. Disturbance in marine boulder fields: the non-equilibrium maintenance of species diversity. *Ecology* 60, 1225–39.

Tansley, A. G. 1914. Presidential address. *J. Ecol.* 2, 194–202.

Tilman, D. 1981. Resources: a graphical–mechanistic approach to competition and predation. *Amer. Natur.* 116, 362–93.

Underwood, A. J. 1980. The effects of grazing by gastropods and physical factors on the upper limits of distribution of intertidal macroalgae. *Oecologia* 46, 201–13.

Underwood, A. J. 1981. Techniques of analysis of variance in experimental marine biology and ecology. *Oceanogr. Mar. Biol. Annu. Rev.* 19, 513–605.

Underwood, A. J. and Denley, E. J. 1984. Paradigms, explanations and generalizations in models for the structure of intertidal communities on rocky shores. In Strong, D., Abele, L., and Simberloff, D. (eds.). *Ecological Communities: Conceptual Issues and the Evidence*, pp. 151–80. Princeton University Monograph, Princeton, N.J.

Underwood, A. J., Denley, E. J., and Moran, M. J. 1983. Experimental analysis of the structure and dynamics of mid-shore rocky intertidal communities in New South Wales. *Oecologia* 56, 202–19.

Underwood, A. J., and Jernakoff, P. 1981. Effects of interactions between algae and grazing gastropods on the structure of a low-shore intertidal algal community. *Oecologia* 48, 221–33.

Vadas, R. L. 1968. "The Ecology of *Agarum* and the Kelp Bed Community." Ph.D. dissertation, University of Washington, Seattle. 282 pp.

Wu, R. S. S., Levings, C. D., and Randall, D. J. 1977. Differences in energy partitioning between crowded and uncrowded individual barnacles (*Balanus glandula* Darwin). *Can. J. Zool.* 55, 643–7.

26: Herbivory

ROBERT L. VADAS

Departments of Botany and Plant Pathology and Zoology, University of Maine, Orono, Maine 04469

CONTENTS

I. Introduction

Approaches and solutions to the problems and measurement of herbivory are strongly influenced by the nature of the questions being asked. Many algal–herbivore interactions have significance in an ecological context only when examined by a whole-community approach, often involving experimental manipulation of seemingly unrelated parameters. These may include attempts to mimic higher-order interactions, stochastic events, or other processes not generally addressed in descriptive algal ecology. Pattern searching, however, is often a valuable first step in recognizing potentially important processes or interactions. Philosophy, rationale, and experimental approach, therefore, are important aspects in formulating and testing hypotheses regarding the influence of herbivores on benthic seaweeds. The tack taken here is to review various methodologies and approaches and combine them with conceptual frameworks, the assumption being that this will encourage rigorous testing of hypotheses and provide meaningful data and generalizations regarding herbivory.

From an ecological and evolutionary perspective, the potential reciprocal effects of herbivores on algae, or the "Arms Race" (Feeny 1975), cannot be ignored. Detailed information on herbivore biology and feeding behavior permits greater understanding of the physical and chemical composition of algal thalli and life history strategies as potential defensive mechanisms. Thus, the rationale and methods for examining the influence of algae on herbivore fitness are included. In turn, these measures may clarify the presence of interesting properties or strategies evolved by seaweeds and advance the form–functional approach to algal biology (Littler and Littler 1980; but see also Vadas 1979; Norton et al. 1982; Steneck and Watling 1982).

Information on the sources and rates of algal mortality are requisite for understanding the ecological and evolutionary constraints on thalli and on the functional roles of algae. Benthic algae are grazed by a variety of animal groups, including prosobranchs (Castenholz 1961; Sutherland 1970; Stimson 1970; Haven 1973; Lubchenco 1978; Underwood 1980; Creese 1980; Steneck 1982), opisthobranchs (Carefoot 1967; Jensen 1983), amphipods (Greze 1968; Brenner et al. 1976; Brawley and Adey 1981), isopods (Nicotri 1980), crabs (Raffaelli 1979; Sousa 1979; Elner 1981; Robles and Cubit 1981), pycnogonids (Bamber and Davis 1982), polychaetes (Lewis and Whitney 1968; Woodin 1977), insects (Robles and Cubit 1981; Robles 1982), sea urchins (Lewis 1958; North 1963; Ogden et al. 1973; Lawrence 1975), starfish (Branch and Branch 1980), fish (Randall 1965; Jones 1968b; Ogden and Lobel 1978; Montgomery 1980b; Choat 1982), turtles (Bjorndal 1980), dugongs and manatees (Heinsohn and Birch 1972; Lipkin 1975), and birds (Jacobs et al. 1981). The dominant grazers in temperate and boreal waters are sea urchins, mollusks, and amphipods. These groups may influence the structure of tropical algal communities but mainly in the absence of grazing fish (Hay 1981a,c; Hay et al. 1983; Choat 1982).

Benthic herbivores can be classified by sizes that range from millimeters to 2 m in length (micro- versus macrograzers), functional feeding groups (grazing, browsing, rasping, or picking), and mobility (sessile, crawling, or swimming). Most of that which is known about the effects of herbivores on algae is derived from studies of macrograzers. However, it is rapidly becoming apparent (Brawley and Adey 1981; Robles and Cubit 1981) that micrograzers play a larger role in structuring benthic communities than was previously suspected. Our ignorance is not surprising given that, in addition to being cryptic, micrograzers are difficult to manipulate experimentally in the field. In fact, the recognition of their potential importance often has resulted from their confounding of experiments designed to study macrograzers, fouling communities, or predators (Brawley and Adey 1981).

Conspicuous macrograzers such as sea urchins commonly reduce the photosynthetic surfaces of adult thalli, shorten life spans, or preclude macroalgal development entirely, whereas small and inconspicuous grazers such as amphipods are likely to be important sources of mortality for germlings, juveniles, and microalgae. Understanding the role and influence of smaller cryptic herbivores on algal associations will continue to require rigorous natural history observations and innovative and meticulous experimental manipulation.

The approaches for investigating algal–herbivore interactions are classified under three broad headings: herbivore ecology and feeding, herbivore biology and fitness, and herbivore effects. These three aspects are obviously interrelated but are delineated for convenience. A limited number of studies and techniques that could be extended to other algal–herbivore interactions are listed in each area. An attempt has been made to provide examples from diverse taxonomic groups.

II. Herbivore ecology and feeding

There are no armchair substitutes for good natural history observations. Not only do they stimulate research, but they are requisite for establishing testable hypotheses and designing experiments. When integrated with theory, they become a powerful, perhaps optimal (Fretwell 1972) strategy for understanding distribution and abundance patterns.

Determining the functional feeding roles of herbivores has been controversial. With herbivorous fish, Bakus (1967) and Jones (1968b) recognized two basic categories: grazers (or raspers) and browsers. Grazers scrape substrata (crusts and basal systems) including calcareous algae, whereas browsers consume "upright" multicellular forms. More recently, Lobel (1981) and Choat (1982) asserted that grazers take benthic organisms nonselectively due to the reduced size of prey relative to the predator, whereas browsers take whole or parts of individually recognizable sessile organisms and therefore are more selective. Similar categories have been used for invertebrates, although there is little consistency between groups. Shepherd (1973) described scraping as browsing, whereas feeding on macroalgae was considered grazing with abalones. Generally, there is a lesser tendency to erect rigid categories with invertebrates, and typically grazing and browsing are treated synonymously (Hughes 1980a, my preference; see also Lubchenco 1979; Hawkins and Hartnoll 1983).

These definitional problems are exacerbated when one attempts to characterize sea urchins, for example, that ingest algae, invertebrates (Ayling 1978; Vance 1979), or sediment (Vadas et al. 1982). This exposes two other problems: failure to recognize diet changes resulting from outgrowing a prey or food limitation. Numerous studies show that animals become more catholic in their diets when food becomes limiting (Emlen 1966; Pyke et al. 1977; Vadas 1977; Hughes 1980a,b). Thus, for some herbivores, grazing, browsing, or other categories may be valid only during periods of plenty. Perhaps rigid designations should be limited to cases involving particular feeding guilds or functional–form groups such as herbivorous gastropods (see Steneck and Watling 1982).

A. Factors affecting assessment of herbivore abundance

One phase of determining herbivore effects on algal communities or species is assessing their abundance. Generalizations about herbivore abundances and effects cannot be made unless one understands processes that affect dispersion and activity patterns. Several behavioral aspects can alter dispersion patterns and modify the potential impact of herbivores on algae. Many herbivores, especially in temperate and boreal latitudes, migrate seasonally between two or more habitats. The intertidal gastropod *Littorina littorea* migrates upward in spring and downward in fall (Williams and Ellis 1975) and affects algal recruitment, development, and zonation over a large portion of the shore (Lubchenco 1980). Seasonal migrations also are evident in sea urchins, which differentially control the vertical distribution of *Alaria* (Himmelman 1980).

Many herbivores use specific movements and dispersion patterns to avoid predators. Two widespread responses include fleeing (Garrity and Levings 1981; Schmitt 1982) and remaining in place (Phillips and Castori 1982), often on feeding or homing scars, which may result in nonrandom distribution and grazing patterns and therefore in patch formation among algae. Fleeing can involve migrating in front of a rising tide (Garrity and Levings 1981) or timing foraging activities to avoid predators. Predation by fish, spiny lobsters, and starfish in temperate latitudes affects where and when sea urchins feed and consequently their diet (Vance and Schmitt 1979; Schroeter et al. 1983). These urchins remain in refuges during the day and feed nocturnally when their major predators are inactive. Abalones and sea urchins remain in refuges and rely primarily on drift algae for food in the presence of sea otters (Lowry and Pearse 1973). Diel grazing by fish (Randall 1965) and nocturnal grazing by sea urchins (Ogden et al. 1973) are examples of fleeing and result in "halos" around Caribbean patch reefs. The avoidance of aggregations of *Purpura* by the grazing snail *Nerita* produces reverse halos, patches of algae around the predator (Garrity and Levings 1981).

Both macro- and micrograzers may have diel or tidal feeding cycles. In many cases, it is not immediately clear what causes a particular pattern. For example, Greze (1968) and Lawrence and Hughes (1972) observed food in the guts of amphipods and sea urchins, respectively, only at night. Similar observations on nocturnal feeding and predator avoidance were made on amphipods in a microcosm (Brawley and Adey 1981). *Littorina irrorata* forages on algal flats during low tide but migrates up the stems of salt marsh grasses during incoming tides to avoid predators (Hamilton, 1977). Daily or nearly continuous feeding also occurs in gastropods (Berg 1975; Hirano 1979), sea urchins (Lawrence 1975), and fish (John and Pople 1973).

Territoriality and agonism also modify dispersion patterns and may result in reduced local grazing pressure and influence the persistence of specific algae. For example, guilds of highly productive filamentous algae are usually selected for by herbivorous fish on coral reefs (Vine 1974; Wanders 1977; Lobel 1980; Hay 1981b). In essence, territorial herbivores maintain algal communities in an early and highly productive stage of succession. Similar but smaller groups (benthic microalgae or juvenile stages) persist within the territory of limpets (Stimson 1970; MacKay and Underwood 1977).

Measurement of abundance. Abundances of herbivores have been estimated or determined in a variety of ways. Most commonly, number or weight per unit area (density or biomass) is used for estimating grazing rates and energetic needs and for predicting effects on algal populations. Abundances should be estimated regularly (weekly, monthly, quarterly, etc.), the time interval depending on the rapidity with which densities change. Densities obtained only during periods of maximum or minimum abundances will bias grazing estimates. Both intertidal and subtidal herbivores form nonrandom distribution patterns that require stratified or plotless sampling (Underwood 1976). These strata should be based on biologically meaningful boundaries such as depth, substrate type, or light level in the subtidal, tidal height in the intertidal, and space where grazers are territorial. If there is no basis for stratification, then sampling should be random from within the whole area. Estimating densities of nonmobile forms can be based on quadrats or clear plastic sheets with random or evenly spaced dots (Dayton 1971; Grant 1977). Line, belt, or cine transects (Ebeling et al. 1980) may be appropriate for highly mobile forms. For specific comments on sampling and statistical procedures, see Underwood (1981) and Zar (1984); for sampling and experimental designs, see Green (1979) and Hurlbert (1984); and for testing spatial patterns and movements, see Underwood (1976) and MacKay and Underwood (1977).

The causal mechanisms for these and similar herbivore patterns can be examined with transplants. Activity patterns such as habitat selection and sources of predation can be determined by the reciprocal transplantation of individuals to substrata in different zones or environments. One can achieve this by tethering herbivores on thin monofilament line (McQuaid 1981) or by gluing them directly to substrata (Levings and Garrity 1983). Direct observations or movies can be made to quantify movements.

Behavioral studies on nocturnal or shy herbivores usually require special monitoring, marked individuals, and frequent observations. Invertebrate shells are easily marked by drying and lightly sanding the spot to be painted with lacquer or marine enamel. For population studies they can be tagged with coded dots or small numbered tags in a spot of clear epoxy cement (Williams and Ellis 1975; Creese 1980; McQuaid 1981; Byers and Mitton 1981; Garrity and Levings 1981). For invertebrates with perforations or special structures on the shell, tags can be fashioned to fit the animal (Hayashi 1980).

Long-term tagging of sea urchins or animals with fragile tests, for example, is difficult since the method itself often increases mortality (Ebert 1965; Olsson and Newton 1978). Soft-bodied invertebrates present additional problems, although Joule (1983) developed a method of tagging polychaetes that involves imbedding binary-coded microwire inside the animal and recording recaptured individuals with a metal detector. In any case, one should anticipate high mortality and mark a large number of animals or conduct adequate preliminary tests to estimate mortality due to methodology.

Long-term or continuous observations can also be conducted with time-lapse movies or video cameras. Ogden et al. (1980) used in situ time-lapse photography with a 30-s intervalometer for undisturbed 72-h periods to record the frequency and duration of turtle feeding in grass beds. Littler et al. (1983b) documented feeding by kyphosid fishes only with time-lapse photography. Kitting (1979) and Steneck (1982) used similar techniques to observe limpet feeding, whereas R. L. Vadas et al. (unpublished data) used infrared light sources and video cameras to study predator avoidance responses of sea urchins at night. Feeding noises and acoustic tags have been used to record and characterize gastropod feeding activity and diets (Kitting 1979, 1980) and to study feeding and activity patterns in green sea turtles (Ogden et al. 1983). Lissner (1980) cleverly characterized activity periods and the effects of storms on the behavior of sea urchins by gluing "trip threads" across burrow entrances and determining the frequency of broken threads diurnally and during different storm conditions.

B. Determining herbivore diets, feeding rates, and food preferences

The foraging strategies of herbivores, and predators in general, have become increasingly of theoretical concern. Not only do foraging theories reveal interesting aspects of food choice, but they also take into account an animal's time and energy budget (Hughes 1980a,b) and efficiency (Lubchenco and Gaines 1981) under various environmental conditions. Optimal foraging theory assumes that natural selection will favor herbivores that maximize their net rate of energy, food, or nutrient intake. For example, optimal patch use, as predicted by the marginal value theorem (Pyke et al. 1977), requires that an herbivore concentrate on the richest patches and remain in them until the net rate of food intake decreases to the average rate for the habitat. The energy maximization premise, from which the marginal value theorem is developed, can also be applied to other aspects of feeding behavior such as searching method and choice of prey within a patch. The optimality of the strategy used in any one circumstance is assumed to be constantly assessed by the animal.

Implicit in optimization strategy is the assumption that selective pressures are acting on the herbivore and that it has been either adaptive to have developed such a strategy or limiting not to have done so. Conceivably, some herbivores (e.g., *Littorina obtusata*, which utilizes fucoids; van Dongen 1955) may have such stereotyped feeding behavior that they have not had the opportunity to become optimal foragers. Herbivores that make higher-level decisions (e.g., *Strongylocentrotus droebachiensis*; Vadas 1977; Larson et al. 1980) would have more opportunity to evolve different strategies. Given these assumptions, optimality theory enables the researcher to predict an animal's optimum behavior, diet, performance, and so on, which can be stated as a null hypothesis and used to test the animal's fitness, performance, and so on, by comparing theoretical optima with real-world data (Townsend and Hughes 1981). In some cases the herbivore may be attempting to maximize some essential nutrient rather than energy per se, an aberration that can also be analyzed with optimal foraging theory.

Not all herbivores should be expected to be energy maximizers. Some may be time minimizers (Emlen 1966; Schoener 1971) and may be required by environmental constraints to forage in relatively set periods of time. This may be due to minimizing risks from predation or from desiccation due to tides, as in the case of gastropods (Hawkins and Hartnoll 1983). The diet width of animals may also be affected by prey availability and predictability (Menge 1972) and thereby influence the development or limitation of an optimal foraging strategy. In environments in which the food resource is controlled largely by stochastic process and in which prey densities are low (actual or perceived), nonselective feeding is likely and would result in a wide diet. In addition, some grazers spend a considerable amount of time sampling before recognizing algae, which would also affect their diet width (Hughes 1980a,b; see also Elner and Hughes 1978; Fairweather and Underwood 1983).

Although the topic is largely unexplored, marine herbivores appear to have a wide range of morphological and behavioral adaptations for exploiting and increasing their utilization of seaweeds. Several anatomical and physiological features in herbivorous fish, for example, contribute to the breakdown of the diverse algal storage and chemical products, which Montgomery and Gerking (1980) and Horn et al. (1982) consider a major problem in digesting seaweeds. Relative to carnivorous species, herbivorous fish have some or all of the following: longer alimentary tracts (Jones 1968b), modified mouth parts including dentition or pharyngeal apparatus (Christensen 1978), modified caeca and lymph system (Montgomery 1977), and acid lysis via gut pH (Lobel 1981). Most invertebrates have one or more enzymes capable of degrading some of the algal polysaccharides, but as Lawrence (1975, 1982) notes for sea urchins, many can use only soluble and not structural carbohydrates. Also, some herbivores have associated bacteria that digest algae, for example, sea urchins (Prim and Lawrence 1975; Guerinot and Patriquin 1981), turtles (Bjorndal 1980), and gastropods (Galli and Giese 1959). In many cases protozoans are also present in the guts of invertebrates, but their contribution to algal degradation is unknown (Lawrence 1975). Radular morphology and function in mollusks appear to have undergone considerable evolution (Steneck and Watling 1982; Steneck 1983), which permits grazing on a wide range of algal forms.

1. Natural diets. Herbivore diets can be determined directly or indirectly and, whenever possible, should be independently corroborated to minimize assumptions and lessen the importance of prey that are inadvertently ingested (see Menzel 1959; Elner 1981). Indirect observations include (1) gut analysis in sea urchins (Lewis 1958; Neill and Larkum 1965; Himmelman and Steele 1971; Lawrence 1975), fish (Menzel 1959; Jones 1968a; Westernhagen 1973; Christensen 1978; Lundberg and Lipkin 1979; Montgomery 1980b; Horn et al. 1982), mollusks (Leighton and Boolootian 1963; Tsuda and Randall 1971; Kitting 1979), amphipods (Brenner et al. 1976), turtles (Bjorndahl 1980), and dugongs (Lipkin 1975); (2) $^{13}C/^{12}C$ ratios (Thayer et al. 1978; Fry and Parker 1979); (3) pigment shifts (van Montfrans et al. 1982); (4) feeding scars in seagrass beds or on coralline algae (Johnstone and Hudson 1981; Ayling 1981); (5) artificial algal surfaces (Ayling 1981); and (6) acoustic patterns (Kitting 1979, 1980).

Of these, gut analysis is the simplest to use and interpret but not necessarily the most reliable (Leighton 1966). Several implicit assumptions are involved in this method, including the following. (1) Relative abundance in the gut depicts selection or availability, (2) prey are not differentially digested, (3) digestion stops when the animal is collected, and (4) all prey are deliberately ingested. To minimize some of these concerns, animals can be injected with a buffered 4% Formalin solution into the gut upon collection (Westernhagen 1973). In addition, corroborative laboratory feeding can be done (Vadas 1977). The quantification of algae in gut analysis is difficult but best done following the randomization procedure of Jones (1968a) for fish and by Tsuda and Randall (1971) for gastropods.

Direct observations of feeding can be made in several ways. Slow-moving or large herbivores can be rapidly overturned and examined for algae in the mouth parts (Moore and McPherson 1965; Vadas 1977; Cowen et al. 1982). For smaller gastropods that feed on algal crusts and diatoms, time-lapse photography can record the frequency and duration of time spent on various algae (Kitting 1979). With territorial and some mobile species, divers can count or film both the prey and the number of bites per unit time (Montgomery 1980b; Lobel and Ogden 1981; Kingett and Choat 1981). One source of bias in direct estimates may be the handling times of different prey. Those items that take a relatively long time may be overestimated (Fairweather and Underwood 1983).

Field diets combined with algal availability (biomass or percentage of cover) may indicate the degree of food selectivity. Several indices have been used with marine herbivores (Vadas 1977; Lundberg and Lipkin 1979; Carpenter 1981; Vadas et al. 1982; Robertson and Lucas 1983), most being based on rankings or Ivlev's (1961) electivity index. These graphic models highlight algal species that are avoided, preferred, or simply taken on the basis of availability. Caution must be used, however, in interpreting selection strictly on the basis of electivity indices, since they may be biased and dependent on sample size. A better alternative is the linear food selection index (Strauss 1979). Also, preferences often depend on the array of food presented and on what the researcher assumes is available to an herbivore. One method of reducing these problems is to use the statistical ranking procedure and test of Johnson (1980), which provides comparable results regardless of whether questionable items have been included. To maximize the benefits of any of these procedures and analyses, one must ensure that adequate habitat samples of algae have been taken (see De Wreede, Chap. 7, and Littler and Littler, Chap. 8).

2. Feeding rates. Feeding rates can be determined in two general ways: measuring herbivore food consumption per unit time or rate of disappearance (net weight or percentage of cover loss) of algae. If the experimenter is concerned with algal defense, then the percentage of thallus lost per unit time may be more meaningful than comparisons of absolute weight loss among different algae (cf. calcified vs. fleshy algae). Algal survival, for example, may be increased more by minimizing losses to the percentage of photosynthetic surface rather than to total mass. The most useful approach from a trophic perspective, however, is to incorporate the algal mass or area ingested per unit time as a function of the organic weight of the herbivore. The latter is rarely attainable in the field, and various approximations can be used, including area of natural substratum grazed or denuded (Forster 1959; Ayling 1981), number, size or frequency of feeding scars on coral (Brock 1979) or grass beds (Ogden et al. 1983), and number of scrapes or bites on algae per unit time (Montgomery 1980b; Lobel and Ogden 1981). When calibrated for diel or other feeding variations (Montgomery 1980b) and for total and organic weight ingested, these measures provide reasonable estimates of feeding rates.

Laboratory studies provide greater detail on feeding rates and behavior but do not necessarily mimic reality. Feeding may not be continuous throughout an experiment. Used in conjunction with field observations and sensible interpretation, laboratory studies can provide support and insight into mechanisms operating in nature (Vadas 1977; Dayton and Oliver 1980). In general, one should conduct pilot studies to determine relative feeding rates, duration of time during which the fronds and animals remain fresh or healthy, and other factors that affect behavior. One means of assessing health is to compare "righting times" (Lowe and Lawrence 1976) or other behavioral traits of experimental animals regularly.

The choice of feeding regime depends on the size of the herbivore and on the question(s) being asked. Typically, fixed but ad libitum amounts of algae are fed to macrograzers at timed intervals. With micrograzers and omnivores, rations can be impregnated with tracers such as dyes (Lewis 1964; Larson et al. 1980), vital fluorescent stains (B. P. Patterson and R. L. Vadas, unpublished data), or fluorescent algae (Vadas 1977), which can be used as time markers or absolute measures when recovered in the gut. Several other decisions must be made regarding single versus multiple diets, pretreatment history of animals (fed vs. starved), and food ration (limited vs. ad libitum).

To minimize pretreatment effects on feeding, animals should be maintained under uniform conditions. Commonly, larger herbivores are starved for 2 to 21 d (Bryan 1975; Boolootian and Lasker 1964) or until the guts are evacuated, whereas micrograzers can be starved for 1 d (Robertson and Lucas 1983). Starvation presumably normalizes all test animals and induces them to feed maximally. These assumptions have rarely been tested, although Bonsdorff (1983) showed that sea urchins fed at maximum rates after 36 h of starvation.

A second method involves feeding with minimal food rations and using species that will not be tested, that are nonpreferred (Bryan 1975), that are preferred (Littler et al. 1983b), or that are not encountered in the environment, for example, lettuce (O. J. McConnell et al., unpublished data). Unless one has pilot data, the first seems preferable since it involves fewer assumptions and therefore fewer complications.

The third aspect involves the amount of algae to be used in an experiment. In most cases an ad libitum ration permits maximum rates to be achieved. However, the nature of the question(s) being asked may alter this strategy. For example, if one is concerned with why an herbivore "switches" (Murdoch 1969) its diet and the value of alternative prey, one might limit the ration of preferred algae. Conversely, to maintain and test the effects of predetermined ratios on growth and reproduction, rations can be regularly adjusted.

3. Food preferences. Food preferences can best be determined when two or more species are present. Gut contents have been used but provide only corroborative data due to the problems already noted and because availability influences field diets. Feeding rate is a better measure, since it is more likely to correlate with herbivore fitness (Carefoot 1967; Vadas 1977). Ideally, chemoreception and gustation can be measured sequentially in the same experiment, thereby providing additional information on preferences. Incorporating gustation, chemoreception, a combination of both, and field diets into such analyses increases robustness (Vadas 1977). Hierarchical feeding processes in marine herbivores have been largely unexplored, although several components are evident, including (1) chemical detection in gastropods (van Dongen 1955) and echinoids (Lawrence 1975), (2) taste testing (edibility) in opisthobranchs (Jahan-Parwar et al. 1969), (3) acceptance or rejection due to internal (buccal) receptors in opisthobranchs (Kriegstein et al. 1974) and fish (Montgomery 1980b), and (4) continued feeding (palatability) in sea urchins (Larson et al. 1980). Generally, adult feeding behavior in marine herbivores can be reduced to appetitive and consummatory components (Kriegstein et al. 1974; Bonsdorff 1983).

The simplest method of testing for differences in feeding rates of various algae is to provide randomly assorted multiple prey. Switching, as a function of time and availability of preferred algae, can also be assessed in the same experiment if one or more algae can be consumed within the allotted time. One design is an equal-spacing arrangement in which one or more herbivores are located in the center of an array of two or more algae. A variation on this theme was used by Nicotri (1980) for amphipods and isopods and consists of a plexiglass chamber with a central release compartment and four peripheral (algal test) compartments. This method enables one to examine attractiveness, including chemoreception, tactile responses, and taste. To distinguish between preferences for attractiveness and edibility, Nicotri (1980) used plastic plants to mimic living algae. The isopod *Idotea* preferred perennial algae in nature but chose the mimics over several live algae, suggesting that morphological form or habitat may be important and may confound interpretations based on food value alone. With this design and adequate space between the algae and the test animal(s), a large number of prey can be tested simultaneously for both chemical attraction and feeding. Subsequent distribution patterns can be tested with χ^2 or preferably a *G*-test (Zar 1984).

A device commonly used in the study of predator and herbivore feeding behavior is a plexiglass Y maze tank (Vadas 1968, Larson et al. 1980). Live algae or extracts are placed in the opposing arms of the Y (inlets), and the animals are positioned at the end of the long arm (drain). The sizes of these tanks are determined by the size and mobility of the herbivore. For large, relatively slow moving sea urchins, tanks 115 × 22 × 22 cm deep are adequate. A modified l-shaped choice chamber was used by Garnick (1978). The duration and final design of these experiments must be determined in pilot studies, since the results can be confounded by nonmovement, reversals, or trail following (Hall 1973). Following trails of feeding conspecifics in preference experiments is a potential problem with all invertebrates but especially mucoid producers. Some mollusks recognize not only trails but also the direction in which the organisms were moving when the trails were produced (Wells and Buckley 1972; Branch 1981). Thus, what might appear to be a preference for an alga may simply be an artifact of behavior. Testing movements in the maze without food and reversing the location of prey items reduces or eliminates this problem. In some cases, elaborate procedures may have to be taken, for example, testing individuals rather than groups of animals and cleaning the maze between individual runs to remove chemical stimuli (Zafiriou 1972). If chemoreception alone is being tested, animals should be removed once they have reached predetermined locations in the arms.

Another potentially useful method involves impregnating agar with homogenized algal extracts (Geiselman and McConnell 1981). The heated agar, containing ethanol or water extracts and appropriate controls, is poured into divided petri dishes, solidified, and exposed to herbivorous gastropods in tanks with running seawater. Preferences are determined on the basis of the percentage of surface area consumed relative to controls. A variation on this theme, involving algal extracts impregnated into plaster, was used by McClintock et al. (1982). An untested assumption of both methods is that these extracts simulate that which the animals detect or respond to in living thalli.

Preferences in aquaria, in running seawater, or in the field can also be determined by attaching algae to coral (Hay 1981a), weighted grids (Littler et al. 1983a,b), or lines suspended on or above the bottom and exposing them to herbivores (Bryan 1975; Hay 1981c; Littler et al. 1983b; Lobel and Ogden 1981). Thought must be given to the type and placement of controls to ensure that the desired response is being tested. Algae suspended in the water column as a control for benthic feeders or drift feeding fish (Littler et al. 1983b) may have to be shaded to minimize depth or cage effects. Also, vertical and horizontal distances from reefs or structures can be used as controls to test the effects of mobile grazers, especially fish (Hay 1981a,c; Hay et al. 1983; Littler et al. 1983a,b).

These laboratory and field tests usually can be ranked and treated with nonparametric statistics. Many behavioral studies do not yield normally distributed data, and the use of transformations or nonparametric models may be necessary.

III. Herbivore biology and fitness

Examining the influences of various algae on the fitness of an herbivore provides a better understanding of their field behavior and potential effects on marine communities. Understanding why an herbivore feeds on a particular alga is not trivial, and interpreting algal distribution and abundance patterns as well as evolutionary developments may be as dependent on herbivore feeding as on any other ecological factor (Lubchenco and Gaines 1981). Fitness may involve any of the ecological aspects discussed earlier, but it is affected primarily by the efficient assimilation of algae and the ensuing growth and reproduction. The relative fitness of herbivores can be assessed by several indirect measures, including survival, relative size, growth rate, life history change, and reproduction.

Reproductive measures can be subdivided into early maturation resulting from rapid growth (Vadas, 1977; Larson et al. 1980; Robertson and Lucas 1983), total gonad or gamete production (Carefoot, 1967; Vadas 1977), offspring or larval production (Robertson and Lucas 1983), and recruit success. The last is difficult to assess but is an important component of an organism's success (Dayton and Oliver, 1980; Choat, 1982). Recruitment, however, is highly variable, and its measurement and use as an indicator of fitness may be risky (Tegner and Dayton 1977).

Only relatively complete examples from different types of herbivores will be used to illustrate the methods and potential value of these measures to algal–herbivore interactions. Few experimental studies have examined feeding behavior concurrently with one or more aspects of fitness. Energetics have been emphasized in a number of studies, but unfortunately not with different diets. Thus, information on the utility of a particular alga to an herbivore is lacking. Although knowing the value or utility of an alga to an herbivore will be stressed, it is important to recognize that at times feeding in nature may involve availability. The latter does not mean that selective feeding is without evolutionary influence. Rather, inferring evolutionary developments strictly on the basis of present conditions or relationships may be misleading, unless that behavior represents a deeper (genetic) adaptation.

A. Food processing

Absorption is dependent on both the digestive mechanisms of an animal and the composition of the alga it consumes. As a result, absorption efficiency is the first stage in which the adaptive significance of selective feeding can be evaluated. Absorption efficiency can be determined by direct or indirect methods and involves the ratio of algae absorbed to the total ingested. Direct assessments involve gravimetric quantification of the dry weight (based on conversions of blotted wet to dry weight) of algae ingested and egested as feces. This technique has been used with fish (Menzel 1959), gastropods (Grahame 1973), sea urchins (Vadas 1977), and amphipods (Robertson and Lucas 1983). Blotted rather than dry algae are used to simulate normal thallus and feeding conditions closely and thereby reduce artifacts in feeding behavior. Dyes, stains, powdered charcoal, or fluorescent algae can also be used to mark the conveyor-belt movement of food and feces through the gut (Larson et al. 1980) or estimate the rate of digestion in amphipods (Greze 1968) or fish (Buddington 1979).

Indirect measures for aquatic herbivores are based on Conover's (1966) ash fraction technique. Lowe and Lawrence (1976) provided a general form of this indirect measure that involves the relative absorption of any material (e.g., carbohydrate, protein, or lipid) that can be determined in the alga and subsequently in the feces. Modifications of this method were used by Bjorndal (1980) on turtles and Montgomery and Gerking (1980) on fish; lignin and ash, assumed not to be digestible, provided an internal marker against which digestion of various nutrient components was evaluated. The major value of indirect measures in an aquatic environment is the elimination of tedium and errors involved in daily collections of feces. One simply feeds the herbivore a specific diet long enough to effect complete evacuation and then collects and analyzes fresh feces at random. Labeling algae with ^{14}C has been used with fish (Bryan 1975), amphipods (Zimmerman et al. 1979), and pycnogonids (Bamber and Davis 1982). However, the use of ^{14}C-labeled algae assumes that there are no respiratory losses of $^{14}CO_2$ from either the alga (alive inside the gut) or animal tissues.

Absorption efficiency data are most useful within herbivore feeding guilds and with algae that are spatially and temporally available. Absorption is complicated by enteric bacteria that are involved in digestion (Prim and Lawrence 1975) and nitrogen fixation (Guerinot and Patriquin 1981). Both may result in underestimates of absorption, especially of nitrogen. Absorption efficiencies of herbivores feeding on algae show a wide range of values (from less than 10 to 90%) and generally are highest with preferred algae (opisthobranchs, Carefoot 1967; sea urchins, Vadas 1977; fish, Edwards and Horn 1982).

B. Growth and reproduction

Many of the concerns directed toward food processing also apply here. Stratification and randomization of experimental animals are necessary to reduce variability and bias. One way to achieve this is to stratify test animals by size, age, weight, collection site, or other features that might increase variability. Size- and age-specific differences in metabolism and growth were observed in sea urchins (Fuji 1967; Ebert 1968) and herbivorous mollusks (Frank 1965; Paine 1971). Most herbivores also have an upper size limit, and growth may depend on both the alga and the growth potential of the animal. Larger animals have a higher inertia to further growth (Vadas 1977; Larson et al. 1980) and are less likely to show differences as a result of food type.

The available pool of animals used for experimental studies must be large enough to avoid the repeated use of individuals. When repeatedly disturbed, herbivorous fish exhibit abnormal behavior and reduce their acid secretions to the gut (Moriarty 1973). Excessive handling of sea urchins alters behavior (Vadas 1977), which could affect feeding, metabolism, and growth. Several growth models are available against which growth on different algae can be assessed for invertebrates (Frank 1965; Fuji 1967; Ebert 1968) and for fish (Ricker 1979).

The effects of algal quality on growth and reproduction can be examined in single, sequential, or combination diets. With macrograzers this may require relatively long term experiments (6–18 mo), especially if reproductive capacity is included (Vadas 1977; Larson et al. 1980). Micrograzers may require considerably less time. Reproduction and life history properties of an amphipod were delineated in as few as 2 to 4 wk (Robertson and Lucas 1983). Simulating field conditions requires fresh algae and frequent cleaning of experimental chambers to reduce fouling, which can be a problem even with micrograzers (Robertson and Lucas 1983). The availability of fresh algae necessitates the design of studies to coincide with seasonal abundance. Seasonal variation in the chemical composition of thalli may be an important source of variation for short-term studies on micrograzers.

Studies involving single diets generally show that animals grow and reproduce best on the most preferred algae [for sea urchins, Fuji (1967); Vadas (1977); Larson et al. (1980); for opisthobranchs, Carefoot (1967); Jensen (1983); and for amphipods, Nicotri (1980); Robertson and Lucas (1983)]. Sequential diets provide an intermediate condition between single and combination diets. In these experiments, differential growth is determined over relatively short feeding intervals, and growth rates are expressed as a function of a specific alga provided during the interval (Leighton and Boolootian 1963). Starvation periods between switches in algal diets are used to normalize animals before each test. There are two problems with this method: the increasing size of test animals and the differential growth on various algae. An analysis of covariance is appropriate when such differences cannot be controlled.

Studies with single-prey species deny the potential benefits of mixed diets, which may be critical for the subtle requirements of good growth. Despite this, few herbivore studies have made use of mixed diets. Larson et al. (1980) tested paired algal combinations and various proportions of two algae on the growth and reproduction of boreal sea urchins. Higher growth but not reproductive rates were achieved in the combinations employing the highest proportions of preferred algae. Gonad production, however, was highest on the most preferred alga. Intuitively, one might anticipate that a mixed diet would be adaptive, especially in nutrient-limited environments such as tropical seagrass lagoons and coral reefs (Lowe and Lawrence 1976; Lobel and Ogden 1981). Attempts to raise herbivorous fish on strict algal diets were unsuccessful (Menzel 1959; Vaughan 1978). These and other tropical fish required animal protein to grow, supporting the need for more inclusive diets in tropical herbivores. Mixed diets were also required for growth in the omnivorous temperate mud snail *Ilyanassa* (Curtis and Hurd 1979).

The influence of a particular alga on growth and reproduction can also be determined in certain field situations in which the animals can be marked and restricted to specific stands of algae. Since many herbivores are zoned vertically by size or age (Paine 1971; McQuaid 1981) and because they can grow at different rates (Paine 1971), stratified designs for size and zone should be used.

IV. Herbivore effects

Numerous studies have shown that herbivores influence not only the structure of algal populations, but also the organization of marine and algal communities. Indirectly, predators such as sea otters (Estes and Palmisano 1974) also influence the structure of algal communities by reducing the density of all or specific herbivores and sessile invertebrates. Both of these indirect effects, the reduction in grazing intensity and competition, have a positive effect on the occupation of space by algae. Similarly, parasites, disease, or other factors inducing mortality or regulating herbivore populations enhance the development and persistence of algae. The value of understanding higher-order interactions is reflected in the effects, causal mechanisms, and questions that have been uncovered by the experimental incorporation of these seemingly peripheral interactions (Dayton 1971; Lubchenco and Menge 1978; Garrity and Levings 1981).

Several excellent reviews (Underwood 1979; Lubchenco and Gaines 1981; Branch 1981; Choat 1982; Gaines and Lubchenco 1982; Hawkins and Hartnoll 1983) discuss the influence of grazers on benthic algae. Herbivore effects are covered here indirectly, the focus being on methods and conceptual approaches to field experiments. Mathematical models and experimental simulations that can be used to assess the potential effects of herbivores on algae are described briefly. The methods, rationale, and types of field experiments used to test herbivore effects in nature are given in more detail.

A. Indirect assessments

1. Simulations. The potential effects of herbivores on algal populations can be estimated from the types of information already discussed. Incorporating information on herbivore densities, age or size structure, and size-specific feeding rates of the herbivore provides an approximation of the algal biomass consumed by an herbivore population. A further refinement is achieved if data on seasonal and diel feeding patterns, food preferences, and algal availability (abundance, patchiness, growth, and mortality) are included. Another approach is to incorporate in models the removal rates of algae by animals in the field.

Few estimates contain this detail, but Seip (1980, 1983) and colleagues have provided extensive models of algal development on rocky shores. Of relevance are the parameters identified and the information generated on algal–herbivore cycles, successional processes, zonation, and maturity of intertidal communities, all of which are affected by herbivores. In addition, Lubchenco and Gaines (1981) provide a useful conceptual framework and model for the expected damage to algae from herbivores. Their mathematical and graphic models attempt to integrate plant and herbivore studies and highlight important facets of the interaction including, for plants, abundance and spatial and temporal distributions and, for animals, density, sensory capabilities, and mobility. Their models deal with individuals, populations, and communities, and an important component is the degree of feeding selectivity by generalist herbivores. Community models on species richness incorporate herbivory as well as competitive interactions. These authors were unable, however, to incorporate the effects of multispecies (herbivore) interactions and higher-order predators because of an insufficiency of studies to draw on. Hawkins and Hartnoll (1983) also provide graphic models based on the classical descriptions of Southward (1956) and Southward and Southward (1978) on the balance between fucoids and limpets.

Graphic models can provide valuable generalizations about complex systems as well as starting points for testing hypotheses in other systems. They, along with computer simulations, have the capacity to test or separate the effects of a large number of parameters with a wide range of conditions on community structure. Despite the simplifying nature of the assumptions inherent in simulations, they can provide further directions for field research by identifying which of the parameters and underlying assumptions are based on data and which require additional support. In fact, their greatest utility may lie in the questions they raise.

Energetic studies on specific herbivores or on portions of food webs (Paine 1971; Miller and Mann 1973) also provide indirect measures of the effects of herbivores on algae. Extrapolations from energetic studies, however, must be interpreted with care because the focus is generally on the total amount of energy required to sustain the biomass of herbivores rather than on the population size of the algae or the functional strength of a particular food web (Paine 1980). Most studies show that less than 20% (often less than 10%) of the available algal biomass goes through a typical herbivore food web and that most of the macroalgae end up in detrital food paths (Miller et al. 1971).

A logical conclusion based on energy flow would suggest that herbivores do not control algal populations. Furthermore, the focus on available energy or descriptive food webs, rather than on the individual or population, reduces the likelihood that the sources of algal mortality will be closely related to those units that are acted on by selection. In other words, natural selection acts on individuals and not on biomass and energy. For example, it is not important to an alga that has been grazed and disrupted (often set adrift by damage to the stipe) whether or not it is eaten, unless drift algae are as successful as attached plants at leaving offspring. Simulations based on population data, including the functional relationships in a food web, provide a more complete understanding of the ecological and evolutionary constraints of herbivores on algal populations. Similar arguments regarding the value of population data have been advanced by Choat and Schiel (1982) (also see Paine 1980). The important message here is to apply caution in the selection of data, because the models themselves are only as good as the data and assumptions on which they are based.

2. *Microcosms.* Empirical simulations in laboratory or outdoor microcosms are another means of assessing the potential influence of herbivores on benthic algae. Broadly defined, microcosms can range from small laboratory aquaria to large outdoor tanks with running seawater. Several microcosm studies have contributed to algal–herbivore and predator interactions. Newell et al. (1971) studied grazing rates in *Littorina* by observing radular movements and rates on the walls of glass aquaria. Nicotri (1977a) examined amphipod grazing on benthic algae but found no major impact. Conversely, Brawley and Adey (1981) showed that amphipods controlled fouling algae and indirectly enhanced the productivity of macroalgae. These experiments and several utilizing fish to control amphipods led them to suggest that micrograzers were controlling algal and community structure on wave-exposed tropical algal ridges where herbivorous fish were unable to feed. A flow-through plexiglass microcosm was used to determine the effects of mud snails on benthic microalgae and community structure (Connor and Teal 1982). At low densities, algal growth was stimulated by increased nitrogen cycling, whereas overgrazing and disturbance occurred at high densities, similar to that observed in nature. Finally, Brock (1979) used well-replicated 550-liter flow-through microcosms to assess fish grazing and refuges on benthic community structure. Increased habitat heterogeneity and intermediate fish density resulted in higher algal and invertebrate diversity.

B. Direct assessments

Several types of "field experiments" (see Connell 1974) can be used to study the effects of herbivores on algae and include (1) herbivore manipulations by removals or additions or by various physical or chemical barriers, (2) predator manipulations or inadvertent removals, and (3) algal manipulations. Although all experiments are susceptible to unforeseen complications ("Murphy's law"), field experiments are particularly prone to being seriously flawed, and extensive consideration must be given to controls and to the potential artifacts induced by the manipulation. A solid understanding of natural history, considerable preliminary work, clearly delineated objectives, properly controlled experiments, and a skeptical perspective are perhaps the best insurance against flawing the design or introducing undesirable side effects. An excellent account of how particular field experiments have been flawed is provided by Dayton and Oliver

(1980), but see also Green (1979), Underwood (1980), Choat (1982), and Hurlbert (1984). The basic design of these experiments is such that all factors but the one under consideration are allowed to vary naturally, whereas the one of interest is controlled (Connell 1974). Although this appears to be simple in practice, it is difficult to maintain "natural" conditions. The preferred experimental protocol is to use the least disruptive method or, if that is not possible, to demonstrate through corroborative studies that other techniques are valid.

1. Manipulations. A variety of techniques have been used to alter the natural abundances of grazers. Generally, the removal of herbivores from large isolated areas or "islands," such as tide pools, rock outcrops on sand or mud substrata, patch or natural reefs, and artificial structures, that have been in place for long periods offers the least disruptive means of controlling grazer levels. The island effect provides herbivore free borders around isolated areas, *cordon-sanitaire* (Hawkins and Hartnoll 1983), and reinvasions are reduced by frequent monitoring. Islands may be inappropriate for highly mobile grazers since reinvasion can be rapid (May et al. 1970).

Hand manipulations have been used effectively with slow-moving herbivores including sea urchins in tide pools (Paine and Vadas, 1969), on patch reefs (Ogden et al., 1973; Sammarco et al., 1974), and in subtidal and intertidal habitats (Kitching and Ebling 1961; Jones and Kain 1967; Vadas 1968; Dayton 1975a; Vance 1979; Sousa et al. 1981; Cowen et al. 1982). Manual removals on intertidal shores include limpets (Jones 1948; Lewis and Bowmann 1975; Underwood and Jernakoff 1981), littorinids (Lubchenco 1978; Moreno and Sutherland 1982), and mixtures of both (May et al. 1970; Black 1974; Luckens 1974; Raffaelli 1979; Menge and Lubchenco 1981). A classic demonstration of the controlling effect of limpets on algal recruitment and community structure is the large intertidal clearings of Jones (1948). Adjacent tide pools, rock outcrops, patch reefs, or substrata with similar algal associations and natural herbivore levels may serve as controls. The placement of replications and controls, however, must be thoroughly considered (Hurlbert 1984). Alternatively, grazers can be added to an area, as a control or to test effects along a gradient of grazer intensity (Lang and Mann 1976; Paine 1977). Herbivores that are removed from one area can be censused, marked, and transplanted, cautiously with animals easily dislodged by waves (Underwood and Jernakoff 1981; Underwood et al. 1983), to control or experimental sites.

Toxic chemicals have also been used to control or eliminate grazers. The potential exists for using chemicals in the field, provided that side effects are controlled or are insignificant. Controls should be identical to treatments except for the active ingredient. Paired tide pools (experimental vs. control) may be useful for preliminary testing of side effects on herbivores and other members of the community. Rotenone, for example, is commonly used on fish (Sale 1980). Sale also described various other techniques for manipulating fish. Sea urchins are readily killed by quicklime (CaO) (Leighton et al. 1966; Paine 1977). A variety of insecticides have been used for controlling crustaceans and gastropods (Shacklock and Croft 1981) and insects (Robles and Cubit 1981).

An assortment of barriers has been used in intertidal and subtidal communities to assess herbivore effects. Barriers are designed to exclude or include specific numbers of particular herbivores (exclusion vs. inclusion experiments). These experiments are usually performed in relatively small, often isolated areas. Such experiments may suffer from increased edge effects, lack of a seed source (spores, zygotes, etc.), and behavioral effects (herbivore avoidance) associated with barriers and small patches (Dayton 1973; Choat 1982). However, different successional stages often co-occur in many marine communities (Grant 1977; Connell and Slatyer 1977; Paine and Levin 1981) and thus exist naturally in a patchy network. One advantage of the small size of barrier experiments is that they are easily replicated.

Some of the physical barriers used to control herbivore densities include plastic or uncoated galvanized steel, wire mesh, hemp or nylon fish netting or screening, fiberglass, stainless steel or plastic cages and fences, platforms and elevated structures, artificial grass, plastic dishes, marine cement, nontoxic underwater epoxy putty [Sea Goin' Poxy Putty for soft applications and Z-Spar A-788 Splash Zone Compound (R. Steneck, personal communication) for hard applications and cold conditions], and urethan foam. These barriers may be effective against some but not all herbivores in an area. To clarify the effects of a particular herbivore, a series of experiments employing a combination of barriers (physical, chemical, or biological) may be required. In other cases, in which the number of herbivore and predator species is high and the difficulty of separating individual effects is increased, it may be more prudent to categorize herbivores by size and treat them as functional consumer groups (Menge and Lubchenco 1981).

Artifacts are a potential problem with barriers, and appropriate controls must include or account for the altered environment created by the barrier. Some of the problems or side effects observed for physical barriers include reduced light intensity, desiccation, temperature, and water flow; increased fouling, debris, sedimentation, and moisture; and greater attraction and protection of micrograzers and juvenile stages of numerous invertebrates. Attempts to analyze some of these effects have been made by Underwood (1980), Moran (1980), Menge and Lubchenco (1981), and Kennelly (1983).

Solid and mesh fences have been commonly used to prevent reinvasion by limpets, gastropods, and chitons (Haven 1973; Slocum 1980; Underwood 1980; Creese 1980; Lein 1980; Hawkins 1981; Underwood and Jernakoff 1981; Foster 1982), whereas Dayton (1971) used plastic dishes to control limpets. Ebert (1977), Pace (1981), Castilla and Moreno (1982), and Keller (1983) constructed subtidal fences or corrals (topless cages) to exclude or include sea urchins. Fences ranged in height from 2 cm for molluskan grazers to 30 cm for sea urchins. One of the larger fence designs is three parallel 20 × 5 m areas enclosed by nylon nets to control large sea urchins (Pace 1981). Movement of grazers over the fence can be discouraged by curling the upper edges of the fence outward for exclosures and inward for enclosures. Mesh sizes have also varied (0.2–5 cm) depending on the size of the herbivore being manipulated. A solid sheet or structure such as Dayton's (1971) dog dishes and Underwood's (1980) galvanized iron fences is a more effective barrier but also potentially increases artifacts such as altered water flow, sedimentation, and edge effects (Dayton 1971). One solution is to enclose larger areas within fenced plots and sample or conduct experiments only in the central portion of the plot, thereby reducing or eliminating edge effects and artifacts associated with the proximity of the barrier (Hawkins and Hartnoll 1983).

Fencing studies commonly employ unfenced, adjacent plots as controls. Some controls may have one side bounded by the barrier, but even then they do not mimic the experimental treatment (Underwood 1980; Hawkins 1981). One alternative is to employ partial fences to mimic the physical barrier but allow the grazers access to the plot. A potential problem with partial fences, however, is the tendency for herbivores to accumulate in them (Menge 1976; Foster 1982). Another alternative (control) is to enclose herbivores in fenced plots at natural densities. This method mimics the physical environment well but may increase or decrease grazing intensity depending on the reaction of the grazers to the barrier (Dayton 1971; Haven 1973; Hawkins 1981). The best solution here might be a combination of several controls to assess both effects and artifacts (Underwood 1976, 1980).

Various cages ranging in size from 100 cm^2 with small mesh (Dayton 1971) to 6.75 m^3 with large mesh (Fitz et al. 1983) have been used to manipulate fish or actively climbing invertebrates. Several groups of mobile grazers have been included or excluded from natural or artificial substrata, including sea urchins (Kitching and Ebling 1961; Vadas 1968; Paine and Vadas 1969; Vine 1974; Wanders 1977; Sammarco 1980, 1982a,b; Keller 1983), gastropods (Castenholz 1961; Luckens 1974; Lubchenco and Menge 1978; Lubchenco and Cubit 1980; Underwood 1980; Underwood and Jernakoff 1981), crabs (Sousa 1979; Menge and Lubchenco 1981), and fish (Stephenson and Searles 1960; Randall 1961, 1965; Bakus 1967; Earle 1972; John and Pople 1973; Vine 1974; Day 1977; Wanders 1977; Montgomery 1980a; Hay 1981c; Hixon and Brostoff 1981; Menge and Lubchenco 1981; Kennelly 1983; Hatcher and Larkum 1983).

The general artifacts or side effects described for barriers also apply to cages, and therefore it is essential that appropriate controls be employed. Again, a combination of controls can be used, including isolated boulders or ledges (Vadas 1968), artificial or elevated substrata (Robles and Cubit 1981), partial cages (Wanders 1977; Fitz et al. 1983; Keller 1983; Petraitis 1983), and enclosure cages either covering natural substrata (Castenholz 1961; Kitching and Ebling 1961; Lubchenco and Menge 1978; Underwood 1980; Hawkins 1981) or containing introduced boulders (Vadas 1968) or artificial substrata (Foster 1975; Kennelly 1983). Intertidal open-sided roofs or covers, used as controls for cages, deserve attention because they are commonly employed to mimic the physical environment, especially shading, and moisture regimes in cages (Connell 1974; Menge 1976; Lubchenco 1978, 1980). Unfortunately, like partial fences, they may shelter an abnormally large number of predators (Menge 1976). Roof effects are not limited to predators and occur with herbivorous gastropods (Underwood 1980; Underwood and Jernakoff 1981; Vadas 1982; Petraitis 1983).

Fouling occurs on most introduced surfaces, and compromises must be made regarding the size of the mesh and the frequency of maintenance or replacement. A large mesh with good water flow is preferable when a choice is possible. When fouling is severe, fences may be more appropriate (Choat 1982). However, the larger the mesh, the more likely it is that micrograzers or juvenile grazers will enter the cage and influence the results. A quick estimate of the grazers present and their sizes, degree of motility, and likelihood of reinvasion between censuses will aid in this decision. For example, the lack of herbivorous fish in colder waters and the general paucity of smaller invertebrate grazers allow larger meshes to be used in subtidal studies. The larger mesh is also desirable for the development of high-light-requiring, opportunistic species.

With herbivore inclusion studies both caged and uncaged (natural) controls are needed to clarify the effects of particular cage and mesh sizes. Subtidal colonization and succession patterns inside of exclusion cages were remarkably similar to natural patterns in Washington (Vadas 1968) and in California (Foster 1975). Subtidal and intertidal cages also suffer from other problems. Like partial fences, they may attract grazers (Luckens 1974), which, if dense enough, may cause the algae to become detached. Moreover, large cages designed to exclude fish may attract snails and other grazers (Kennelly 1983). Uncaged plants also suffer catastrophic mortality from episodic invasions by smaller grazers (Vadas 1968; Thomas and Page 1983).

Other types of experimental and cage designs can be used for particular situations. For example, using both covered and topless cages may separate the effects of two different herbivores (Keller 1983). A larger mesh can be used on the top or upper portions of the cage to increase light penetration where grazers are not apt to climb. In nonturbulent environments, smaller cages can be sequentially replaced with larger cages to reduce crowding and competition among algae (Vadas 1982). In the latter study stainless steel cages 20 × 20 × 5 cm high were replaced by cages 10 and 20 cm high or removed entirely. Reduced algal growth, especially of *Fucus* spp., and crowding occurred in all cases except when the cages were removed entirely, not unlike an effect observed by Foster (1975). The point here is that, the smaller the cage and mesh size, the more likely, it is that algal growth will be abnormal and the experiment flawed. Even in the 20 × 20 cm cages the algal thalli appear to be "stuffed" into the cages, which likely reduces the light reaching the understory and the flow of water through the cage (see Foster and Sousa, Chap. 13, for details on anchoring structures).

Chemical barriers provide another means of regulating grazing and have the advantage (Robles and Cubit 1981) of minimizing but not eliminating (Little and Smith 1980) the physical and biological artifacts encountered with fences and cages. Two types of barriers have been utilized: one a paint that consists of a slurry of copper powder and a copper-based lacquer or enamel, the other a sticky insecticidal strip that is glued to the substratum (Sousa 1979; Kitting 1980; Paine 1980; Robles and Cubit 1981). In both, care must be taken in the placement of plots to avoid runoff or standing water and to remove fouling organisms. Benthic diatoms can form extensive mats over the copper paint, which allows limpets to migrate into the plot (Little and Smith 1980). Controls for experimental plots should include adjacent (natural) plots and partial (small openings in strip) chemical barriers.

Another potential means of regulating grazing activity and assessing the role of herbivores is the manipulation of animals that prevent or reduce access by other animals to substrata colonized by algae.

The beneficial effects on algae of top predators feeding on herbivores was noted by Lubchenco and Gaines (1981). Predators that exert a controlling effect on functionally important herbivores, such as sea otters and fish (Estes et al. 1978; Cowen 1983), and on territorial herbivores, such as limpets and fish (Stimson 1970; Brawley and Adey 1977), can be removed or excluded from specific habitats and the effects assessed. A severe reduction or elimination of predators that control important herbivores (e.g., sea otters) can result in significant decreases in the dominant algal forms and associated fauna (Estes et al. 1978). A variety of methods ranging from manual removals to selective trapping may be required to control or eliminate specific predators. Removals should focus on keystone species (Paine 1969) or those suspected of having stronger linkage strengths on herbivores within the community (Paine 1980). See also Pringle et al. (1982), Edwards et al. (1982), and Bradley (1983) for criticism of unsupported hypotheses regarding the use of keystone concepts when complex and multispecies assemblages may be important.

An alternative strategy in situations in which other barriers are ineffective or cannot be deployed is to increase the density of functionally important predators by providing artificial shelters or burrows. Sheehy (1976), for instance, constructed artificial lobster shelters from cement blocks, which quickly attained a high rate of occupancy. Ceramic drain tiles and similar structures can be used for crustacean or other predators. Biological barriers may be effective when predators or the lack of refuges restrict the movements and foraging activity of grazers – for example, for sea urchins: sea otters, fish, and lobsters (Lowry and Pearse 1973; Vance and Schmitt 1979; Tegner and Levin 1983); for gastropods: predatory gastropods (Garrity and Levings 1981).

Inadvertent removals and natural phenomena also provide opportunities for investigating the roles of herbivores on algae. In several areas, but especially on Carribean reefs, humans have been active predators on marine herbivores (e.g., fish, conchs, and turtles), which has altered natural relationships in these algal–seagrass communities (Bjorndal 1980). When experimental manipulations can be performed, human effects may be distinguished from historical or other accidents. When aspects of the biology and ecology of herbivores or predators are differentially affected by such phenomena as volcanic activity, landslides, oil pollution, or disease, the ecological effects can be compared with unaffected areas (Southward and Southward 1978; Hawkins and Hartnoll 1983; Pearse and Hines 1979). Such comparisons, however, must be interpreted with caution unless adequate preliminary sampling is available for these areas. The lack of such data and testable hypotheses necessitates the use of a posteriori assumptions that are not always valid (Green 1979).

Algal transplants, removals, or manipulations of structures on which algae grow provide another means of evaluating the influence of grazers. Transplants of rocks or boulders with attached algae from natural areas to sites suspected of being influenced by grazers can be used to identify herbivore effects (John and Pople 1973). Coupled with predator exclusion cages, transplant experiments can also be used in natural (control) areas to determine whether predators are affecting herbivore levels. Also, artificial substrata such as clay tiles can be introduced into herbivore-free areas to allow algal development to proceed to a desired stage before the algae are exposed to herbivores (Nicotri 1977b). With some algae, such as corallines, the thallus can be glued with epoxy putty to polyvinyl chloride or other substrata. If threaded pipes and fittings are used, the various pieces with algae can be readily transplanted to a variety of habitats. The latter method was used to assess grazing by parrotfish on algal reefs (Steneck and Adey 1976).

2. Complications and considerations. Experimental designs, factor interactions, and paradigms in science are briefly explored as a reminder that variability and deterministic processes are part of biological interactions. To test hypotheses and demonstrate generality clearly, serious consideration must be given to the problems of replication, especially in field trials (see Hurlbert 1984). On the one hand, for example, are paired experiments (one control and one experimental treatment), which have been utilized in some studies. When differences are clear-cut, data from paired experiments are useful but limited since variability cannot be assessed. On the other hand, well-replicated experiments with a high degree of similarity within treatments allow for rigorous hypothesis testing and provide robust interpretations and generalizations. Of even greater impact is the similarity of results from two or more independently run experiments.

There is no easy answer to the question of how much replication is enough. This depends on the natural variability of the system under study and numerous other complications. In many field situations interactions between one or more factors may be important, for example, community structure and organization in intertidal habitats (Dayton 1971; Underwood 1980) and subtidal habitats (Dayton 1975b; Cowen et al. 1982). Resolution of these interactions may require multivariate or factorial analysis (Green 1979; Underwood 1980). Highly variable systems or results require greater replication. Pilot studies can provide an indication of the variability, which in turn can be used to estimate acceptable variance levels and hence the replications needed (Green 1979; Zar 1984).

Throughout this chapter, I expressed skepticism about the merit of any particular approach or concept. Unfortunately, the uncritical acceptance of singular concepts leads to paradigms, which are views so dominant that they make other approaches to a discipline seem irrelevant (Strong 1980). Subsequently, ecological paradigms, like others, are extremely difficult to overthrow once they have become firmly established (Simberloff 1980). Therefore, it is essential to state and test hypotheses explicitly, and as Dayton and Oliver (1980) note, one can escape the trap of ambiguous testing and failure to falsify hypotheses by vigorous confrontation of preconceptions and paradigms. Ensuring that the questions or hypotheses have been properly framed, that one's own biases have been minimized, and that the experimental design has provided for adequate replication and random allocation will ensure the continuing development of algal ecology into a mature, quantifiable science.

V. Acknowledgments

I thank Dr. C. Campbell, Dr. R. Elner, and Dr. G. Jacobson for reviewing all or parts of the chapter. I also appreciate the criticism of my graduate students, especially I. Babb, P. Garwood, and F. P. Ojeda, who were coerced into reading several drafts and sharing their knowledge of the literature with me. I gratefully acknowledge the secretarial assistance of Jean Ketch and Donna Wilbur, the computer and bibliographic work of J. Breton, P. O'Connell, and W. Wright, and the patience of my family.

VI. References

Ayling, A. L. 1978. The relation of food availability and food preferences to the field diet of an echinoid *Evechinus chloroticus* (Valenciennes). *J. Exp. Mar. Biol. Ecol.* 33, 223–35.

Ayling, A. L. 1981. The role of biological disturbance in temperate subtidal encrusting communities. *Ecology* 62, 830–47.

Bakus, G. J. 1967. The feeding habits of fishes and primary production at Eniwetok, Marshall Islands. *Micronesica* 3, 135–49.

Bamber, R. N., and Davis, M. H. 1982. Feeding of *Achelia echinata* Hodge (Pycnogonida) on marine algae. *J. Exp. Mar. Biol. Ecol.* 60, 181–7.

Berg, C. J. Jr., 1975. Behavior and ecology of conch (superfamily Strombaceae) on a deep subtidal algal plain. *Bull. Mar. Sci.* 25, 307–17.

Bjorndal, K. A. 1980. Nutrition and grazing behavior of the green turtle *Chelonia mydas*. *Mar. Biol.* 56, 147–54.

Black, R. 1974 Some biological interactions affecting intertidal populations of the kelp *Egregia laevigata. Mar. Biol.* 28, 189–98.

Bonsdorff, E. 1983. Appetite and food consumption in the sea urchin *Echinus esculentus* L. *Sarsia* 68, 25–7.

Boolootian, R. A., and Lasker, R. 1964. Digestion of brown algae and the distribution of nutrients in the purple sea urchin *Strongylocentrotus purpuratus. Comp. Biochem. Physiol.* 11, 273–89.

Bradley, R. A. 1983. Complex food webs and manipulative experiments in ecology. *Oikos* 41, 150–2.

Branch, G. M. 1981. The biology of limpets: physical factors, energy flow, and ecological interactions. *Oceanogr. Mar. Biol. Annu. Rev.* 19, 235–380.

Branch, G. M., and Branch, M. L. 1980. Competition between *Cellana tramoserica* (Sowerby) (Gastropoda) and *Patiriella exiqua* (Lamarck) (Asteroidea), and their influence on algal standing stocks. *J. Exp. Mar. Biol. Ecol.* 48, 35–49.

Brawley, S. H., and Adey, W. H. 1977. Territorial behavior of threespot damselfish (*Eupomacentrus planifrons*) increases reef algal biomass and productivity. *Environ. Biol. Fish.* 2, 45–51.

Brawley, S. H., and Adey, W. H. 1981. The effect of micrograzers on algal community structure in a coral reef microcosm. *Mar. Biol.* 61. 167–77.

Brenner, D., Valiela, I., and van Raalte, C. D. 1976. Grazing by *Talorchestia longicornis* on an algal mat in a New England salt marsh. *J. Exp. Mar. Biol. Ecol.* 22, 161–9.

Brock, R. E. 1979. An experimental study on the effects of grazing by parrotfishes and role of refuges in benthic community structure. *Mar. Biol.* 51, 381–8.

Bryan, P. G. 1975. Food habits, functional digestive morphology, and assimilation efficiency of the rabbitfish *Siganus spinus* (Pisces, Siganidae) on Guam. *Pac. Sci.* 29, 269–77.

Buddington, R. K. 1979. Digestion of an aquatic macrophyte by *Tilapia zillii* (Gervais). *J. Fish. Biol.* 15, 449–55.

Byers, B. A., and Mitton, J. B. 1981. Habitat choice in the intertidal snail *Tegula funebralis. Mar. Biol.* 65, 149–54.

Carefoot, T. H. 1967. Growth and nutrition of *Aplysia punctata* feeding on a variety of marine algae. *J. Mar. Biol. Assoc. U.K.* 47, 565–90.

Carpenter, R. C. 1981. Grazing by *Diadema antillarum* (Philippi) and its effects on the benthic algal community. *J. Mar. Res.* 39, 749–65.

Castenholz, R. W. 1961. The effect of grazing on marine littoral diatom populations. *Ecology* 42, 783–94.

Castilla, J. C., and Moreno, C. A. 1982. Sea urchins and *Macrocystis pyrifera*: experimental test of their ecological relations in southern Chile. In Lawrence, J. M. (ed.), *Echinoderms: Proceedings of the International Conference, Tampa Bay*, pp. 257–63. Balkema, Rotterdam.

Choat, J. H. 1982. Fish feeding and the structure of benthic communities in temperate waters. *Annu. Rev. Ecol. Syst.* 13, 423–49.

Choat, J. H., and Schiel, D. R. 1982. Patterns of distribution and abundance of large brown algae and invertebrate herbivores in subtidal regions of northern New Zealand. *J. Exp. Mar. Biol. Ecol.* 60, 129–62.

Christensen, M. S. 1978. Trophic relationships in juveniles of three species of sparid fishes in the South African marine littoral. *Fish. Bull.* 76, 389–401.

Connell, J. H. 1974. Ecology: field experiments in marine ecology. In Mariscal, R. N. (ed.), *Experimental Marine Biology*, pp. 21–54. Academic Press, New York.

Connell, J. H., and Slatyer, R. O. 1977. Mechanisms of succession in natural communities and their role in community stability and organization. *Amer. Natur.* 111, 1119–44.

Connor, M. S., and Teal, J. M. 1982. The effect of feeding by mud snails, *Ilyanassa obsoleta* (Say), on the structure and metabolism of a laboratory benthic algal community. *J. Exp. Mar. Biol. Ecol.* 65, 29–45.

Conover, R. J. 1966. Assimilation of organic matter by zooplankton. *Limnol. Oceanogr.* 11, 338–45.

Cowen, R. K. 1983. The effect of sheephead (*Semicossypus pulcher*) predation on red sea urchin (*Strongylocentrotus franciscanus*) populations: an experimental analysis. *Oecologia* 58, 249–55.

Cowen, R. K., Agegian, C. R., and Foster, M. S. 1982. The maintenance of community structure in a central California giant kelp forest. *J. Exp. Mar. Biol. Ecol.* 64, 189–201.

Creese, R. G. 1980. An analysis of distribution and abundance of populations of the high-shore limpet, *Notoacmea petterdi* (Tenison-Woods). *Oecologia* 45, 252–60.

Curtis, L. A., and Hurd, L. E. 1979. On the broad nutritional requirements of the mud snail, *Ilyanassa* (*Nassarius*) *obsoleta* (Say), and its polytrophic role in the food web. *J. Exp. Mar. Biol. Ecol.* 41, 289–97.

Day, R. W. 1977. Two contrasting effects of predation on species richness in coral reef habitats. *Mar. Biol.* 44, 1–5.

Dayton, P. K. 1971. Competition, disturbance, and community organization: the provision and subsequent utilization of space in a rocky intertidal community. *Ecol. Monogr.* 41, 351–89.

Dayton, P. K. 1973. Dispersion, dispersal, and persistence of the annual intertidal alga, *Postelsia palmaeformis* Ruprecht. *Ecology* 54, 433–8.

Dayton, P. K. 1975a. Experimental evaluation of ecological dominance in a rocky intertidal algal community. *Ecol. Monogr.* 45, 137–59.

Dayton, P. K. 1975b. Experimental studies of algal canopy interactions in a sea otter-dominated kelp community at Amchitka Island, Alaska. *Fish. Bull.* 73, 230–7.

Dayton, P. K., and Oliver, J. S. 1980. An evaluation of experimental analyses of population and community patterns in benthic marine environments. In Tenore, K. R., and Coull, B. C. (eds.), *Marine Benthic Dynamics*, pp. 93–120. Belle W. Baruch Library of Marine Science Publ. 11, University of South Carolina Press, Columbia.

Earle, S. A. 1972. The influence of herbivores on the marine plants of Great Lameshur Bay, with an annotated list of plants. *Sci. Bull. Natur. Hist. Mus. Los Angeles* 14, 17–44.

Ebeling, A. W., Larson, R. J., Alevizon, W. S., and Bray, R. N. 1980. Annual variability of reef-fish assemblages in kelp forests off Santa Barbara, California. *Fish. Bull.* 78, 361–77.

Ebert, T. A. 1965. A technique for the individual marking of sea urchins. *Ecology* 46, 193–4.

Ebert, T. A. 1968. Growth rates of the sea urchin *Strongylocentrotus purpuratus* related to food availability and spine abrasion. *Ecology* 49, 1075–91.

Ebert, T. A. 1977. An experimental analysis of sea urchin dynamics and community interactions on a rock jetty. *J. Exp. Mar. Biol. Ecol.* 27, 1–22.

Edwards, D. C., Conover, D. O., and Sutter, F. 1982. Mobile predators and the structure of marine intertidal communities. *Ecology* 63, 1175–80.

Edwards, T. W., and Horn, M. H. 1982. Assimilation efficiency of a temperate-zone intertidal fish (*Cebidichthys violaceus*) fed diets of macroalgae. *Mar. Biol.* 67, 247–53.

Elner, R. W. 1981. Diet of green crab *Carcinus maenas* (L.) from Port Herbert, southwestern Nova Scotia. *J. Shellfish Res.* 4, 89–94.

Elner, R. W., and Hughes, R. N. 1978. Energy maximization in the diet of the shore crab, *Carcinus maenas*. *J. Anim. Ecol.* 47, 103–16.

Emlen, J. M. 1966. The role of time and energy in food preference. *Amer. Natur.* 100, 611–17.

Estes, J. A., and Palmisano, J. F. 1974. Sea otters: their role in structuring nearshore communities. *Science* 185, 1058–60.

Estes, J. A., Smith, N. S., and Palmisano, J. F. 1978. Sea otter predation and community organization in the western Aleutian Islands, Alaska. *Ecology* 59, 822–33.

Fairweather, P. G., and Underwood, A. J. 1983. The apparent diet of predators and biases due to different handling times of their prey. *Oecologia* 56, 169–79.

Feeny, P. 1975. Biochemical coevolution between plants and their insect herbivores. In Gilbert, L. E., and Raven, P. H. (eds.), *Coevolution of Animals and Plants*, pp. 3–19. University of Texas Press, Austin.

Fitz, H. C., Reaka, M. L., Bermingham, E., and Wolf, N. G. 1983. Coral recruitment at moderate depths: the influence of grazing. In Reaka, M. L. (ed.), *The Ecology of Deep and Shallow Coral Reefs*, vol. 1, no. 1 pp. 89–96. NOAA's Undersea Research Program, Washington, D.C.

Forster, G. 1959. The ecology of *Echinus esculentus* L.: quantitative distribution and the rate of feeding. *J. Mar. Biol. Assoc. U.K.* 38, 361–7.

Foster, M. S. 1975. Regulation of algal community development in a *Macrocystis pyrifera* forest. *Mar. Biol.* 32, 331–42.

Foster, M. S. 1982. Factors controlling the intertidal zonation of *Iridaea flaccida* (Rhodophyta). *J. Phycol.* 18, 285–94.

Frank, P. W. 1965. Shell growth in a natural population of the turban snail, *Tegula funebralis*. *Growth* 29, 395–403.

Fretwell, S. D. 1972. *Populations in a Seasonal Environment*. Princeton University Press, Princeton, N. J. 217 pp.

Fry, B., and Parker, P. L. 1979. Animal diet in Texas seagrass meadows: δ^{13}C evidence for the importance of benthic plants. *Est. Coast. Mar. Sci.* 8, 499–509.

Fuji, A. 1967. Ecological studies on the growth and food consumption of Japanese common littoral sea urchin, *Strongylocentrotus intermedius* (A. Agassiz). *Mem. Fac. Fish.* 15, 83–160.

Gaines, S. D., and Lubchenco, J. 1982. A unified approach to marine plant–herbivore interactions. 2. Biogeography. *Annu. Rev. Ecol. Syst.* 13, 111–38.

Galli, D. R., and Giese, A. C. 1959. Carbohydrate digestion in a herbivorous snail, *Tegula funebralis*. *J. Exp. Zool.* 140, 415–40.

Garnick, E. 1978. Behavioral ecology of *Strongylocentrotus droebachiensis* (Muller) (Echinodermata: Echinoidae) aggregating behavior in chemotaxis. *Oecologia* 37, 77–84.

Garrity, S. D., and Levings, S. C. 1981. A predator–prey interaction between two physically and biologically constrained tropical rocky shore gastropods: direct, indirect and community effects. *Ecol. Monogr.* 51, 267–86.

Geiselman, J. A., and McConnell, O. J. 1981. Polyphenols in brown algae *Fucus vesiculosus* and *Ascophyllum nodosum:* chemical defenses against the marine herbivorous snail, *Littorina littorea*. *J. Chem. Ecol.* 7, 1115–33.

Grahame, J. 1973. Assimilation efficiency of *Littorina littorea* (L.) (Gastropoda: Prosobranchiata). *J. Anim. Ecol.* 4, 383–9.

Grant, W. S. 1977. High intertidal community organization on a rocky headland in Maine, USA. *Mar. Biol.* 44, 15–25.

Green, R. H. 1979. *Sampling Design and Statistical Methods for Environmental Biologists*. Wiley, New York, 257 pp.

Greze, I. I. 1968. Feeding habits and food requirements of some amphipods in the Black Sea. *Mar. Biol.* 1, 316–21.

Guerinot, M. L., and Patriquin, D. G. 1981. The association of N_2-fixing bacteria with sea urchins. *Mar. Biol.* 62, 197–207.

Hall, J. R. 1973. Intraspecific trail-following in the marsh periwinkle, *Littorina irrorata* Say. *Veliger* 16, 72–5.

Hamilton, P. V. 1977. Daily movements and visual location of plant stems by *Littorina irrorata* (Mollusca, Gastropoda). *Mar. Behav. Physiol.* 4, 293–304.

Hatcher, B. G., and Larkum, A. W. D. 1983. An experimental analysis of factors controlling the standing crop of the epilithic algal community on a coral reef. *J. Exp. Mar. Biol. Ecol.* 69, 61–84.

Haven, S. B. 1973. Competition for food between the intertidal gastropods *Acmaea scrabra* and *Acmaea digitalis*. *Ecology* 54, 143–51.

Hawkins, S. J. 1981. The influence of *Patella* grazing on the fucoid barnacle mosaic on moderately exposed rocky shores. *Kieler Meeresforsch. Sonderh.* 5, 537–43.

Hawkins, S. J., and Hartnoll, R. G. 1983. Grazing of intertidal algae by marine invertebrates. *Oceanogr. Mar. Biol. Annu. Rev.* 21, 195–282.

Hay, M. E. 1981a. Spatial patterns of grazing intensity on a Caribbean barrier reef: herbivory and algal distribution. *Aquat. Bot.* 11, 97–109.

Hay, M. E. 1981b. The functional morphology of turf-forming seaweeds: persistence in stressful marine habitats. *Ecology* 62, 739–50.

Hay, M. E. 1981c. Herbivory, algal distribution, and the maintenance of between-habitat diversity on a tropical fringing reef. *Amer. Natur.* 118, 520–40.

Hay, M. E., Colburn, T., and Downing, D. 1983. Spatial and temporal patterns in herbivory on a Caribbean fringing reef: the effects on plant distribution. *Oecologia* 58, 299–308.

Hayashi, I. 1980. Structure and growth of a shore population of the ormer, *Haliotis tuberculata. J. Mar. Biol. Assoc. U.K.* 60, 431–7.

Heinsohn, G. E., and Birch, W. R. 1972. Food and feeding habits of the dugong *Dugong dugong* (Erxleben), in Northern Queensland, Australia. *Mammalia* 36, 414–22.

Himmelman, J. H. 1980. The role of the green sea urchin, *Strongylocentrotus droebachiensis,* in the rocky subtidal region of Newfoundland. In Pringle, J. D., Sharp, G. J., and Caddy, J. F. (eds.), *Proceedings of a Workshop on the Relationship betwen Sea Urchin Grazing and Commercial Plant/Animal Harvesting,* pp. 92–119. Candian Technical Report on Fisheries and Aquatic Sciences no. 954, Government of Canada, Department of Fisheries and Oceans, Winnipeg, Manitoba.

Himmelman, J. H., and Steele, D. H. 1971. Food and predators of the green sea urchin *Strongylocentrotus droebachiensis* in Newfoundland waters. *Mar. Biol.* 9, 315–22.

Hirano, Y. 1979. Studies on activity pattern of the patellid limpet *Cellana toreuma* (Reeve). *J. Exp. Mar. Biol. Ecol.* 40, 137–48.

Hixon, M. A., and Brostoff, W. N. 1981. Fish grazing and community structure of Hawaiian reef algae. In Gomez, E. D., Birkelaris, C. E., Buddemares, R. W., Johannes, R. E., Marsh, J. A., Jr., and Tsuda, R. T. (eds.), *Proceedings of the Fourth International Coral Reef Symposium,* vol. 2, pp. 507–14. Marine Sciences Center, University of the Philippines, Quezon City, Philippines.

Horn, M. H., Murray, S. N., and Edwards, T. W. 1982. Dietary selectivity in the field and food preferences in the laboratory for two herbivorous fishes (*Cebidichthys violaceus* and *Xiphister mucosus*) from a temperate intertidal zone. *Mar. Biol.* 67, 237–46.

Hughes, R. N. 1980a. Predation and community structure. In Price, J. H., Irvine, D. E. G., and Farnham, W. F. (eds.), *The Shore Environment,* vol. 2: *Ecosystems.* pp. 699–728. Systematics Association Special Volume 17(b), Academic Press, London.

Hughes, R. N. 1980b. Optimal foraging theory in the marine context. *Oceanogr. Mar. Biol. Annu. Rev.* 18, 423–81.

Hurlbert, S. H., 1984. Pseudoreplication and the design of ecological field experiments. *Ecol. Monogr.* 54, 187–211.

Ivlev, V. S. 1961. *Experimental Ecology of the Feeding of Fishes.* Yale University Press, New Haven, Conn. 302 pp.

Jacobs, R. P. W. M., den Hartog, C., Braster, B. F., and Carriere, F. C. 1981. Grazing of the seagrass *Zostera noltii* by birds at Terschelling (Dutch Wadden Sea). *Aquat. Bot.* 10, 241–59.

Jahan-Parwar, B., Smith, M., and von Baumgarten, R. 1969. Activation of neurosecretory cells in *Aplysia* by osphradial stimulation. *Amer. J. Physiol.* 216, 1246–57.

Jensen, K. R. 1983. Factors affecting feeding selectivity in herbivorous ascoglossa (Mollusca, Opisthobranchia). *J. Exp. Mar. Biol. Ecol.* 66, 135–48.

John, D. M., and Pople, W. 1973. The fish grazing of rocky shore algae in the Gulf of Guinea. *J. Exp. Mar. Biol. Ecol.* 11, 81–90.

Johnson, D. H. 1980. The comparison of usage and availability measurements for evaluating resource preference. *Ecology* 61, 65–71.

Johnstone, I. M., and Hudson, E. T. 1981. The dugong diet: mouth sample and analysis. *Bull. Mar. Sci.* 34, 681–90.

Jones, N. S. 1948. Observations and experiments on the biology of *Patella vulgata* at Port St. Mary, Isle of Man. *Proc. Trans. Liverpool Biol. Soc.* 56, 60–77.

Jones, N. S., and Kain, J. M. 1967. Subtidal algal colonization following the removal of *Echinus*. *Helgo. Meeres.* 15, 460–6.

Jones, R. S. 1968a. A suggested method for quantifying gut contents in herbivorous fish. *Micronesica* 4, 369–71.

Jones, R. S. 1968b. Ecological relationships in Hawaiian and Johnston Island Acanthuridae (surgeonfishes). *Micronesica* 4, 309–61.

Joule, B. J. 1983. An effective method for tagging marine polychaetes. *Can. J. Fish. Aquat. Sci.* 40, 540–1.

Keller, B. D. 1983. Coexistence of sea urchins in seagrass meadows: an experimental analysis of competition and predation. *Ecology* 64, 1581–98.

Kennelly, S. J. 1983. An experimental approach to the study of factors affecting algal colonization in a sublittoral kelp forest. *J. Exp. Mar. Biol. Ecol.* 68, 257–76.

Kingett, P. D., and Choat, J. H. 1981. Analysis of density and distribution patterns in *Chrysophrys auratus* (Pisces, Sparidae) within a reef environment: an experimental approach. *Mar. Ecol. Progr. Ser.* 5, 283–90.

Kitching, J. A., and Ebling, F. J. 1961. The ecology of Lough Ine. 11. The control of algae by *Paracentrotus lividus* (Echinoidea). *J. Anim. Ecol.* 30, 373–83.

Kitting, C. L. 1979. The use of feeding noises to determine the algal foods being consumed by intertidal molluscs. *Oecologia* 40, 1–17.

Kitting, C. L. 1980. Herbivore–plant interactions of individual limpets maintaining a mixed diet of intertidal marine algae. *Ecol. Monogr.* 50, 527–50.

Kriegstein, A. R., Castellucci, V. C., and Kandel, E. R. 1974. Metamorphosis of *Aplysia californica* in laboratory culture. *Proc. Nat. Acad. Sci. U.S.* 71, 3654–8.

Lang, C., and Mann, K. H. 1976. Changes in sea urchin populations after the destruction of kelp beds. *Mar. Biol.* 36, 321–6.

Larson, B. R., Vadas, R. L., and Keser, M. 1980. Feeding and nutritional ecology of the sea urchin *Strongylocentrotus droebachiensis* in Maine, U.S.A. *Mar. Biol.* 59, 49–62.

Lawrence, J. M. 1975. On the relationship between marine plants and sea urchins. *Annu. Rev. Oceanogr. Mar. Biol.* 13, 213–86.

Lawrence, J. M. 1982. Digestion. In Jangoux, M., and Lawrence, J. M. (eds.), *Echinoderm Nutrition*, pp. 283–316. Balkema, Rotterdam.

Lawrence, J. M., and Hughes, L. 1972. The diurnal rhythm of feeding and passage of food through the gut of *Diadema setosum* (Echinodermata, Echinoidea). *Isr. J. Zool.* 21, 13–16.

Leighton, D. L. 1966. Studies of food preference of algivorous invertebrates of southern California kelp beds. *Pac. Sci.* 20, 104–13.

Leighton, D. L., and Boolootian, R. A. 1963. Diet and growth in the black abalone, *Haliotis cracherodii. Ecology* 44, 227–38.

Leighton, D. L., Jones, L. G., and North, W. J. 1966. Ecological relationships between giant kelp and sea urchins in southern California. In Young, E. G., and McLachlan, J. L. (eds), *Proceedings of the Fifth International Seaweed Symposium*, pp. 141–53. Pergamon Press, Oxford.

Lein, T. E. 1980. The effects of *Littorina littorea* L. (Gastropoda) grazing on littoral green algae in the inner Oslofjord, Norway. *Sarsia* 65, 87–92.

Levings, S. C., and Garrity, S. D. 1983. Diel and tidal movement of two co-occurring neritid snails: differences in grazing patterns on a tropical rocky shore. *J. Exp. Mar. Biol. Ecol.* 67, 261–78.

Lewis, D. B., and Whitney, P. J. 1968. Cellulase in *Nereis virens. Nature* 220, 603–4.

Lewis, J. B. 1958. The biology of the tropical sea urchin, *Tripneustes esculentus* Leske in Barbados, British West Indies. *Can. J. Zool.* 36, 607–21.

Lewis, J. B. 1964. Feeding and digestion in the tropical sea urchin *Diadema antillarum* Philippi. *Can. J. Zool,* 42, 549–57.

Lewis, J. R., and Bowman, R. S. 1975. Local habitat-induced variations in the population dynamics of *Patella vulgata* L. *J. Exp. Mar. Biol. Ecol.* 17, 165–203.

Lipkin, Y. 1975. Food of the Red Sea dugong (Mammalia, Sirenia) from Sinai. *Isr. J. Zool.* 24, 81–98.

Lissner, A. L. 1980. Some effects of turbulence on the activity of the sea urchin *Centrostephanus coronatus* Verrill. *J. Exp. Mar. Biol. Ecol.* 48, 185–93.

Little, C., and Smith, L. P. 1980. Vertical zonation on rocky shores in the Severn Estuary. *Est. Coast. Mar. Sci.* 11, 651–69.

Littler, M. M., and Littler, D. S. 1980. The evolution of thallus form and survival strategies in benthic marine macroalgae: field and laboratory tests of a functional form model. *Amer. Natur.* 116, 25–44.

Littler, M. M., Littler, D. S., and Taylor, P. R. 1983a. Evolutionary strategies in a tropical barrier reef system: functional-form groups of marine macroalgae. *J. Phycol.* 19, 229–37.

Littler, M. M., Taylor, P. R., and Littler, D. S. 1983b. Algal resistance to herbivory on a Caribbean barrier reef. *Coral Reefs* 2, 111–18.

Lobel, P. S. 1980. Herbivory by damselfishes and their role in coral reef ecology. *Bull. Mar. Sci.* 30, 273–89.

Lobel, P. S. 1981. Trophic biology of herbivorous reef fishes: alimentary pH and digestive capabilities. *J. Fish. Biol.* 19, 365–97.

Lobel, P. S., and Ogden, J. C. 1981. Foraging by the herbivorous parrotfish *Sparisoma radians. Mar. Biol.* 64, 173–83.

Lowe, E. F., and Lawrence, J. M. 1976. Absorption efficiencies of *Lytechinus variegatus* (Lamarck) (Echinodermata, Echinoidea) for selected marine plants. *J. Exp. Mar. Biol. Ecol.* 21, 223–34.

Lowry, L. F., and Pearse, J. S. 1973. Abalones and sea urchins in an area inhabited by sea otters. *Mar. Biol.* 23, 213–19.

Lubchenco, J. 1978. Plant species diversity in a marine intertidal community: importance of herbivore food preference and algal competitive abilities. *Amer. Natur.* 1, 23–39.

Lubchenco, J. 1979. Consumer terms and concepts. *Amer. Natur.* 113, 315–17.

Lubchenco, J. 1980. Algal zonation in the New England rocky subtidal community: an experimental analysis. *Ecology* 61, 333–44.

Lubchenco, J., and Cubit, J. 1980. Heteromorphic life histories of certain marine algae as adaptations to variations in herbivory. *Ecology* 61, 676–87.

Lubchenco, J., and Gaines, S. D. 1981. A unified approach to marine plant–herbivore interactions. 1. Populations and communities. *Annu. Rev. Ecol. Syst.* 12, 405–37.

Lubchenco, J., and Menge, B. A. 1978. Community development and persistence in a low rocky intertidal zone. *Ecol. Monogr.* 48, 67–94.

Luckens, P. A. 1974. Removal of intertidal algae by herbivores in experimental frames and on shores near Auckland, New Zealand. *N.Z. J. Mar. Freshw. Res.* 8, 637–54.

Lundberg, B., and Lipkin, Y. 1979. Natural food of the herbivorous rabbitfish (*Siganus* spp.) in northern Red Sea. *Bot. Mar.* 22, 173–81.

McClintock, J. B., Klinger, T. S., and Lawrence, J. M. 1982. Feeding preferences of echinoids for plant and animal food models. *Bull. Mar. Sci.* 32, 365–69.

MacKay, D. A., and Underwood, A. J. 1977. Experimental studies on homing in the intertidal patellid limpet *Cellana tramoserica* (Sowerby). *Oecologia* 30, 215–37.

McQuaid, C. D. 1981. The establishment and maintenance of vertical size gradients in populations of *Littorina africana knysnaensis* (Philippi) on an exposed rocky shore. *J. Exp. Mar. Biol. Ecol.* 54, 77–89.

May, V., Bennett, I., and Thompson, T. E. 1970. Herbivore–algal relationships on a coastal rock platform (Cape Banks, N.S.W.). *Oecologia* 6, 1–14.

Menge, B. A. 1972. Foraging strategy of a starfish in relation to actual prey availability and environmental predictability. *Ecol. Monogr.* 42, 25–50.

Menge, B. A. 1976. Organization of the New England rocky intertidal community: role of predation, competition and environmental heterogeneity. *Ecol. Monogr.* 46, 355–93.

Menge, B. A., and Lubchenco, J. 1981. Community organization in temperate and tropical rocky intertidal habitats: prey refuges in relation to consumer pressure gradients. *Ecol. Monogr.* 51, 429–50.

Menzel, D. W. 1959. Utilization of algae for growth by the angelfish. *Holacanthus bermudensis. J. Cons. Int. Expl. Mer* 24, 308–13.

Miller, R. J., and Mann, K. H. 1973. Ecological energetics of the seaweed zone in a marine bay on the Atlantic coast of Canada. 3. Energy transformations by sea urchins. *Mar. Biol.* 18, 99–114.

Miller, R. J., Mann, K. H., and Scarratt, D. J. 1971. Production potential of a seaweed–lobster community in eastern Canada. *J. Fish. Res. Bd. Can.* 28, 1733–8.

Montgomery, W. L. 1977. Diet and gut morphology in fishes, with special reference to the monkeyface prickleback, *Cebidichthys violaceus* (Haeidae, Blennioidei). *Copeia* 1, 179–82.

Montgomery, W. L. 1980a. The impact of non-selective grazing by the giant blue damselfish, *Microspathodon dorsalis*, on algal communities in the Gulf of California, Mexico. *Bull. Mar. Sci.* 30, 290–303.

Montgomery, W. L. 1980b. Comparative feeding ecology of two herbivorous damselfishes (Pomacentridae: Teleostei) from the Gulf of California, Mexico. *J. Exp. Mar. Biol. Ecol.* 47, 9–24.

Montgomery, W. L., and Gerking, S. D. 1980. Marine macroalgae as foods for fishes: an evaluation of potential food quality. *Environ Biol. Fish.* 5, 143–53.

Moore, H. B., and McPherson, B. F. 1965. A contribution to the study of the productivity of the urchins *Tripneustes esculentus* and *Lytechinus variegatus*. *Bull. Mar. Sci.* 15, 855–71.

Moran, P. J. 1980. Natural physical disturbance and predation, their importance in structuring a marine sessile community. *Aust. J. Ecol.* 5, 193–200.

Moreno, C. A., and Sutherland, J. P. 1982. Physical and biological processes in a *Macrocystis pyrifera* community near Valdiva, Chile. *Oecologia* 55, 1–6.

Moriarty, D. J. W. 1973. The physiology of digestion of blue-green algae in the cichlid fish, *Tilapia nilotica*. *J. Zool. Lond.* 171, 25–39.

Murdoch, W. W. 1969. Switching in general predators: experiments on predator specificity and stability of prey populations. *Ecol. Monogr.* 39, 335–54.

Neill, S. R. St. J., and Larkum, A. H. 1965. *Ecology of Some Echinoderms in Maltese Waters*. Symposium of the Underwater Association of Malta, pp. 51–55.

Newell, R. C., Pye, V. I., and Absanullah, M. 1971. Factors affecting the feeding rate of the winkle *Littorina littorea*. *Mar. Biol.* 9, 1020–32.

Nicotri, M. E. 1977a. The impact of crustacean herbivores on cultured seaweed populations. *Aquaculture* 12, 127–36.

Nicotri, M. E. 1977b. Grazing effects of four marine intertidal herbivores on the microflora. *Ecology* 58, 1020–32.

Nicotri, M. E. 1980. Factors involved in herbivore food preference. *J. Exp. Mar. Biol. Ecol.* 42, 13–26.

North, W. J. 1963. *Kelp Habitat Improvement Project: Final Report*. University of California Institute of Marine Resources, Imr 63-13, Davis. 123 pp.

Norton, T. A., Mathieson, A. C., and Neushul, A. C. 1982. A review of some aspects of form and function in seaweeds. *Bot. Mar.* 25, 501–10.

Ogden, J. C., Brown, R. A., and Salesky, N. 1973. Grazing by the echinoid *Diadema antillarum* Phillippi: formation of halos around West Indian patch reefs. *Science* 182, 715–17.

Ogden, J. C., and Lobel, P. S. 1978. The role of herbivorous fishes and urchins in coral reef communities. *Environ. Biol. Fish.* 3, 49–63.

Ogden, J. C., Robinson, L., Whitlock, K., Daganhardt, H., and Cebula, R.

1983. Diel foraging patterns in juvenile green turtles (*Chelonia mydas* L.) in St. Croix, United States Virgin Islands. *J. Exp. Mar. Biol. Ecol.* 66, 199–205.

Ogden, J. C., Tighe, S., and Miller, S. 1980. Grazing of seagrasses by large herbivores in the Caribbean. *Amer. Zool.* 20, 949 (abstr.).

Olsson, M., and Newton, G. 1978. A simple, rapid method for marking individual sea urchins. *Calif. Fish Game* 1, 58–62.

Pace, D. 1981. Kelp community development in Barkley Sound, British Columbia following sea urchin removal. In Fogg and Jones (eds.), *Proceedings of the Eighth International Seaweed Symposium.* pp. 457–463. The Marine Science Laboratories, Menai Bridge.

Paine, R. T. 1969. A note on trophic complexity and community stability. *Amer. Natur.* 103, 91–3.

Paine, R. T. 1971. Energy flow in a natural population of the herbivorous gastropod *Tegula funebralis. Limnol. Oceanogr.* 16, 86–98.

Paine, R. T. 1977. Controlled manipulations in the marine intertidal zone, and their contributions to ecological theory, pp. 245–70. Academy of Natural Sciences, Special Publ. 12, Philadelphia.

Paine, R. T. 1980. Food webs: linkage, interaction strength, and community infrastructure. *J. Anim. Ecol.* 49, 667–85.

Paine, R. T., and Levin, S. A. 1981. Intertidal landscapes: disturbance and the dynamics of pattern. *Ecol. Monogr.* 51, 145–78.

Paine, R. T., and Vadas, R. L. 1969. The effects of grazing by sea urchins, *Strongylocentrotus* spp., on benthic algal populations. *Limnol. Oceanogr.* 14, 710–19.

Pearse, J. S., and Hines, A. H. 1979. Expansion of a central California kelp forest following the mass mortality of sea urchins. *Mar. Biol.* 51, 83–91.

Petraitis, P. S. 1983. Grazing patterns of the periwinkle and their effect on sessile intertidal organisms. *Ecology* 64, 522–33.

Phillips, D. W., and Castori, P. 1982. Defensive responses to predatory seastars by two specialist limpets, *Notoacmea insessa* (Hinds) and *Collisella instabilis* (Gould), associated with marine algae. *J. Exp. Mar. Biol. Ecol.* 59, 23–30.

Prim, P., and Lawrence, J. M. 1975. Utilization of marine plants and their constituents by bacteria isolated from the gut of echinoids (Echinodermata). *Mar. Biol.* 33, 167–73.

Pringle, J. D., Sharp, G. J., and Caddy, J. F. 1982. *Interactions in Kelp Bed Ecosystems in the Northwest Atlantic: Review of a Workshop.* Canadian Special Publication of Fisheries and Aquatic Sciences no. 59, 169 pp.

Pyke, G. H., Pulliam, H. R., and Charnov, E. I. 1977. Optimal foraging: a selective review of theory and tests. *Quart. Rev. Biol.* 52, 137–54.

Raffaelli, D. 1979. The grazer–algae interaction in the intertidal zone on New Zealand rocky shores. *J. Exp. Mar. Biol. Ecol.* 38, 81–100.

Randall, J. E. 1961. Overgrazing of algae by herbivorous marine fishes. *Ecology* 42, 812.

Randall, J. E. 1965. Grazing effect on sea grasses by herbivorous reef fishes in the West Indies. *Ecology* 46, 255–60.

Ricker, W. E. 1979. Growth rates and models. In Hoar, W. S., Randa, D. J., and Brett, J. R. (eds.), *Fish Physiology*, vol. 8: *Bioenergetics and Growth*, pp. 677–743. Academic Press, New York.

Robertson, A. I., and Lucas, J. S. 1983. Food choice, feeding rates, and the turnover of macrophyte biomass by a surf-zone inhabiting amphipod. *J. Exp. Mar. Biol. Ecol.* 72, 99–124.

Robles, C. 1982. Disturbance and predation in an assemblage of herbivorous diptera and algae on rocky shores. *Oecologia* 54, 23–31.

Robles C. D., and Cubit, J. 1981. Influence of biotic factors in an upper intertidal community; dipteran larvae grazing on algae. *Ecology* 62, 1536–47.

Sale, P. F. 1980. The ecology of fishes on coral reefs. *Oceanogr. Mar. Biol. Annu. Rev.* 18, 367–421.

Sammarco, P. W. 1980. *Diadema* and its relationship to coral spot mortality: grazing, competition, and biological disturbance. *J. Exp. Mar. Biol. Ecol.* 45, 245–72.

Sammarco, P. W. 1982a. Echinoid grazing as a structuring force in coral communities: whole reef manipulations, *J. Exp. Mar. Biol. Ecol.* 61, 31–55.

Sammarco, P. W. 1982b. Effects of grazing by *Diadema antillarum* Philippi (Echinodermata: Echinoidea) on algal diversity and community structure. *J. Exp. Mar. Biol. Ecol.* 65, 83–105.

Sammarco, P. W., Levinton, J. S., and Ogden, J. C. 1974. Grazing and control of coral reef community structure by *Diadema antillarum* Philippi (Echinodermata, Echinoidea): a preliminary study. *J. Mar. Res.* 32, 47–53.

Schmitt, R. J. 1982. Consequences of dissimilar defenses against predation in a subtidal marine community. *Ecology* 63, 1588–1601.

Schoener, T. W. 1971. Theory of feeding strategies. *Annu. Rev. Ecol. Syst.* 2, 369–404.

Schroeter, S. C., Dixon, J., and Kastendiek, J. 1983. Effects of the starfish *Patiria miniata* on the distribution of the sea urchin *Lytechinus anamesus* in a southern California kelp forest. *Oecologia* 56, 141–7.

Seip, K. L. 1980. A mathematical model of predation in the rocky shore community with some implications for the recovery after oil pollution. *Isem J.* 2, 3–53.

Seip, K. L. 1983. Mathematical models of rocky shore ecosystem, chap. 13. In Mitchel, W. J. and Jorgensen, S. E. (eds.), *Application of Ecological Modelling in Environmental Management,* part B. Elsevier, Amsterdam.

Shacklock, P. F., and Croft, G. B. 1981. Effect of grazers on *Chondrus crispus* in culture. *Aquaculture* 22, 331–42.

Sheehy, D. J. 1976. Utilization of artificial shelters by the American Lobster (*Homarus americanus*). *J. Fish. Res. Bd. Can.* 33, 1615–22.

Shepherd, S. A. 1973. Studies on southern Australian abalone (genus *Haliotis*). 1. Ecology of five sympatric species. *Aust. J. Mar. Freshw. Res.* 24, 217–57.

Simberloff, D. 1980. A succession of paradigms in ecology: essentialism to materialism and probabilism. *Synthese* 43, 3–39.

Slocum, C. J. 1980. Differential susceptibility to grazers in two phases of an intertidal alga: advantages of heteromorphic generations. *J. Exp. Mar. Biol. Ecol.* 46, 99–110.

Sousa, W. P. 1979. Experimental investigations of disturbance and ecological succession in a rocky intertidal algal community. *Ecol. Monogr.* 49, 227–54.

Sousa, W. P., Schroeter, S. C., and Gaines, S. D. 1981. Latitudinal variation in intertidal algal community structure: the influence of grazing and vegetative propagation. *Oecologia* 48, 297–307.

Southward, A. J. 1956. The population balance between limpets and seaweeds on wave-beaten rocky shores. *Rep. Mar. Biol. St. Port Erin* 86, 20–9.

Southward, A. J., and Southward, E. C. 1978. Recolonization of rocky shores in Cornwall after use of toxic dispersants to clean up the Torrey Canyon spill. *J. Fish. Res. Bd. Can.* 35, 682–706.

Steneck, R. S. 1982. A limpet–coralline alga association: adaptations and defenses between a selective herbivore and its prey. *Ecology* 63, 507–22.

Steneck, R. S. 1983. Escalating herbivory and resulting adaptive trends in calcareous algal crusts. *Paleobiology* 9, 44–61.

Steneck, R. S., and Adey, W. H. 1976. The role of environment in control of morphology in *Lithophyllum congestum*, a Caribbean algal ridge builder. *Bot. Mar.* 19, 197–215.

Steneck, R. S., and Watling, L. 1982. Feeding capabilities and limitations of herbivorous molluscs: a functional group approach. *Mar. Biol.* 68, 299–319.

Stephenson, W., and Searles, R. B. 1960. Experimental studies on the ecology of intertidal environments at Heron Island. 1. Exclusions of fish from beach rock. *Aust. J. Mar. Freshw. Res.* 11, 241–67.

Stimson, J. S. 1970. Territorial behavior in the owl limpet, *Lottia gigantea*. *Ecology* 51, 113–18.

Strauss, R. E. 1979. Reliability estimates of Ivlev's electivity index, the forage ratio, and a proposed linear index of food selection. *Trans. Amer. Fish. Soc.* 108, 344–52.

Strong, D. R., Jr. 1980. Null hypotheses in ecology. *Synthese* 43, 271–85.

Sutherland, J. P. 1970. Dynamics of high and low populations of the limpet *Acmaea scabra* (Gould). *Ecol. Monogr.* 40, 169–88.

Tegner, M. J., and Dayton, P. K. 1977. Sea urchin recruitment patterns and implications of commercial fishing. *Science* 196, 324–6.

Tegner, M. J., and Levin, L. A. 1983. Spiny lobsters and sea-urchins: analysis of a predator–prey interaction. *J. Exp. Mar. Biol. Ecol.* 73, 125–50.

Thayer, G. W., Parker, P. L., LaCroix, M. W., and Fry, B. 1978. The stable carbon isotope ratio of some components of an eelgrass, *Zostera marina*, bed. *Oecologia* 35, 1–12.

Thomas, M. L. H., and Page, F. H. 1983. Grazing by the gastropod, *Lacuna vincta*, in the lower intertidal area at Musquash Head, New Brunswick, Canada. *J. Mar. Biol. Assoc. U.K.* 63, 725–36.

Towsend, C. R., and Hughes, R. N. 1981. Maximizing net energy returns from foraging. In Towsend, C. R., and Calow, P. (eds.), *Physiological Ecology*, pp. 86–108. Sinauer, Sunderland, Mass.

Tsuda, R. T., and Randall, J. E. 1971. Food habits of the gastropods *Turbo argyrostoma* and *T. setosus*, reported as toxic from the tropical Pacific. *Micronesica* 7, 153–62.

Underwood, A. J. 1976. Nearest neighbor analysis of spacial dispersion of intertidal prosobranch gastropods within two substrata. *Oecologia* 26, 257–66.

Underwood, A. J. 1979. Ecology of intertidal gastropods. *Adv. Mar. Biol.* 16, 111–210.

Underwood, A. J. 1980. The effects of grazing by gastropods and physical factors on the upper limits of distribution of intertidal macroalgae. *Oecologia* 46, 201–13.

Underwood, A. J. 1981. Techniques of analysis of variance in experimental marine biology and ecology. *Oceanogr. Mar. Biol. Annu. Rev.* 19, 513–605.

Underwood, A. J., Denley, E. J., and Moran, M. J. 1983. Experimental analyses of the structure and dynamics of mid-shore rocky intertidal communities in New South Wales. *Oecologia* 56, 202–19.

Underwood, A. J., and Jernakoff, P. 1981. Effect of interactions between algae and grazing gastropods on the structure of a low-shore intertidal algal community. *Oecologia* 48, 221–33.

Vadas, R. L. 1968. "The Ecology of *Agarum* and the Kelp Bed Community." Ph.D. dissertation, University of Washington, Seattle. 280 pp.

Vadas, R. L. 1977. Preferential feeding: an optimization strategy in sea urchins. *Ecol. Monogr.* 47, 337–71.

Vadas, R. L. 1979. Seaweeds: an overview; ecological and economic importance. *Experientia* 35, 429–32.

Vadas, R. L. 1982. The role of herbivores and stochastic processes on intertidal communities in Maine. In *First International Phycology Congress. St. Johns, Newfoundland,* p. a51. International Phycological Society.

Vadas, R. L., Fenchel, T., and Ogden, J. C. 1982. Ecological studies on the sea urchin, *Lytechinus variegatus,* and the algal-seagrass communities of the Miskito Cays, Nicaragua. *Aquat. Bot.* 14, 109–25.

Vance, R. R. 1979. Effects of grazing by the sea urchin, *Centrostephanus coronatus* on prey community composition. *Ecology* 60, 537–46.

Vance, R. R., and Schmitt, R. J. 1979. The effect of the predator-avoidance behavior of the sea urchin (*Centrostephanus coronatus*) on the breadth of its diet. *Oecologia* 44, 21–5.

van Dongen, A. 1955. The preference of *Littorina obtusata* for Fucaceae. *Arch. Neerl. Zool.* 2, 373–86.

van Montfrans, J., Orth, R. J., and Vay, S. A. 1982. Preliminary studies of grazing by *Bittium varium* on eelgrass periphyton. *Aquat. Bot.* 14, 75–89.

Vaughan, F. A. 1978. Food habits of the sea bream, *Archosargus rhomboidalis* (Linnaeus), and comparative growth on plant and animal food. *Bull. Mar. Sci.* 28, 527–36.

Vine, P. J. 1974. Effects of algal grazing and aggressive behavior of the fishes *Pomacentrus lividus* and *Acanthurus sohal* on coral-reef ecology. *Mar. Biol.* 24, 131–6.

Wanders, B. W. 1977. The role of benthic algae in the shallow reef of Curaçao (Netherlands Antilles). 3. The significance of grazing. *Aquat. Bot.* 3, 357–90.

Wells, M. J., and Buckley, S. K. L. 1972. Snails and trails. *Anim. Behav.* 20, 345–55.

Westernhagen, H. von. 1973. The natural food of the rabbitfish *Siganus oramin* and *S. striolata*. *Mar. Biol.* 22, 367–70.

Williams, I. C., and Ellis, C. 1975. Movements of the common periwinkle, *Littorina littorea* (L.), on the Yorkshire coast in winter and the influence of infection with larval *Digenea*. *J. Exp. Mar. Biol. Ecol.* 17, 47–58.

Woodin, S. A. 1977. Algal "gardening" behavior by nereid polychaetes: effects on soft-bottom community structure. *Mar. Biol.* 44, 39–42.

Zafiriou, O. 1972. Response of *Asterias vulgaris* to chemical stimuli. *Mar. Biol.* 17, 100–7.

Zar, J. H. 1984. *Biostatistical Analysis*, 2nd ed. Prentice-Hall, Englewood Cliffs, N.J. 718 pp.

Zimmerman, R., Gibson, R., and Harrington, J. 1979. Herbivory and detritivory among gammaridean amphipods from a Florida seagrass community. *Mar. Biol.* 54, 41–7.

27: Pathology

JOHN H. ANDREWS

Department of Plant Pathology,
University of Wisconsin, Madison, Wisconsin 53706

LYNDA J. GOFF

Department of Biology and Center for Coastal Marine Studies,
University of California, Santa Cruz, California 95064

CONTENTS

[573]

I. Introduction

Disease is one kind of impairment of the normal state of an organism. The purpose of this chapter on the pathology of macroalgae is to provide guidelines for assessing either the cause or extent of the impairment. The distinction between a "healthy" and a "diseased" individual cannot be clearly drawn, and there is considerable debate on the nature of disease (Wheeler 1975; Grogan 1981). Without embellishing previous definitions, we emphasize that any concept must at least be broad enough to encompass both biotic (infectious) and abiotic (noninfectious) factors. It is often overlooked that abiotic anomalies induced by such agents as pollutants, desiccation, and extremes of temperature or salinity constitute disease (Andrews 1976; Andrews et al. 1979; Goff and Glasgow 1980). Most of this chapter focuses on methods pertinent to infectious diseases.

Knowledge of seaweed pathology is recent, despite long-standing observations of the association of seaweeds with various microorganisms presumed to be parasitic. In the few detailed investigations, methods typically have been improvised or adapted in an ad hoc fashion from terrestrial pathology. Extensive research is needed to establish the best approaches. Wherever possible, we outline procedures used successfully for macroalgae. If data are lacking, we propose methods that, from analogous situations, appear to be feasible. We consider, first, standard (primarily laboratory) approaches to determining the cause or etiology of marine benthic algal disease and, second, techniques for investigating various aspects of host–pathogen interactions. Field-oriented methodologies pertaining to epidemiology or disease loss assessment are insufficiently developed to warrant inclusion.

II. Diagnosis and etiology

A. Symptomatology

The diagnosis of disease is both an art and a science. In essence, it is an exercise in sleuthing and multiple hypothesis testing. The concept of diagnosis and a detailed assessment of diagnostic proce-

[574]

dures derived from the concept are beyond the scope of this chapter (see Streets 1969; McIntyre and Sands 1977; Grogan 1981). Although some attempt has been made to categorize seaweed diseases by symptoms (Suto et al. 1972; Kohlmeyer 1974), symptomatology analogous to that existing in terrestial pathology does not yet exist. This situation greatly handicaps the diagnosis of aquatic plant diseases.

Diagnosis should not be attempted without familiarity with the algal host affected, preferably by personal experience or, as a minimum, by the study of publications on the growth of the plant and the biotic and abiotic factors affecting its development. Among many considerations, the investigator must realize the following: (1) the direct or indirect role of environmental factors in disease development; (2) that cases in which a primary pathogen is involved such that the disease can be relatively easily diagnosed by symptoms (expressions of the host) or signs (visible evidence of the pathogen) will likely be rare; (3) that several different diseases may occur concurrently; and (4) that visual diagnosis is limited because similar symptoms may result from different causal agents and, conversely, the same agent (particularly if abiotic) may induce different symptoms in different algae.

B. Field and laboratory examination

The objective of field examination is to establish a case history. Remove the plant intact and study it completely. Compare its condition (e.g., growth characteristics and symptom expression) with both healthy and other diseased specimens at the same general location. Record pertinent habitat characteristics such as substratum type, water temperature, and salinity. Patterns of disease in the community are often a valuable diagnostic clue. For example, is more than one species affected? Do symptomatic plants occur in a random or patchy array? Irregular or localized distribution may be related, for example, to a specific region of the intertidal zone or to a sewage outfall. When possible, as in commercial or experimental aquaculture operations, also record the observations of the responsible grower or operator. This should establish, among other things, when the problem was first observed, whether it appeared in previous years, which cultural practices were used, and whether unusual environmental conditions occurred.

Laboratory examination is necessary to confirm suspicions aroused by field observations and especially to determine whether an infectious agent is involved (Grogan 1981). Keep specimens refrigerated pending examination and isolation of microorganisms. This should be done as soon as possible. Representative healthy and pathological material should be preserved (dry or wet) in some suitable manner. The simplest and most common preservatives are buffered 5% Formalin in seawater and modified Karpochenko's fixative (Papenfuss 1946). Natural colors of plants can be preserved by the addition of copper sulfate (0.2%, w/v) to one of several formulas (Berlyn and Miksche 1976: 122).

Examine algal surfaces with a dissecting microscope to ascertain the specific areas affected and whether fungi, nematodes, or bacterial ooze occur. Then make freehand or cryostat sections (Streets 1969), stain with aniline blue-HCl (Goff and Cole 1975) or lactophenol cotton blue (Berlyn and Miksche 1976), mount for a semipermanent preparation in a suitable medium such as 60–80% Karo (Abbott 1971), and examine with a compound microscope at low magnification. Observation of fungal structures in algal cells routinely provides essential information on host–parasite interaction (Kohlmeyer 1979). Light microscopy of epidermal strips stained to reveal viral inclusions (Christie 1967) and electron microscopy of negatively stained plant extracts have proved to be rapid, reliable diagnostic procedures for higher plant viruses. By analogy, these methods should be generally applicable to algae. Microscopic study, which unfortunately tends to be underemphasized in diagnosis, should precede or accompany efforts to culture pathogens from symptomatic plants (Streets 1969; Grogan 1981). Microscopy provides information on the nature and location of the pathogen in host tissues, and it avoids the selectivity inherent in all surface sterilization and cultural procedures. Bacteria epiphytic on aquatic plants have been enumerated in situ by epifluorescence microscopy or, after staining of leaves in phenolic aniline blue, by bright-field microscopy (Fry and Humphrey 1978).

C. Establishing causality of biotic disease: completion of Koch's postulates

Examination of symptomatic specimens can provide evidence for association of a presumed causal agent with disease. The protocol developed by Koch from his classic studies of anthrax provides rules of proof necessary to establish causality. Briefly, the criteria are as follows. (1) The causal agent must be associated in every case with the disease under natural conditions and, conversely, the disease must not appear in the absence of the agent; (2) the causal agent must be isolated in pure culture and characterized; (3) typical symptoms must develop when the host is inoculated with the agent under suitable conditions, and the appropriate control inoculations must be made concurrently; and (4) the causal agent must be reisolated and demonstrated to be identical to the agent isolated originally. Pertinent aspects of these postulates, including their modification and experimental errors in their execution, are outlined in the following paragraphs. When several factors contribute substantially to a disease, diagnosis is considerably more complicated, involving correlation matrices and multiple regression analyses. Consult Wallace (1978) for approaches.

1. Association. Of Koch's four postulates, evidence bearing on the first is the most fundamental and the easiest to obtain. Association can usually be demonstrated by the routine field and laboratory examinations already described. Various manipulations may be necessary to facilitate observations. For example, fungi may be induced to sporulate if pieces of algae are washed vigorously for 1 to 12 h in running water (preferably salt water if the host is marine) to remove extraneous epiflora and then partially immersed in sterile seawater in a petri dish. Motile nematodes suspected of residing in holdfast interstices (Moore 1971) or other inaccessible areas can be induced to swim out of crevices by the addition of a small volume of Formalin to the sample (add drops of 10% Formalin until the desired effect is achieved). Obviously, the microorganisms observed may be the causal agents, especially if there is a large number of them. They may also be epiphytes or secondary invaders. Thus, observation of association provides essential, but by itself insufficient evidence for causality.

2. Isolation. Consistent isolation of a microbe in pure culture from diseased tissue provides additional evidence that it may be the causal agent and also permits the isolate to be characterized for subsequent comparison with isolates obtained from the reinoculated host. Unfortunately, certain pathogens (i.e., physiologically obligate parasites such as viruses, many nematodes, fungi, and algae) are difficult or impossible to culture and so this step must frequently be modified. Occasionally, one must settle for less rigorous, indirect evidence that an organism is implicated in the disease. For example, transmitting the agent directly from diseased to healthy plants or curing diseased specimens by treating with specific chemicals (e.g., nematocides) is possible.

Reduced to its simplest form, isolation involves separation of the desired agent from extraneous living and nonliving material. The diverse physical, chemical, and biological methods (Durbin 1961; Tuite 1969; Willoughby 1978) all share the operational premise of enrichment, that is, exploiting the distinctive properties of the desired agent to facilitate isolation, and each includes some kind of monitoring scheme to assess the presence and relative purity of the agent during the isolation process. A detailed discussion of only a few of the numerous considerations in isolating a presumed pathogen is presented here. Among these are the probable type of pathogen (virus, bacterium, fungus, etc.), the type of sample (host tissue, water, or sediment), the organ or tissue involved if the isolation is to be made from the host, treatment of the infected tissue, selection of a culture medium, and plating or other final isolation procedure. Developing credible working hypotheses from an examination of specimens early on will enable one to make an intelligent choice of medium, appropriate tissue for culture, and method of plating, and it will suggest whether, indeed, culturing is even feasible.

As a general rule, isolations for fungi and bacteria should be made from the margins of expanding lesions, and at least some samples should be surface-sterilized before plating. These techniques reduce contamination by secondary or saprophytic microorganisms and enhance the probability of recovery of the pathogen in pure culture. Numerous surface-sterilizing agents are available (Tuite 1969; Booth 1971). Among the more common of these are sodium hypochlorite (10% commercial bleach) or 3% hydrogen peroxide, used by Andrews (1977) in isolating microbes from diseased seaweeds. Normally, tissue pieces are immersed for 1 to 5 min and then rinsed in sterile seawater to remove the chemical before plating. Isolation procedures for specific classes of pathogens follow; for clarity they are described separately, although in practice the investigator pursues several simultaneously.

a. Bacteria. Marine bacteria should be isolated on a seawater-based medium. Natural seawater, recently collected or aged, should be used whenever possible; for details on its collection, preparation, and storage, consult Johnson and Sparrow (1961: 27), McLachlan (1973: 29–31), and Collins et al. (1973). Artificial seawater formulated in the laboratory or prepared from commercial mixes may be substituted. To minimize precipitation during autoclaving, the seawater can be diluted to a salinity of ~25°/oo, which is usually sufficient to support growth (Collins et al. 1973). Alternatively, the seawater can be sterilized by passage through a membrane filter (porosity, 0.2 μm) after prefiltering to remove larger particles.

Numerous general media for marine bacteria exist. Medium 2216 (Zobell 1941) is well known; per liter of aged seawater the ingredients are Bacto-peptone, 5.0 g; $FePO_4$, 0.1 g; and Bacto-agar 15.0 g (final pH, 7.5–7.6). A slightly modified version (Marine Agar 2216) containing yeast extract, ferric citrate in place of $FePO_4$, and various salts to simulate seawater is available from Difco Laboratories in dehydrated form for rehydration with distilled water. Bacteriological media common in terrestial studies such as nutrient agar and Trypticase soy agar (Lennette 1980) have been modified for marine use by substitution of seawater for distilled water (Chan and McManus 1969; Andrews 1977). Notwithstanding the "general" nature of these formulations, all media are to some extent selective. Thus, they may not be adequate for the pathogen, and viable plate counts will always lead to an underestimate of the actual population.

Various selective media, including one that favors alginate-decomposing bacteria, are available, as are differential media such as the marine version (Collins et al. 1973) of King's medium B for detecting fluorescent pseudomonads. However, despite great potential, these have rarely been used in seaweed pathology. Consult Elliott and Georgala (1969) and Bridson and Brecker (1970) for standard procedures regarding the preparation and storage of media.

Bacteria in undiluted samples produce a large number of colony-forming units and are therefore most efficiently isolated by standard serial dilution and streak- or spread-plate techniques, rather than by culturing intact pieces of tissue. Use a sterile glass rod to grind small pieces of the surface-sterilized lesion aseptically in a test tube containing 3–5 ml sterile seawater. Agitate thoroughly with a vortex-type mixer (Vortex Genie, Scientific Products). and dilute if necessary. Use standard bacteriological methods (Collins and Lyne 1976) to streak duplicate or triplicate plates of medium with an inoculating loop. When well-isolated colonies develop, pick a typical colony and restreak. Instead of streaking, 0.1 ml of inoculum can be distributed uniformly over the agar with an alcohol flame-sterilized, L-shaped glass spreader or, alternatively, incorporated into the molten medium, which is poured as a soft (1%) agar overlay above an underlay containing 1.5% agar.

For initial isolations, use relatively nonselective, "general" media (see Sec. II.C.2.a). If these media are also differential (i.e., allow the differentiation by pigmentation or other features of microbes of interest from the background population), so much the better. Selective media can be included if the investigator is sufficiently familiar with the presumed etiology to strongly suspect a specific pathogen. Plates are routinely incubated inverted at 22 to 25°C for 3 to 5 d; occasionally, lower temperatures and correspondingly longer incubation periods are used (Chan and McManus 1969; Laycock 1974).

If the plates display a variety of colonies, little can be concluded and the isolation must be repeated. Consistent isolation of a single or highly predominant colony type suggests that it is the pathogen. To confirm or rule out this possibility, proceed with Koch's protocol. Characterize the isolate by standard bacteriological criteria including Gram's stain, motility, and numerous physiological tests (e.g., see Chan and McManus 1969; Cowan 1974; Lennette 1980). These procedures require proper controls and should be performed by or at least in consultation with a bacteriologist.

b. Fungi and mycetozoans. Numerous marine media have been used for the isolation of fungi (Johnson and Sparrow 1961; Jones 1971). These range from seawater solidified with agar to potato dextrose (Andrews 1977) or cornmeal (Sasaki and Sato 1969) agars prepared with seawater instead of distilled water. (Note comments on seawater in Sec. II.C.2.a on bacteria). A common medium is glucose/yeast extract/seawater, prepared as follows (Johnson and Sparrow 1961): glucose, 1 g; Bacto-yeast extract, 0.1 g; agar, 18 g; aged seawater, 1 liter. For labyrinthulas, more complex media based on gelatin hydrolysate and serum are used (Olive 1975). Antibiotics such as tetracycline (500 mg·liter^{-1}) or penicillin G (100 mg·liter^{-1}) plus streptomycin (50 mg·liter^{-1}) should be added to inhibit bacteria. Consult Tuite (1969: ch. 3) or Guillard (1973; 81) for details on types and preparation of antibiotics. Media should not be acidified to depress bacterial growth, as is done frequently to isolate terrestrial or freshwater fungi, because marine Ascomycetes and fungi imperfecti are characteristically inhibited by acid (Johnson and Sparrow 1961).

There are basically four techniques for isolating algicolous fungi and mycetozoans: (1) induction of sporulation, (2) induction of mycelial growth, (3) dilution plating, and (4) direct isolation. To induce sporulation, incubate pieces of rinsed or surface-sterilized lesions in moist chambers, as described in Sec. II.C.1. Transfer developing mycelium or propagules to media. To induce mycelial growth, aseptically place surface-sterilized sections of infected tissue (three or four pieces per plate) on media. Generally, both a basal medium such as seawater agar and a more nutritious medium should be used. After growth occurs, transfer mycelium from the edges of developing colonies to test tube slants of media. Also examine plates for sporulation on host tissue. Dilution plating involves spreading aliquots of seawater or finely minced infected tissue onto the surface of solid media (Sec.II.C.2.a) or incorporating inocula into cooled but still molten agar medium (40–50°C) before pouring plates. Direct isolation is the method of choice when an organism is fruiting on diseased material. Remove spores or fruiting bodies aseptically with a dissecting needle or other suitable instrument. Streak these directly onto the medium or transfer them to a sterile 1- to 2-ml water blank and inoculate plates by spreading as for bacteria.

As a general rule, incubate plates at 15 to 25°C for 7 to 14 d (although there are some exceptions; Kohlmeyer and Kohlmeyer 1979). After consistent isolation of particular forms suggestive of a pathogen (Sec. II.C.2.a), individual lines should be established by standard hyphal-tipping or single-sporing techniques (Booth 1971) for maintenance. Consult McGinnis (1980) for purification of mixed cultures and proper storage procedures. Two of the best general references for identification are Johnson and Sparrow (1961) and Kohlmeyer and Kohlmeyer (1979).

As in the case of many fungal pathogens of terrestrial plants (Allen 1976), the spores of many marine fungal parasites (obligate parasites) of algae appear to require the presence of a suitable host to elicit spore germination (L. J. Goff, unpublished observations). This mechanism ensures that a spore of an obligate parasite will not be committed to germinate except in situations that are favorable to its continued development. Isolation and subsequent growth of such obligate fungal parasites in pure culture (away from the host) are generally not possible. Therefore, to be cultured, the fungus must be grown in association with its host. The fungal spores only may be isolated (Koch's postulate 2), and these must be inoculated (postulate 3) immediately into a culture containing the appropriate host.

The establishment and maintenance of obligately parasitic marine fungi and their algal hosts in culture have not yet been reported. However, the techniques used to isolate and maintain obligately parasitic red algae (see next section) and their hosts may be applicable to studies of such fungi.

c. Parasitic algae. The spores (mitospores or meiospores) of epiphytic, endophytic, or parasitic algae can be isolated by standard algal culturing techniques (Chapman 1973). After removing the larger surface contaminants from the reproductive thallus by gentle brushing and/or sonication (Polne et al. 1980), surface-sterilize the thallus in a 1 to 5% solution of sodium hypochorite in sterile seawater (1–10 min). Antibiotics may also be useful as surface-sterilizing agents (Boalch 1961; Druel and Hsiao 1969; Chapman 1973), as may molecular iodine (Jodopax–0.01% iodine, 2 min) (Fries 1963; A. Gibor, personal communication).

Place small pieces of the cleaned, surface-sterilized reproductive thallus in a crystallizing dish containing sterilized seawater. Incubate sporulating tissues under temperature and photoperiodic conditions simulating the natural environment. Zoospores can be concentrated by gentle centrifugation (as a starting point try 100–500 g for 3 to 5 min) or filtration (membrane filters or fine-porosity depth filters), or in the case of phototactic spores, the phototactic response itself can be employed to concentrate the swarmers.

Flagellated and nonflagellated spores must be transferred from the sporulation dish before they begin to adhere. To accomplish this, use a finely drawn sterile Pasteur-type pipette or micropipette with a bore diameter just slightly larger than the spore diameter. Wash the spores several times in sterile seawater using standard techniques

(Hoshaw and Rosowski 1973). Inoculate the washed spores into a sterile dish containing the appropriate host alga and sterile seawater or other culture medium. A gyratory shaker operating at ~100 rpm can be used to mix and concentrate the spores and the host material, thereby maximizing the chances of host–parasite contact.

In the case of parasitic red algae, host penetration is generally accomplished within 48 h of inoculation (Goff 1982). At this time, the infected host plants should be transferred to fresh, sterilized seawater or to the appropriate culture medium and cultured under conditions that optimize the progress of the disease cycle.

d. Viruses. The isolation of viruses involves concentrating and subsequently characterizing the particles by several complex biophysical and biochemical techniques (e.g., Kado and Agrawal 1972), a discussion of which is beyond the scope of this chapter. Although numerous presumed viral infections of algae exist (Andrews 1976), only rarely (Gibbs et al. 1975; Cole et al. 1980) have investigations proceeded beyond the stage of observing viruslike particles in host cytoplasm by electron microscopy. The complexity of viral manipulations and the virtual absence of information concerning appropriate procedures for macroalgae make it imperative that a virologist join the investigation if a viral etiology is suspected.

e. Nematodes. There is extensive information on collecting nematodes from soil or roots, somewhat less about marine free-living forms, and virtually none regarding algal pathogens. Most extraction methods depend either on the principle of mobility to separate nematodes from inert materials (the Baermann funnel and its modifications) or on differential size or density, whereby nematodes are concentrated by sieving (the Cobb method and its variations). The method recommended for collecting nematodes from seaweeds or sediment by the late Dr. M. W. Allen (personal communication) is as follows. (1) Place material in a container of seawater, (2) mix thoroughly, (3) let contents settle for a few seconds, (4) decant into a second container, and (5) pass contents of the second container through the standard sieve series (Southey 1970). The funnel technique should be avoided. Use only seawater as the wash medium. The screening procedure should follow basically the same pattern as for standard soil samples, and all residue should be kept from the screens.

Concentrate nematodes into a small volume of seawater in a watch glass, where they can be killed by the addition of an equal volume of boiling seawater with rapid stirring. To fix, add an equal volume of 5% Formalin in seawater. Consult Southey (1970) or Dropkin (1980) for methods of examination. Preserved specimens should be sent to taxonomic authorities for identification.

3. Inoculation. Experimental reproduction of the disease by inoculation of plants with a pure culture of the presumed pathogen is the most critical step in Koch's protocol for two reasons. First, it enables one to determine whether organisms associated with the disease (postulate 1) and isolated from the host (postulate 2) are in fact pathogenic. Second, it requires the greatest knowledge and skill to execute correctly. The investigator must not only choose suitable inoculum, but also should know how the pathogen spreads, how infection occurs, how the environment predisposes the host and subsequently affects disease development, and whether symptoms on inoculated plants duplicate those of the disease in nature. For example, if unreasonably high inoculum levels are used, if the inoculum is supplied with an unusually nutritious food base, or if the plant is severely stressed, the development of symptoms may be meaningless (falsely positive). Conversely, if the organism attacks only senescent or damaged parts, symptoms will probably not develop if nonsenescent or intact tissues are inoculated erroneously (falsely negative).

The key operational guidelines are to include proper controls (including positive controls of isolates of the pathogen, if these exist) and to simulate field conditions as much as possible. Controlled experiments are feasible if aquaria supplied with aerated and recirculated or, preferably, running seawater are available (Spotte 1974; Andrews 1977). To simulate natural conditions, inoculations should ultimately be done in the field. These tests may be facilitated by confinement within sheltered coves or large plastic bags (Strickland and Terhune 1961). The use of portable plexiglass fouling plates attached to the sea floor (Neushul et al. 1976) will make it possible to remove plants periodically for observation and manipulation. Consult Connell (1974), Foster and Sousa (Chap. 13), Denley and Dayton (Chap. 25), and Vadas (Chap. 26) for good discussions of the larger issue of field experiments in marine biology, particularly with respect to experimental ecology.

After considering the major requirements for establishing disease, the investigator must select an appropriate inoculation technique (for details see Tuite 1969: ch. 6). There appear to be four possibilities for seaweeds: (1) adding inoculum to the water column, (2) exposing external surfaces of emersed plants to inoculum, (3) applying inoculum to inaccesible (internal) infection sites, and (4) infesting sediments (because most seaweeds adhere to rocky substrata, this approach has limited application). Which method is chosen will depend on the class of pathogen and the infection process in nature. For example, fungi grow actively into their hosts; bacteria enter plants only through natural openings or wounds; nematodes infect directly or occasionally through openings; viruses, viroids, mycoplasma-like organisms, and rickettsia-like bacteria must be introduced either mechanically or by vectors. Andrews and Hecht (1981) experimented with several forms of inoculation of Eurasian water milfoil, and these should also be applicable to seaweeds.

The investigator should select appropriate criteria for demonstrating pathogenicity. The most important of these is duplication of the natural disease syndrome. Growth of the microbe within the host, especially in the case of bacteria, is often also used. It is worthwhile to quantify the impact of the pathogen on the host by one or more assessments, namely, extent of discoloration, dry weight reduction, chlorophyll synthesis, and photosynthetic capacity.

In seaweed pathology, no area of greater ignorance exists than that centering on Koch's third postulate. There is tremendous scope for innovative approaches to manipulations for disease induction. These must be developed before experimental procedures can begin to compare favorably with those existing in terrestrial pathology.

4. Reisolation. This final stage in the demonstration of causality involves the recovery of the inoculated organism from diseased tissues. This step should not be disregarded, for without it the series of evidence is inconclusive. There are two primary considerations: (1) It is desirable to use the same technique for reisolation as was used for the initial isolation. (2) It must be shown that the organism reisolated is in fact identical to the species isolated initially – hence, the importance of correct taxonomic procedures and thorough characterization of both isolates.

III. Assessing the extent of host–pathogen interaction

A. Types of interactions

A broad spectrum of interactions is evident in the associations that have evolved between marine plant pathogens and their algal hosts (Goff 1983), and it is important to understand the nature of a particular interaction to determine what steps (if any) should be taken to control the "pathogen." In general, the effect of most marine pathogens on their algal hosts is highly localized, with few external symptoms of disease. Invasion by fungi, bacteria, viruses, copepods, nematodes, protozoans, and other algae may cause (1) no external symptoms ("symptomless parasitism," Lewis 1973); (2) localized cell hyperplasia, resulting in the formation of galls or aberrant host growth; (3) localized tissue necrosis, which may permit the invasion of secondary perthophytic or saprobic fungi (Kohlmeyer 1979); or rarely, (4) extensive host tissue necrosis and malformation.

Table 27–1. *Characteristics of necrotrophic and biotrophic parasites*

Necrotroph	Biotroph
Host penetration through natural openings or wounds	Host penetration direct or through natural openings
No special parasitic structures formed	Special parasitic structures (e.g., haustoria)
Host cells rapidly killed	Host cells are not rapidly killed
Toxins and cytolytic enzymes produced	Few or no toxins or cytolytic enzymes produced
Wide host range	Narrow host range
Able to grow saprophytically away from host	Unable to grow away from living host

Source: Modified primarily after Dickinson and Lucas (1977); see also Lewis (1973) and Goff (in press). This summary is considerably simplified, and there are exceptions to each correlate.

Many of the pathogens that have localized effects on their hosts are best described as obligately biotrophic parasites, that is, organisms that do not kill their host overtly, but rather depend on a living host and obtain energy from living host cells (Lewis 1973; and Table 27–1). In contrast to biotrophic parasites, necrotrophic parasites (Table 27–1) kill part or all of their host and obtain energy predominantly from cells that they have killed. Generally, the effects of necrotrophic parasites on their hosts are much more extensive and may result in damaging epidemics within natural populations or monocultures of a host species. The fungus *Pythium porphyrae* is an example of one such necrotrophic parasite. It has caused extensive destruction and economic loss in cultivated *Porphyra tenera* and *P. yezoensis* (Kazama 1979), and certain environmental conditions (warm seawater temperature, low salinity, and plant overcrowding) favor the occurrence, severity, and spread of the economically costly disease (Arasaki 1947, 1962).

B. Determination of interaction types

One can determine the type of interaction between a pathogen and its marine algal host by examining features of the interaction in culture and microscopically. Culture methods can be employed to examine the interaction of the pathogen and its host at each step of the disease cycle (i.e., inoculation, penetration, infection, growth, reproduction, and dissemination) as well as the host range of a pathogen. In addition, culturing a pathogen and host through the infection cycle enables one to study the effect of specific environmental parameters on the severity and spread of the disease.

Microscopy of infected tissues, even without supplementary enzyme histochemistry, autoradiography, or fluorescent antibody techniques can provide presumptive information on the type of interaction. For example, simple aniline blue-HCl-stained cryostat or hand sections (Goff and Cole 1975) of host tissue inoculated by a particular pathogen will indicate whether penetration is restricted to natural openings or wounds (most necrotrophic parasites) or whether the pathogen is able to penetrate directly the intact outer surface of the host (many biotrophic parasites). Upon entering the tissues of the host, biotrophic parasites grow predominantly in the intercellular region, although most have evolved modified structures (haustoria) that bring the parasite into more intimate contact with the host cell. The presence of such structures and the type of internal development (intercellular vs. intracellular growth), may provide some indication of whether the interaction is biotrophic or necrotrophic.

The effect of the infection on the host may also indicate the type of interaction, and many of these features are readily seen in stained cryostat or hand sections of fixed material or in plastic-embedded, sectioned, and stained tissues (McCully et al. 1980). In necrotrophic associations, host cells are rapidly killed, and such cells appear to be collapsed and often very dense and granulated. In living tissues, the stain Evans' blue B (0.5%, w/v) can be used to distinguish dead cells (Gaff and OKong'O-Ogola 1971), since this stain is incorporated only into cells that have lost their selective permeability.

Cellular changes are less apparent in host cells affected by biotrophic parasites. Occasionally, the host cells may become plasmolyzed, and the amount of storage inclusions may change, as may the cytoplasmic density of the cell (Kohlmeyer and Kohlmeyer 1975; Goff 1982). In these associations, host cells are rarely killed. At the tissue level, damage inflicted by a necrotrophic pathogen may also be much more extensive than in biotrophic interactions, because necrotrophic parasites often kill their host cells via toxins and/or cytolytic enzymes that are released by the parasite into the host tissues.

C. General considerations

At the populational level, most of the diseases that have been reported to affect marine plants appear to be endemic rather than epidemic. In the case of marine higher fungal parasites of algae, apparently none of the known 36 taxa cause epidemics (Kohlmeyer 1979), and most appear to be biotrophic parasites of their hosts. This is in sharp contrast to the terrestial higher fungal parasites, many of which cause extensive and highly destructive epidemics.

The relative infrequency of epidemics and necrotrophic interac-

tions in marine plants may merely reflect the genetic diversity of natural marine plant communities. By the establishment of mono-cultures of economically valuable races of marine plants, the genetic diversity of a population is reduced, thereby providing the conditions necessary for the epidemic spread of a pathogen. In addition, the physiological stress of marine plants subjected to suboptimal growth conditions (e.g., crowding, pollution, and thermal stress) may make them even more susceptible to attack by pathogens or change the ability of a host to control normally biotrophic parasites. Some of these problems have been experienced in the cultivation of *Porphyra* (Kazama 1979) and *Laminaria* (C. K. Tseng, personal communication), and there remains an obvious need for an acceleration of both basic and applied research of marine algal pathology. Much of this research should deal with appropriate methodologies for pathological studies. As is apparent from this chapter, there is essentially no field information available even to guide the researcher attempting to determine the cause of stress symptoms apparent in a plant, much less a body of information on symptomatology, predisposition, epi-demiology, disease loss assessment, or ecology of the pathogens involved. The laboratory methods discussed have a key, and perhaps major, role in aquatic pathology, as in terrestrial research. However, the really novel and exciting advances to be made are those involving host–parasite interactions in the ocean.

IV. References

Abbott, I. A. 1971. On the species of *Iridaea* (Rhodophyta) from the Pacific coast of North America. *Syesis* 4, 51–72.

Allen, P. J. 1976. Control of spore germination and infection structure formation in the fungi. In Heitefuss, R., and Williams, P. H. (eds.), *Physiological Plant Pathology*, pp. 51–85. Springer-Verlag, Berlin.

Andrews, J. H. 1976. The pathology of marine algae. *Biol. Rev. Cambr. Philos. Soc.* 51, 211–53.

Andrews, J. H. 1977. Observations on the pathology of seaweeds in the Pacific Northwest. *Can. J. Bot.* 55, 1019–27.

Andrews, J. H., and Hecht, E. P. 1981. Evidence for pathogenicity of *Fusarium sporotrichioides* to Eurasian water milfoil, *Myriophyllum spicatum*. *Can. J. Bot.* 59, 1069–77.

Andrews, J. H., Vadas, R. L., Wheeler, W. N., Neushul, M., Woessner, J. W., Kohlmeyer, J., Dodds, J. A., Kazama, F. Y., and North, W. J. 1979. Pathology of seaweeds: current status and future prospects. *Experientia* 35, 429–50.

Arasaki, S. 1947. Studies on the deterioration of *Porphyra tenera*. *Bull. Jap. Soc. Sci. Fish.* 13, 74–90.

Arasaki, S. 1962. Studies on the artificial culture of *Porphyra tenera* Kjellm. *J. Agric. Lab. Abiko* 3, 87–93.

Berlyn, G. P., and Miksche, J. P. 1976. *Botanical Microtechnique and Cytochemistry.* Iowa State University Press, Ames. 326 pp.

Boalch, G. T. 1961. Studies on *Ectocarpus* in culture. 1. Introduction and methods of obtaining uni-algal and bacteria-free cultures. *J. Mar. Biol. Assoc. U.K.* 41, 279–86.

Booth, C. 1971. Introduction to general methods. In Booth, C. (ed.), *Methods in Microbiology,* vol. 4, pp. 1–47. Academic Press, New York.

Bridson, E. Y., and Brecker, A. 1970. Design and formulation of microbial culture media. In Booth, C. (ed.), *Methods in Microbiology,* vol. 4, pp. 49–94. Academic Press, New York.

Chan, E. C. S., and McManus, E. A. 1969. Distribution, characterization and nutrition of marine microorganisms from the algae *Polysiphonia lanosa* and *Ascophyllum nodosum. Can. J. Microbiol.* 15, 409–20.

Chapman, A. R. O. 1973. Methods for macroscopic algae. In Stein, J. R. (ed.), *Handbook of Phycological Methods: Culture Methods and Growth Measurements,* pp. 87–104. Cambridge University Press, Cambridge.

Christie, R. G. 1967. Rapid staining procedures for differentiating plant virus inclusions in epidermal strips. *Virology* 31, 268–71.

Cole, A., Dodds, J. A., and Hamilton, R. I. 1980. Purification and some properties of a double-stranded DNA-containing virus-like particle from *Uronema gigas,* a filamentous eucaryotic green alga. *Virology* 100, 166–74.

Collins, C. H., and Lyne, P. M. (eds.). 1976. *Microbiological Methods,* 4th ed. University Park Press, Baltimore, Md. 521 pp.

Collins, V. G., Jones, J. G., Hendrie, M. S., Sewan, J. M., Wynn-Williams, D. P., and Rhodes, M. E. 1973. Sampling and estimation of bacterial populations in the aquatic environment. In Board, R. G., and Lovelock, D. W. (eds.), *Sampling: Microbiological Monitoring of Environments,* pp. 77–110. Academic Press, New York.

Connell, J. H. 1974. Ecology: field experiments in marine biology. In Mariscal, R. N. (ed.), *Experimental Marine Biology,* pp. 21–54. Academic Press, New York.

Cowan, S. T. 1974. *Cowan and Steel's Manual for the Identification of Medical Bacteria,* 2nd ed. Cambridge University Press, Cambridge. 238 pp.

Dickinson, C. H., and Lucas, J. A. 1977. *Plant Pathology and Plant Pathogens.* Wiley, New York. 161 pp.

Dropkin, V. H. 1980. *Introduction to Plant Nematology.* Wiley, New York. 293 pp.

Druel, L. D., and Hsiao, S. I. C. 1969. Axenic cultures of Laminariales in defined media. *Phycologia* 8, 47–9.

Durbin, R. D. 1961. Techniques for the observation and isolation of soil microorganisms. *Bot. Rev.* 27, 522–60.

Elliott, E. C., and Georgala, D. C. 1969. Sources, handling, and storage of media and equipment. In Norris, J. R., and Ribbons, D. W. (eds.), *Methods in Microbiology,* vol. 1, pp. 1–20. Academic Press, New York.

Fries, L. 1963. On the cultivation of axenic red algae. *Physiol. Plant.* 16, 695–708.

Fry, J. C., and Humphrey, N. C. B. 1978. Techniques for the study of bacteria epiphytic on aquatic macrophytes. In Lovelock, D. W., and Davies,

R. (eds.), *Techniques for One Study of Mixed Populations*, pp. 1–29. Academic Press, New York.

Gaff, D. F., and OKong'O-Ogola, O. 1971. The use of nonpermeating pigments for testing the survival of cells. *J. Exp. Bot.* 22, 756–8.

Gibbs, A., Skotnicki, A. H., Gardiner, J. E., and Walker, E. S. 1975. A tobamovirus of a green alga. *Virology* 64, 571–4.

Goff, L. J. 1982. The biology of parasitic red algae. In Round, F., and Chapman, D. (eds.), *Progress in Phycological Research*, vol. 1, pp. 289–369. Elsevier/Biomedical, Holland.

Goff, L. J. 1983. Marine algal interactions: epibiosis, endobiosis, parasitism and disease. In Tseng, C. K., and Starr, R. (eds.), *Proceedings of the Joint U.S.–China Phycological Symposium*, pp. 221–74. Science Press, Beijing, People's Republic of China.

Goff, L. J., and Cole, K. 1975. The biology of *Harveyella mirabilis* (Cryptonemiales, Choreocolaceae). 2. Carposporophyte development as related to the taxonomic affiliation of the parasitic red alga *Harveyella mirabilis*. *Phycologica* 14, 227–38.

Goff, L. J., and Glasgow, J. 1980. *Pathogens of Marine Algae*. Special Report 7, California Sea Grant College Program, La Jolla. 236 pp.

Grogan, R. G. 1981. The science and art of plant-disease diagnosis. *Annu. Rev. Phytopathol.* 19, 333–51.

Guillard, R. R. L. 1973. Methods for microflagellates and nannoplankton. In Stein, J. R. (ed.), *Handbook of Phycological Methods: Culture Methods and Growth Measurements*, pp. 69–85. Cambridge University Press, Cambridge.

Hoshaw, R. W., and Rosowski, J. R. 1973. Methods for microscopic algae. In Stein, J. R. (ed.), *Handbook of Phycological Methods: Culture Methods and Growth Measurements*, pp. 53–68. Cambridge University Press, Cambridge.

Johnson, T. W., Jr., and Sparrow, F. K., Jr. 1961. *Fungi in Oceans and Estuaries*. Hafner Press, New York, 668 pp.

Jones, E. B. G. 1971. Aquatic fungi. In Booth, C. (ed.), *Methods in Microbiology*, vol. 4, pp. 335–65. Academic Press, New York.

Kado, C. I., and Agrawal, H. O. (eds.). 1972. *Principles and Techniques in Plant Virology*. Van Nostrand Reinhold, New York. 688 pp.

Kazama, F. Y. 1979. *Pythium* 'red rot disease' of *Porphyra*. *Experientia* 35, 443–4.

Kohlmeyer, J. 1974. Higher fungi as parasites and symbionts of algae. *Veroff. Inst. Meeresforsch. Bremerh.* 5(suppl.) 399–56.

Kohlmeyer, J. 1979. Marine fungal pathogens among Ascomycetes and Deuteromycetes. *Experientia* 35, 437–9.

Kohlmeyer, J., and Kohlmeyer, E. 1975. Biology and geographical distribution of *Spathulospora* species. *Mycologia* 67, 629–37.

Kohlmeyer, J., and Kohlmeyer, E. 1979. *Marine Mycology: The Higher Fungi*. Academic Press, New York. 690 pp.

Laycock, R. A. 1974. The detrital food chain based on seaweeds. 1. Bacteria associated with the surface of *Laminaria* fronds. *Mar. Biol.* 25, 223–31.

Lennette, E. H. (ed.). 1980. *Manual of Clinical Microbiology*, 3rd ed. American Society for Microbiology, Washington, D. C. 1044 p.

Lewis, D. H. 1973. Concepts in fungal nutrition and the origin of biotrophy. *Biol. Rev.* 48, 261–78.

McCully, M. W., Goff, L. J., and Adshead, P. 1980. Preparation of algae for light microscopy. In Gantt, E. (ed.), *Handbook of Phycological Methods: Developmental and Cytological Methods*, pp. 263–83. Cambridge University Press, Cambridge.

McGinnis, M. R. 1980. *Laboratory Handbook of Medical Mycology.* Academic Press, New York. 662 pp.

McIntyre, J. L., and Sands, D. C. 1977. How disease is diagnosed. In Horsfall, J. G., and Cowling, E. B. (eds.), *Plant Disease: An Advanced Treatise,* vol. 1, pp 35–53. Academic Press, New York.

McLachlan, J. 1973. Growth media: marine. In Stein, J. R. (ed.), *Handbook of Phycological Methods: Culture Methods and Growth Measurements,* pp. 25–51. Cambridge University Press, Cambridge.

Moore, P. G. 1971. The nematode fauna associated with holdfasts of kelp (*Laminaria hyperborea*) in North-East Britain. *J. Mar. Biol. Assoc. U. K.* 51, 589–604.

Neushul, M., Foster, M. S., Coon, D. A., Woessner, J. W., and Harger, B. W. W. 1976. An *in situ* study of recruitment, growth and survival of subtidal marine algae: techniques and preliminary results. *J. Phycol.* 12, 397–408.

Olive, L. S. 1975. *The Mycetozoans.* Academic Press, New York, 293 pp.

Papenfuss, G. F. 1946. Structure and reproduction of *Trichogloea requienii*, with a comparison of the genera of Helminthocladiaceae. *Bull. Torrey Bot. Club* 73, 319–437.

Polne, M., Gibor, A., and Neushul, M. 1980. The use of ultrasound for the removal of macroalgal epiphytes. *Bot. Mar.* 34, 731–4.

Sasaki, M., and Sato, S. 1969. Composition of medium and cultural temperature of *Pythium* sp., a pathogenic fungus, of the "Akagusare" disease of cultivated *Porphyra*. *Bull. Tohoku Reg. Fish. Res. Lab.* 29, 125–32. (In Japanese; English summary.)

Southey, J. F. (ed.). 1970. *Laboratory Methods for Work with Plant and Soil Nematodes.* Ministry of Agriculture, Fisheries and Food, Technical Bulletin no. 2, Her Majesty's Stationery Office, London. 148 pp.

Spotte, S. 1974. Aquarium techniques: closed system marine aquariums. In Mariscol, R. N. (ed.), *Experimental Marine Biology,* pp. 1–19. Academic Press, New York.

Streets, R. B., Sr. 1969. *The Diagnosis of Plant Diseases.* Extension Service, Agricultural Experiment Station, University of Arizona, Tucson, 134 pp.

Strickland, J. D. H., and Terhune, L. D. B. 1961. The study of *in situ* marine photosynthesis using a large plastic bag. *Limnol. Oceanogr.* 6, 93–6.

Suto, S., Saito, Y., Akiyama, K., and Umebayashi, O. 1972. *Text Book of Diseases and Their Symptoms in Porphyra.* Contribution E, no. 18, Tokai Regional Fisheries Research Laboratory, Japan. 37 pp. (In Japanese.)

Tuite, J. 1969. *Plant Pathological Methods: Fungi and Bacteria.* Burgess, Minneapolis. 239 pp.

Wallace, H. R. 1978. The diagnosis of plant diseases of complex etiology. *Annu. Rev. Phytopathol.* 16, 379–402.

Wheeler, H. 1975. *Plant Pathogenesis.* Springer-Verlag, New York. 106 pp.

Willoughby, L. G. 1978. Methods for studying micro-organisms in decaying leaves and wood in freshwater. In Lovelock, D. W., and Davies, R. (eds.), *Techniques for the Study of Mixed Populations,* pp. 31–50. Academic Press, New York.

Zobell, C. E. 1941. Studies on marine bacteria. 1. The cultural requirements of heterotrophic aerobes. *J. Mar. Res.* 41, 42–75.

Section V

Appendix

Suppliers

Louis Adamo, Inc., P.O. Box L, Solana Beach, CA 92075

Allied Trading Company, P.O. Box 2578, Secaucus, NJ 07094

American Hospital Supply Corporation, 1 American Plaza, Evanston, IL 60201

American Optical Corporation (now AO Reichert), Scientific Instruments, P.O. Box 123, Buffalo, NY 14240

American Scientific Products Div., American Hospital Supply Corporation, 1430 Waukegan Rd., McGaw Park, IL 60085

American Type Culture Collection (ATCC), 12301 Parklawn Drive, Rockville, MD 20852

Amersham, 2636 S. Clearbrook, Arlington Heights, IL 60005

Analog Devices, Inc., P.O. Box 280, Rte. 1 Industrial Park, Norwood, MA 02062

Analytical Development Co., Ltd., Pindar Rd., Hoddesdon, Herts ENll OAQ, United Kingdom

Anarad Inc., 534 E. Ontario St., Santa Barbara, CA 93103

Beckman Instruments, Inc., Scientific Instruments Division, Campus Dr. at Jamboree Blvd., P.O. Box C-196001, Irvine CA 92713

Ben Meadows Co., P.O. Box 80549, Chamblee, GA 30366

Bendix Corp., Environmental Sciences Division (now Environmental and Process Instrument Div.), 1400 Taylor Ave., Baltimore, MD 21204

Bio Quip Products, Inc., P.O. Box 61, Santa Monica, CA 90406

Biospherical Instruments, 4901 Morena Blvd., #1003, San Diego, CA 92117

Black and Decker Manufacturing Co., 626 Hanover Pike Dr., Hampstead, MD 21074

Broadley-James Corporation, 1714 S. Lyon, Santa Ana, CA 92705

Carl Zeiss, Inc., 1 Zeiss Dr., Thornwood, NY 10594

Carpenter/Offutt Paper, Inc., Herbarium Dept., P.O. Box 3333, S. San Francisco, CA 94080

Chevron Chemical Company, Ortho Division, 940 Hensley St., Richmond, CA 94804

Clay Adams (Division of Becton, Dickerson and Co.), 299 Webro Rd., Parsipanny, NJ 07054

Curtin Matheson Scientific, Inc., 10727 Tucker St., Beltsville, MD 20705

Danish Hydraulic Institute, Agern Alle 5, DK-2970, Horsholm, Denmark

A. D. Data Systems, Inc., 200 Commerce Dr., Henrietta, NY 14623

Datel Intersil, 11 Cabot Blvd., Mansfield, MA 02043

Deep Ocean Work Systems, Inc., 775 Battery St., San Pedro, CA 90731

[595]

Difco Laboratories, Inc., P.O. Box 1058A, Detroit, MI 48232

DISA Electronics, Inc. (now Dantec), (Li-Cor) 779 Susquebanna Ave., Franklin Lakes, NJ 07147

Eastman Kodak Co., 343 State St., Rochester, NY 14650

ECO Systems Management Associates, Inc., Suite 119, 531 Encinitas Blvd., Encinitas, CA 92024

Edmund Scientific Co., 7082 Edscorp Bldg., Barrington, NJ 08007

EFCOM Communication Systems, Inc., 1540 S. Lyon St., Santa Ana, CA 92705

EG&G, Electro-Optics Division, 35 Congress St., Salem, MA 01970

Ehrenreich Photo Optical Inc., Garden City, NY 11530

Eppley Laboratory, Inc., 12 Sheffield Ave., Newport, RI 02840

Erba Instruments, Inc., 4 Doulton Pl., Peabody, MA 01960

EROS Data Center, NOAA/NESDIS Landsat, Customer Services, Sioux Falls, SD 57198

Fisher Scientific Co., Inc., Allied Corporation, 711 Forbes Ave., Pittsburgh, PA 15219

Floy Tag and Manufacturing Co., Inc., P.O. Box 85357, Seattle, WA 98145-1357

Forestry Suppliers, Inc., 205 West Rankin St., Box 8397, Jackson, MS 39204

Gelman Sciences, Inc., 600 S. Wagner Rd., Ann Arbor, MI 48106

Gilson Medical Electronics Inc., P.O. Box 27, Middleton, WI 53562

General Oceanics Inc., 1295 N.W. 163rd St., Miami, FL 33169

GFS Chemicals, P.O. Box 23214, Columbus, OH 43223

Helle Engineering, Inc., 7198 Convoy Court, San Diego, CA 92111

Ikelite Underwater Systems, P.O. Box 88100, Indianapolis, IN 46208

Illinois Bronze and Paint Company, 300 East Main St., Lake Zurich, IL 60047

Industrial Formulators of Canada, Ltd., 3824 William St., Burnaby, British Columbia, Canada V5C 3H9

Kahl Scientific Instruments Corporation, P.O. Box 1166, El Cajon, CA 92022

Koppers Co., Inc., P.O. Box 911041, Commerce, CA 90011

Laboratory for Computer Graphics and Spatial Analysis, 48 Quincy St., Harvard University, Cambridge, MA 02138

Lambda Instruments, Inc. (now LI-COR, Inc.), 4421 Superior St. Lincoln, NB 68504.

Leybold-Heraeus GMBH, P.O. Box 51 07 60, 5000 Cologne 51, West Germany (USA branch: Leybold-Heraeus Vacuum Products, Inc., 5700 Mellon Rd., Export PA 15632)

M and E Marine Supply, P.O. Box 601, Camden, NJ 08101

McMaster-Carr Supply Co., P.O. Box 54960, Los Angeles, CA 90054

Markson Science, Inc., P.O. Box 8017, Phoenix, AZ 85040

Marsh-McBirney, Inc., 8595 Grovemont Circle, Gaithersburg, MD 20877

Measurements Group, Vishay, P.O. Box 27777, Raleigh, NC 27611

Motorola, Inc., Position Determining Systems, P.O. Box 2606, Scottsdale, AZ 85252

Nalge Co., Division of Sybron Corporation, P.O. Box 365, Rochester, NY 14602

National Ocean Survey (should read: National Ocean Service, Distribution Branch N/CG 33, 6501 Lafayette Ave., Riverdale, MD 20737)

National Oceanographic Data Service (should read: National Ocean Service, Tide and Current Prediction Section N/OMS, 6001 Executive Blvd., Rockville, MD 20852)

National Semiconductor Corp., 2900 Semiconductor Dr., Santa Clara, CA 95051

Neil Brown Instrument Systems, Inc., P.O. Box 498, Cataumet, MA 02534

New Brunswick Scientific Co., Inc., 44 Talmadge Rd., Edison, NJ 08818

New England Nuclear, 549 Albany St., Boston, MA 02118

Ohaus Scale Corp., 29 Hanover Rd., Florham Park, NJ 07932

Orbisphere Laboratories, 287 Lackawanna Ave., West Paterson, NJ 07424

Orion Research Inc., 840 Memorial Dr., Cambridge, MA 02139

Parr Instrument Co., 211 Fifty-third Street, Moline, IL 61265

Permalite Plastics Corporation, 1537 Monrovia Ave., Newport Beach, CA 92663

Pierce Chemical Co., P.O. Box 117, Rockford, IL 61105

Proven Pumps Corporation, 1440 N. Spring St., Los Angeles, CA 90012

Ramset Fastening Systems (now Northern Contractors and Industrial Supply), P.O. Box 500, Fairhaven Station, New Haven, CT 06513

Ryan Instruments, Peabody Ryan, Inc., P.O. Box 599, Kirkland, WA 98033

Sea Tech, Inc., P.O. Box 779, Corvallis, OR 97339

Sierra Chemical Co., 1001 Yosemite Dr., Milpitas, CA 95035

Sound-Wave Systems, Inc., 3001 Red Hill, Bldg. 1, Suite 102, Costa Mesa, CA 92626

SPOT IMAGE Corporation, 1150 17th Street N.W., Washington, DC 20036

Stanly Hydraulic Tools, 3810 S.E. Naef Rd., Milwaukee, WI 97222

Star Expansion, 1061 The Queensway, Toronto, Ontario, Canada M8Z 1R3

Surveyors Service Co., 2942 Century Place, Costa Mesa, CA 92626

Swoffer Instruments, Inc., 1048 Industry Dr., Seattle, WA 98188

Technical Associates, 7051 Eton Ave., Canoga Park, CA

Tec Quipment Inc., Box 1074, Acton, MA 01720

A. H. Thomas Scientific Company, Philadelphia, PA 19105

Tri Hawk International, 8230 Mayrand St., Montreal, Canada H4P 2C6

TSI, Inc., P.O. Box 43394, St. Paul, MN 55164

Turner Designs, 2247 Old Middlefield Way, Mountain View, CA 94043-2489

Turtox Biological Supplies (Now Turtox, Inc.), 5000 W. 128th Pl., Alsip, IL 60658

Leo Ulfelder Co., 420 South Fulton Ave., Mount Vernon, NY 10553

University of Texas Culture Collection (UTEX) [should read Culture Collection of Algae (UTEX)], Department of Botany, University of Texas, Austin, TX 78712

VWR Scientific Inc., subsidiary of Univar Corp., P.O. Box 7900, San Francisco, CA 94120

Ward's Natural Science Establishment, Inc., P.O. Box 9912, Rochester, NY 14692-9012

William T. Bear, Inc. 18915, Grand River Ave., Detroit, MI 48223

Yellow Springs Instrument Co., P.O. Box 279, Yellow Springs, OH 45387

Section VI

Indexes

Author index

(Italic numbers indicate pages with reference citations in full; * indicates chapter author.)

[601]

Taxonomic index